Examining Biological Relevance of Fused S-Heterocycles

Shrikaant Kulkarni
Victorian Institute of Technology, Melbourne, Australia

Hemantkumar Akolkar
Rayat Shikshan Sanstha's Abasaheb Marathe Arts and New Commerce Science College, India

Vijay M. Khedkar
Vishwakarma University, India

Nirmala R. Darekar
Rayat Shikshan Sanstha's Radhabai Kale Mahila Mahavidyalaya, India

IGI Global
Scientific Publishing
Publishing Tomorrow's Research Today

Vice President of Editorial	Melissa Wagner
Managing Editor of Acquisitions	Mikaela Felty
Managing Editor of Book Development	Jocelynn Hessler
Production Manager	Mike Brehm
Cover Design	Phillip Shickler

Published in the United States of America by
IGI Global Scientific Publishing
701 East Chocolate Avenue
Hershey, PA, 17033, USA
Tel: 717-533-8845
Fax: 717-533-8661
Website: https://www.igi-global.com E-mail: cust@igi-global.com

Library of Congress Cataloging-in-Publication Data

Names: Kulkarni, Shrikaant, editor.
Title: Examining biological relevance of fused S-heterocycles / edited by Shrikaant Kulkarni, Hemantkumar Akolkar, Vijay Khedkar, Nirmala Darekar.

Description: Hershey, PA : IGI Global Scientific Publishing, 2025. | Includes bibliographical references and index. | Summary: "The objective of the book is to provide a thorough examination of the biological relevance of fused s-heterocycles, aiming to uncover their potential applications in various fields such as medicine and agriculture" -- Provided by publisher.
Identifiers: LCCN 2024049301 (print) | LCCN 2024049302 (ebook) | ISBN 9798369375204 (h/c) | ISBN 9798369375211 (s/c) | ISBN 9798369375228 (eISBN)
Subjects: LCSH: Sulfur compounds--Physiological effect. | Heterocyclic compounds--Physiological effect. | Organosulfur compounds--Physiological effect.
Classification: LCC QP535.S1 E93 2025 (print) | LCC QP535.S1 (ebook) | DDC 572/.5544--dc23/eng/20250109
LC record available at https://lccn.loc.gov/2024049301
LC ebook record available at https://lccn.loc.gov/2024049302

British Cataloguing in Publication Data
A Cataloguing in Publication record for this book is available from the British Library.

Table of Contents

Detailed Table of Contents

Heterocyclic compounds are essential in the metabolism of all living cells, and most nitrogen-containing fused heterocyclic compounds with five-membered rings are biologically active. Nitrogen and sulphur-containing heteroaromatic compounds, such as imidazothiazoles and their derivatives, are acknowledged for their biological activity and substantial contributions to medicinal chemistry. Imidazothiazoles are defined by their fused bicyclic structure, merging an imidazole ring (a five-membered ring with two non-adjacent nitrogen atoms) with a thiazole ring (a five-membered ring containing both sulphur and nitrogen atoms). This distinctive structure plays a crucial role in their biological activities and chemical stability. Imidazothiazoles display a wide range of biological activities, encompassing antimicrobial, antiviral, antifungal, anticancer, and anti-inflammatory properties.

Fused sulfur heterocycles, such as triazolothiadiazoles, contain four nitrogen atoms and one sulfur atom as heteroatoms in their structure. These elements impart electron-donating and electron-withdrawing properties, influencing the reactivity and chemical behaviour of the compounds. Such effects contribute to various chemical interactions, including coordination with metals and hydrogen bonding. For example, in drug discovery, these interactions play a role in binding with proteins through non-covalent interactions. Additionally, the electronic properties of this heterocycle can interact with plant pathogens, showing potential for applications in the agricultural sector. This chapter discusses the structural significance of triazolothiadiazoles and how they affect their reactivity, as well as their biological importance in medicinal and agricultural contexts. The triazolothiadiazole heterocycle shows significant promise for medicinal as well as agricultural applications, warranting further exploration.

Recent research on cancer therapy has revealed that the emergence of malignant cells is caused by several enzyme alterations. such include kinase inhibitors of ErB4 (HER4), B-Raf kinase, and indoleamine 2,3-dioxygenase 1 (IDO1). Researchers discovered that drugs based on imidazothiazoles, and their analogues exhibit more robust anticancer action by inhibiting these enzymatic receptors. In recent times, imidazothiazole scaffold is broadly explored for its anticancer activity, which acts through various mechanisms such as EGFR, B-RAF, DHFR kinase inhibition and tubulin polymerization inhibition and other molecular mechanisms of action. Due to their feasible synthetic accessibility and promising pharmacological profile, it has attracted various medicinal chemists to explore and develop imidazothiazole derivatives as potent and safe anticancer agents. In the present book chapter, we have reviewed various potent imidazothiazole scaffold-based derivatives reported as anticancer agents, their synthetic strategies, with their future perspective.

Thienopyrroles, characterized by a thiophene ring fused to a pyrrole ring, have garnered significant interest in medicinal chemistry for their diverse therapeutic properties. This review explores the recent advancements in the medical applications of thienopyrroles, highlighting their roles as anticancer, antimicrobial, anti-inflammatory, antiviral, neuroprotective, analgesic, and antidiabetic agents. Through examining current research, we discuss the mechanisms by which thienopyrroles exert their

Thienopyridine derivatives are important sulfur-containing compounds with a unique structure. There are six different Thienopyridine analogues that have shown various biological effects. These compounds are widely used in medicinal chemistry and some have been approved by the FDA for use as drugs. The most well-known applications of thienopyridine-based drugs are in treating cardiovascular and central nervous system (CNS) diseases. These compounds also have anti-inflammatory, anti-infective, antiviral, antiproliferative, antimicrobial, and antitumor properties. Thienopyridine derivatives also have positive effects on important characteristics of the compounds they are part of, such as solubility, lipophilicity, and hydrogen bonding capacity. This chapter will provide a brief overview of the various bioactivities of Thienopyridine derivatives.

Heterocyclic compounds are ubiquitous in biologically active compounds, including many drugs. Benzothiadiazine is a class of sulphur and nitrogen-containing heterocyclic compounds comprised of a benzene ring fused with thiadiazine. Benzothiadiazine rings appear in many marketed drugs, such as bendroflumethiazide, chlorothiazide, cyclothiazide, hydrochlorothiazide, and diazoxide, which are popularly known for their diuretic activity. The ring has diverse pharmacological activity, including anti-cancer, antifungal, antimicrobial, anthelmintic, hypolipidaemic, and antidiabetic activity. This chapter focuses on the chemistry, synthetic methods, and bioactivity of different benzothiadiazine. Furthermore, the updates on patents and clinical trials have also been summarized in the present chapter.

Preface

Fused S-heterocycles, molecules defined by the amalgamation of sulfur-containing heterocyclic rings, have attracted considerable interest in recent years owing to their substantial biological significance. These compounds have shown significant potential in tackling several therapeutic difficulties, including infectious diseases, cancer, and neurodegenerative disorders.

The book, "Examining Biological Relevance of Fused S-Heterocycles," seeks to deliver a thorough overview of recent advancements in the field, elucidating the synthesis, characteristics, and biological uses of these intriguing molecules. This volume, authored by prominent experts in medicinal chemistry, pharmacology, and molecular biology, is a significant resource for academics, scholars, and clinicians interested in investigating the therapeutic potential of fused S-heterocycles.

ORGANIZATION OF THE BOOK

The book comprises four sections and thirteen chapters, each concentrating on a distinct facet of fused S-heterocycles, encompassing; synthesis and structural heterogeneity, pharmacological assessment and mechanism of action, antimicrobial, anticancer, and neuroprotective attributes, computational modeling and structural analysis structure-activity relationships, Present obstacles and prospective trajectories'.

Chapter 1: Biological Relevance of Five-Membered Fused Heterocycle Imidazothiazole

In this chapter, Dr. Sandeep Chandrashekharappa, Subrata Barick, and Kaustav Dhara examine the pivotal role of imidazothiazoles—five-membered fused heterocyclic compounds that are biologically significant for various medicinal applications. By combining nitrogen and sulfur in their structure, imidazothiazoles exhibit a unique stability and activity profile, with potential applications in antimicrobial, antiviral, antifungal, anticancer, and anti-inflammatory therapies. This analysis provides an in-depth understanding of imidazothiazoles' chemical structure and broad biological functions, positioning them as critical components within medicinal chemistry.

Chapter 2: Biological Relevance and Applications of Triazolothiadiazole – Exploring Interactions, Synthesis, and Mechanisms

Authored by Biki Hazarika, Lal Ropuia, and Ved Singh, this chapter delves into the structure and significance of triazolothiadiazoles, which contain a unique arrangement of nitrogen and sulfur atoms that modulate their reactivity and interactions. The chapter explores how these compounds engage in non-covalent bonding with proteins, essential for drug discovery, and demonstrate promise in agricultural applications for pathogen control. Through a detailed analysis of their electronic properties and structural relevance, this chapter highlights triazolothiadiazoles as potential multi-functional agents in both medicinal and agricultural domains.

Chapter 3: Imidazothiazole – Different Synthetic Approaches and Their Anticancer Activity

Shalini Jaiswal investigates the synthetic versatility and anticancer efficacy of imidazothiazoles in this chapter. Highlighting the scaffold's inhibitory effects on key enzymes like EGFR and B-Raf, imidazothiazoles are presented as potent agents in cancer therapy, with synthetic approaches yielding diverse anticancer compounds. The chapter offers a thorough review of synthetic strategies, exploring future directions for developing imidazothiazole-based treatments that can effectively inhibit cancer progression.

Chapter 4: Emerging Therapeutic Agents in Modern Medicine – A Review of Medical Applications

Authors Prashanth G.K., Llalithamba H.S., Srilatha Rao, Rashmi K.V., Bhagya N.P., and Manoj Gadewar provide a comprehensive overview of thienopyrroles, focusing on their therapeutic potential across multiple domains. With effects ranging from apoptosis induction to neuroprotection, thienopyrroles are examined as promising scaffolds for treating cancer, microbial infections, and neurodegenerative diseases. This chapter underscores the versatility of thienopyrroles and emphasizes the need for further research to confirm their efficacy and safety in clinical applications.

Chapter 5: Examining Biological Relevance of Thiazolo Thiadiazole and Thiazolo Thiazole Fused Heterocycles

Vikrant Palekar and Kulbhushan Sasane explore the bioactive behavior of nitrogen- and sulfur-containing fused heterocycles, specifically focusing on thiazolo thiadiazole and thiazolo thiazole. Known for their extensive applications in drug discovery, these compounds display notable anticancer, antimicrobial, and antitubercular activities. This chapter examines the structural basis of their biological activity and potential as therapeutic agents, offering insights into their application in the pharmaceutical industry.

Chapter 6: Potential Impact on Science and Society – Key Findings of Thienothiophene

Hemchandra Chaudhari, Sujal Mondal, Sankalp Salunkhe, Mudassir Nagori, and Kulbhushan Sasane discuss the structural and functional characteristics of thienothiophene, a fused ring structure known for its utility in organic semiconductors. Highlighting applications from pharmacology to optoelectronics, the chapter details thienothiophene's contributions to drug development and its promising role in materials science, emphasizing both therapeutic and technological potential.

Chapter 7: Synthesis, Properties, and Bioactivity of Thiazolopyrimidines – Preparation and Biological Activities

Navnath Hatvate and Mukta Lele provide an in-depth review of thiazolopyrimidines, heterocycles with significant applications in pharmaceuticals and agriculture. The chapter outlines their synthetic routes and bioactivities, including analgesic, antitumor, and antimicrobial properties, and reviews recent clinical trials and patents, underscoring the compound's widespread utility.

Chapter 8: Thienopyrimidines – Exploring the Chemistry and Bioactivity

This chapter by Navnath Hatvate, Trupti Shevkar, and Hemantkumar Akolkar explores thienopyrimidine derivatives, known for their broad spectrum of pharmacological activities such as anticancer and anti-Alzheimer effects. The authors discuss synthetic approaches, bioactivity, and the latest clinical developments, presenting thienopyrimidines as valuable assets in drug discovery.

Chapter 9: Thienopyridine Analogues – A Versatile Scaffold in Medicinal Chemistry for Decades

Bonny Patel and Khushal Kapadiya delve into thienopyridine derivatives, highlighting their extensive use in treating cardiovascular and CNS diseases. The chapter reviews their bioactivity, structural benefits, and the pharmacological roles that make thienopyridines foundational in medicinal chemistry.

Chapter 10: Synthesis, Properties, and Biological Activity of Benzothiadiazine

Navnath Hatvate, Rutuja Ukhade, and Hemantkumar Akolkar present benzothiadiazine, a heterocycle with significant applications as diuretics and antidiabetic agents. The authors review synthetic methods, pharmacological activity, and current patents and trials, reinforcing benzothiadiazine's pharmaceutical relevance.

Chapter 11: Benzothiazepine – Recent Advances in Synthesis and Biological Application

Sarita Kundlikar, Hemantkumar Akolkar, Dilip Thube, Nirmala Darekar, and Vijay Khedkar examine benzothiazepine derivatives, focusing on their application as calcium channel blockers and antidepressants. The chapter provides insights into synthetic methodologies and the bioactive roles of benzothiazepines, highlighting their relevance in drug development.

Chapter 12: Exploring the Biological Relevance of Fused S-Heterocycle Phenothiazine Derivatives

Authors Bhupinder Singh, Christian Kaunert, Sahil Lal, Hind Hammouch, and Manmeet Kaur Arora explore the pharmacological potential of phenothiazine derivatives, characterized by their sulfur-rich fused structure. Known for antipsychotic and anti-inflammatory properties, phenothiazines are reviewed as versatile therapeutic agents, with the chapter discussing synthetic variations and potential clinical applications.

Chapter 13: Harnessing Fused S-Heterocycles for Advanced Pollutant Degradation, Remediation Techniques, and Sustainability: Novel Catalytic Systems for Efficient Pollutant Removal

The last chapter delves into the pivotal role of fused S-heterocycles in environmental remediation efforts, particularly for water and soil. Authors Dr. Hari Shankar Biswas, Amit Kundu, and Prasenjit Mandal analyze how these compounds, known for their unique structural and electronic properties, enhance biological, chemical, and physical pollutant degradation methods.

This book aims to deliver a comprehensive comprehension of the biological significance of fused S-heterocycles, emphasize the most recent procedures and techniques in synthesis and characterization, promote collaboration and knowledge exchange between researchers and physicians, foster innovative solutions for urgent healthcare issues.

We anticipate that this book will act as a catalyst for additional study and discovery, ultimately aiding in advancing innovative medicines and enhancing patient outcomes.

Shrikaant Kulkarni
Victorian Institute of Technology, Melbourne, Australia

Hemantkumar Akolkar
Rayat Shikshan Sanstha's Abasaheb Marathe Arts and New Commerce Science College, India

Vijay M. Khedkar
Vishwakarma University, India

Nirmala R. Darekar
Rayat Shikshan Sanstha's Radhabai Kale Mahila Mahavidyalaya, India

Chapter 1
Biological Relevance of Five Membered Fused Heterocycle Imidazothiazole

Sandeep Chandrashekharappa
https://orcid.org/0000-0003-2475-7270
NIPER Raebareli, India

Subrata Barick
https://orcid.org/0009-0008-4411-1956
NIPER Raebareli, India

Kaustav Dhara
https://orcid.org/0009-0006-2821-9567
NIPER Raebareli, India

ABSTRACT

Heterocyclic compounds are essential in the metabolism of all living cells, and most nitrogen-containing fused heterocyclic compounds with five-membered rings are biologically active. Nitrogen and sulphur-containing heteroaromatic compounds, such as imidazothiazoles and their derivatives, are acknowledged for their biological activity and substantial contributions to medicinal chemistry. Imidazothiazoles are defined by their fused bicyclic structure, merging an imidazole ring (a five-membered ring with two non-adjacent nitrogen atoms) with a thiazole ring (a five-membered ring containing both sulphur and nitrogen atoms). This distinctive structure plays a crucial role in their biological activities and chemical stability. Imidazothiazoles display a wide range of biological activities, encompassing antimicrobial, antiviral,

DOI: 10.4018/979-8-3693-7520-4.ch001

antifungal, anticancer, and anti-inflammatory properties.

INTRODUCTION

Heterocyclic compounds play vital roles in the metabolism of all living cells, and many of the nitrogen-containing fused heterocyclic compounds with five-membered rings are biologically active, (Al-Mulla, 2017) (Kabir & Uzzaman, 2022) (Sandeep *et al.*, 2014) (Dorai *et al.*, 2023) (Uppar *et al.*, 2020a) (Uppar *et al.*, 2020b) (Sandeep, Rashmi, & Kulkarni, 2013) (Venugopala *et al.*, 2019) (Siddesh *et al.*, 2014)(Ningegowda *et al.*, 2020) (Nagfesh *et al.*, 2014) (Venugopala *et al.*, 2020) (Venugopala *et al.*, 2018). Among nitrogen and sulfur-containing heteroaromatic compounds, imidazothiazoles are one of the major calss of compounds having various biological and medicinal applications, (García-Valverde & Torroba, 2005), (Sandeep *et al.*, 2017) (Sumanth *et al.*, 2023). These are fused heterocyclic system containing N and S as heteroatom, formed by joining five membered thiazole and imidazole rings. Currently, there have been five distinct imidazothiazoles reported in the literature depending on the position of N and S atoms in the heterocycle and varying degrees of saturation: imidazo[2,1-*b*]thiazoles (**A**), imidazo[5,1-*b*]-thiazoles **(B)**, imidazo[1,5-*c*]thiazoles **(C)**, imidazo[1,2-*c*]- thiazoles **(D)**, imidazo[4,5-*d*]thiazoles **(E)**(**Figure 1.**), (Saliyeva *et al.*, 2020). Interest towards imidazothiazoles have been increased after the discovery of levamisole, (Amery, Bruynseels, & Levamisole, 1992), a hydrogenated imidazothiazole derivative bearing anthelmintic, (Harrow & Gration, 1985), immunomodulatory, (Kajita *et al.*, 1990), and moderate immune suppressive properties, (Cabaj *et al.*, 1995). Furthermore, the discovery of WAY-181187, (Liu *et al.*, 2010) and pifithrin-β, (Da Pozzo *et al.*, 2014) having anxiolytic and anti-neoplastic properties respectively, attracted researchers to work on these cores. These compounds have demonstrated effectiveness in treating various diseases, highlighting their importance as therapeutic agents. These derivatives demonstrate a broad spectrum of biological activities and pharmacological diversity, making them highly attractive in medicinal chemistry, (Kumar *et al.*, 2022). The imidazothiazole core is well-known for its therapeutic properties, including anticancer, (Kamboj *et al.*, 2023), anthelminthic, (Shetty, Khazi, & Ahn, 2010), fungicide, (Mahfouz & Elhabashy, 1990), cardiotonic, (Bhongade *et al.*, 2016), anti-inflammatory, (Soyer *et al.*, 2022), antibacterial, (Shahin *et al.*, 2023), antiparasitic, (Siddiqui *et al.*, 2022), antiviral, (Chirra et al., 2024), and antioxidant activities, (Dincel *et al.*, 2022).

Figure 1. Different analogues of imidazothiazoles

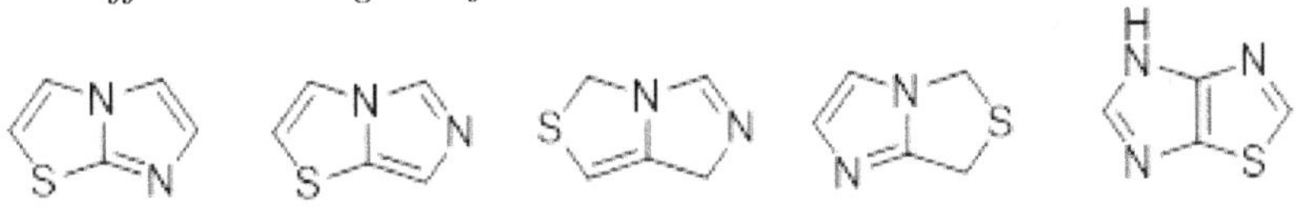

A B C D E

Many synthetic routes have been developed by researchers and few of them has briefly described here, such as, Koudad and coworkers (2020) synthesized imidazothiazole derivatives via the Claisen-Schmidt condensation of aldehydes with various methyl ketones, (Koudad *et al.*, 2020). Substituted acetophenone and aminothiazole were used as substrates in the presence of ethanol. The resulting product underwent a subsequent Vilsmeier-Haack reaction in dimethylformamide (DMF) with phosphoryl chloride ($POCl_3$), yielding imidazothiazole-chalcones. Mishra and colleagues (2016) developed a synthetic approach which is visible light-triggered, and the reaction is free of catalyst for the synthesis of imidazo[2,1-*b*] thiazole in a green medium of ethanol/water. Ewida *et al.* (2018) developed a methodology to synthesize ethyl 5-(4-substituted-phenyl) imidazo[2,1-*b*] thiazole-3-carboxylate by reacting ethyl 2-aminothiazole-4-carboxylate with various derivatives of 2-bromo-1-(4-substituted-phenyl)-ethanone in the presence of acetone at 60 °C for 24 hours. Vekariya *et al.* (2017) developed an efficient one-pot, microwave-assisted green synthesis method for producing new imidazo[2,1-*b*] thiazole derivatives. This method involves reacting 1-(2-amino-4-methylthiazole-5-yl) ethanone or ethyl 2-amino-4-methylthiazole-5-carboxylate with α-bromo aralkyl ketones (phenacyl bromides) in the presence of a water and polyethylene glycol-400 (PEG-400) mixture (in a 1:1 ratio). The use of microwave irradiation accelerates the reaction, while the aqueous PEG-400 solvent provides a more environmentally friendly alternative to traditional organic solvents. This approach allows for the efficient synthesis of imidazo[2,1-*b*] thiazole derivatives. Karaman and colleagues (2016) devised a methodology to synthesize imidazo[2,1-*b*] thiazole derivatives starting from [6-(4-chlorophenyl) imidazo[2,1-*b*] thiazole-3-yl] acetic acid hydrazide. A series of arylidenehydrazide compounds were synthesized by Karaman *et al.* (2016) starting from [6-(4-chlorophenyl) imidazo[2,1-*b*] thiazole-3-yl] acetic acid hydrazide which has shown potent anticancer activity and so on synthetic strategy has been developed. Hydrogenated derivatives of imidazothiazoles can be prepared by various methods, *e.g* reaction of 2-aminothiazole with ethyl chloroacetate in presence of NaH as base and DMF as solvent. Elctrophilic cyclization by Br_2 and I_2 is also a convenient method for the preparation of hydrogenated imidazothiazoles, (Saliyeva *et al.*, 2018).

Other imidazothiazole analogues like, imidazo[5,1-*b*]thiazoles can be synthesized from thiazole derivative by using trifluoroacetic anhydride as an acylating agent, (Schafer, Ahmetovic, & Abele, 2017). In another methodology for the synthesis of imidazo[5,1-*b*]thiazoles, 2-aminomethylthiazole and derivatives of thiazolyl-2-acetic acids served as substrate for synthesizing imidazo[5,1-*b*]thiazoles, (Castanedo *et al.*, 2016). Mostly used method for synthesis of imidazo[1,5-*c*]thiazole involves the cyclocondensation reaction of L-cysteine hydrochloride hydrate with benzaldehyde, (Saliyeva *et al.,* 2020). A convenient method for the synthesis of imidazo[1,2-*c*] thiazole is by bicyclization reaction of isonitriles with thioamides in presence of Pd catalyst, (Tong *et al.,* 2018).

The primary application of imidazothiazole compounds is based on their anthelmintic activity. For example, the L-isomer of Tetramisole, known as levamisole, possesses anthelmintic properties and is marketed under the trade name Ergamisole, (Saliyeva *et al.,* 2020). Levamisole is widely used in livestock, including cattle, sheep, goats, pigs, poultry, dogs, and cats, to control parasitic roundworms. In addition to their anthelmintic activity, imidazothiazole compounds also exhibit a stimulating effect on the immune system and have been used in the treatment of certain tumours. Quizartinib, a benzoimidazothiazole compound, is classified as a phenylurea and specifically inhibits class III receptor tyrosine kinases. These include FMS-related tyrosine kinase 3 (FLT3/STK1), colony-stimulating factor 1 receptor (CSF1R/FMS), stem cell factor receptor (SCFR/KIT), and platelet-derived growth factor receptors (PDGFRs), (Abdullah *et al.,* 2020). This inhibition results in the suppression of ligand-independent leukemic cell proliferation and induces apoptosis. Pifithrin-β, or 2-(4-methylphenyl)-5,6,7,8-tetrahydroimidazo[2,1-*b*] [1,3] benzothiazole is an imidazo[2,1-*b*] benzothiazole derivative that plays a significant role in the treatment of Alzheimer's disease. WAY-181187 is a potent and selective full agonist of the 5-HT_1A receptor. It significantly increases extracellular GABA levels in several brain regions such as the frontal cortex, hippocampus, striatum, and amygdala in rats. Importantly, it does not affect GABA levels in the nucleus accumbens or thalamus. Furthermore, WAY-181187 has minimal to no impact on neurotransmitters like norepinephrine, serotonin, dopamine, or glutamate in these brain regions. In preclinical studies using rodent models, WAY-181187 has shown efficacy in treating depression, (Schechter *et al.,* 2008).

BIOLOGICAL POTENTIAL OF IMIDAZOTHIAZOLE DERIVATIVES

Fused heterocyclic systems are of significant interest in medicinal chemistry due to their wide range of physiological activities. These include anticancer, anti-inflammatory, antioxidant, antiviral, and antimicrobial properties.

Levamisole, one of the well-known imidazothiazole derivatives is known for its anthelmintic properties. In addition to its anthelmintic properties, levamisole also has some more biologically important properties like immune system modulator and immuno-stimulator properties and is also utilized in cancer adjuvant therapy. It is currently prescribed in combination with 5-fluorouracil (5-FU) as an adjuvant therapy after the surgical removal of stage TNM 3 or Duke's C colon cancer, for one-year post-surgery. Additionally, it is used alongside radiation therapy for Duke's stage B2 and TNM stage 4 cancers. Levamisole's biological significance has piqued the interest of numerous researchers, leading them to study structurally related aromatic heterocycles, such as imidazothiazoles, (Kamal *et al.,* 2005).

Imidazothiazoles have garnered significant interest in the past years as a robust platform for exploring new pharmacologically active substances. The biological properties of these compounds have been studied extensively, leading to promising discoveries in the field of drug development. These heterocyclic compounds are known for their diverse biological activities, including antimicrobial, anti-inflammatory, anticancer, and antiviral properties. The extensive array of bioactive compounds incorporating these fused heterocycles, along with the substantial volume of recent papers and reviews, demonstrates the scientific community's keen interest in this field.

Figure 2. Structure of Levamisole

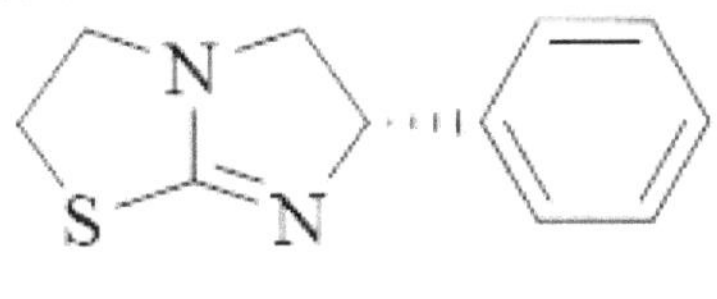

Anticancer Activity

Many imidazothiazole compounds are reported to have therapeutic activity against various types of cancers. Andreani et al. reported on the synthesis and anticancer properties of imidazo[2,1-*b*] thiazolic rings that were replaced with polyphenol groups. Compounds **1** and **2** (**Figure 3**) shown the capacity to hinder the activity of NFκB, a pivotal transcription factor involved in the growth and advancement of cancer. Compound **1** and **2** have demonstrated more effectiveness than their positive controls, with IC_{50} values of 0.36 ± 0.42 μM and 0.53 ± 0.4 μM, respectively. These results were compared to the positive controls, tosyl phenylalanyl chloromethyl ketone (with an IC_{50} value of 3.8 ± 1.1 μM) and BAY-11 (with an IC_{50} value of 2.0 ± 0.54 μM). Both chemicals function by enhancing the activity of a cytoprotective enzyme called NAD(P) Quinone reductase 1 (QR1), which safeguards against the development of cancer by neutralizing and removing cancer-causing substances.

Figure 3. Imidazothiazoles containing polyphenolic structures

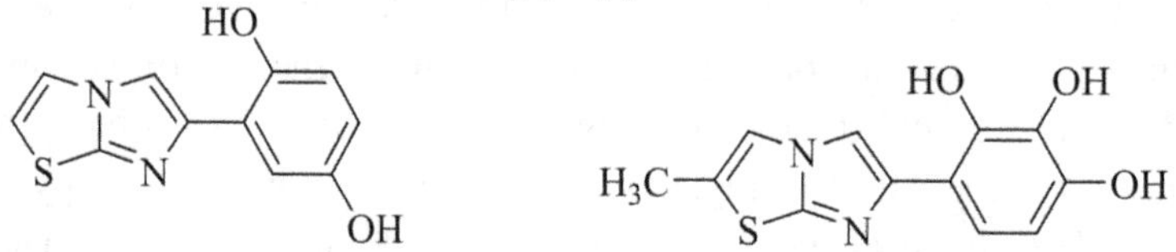

Among the wide variety of imidazothiazoles with antitumour properties the imidazothiazoles featuring a pyrazole ring received thorough examination. Ali et al. documented their synthesis of a series of derivatives of these compounds. Compounds incorporating pyrazolidine-3,5-dione (**3a-c**), 3-methyl-1*H*-pyrazol-5(4*H*)-one (**4a-c**), or 3,5-dimethyl-1H-pyrazoles (**5a-c**) (**Figure 4**) as part of their structure have displayed encouraging potential for advancing as anticancer drug candidates.

Figure 4. The derivatives of imidazothiazole pyrazoles synthesized by Ali

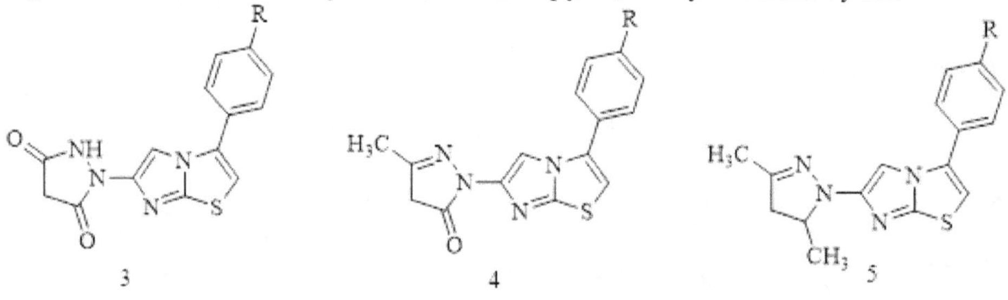

a:R = H b: b:R = Cl; c:R = CH_3

Andreani *et al.* (2012) reported on the synthesis of imidazo[2,1-*b*] thiazole guanylhydrazone derivatives that possess both anticancer and cardiotonic properties. The scientists observed that including a 3- or 4-nitrophenyl group (**Figure 4**) as a pharmacophore in the imidazothiazole guanyl hydrazone series resulted in the creation of three exceptionally potent antitumoral drugs (**6-8**). Compound **8** exhibited a more favorable pharmacological profile compared to other compounds.

It demonstrated substantial differences between cytostatic and cytotoxic dosages in six cellular lines, including leukemia, non-small cell lung cancer, colon cancer, melanoma, ovarian cancer, and breast cancer. Although the exact mechanism of action is unknown, two of the substances examined exhibited a noteworthy supplementary cardiotonic impact. These two functions have the ability to shield cardiac myocytes against apoptosis triggered by anticancer therapies.

Figure 5. Imidazo[2,1-b] thiazole guanyl hydrazones

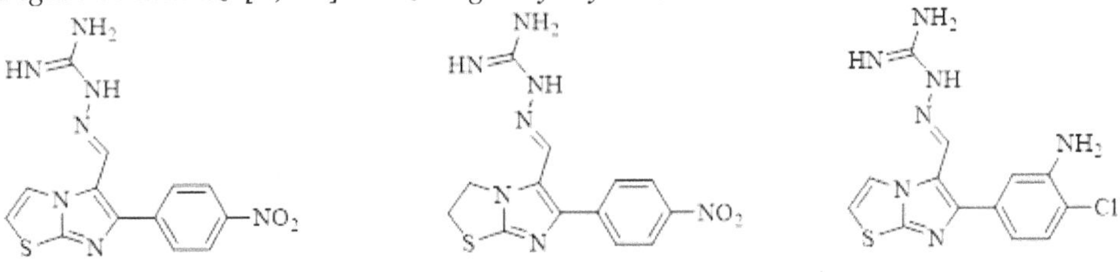

6 7 8

In a recent study, Andreani et al. investigated a new series of imidazo[2,1-*b*] thiazole guanyl hydrazones both as cytotoxic agents and inhibitors of RSK2 kinase. Specifically, analogues of compound **6**, such as 2-F, 2-Br, and 2-Cl-7-(4-nitrophenyl)-imidazo[2,1-*b*] thiazole guanyl hydrazones, demonstrated significant selectivity in inhibiting RSK2 kinase compared to a range of other related kinases. These findings highlight the potential of these compounds as promising candidates for developing novel drugs aimed at interfering with the growth and spread of tumour cells.

Moreover, pyrimidinyl-substituted imidazo[2,1-*b*] thiazole derivatives have been identified as inhibitors of RAF kinases or Rapidly Accelerated Fibrosarcoma kinases, known for their effectiveness against various cancers, including melanoma. Specifically, compounds **9** and **10**, illustrated in **Figure 6**, showed heightened selectivity for melanoma cell lines over other cancer types. They demonstrated sub-micromolar IC_{50} values across six melanoma cell lines (MALME-3M, M14, MDA-MB-435, SK-MEL-28, UACC-257, UACC-62).

Figure 6. Structure of pyrimidinyl subsituted imidazo[2,1-b] thiazoles

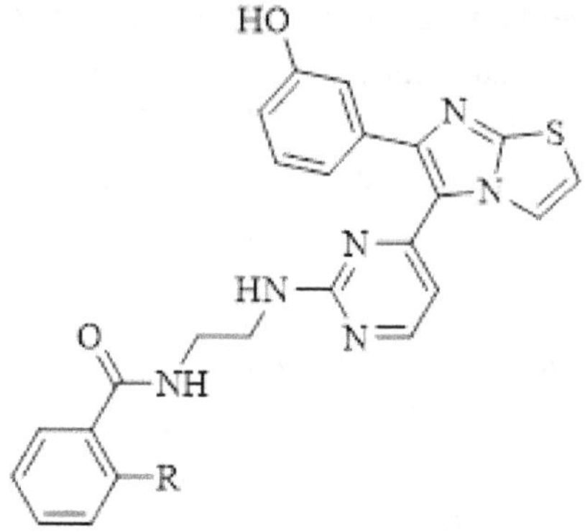

9. R = H
10 R = OH

The 2,5-disubstituted imidazothiazole derivatives SRT1720 (**11**), SRT1460 (**12**), and SRT2183 (**13**) (**Figure 7**) are notable for their role in activating SIRT1 which are among the first synthetic Sirtuin 1 activating compounds (STACs). Sirtuin 1 (SIRT1) is an enzyme that plays a critical role in cellular regulation by removing acetyl groups from various proteins. This deacetylation process influences numerous cellular functions, including gene expression, DNA repair, metabolism, and stress resistance. SIRT1 belongs to the sirtuin family of proteins, which are NAD+-dependent deacetylases. By removing acetyl groups from proteins, SIRT1 alters their function, stability, and interaction with other molecules. Research involving transgenic mice that overexpress SIRT1 has provided valuable insights into the enzyme's functions such as

Leanness and Metabolic Profile: Studies have demonstrated that transgenic mice with elevated levels of SIRT1 are leaner compared to their control littermates. This leanness is attributed to improved metabolic regulation, including enhanced fat oxidation and energy expenditure.

Improved Insulin Sensitivity: These mice show better insulin sensitivity, which is crucial for maintaining normal blood glucose levels. Enhanced insulin sensitivity helps prevent metabolic disorders such as type 2 diabetes.

Increased Mitochondrial Function: Overexpression of SIRT1 has been linked to increased mitochondrial function and biogenesis, leading to better energy production and overall metabolic health.

The upregulation of SIRT1 is associated with a protective effect against a range of age-related diseases, including cancer, Alzheimer's disease, and type 2 diabetes. Its ability to regulate key cellular processes, such as DNA repair, apoptosis, and metabolism, underscores its therapeutic potential. Ongoing research into SIRT1 and its activators continues to explore new avenues for treating these and other age-related conditions. Several studies have highlighted the protective role of SIRT1 in various cancers like Colon Cancer by deacetylating and regulating proteins involved in cell proliferation and apoptosis. This can lead to a reduction in tumour size and number; Prostate Cancer SIRT1 influences the expression of genes involved in tumour suppression and cell survival. Its activation has been linked to decreased tumour growth and improved outcomes in animal models; DNA Damage-Induced Thymic Lymphomas: SIRT1 helps repair DNA damage and maintain genomic stability, which is crucial in preventing the development of thymic lymphomas. By enhancing DNA repair mechanisms, SIRT1 reduces the incidence and growth of these tumours. Two opposing mechanisms have been proposed to explain this activity: (i) direct allosteric activation of Sir2/SIRT1 where STACs directly bind to SIRT1, inducing a conformational change that increases the enzyme's activity. This allosteric activation enhances the enzyme's ability to deacetylate its target proteins, leading to increased SIRT1 activity. Whereas in (ii) indirect activation resulting from off-target effects suggests that the observed increase in SIRT1 activity is due to the indirect effects of STACs on other cellular components, rather than direct binding to SIRT1. These off-target effects might influence pathways that, in turn, lead to an upregulation of SIRT1 activity.

Figure 7. Sirtuin 1 activating structures

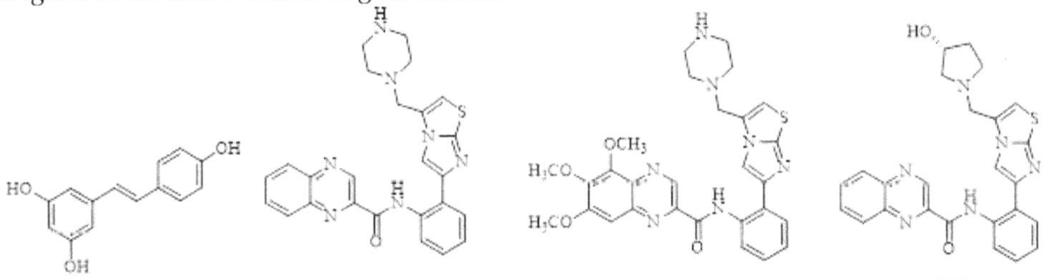

Resveratrol 11. SRT1720 12 SRT1460 13 SRT2183

Hubbard et al. introduced a nuanced mechanism known as assisted-allosteric activation (AAA) which is STACs bind to a specific site on SIRT1 that becomes exposed upon substrate binding. This binding is not to the active site directly but to an allosteric site influenced by substrate presence. The binding of STACs to this allosteric site stabilizes the interaction between SIRT1 and its substrate. This stabilization enhances the enzyme's ability to bind and deacetylate the substrate efficiently. As a result of this stabilization, the rate and efficiency of substrate deacetylation by SIRT1 are significantly increased. This assisted-allosteric activation ensures that STACs effectively boost SIRT1 activity by enhancing its natural substrate interactions.

While resveratrol (3,5,4′-trihydroxystilbene) (**Figure 7**) is recognized as the most potent natural activator of SIRT1, synthetic imidazothiazoles that activate SIRT1 exhibit a comparable response but at significantly lower concentrations.

The drugs were evaluated based on their potency, measured by the concentration required to increase enzyme activity by 50% (EC1.5) and the maximum activation achieved at the highest tested doses. Resveratrol showed an EC1.5 of 46.2 μM and achieved a maximum activation of 201%. In contrast, the imidazothiazoles SRT1460, SRT2183, and SRT1720 demonstrated EC1.5 values of 2.9 μM, 0.36 μM, and 0.16 μM, respectively, with maximum activations of 447%, 296%, and 781%. This highlights the significantly greater potency of the imidazothiazoles in activating SIRT1 compared to resveratrol.

RAF kinases, specifically ARAF, BRAF, and CRAF, play a crucial role in the MAPK (mitogen-activated protein kinase) signalling pathway. This pathway is essential for regulating various cellular processes, including growth, proliferation, differentiation, and survival.

Mutations in RAF kinases, particularly in BRAF, are frequently associated with cancer. BRAF, being a major activator of the MAPK signalling pathway, plays a pivotal role in transmitting signals from cell surface receptors to the nucleus, leading to changes in gene expression that drive cancer progression. As a result, BRAF has become a significant target for cancer therapy, with several BRAF inhibitors developed to specifically target and inhibit the activity of mutant BRAF, thus blocking the aberrant signalling and potentially controlling cancer growth.

Abdal Maksud *et al.* in 2019 has designed and synthesized a new series of imidazo[2,1-*b*] thiazole derivatives. These new compounds feature a 3-fluorophenyl group at the 6-position of the imidazo[2,1-*b*] thiazole ring and a pyrimidine ring at the 5-position. The pyrimidine ring includes either an amide or sulphonamide group, attached via an ethyl or propyl linker at the 2-position of the pyrimidine ring. Most of the derivatives exhibited cytotoxic activity against colon cancer and melanoma cell lines. Furthermore, the IC_{50} values of the target compounds were measured in A375 and SK-MEL-28 cell lines, with sorafenib used as the positive control. The compounds demonstrated superior activity compared to sorafenib. The most potent ones were tested against wild type BRAF, V600E BRAF, and CRAF. Compound **14** showed a potential inhibitory effect against V600E BRAF, with an IC_{50} of 9.3 nM.

Figure 8. Imidazothiazole derivative showing BRAF inhibitory potency

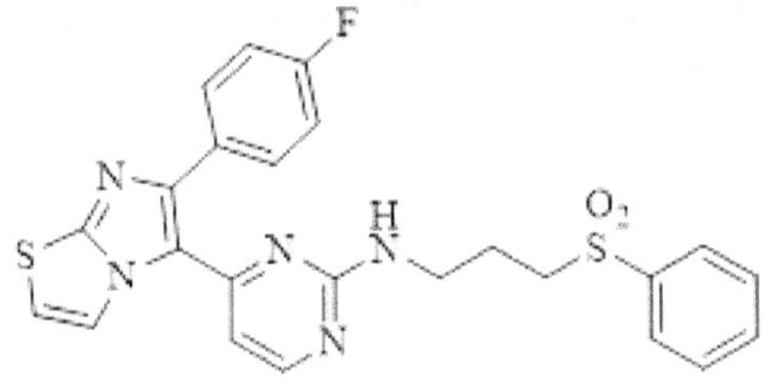

14

M. Feroz. Baig *et al.* has synthesized a series of imidazo[2,1-*b*] thiazole-benzimidazole conjugates and assessed their antiproliferative activity against four human cancer cell lines: HeLa (cervical), A549 (lung), MCF-7 (breast), and DU-145 (prostate), in addition to the normal HEK-293 cell line. Compound **15** showed notable cytotoxicity against the human lung cancer cell line A549, with an IC_{50} value of 1.08 μM. Moreover, cell cycle analysis demonstrated that this compound arrested the cell cycle at the G2/M phase in A549 cells. Additionally, the tubulin polymerization assay results suggest that compound 15 significantly inhibits tubulin assembly, with an IC_{50} value of 1.68 μM.

Figure 9. Imidazothiazole derivative active against cancer

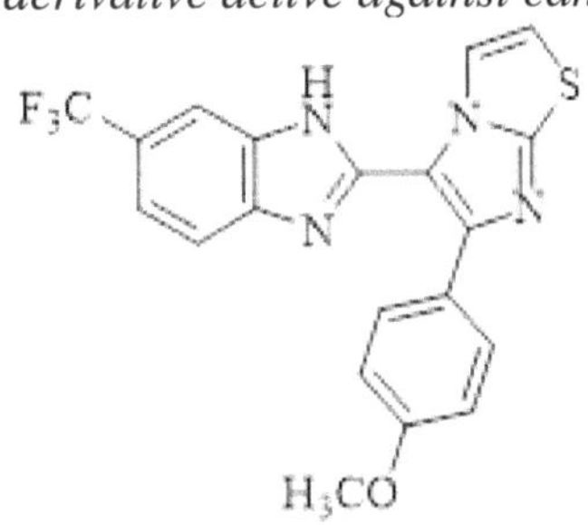

15

A new series of substituted 3-(5-imidazo[2,1-*b*] thiazolylmethylene)-2-indolinones has been identified as effective anticancer agents, (Andreani *et al.*, 2012). Among them, compounds 16, 17, and 18 show significant growth inhibition at sub-micromolar concentrations. These compounds cause apoptosis in HT-29 colon adenocarcinoma cells through caspase activation and phosphatidylserine externalization, although their effects on tubulin are relatively modest.

Notably, compounds 19 and 20 are the most potent, strongly inhibiting the activation of the kinase Akt and demonstrating significant anticancer activity. Further studies in HT-29 cells reveal that these compounds can block cells in the M phase of the cell cycle without affecting microtubule dynamics. This suggests that these substituted indolinones have a unique mechanism of action and significant potential as cancer therapeutics, particularly in targeting Akt signalling pathways and causing cell cycle arrest.

Figure 10. A new series of substituted 3-(5-imidazo[2,1-b] thiazolylmethylene)-2-indolinones

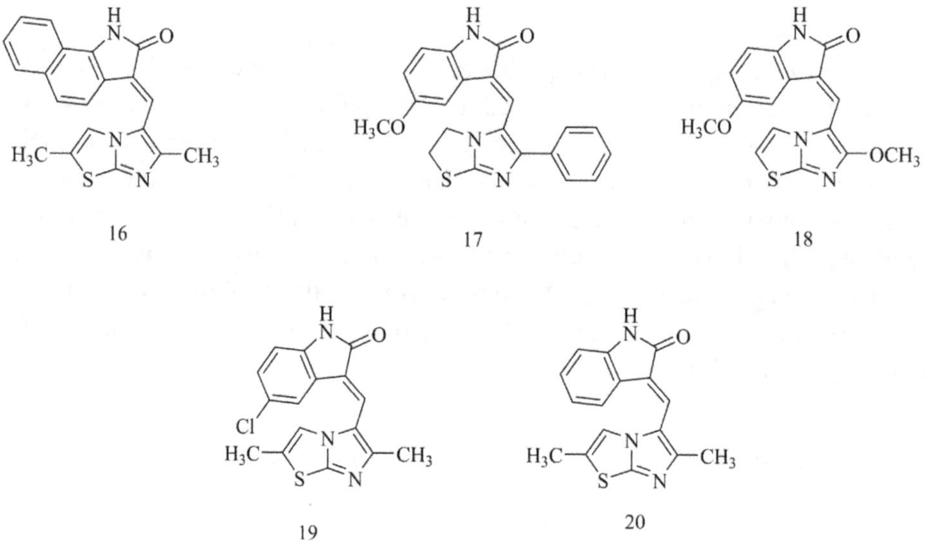

Buttar *et al.* (2008) investigated a series of imidazole-vinyl-pyrimidine derivatives for their potential as anticancer agents. In a phospho-Tie-2 cell-based ELISA assay, compounds **19** and **20** had IC_{50} values of 2.9 μM and 0.33 μM, respectively. Compound **19** displayed good selectivity in a wide range of kinase assays, inhibiting only three other enzymes with IC_{50} values below 10 μM (p38 at 1.6 μM, Flt-4 at 5.5 μM, and KDR at 7.0 μM). However, compound **19** was inactive in vascular endothelial growth factor receptor (VEGFr) and p38 cellular assays, and it showed better oral bioavailability in a mouse cassette dosing experiment (C_{max} = 0.29 μM after a 1 mg/kg dose).

Park *et al.* (2011) synthesized and evaluated the anticancer activity of a series of imidazo[2,1-*b*] thiazole derivatives. These compounds were tested for their in vitro antiproliferative activities against the A375P human melanoma cell line and the NCI-60 cell line panel. Compounds **21**, **22**, and **23** demonstrated superior activity compared to sorafenib against the A375P cell line. Notably, compounds **22** and **23** exhibited greater selectivity for melanoma cell lines over other cancer types (**Figure 11**).

Figure 11. Structure of imidazothiazole conjugates

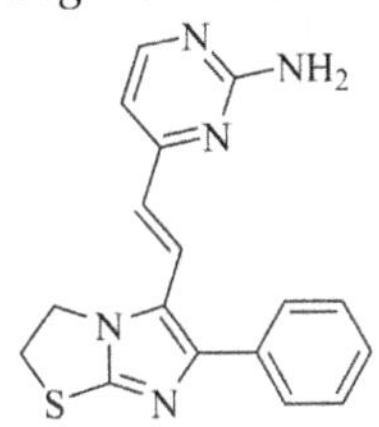

19

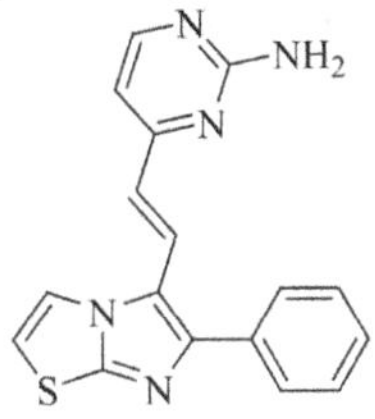

20

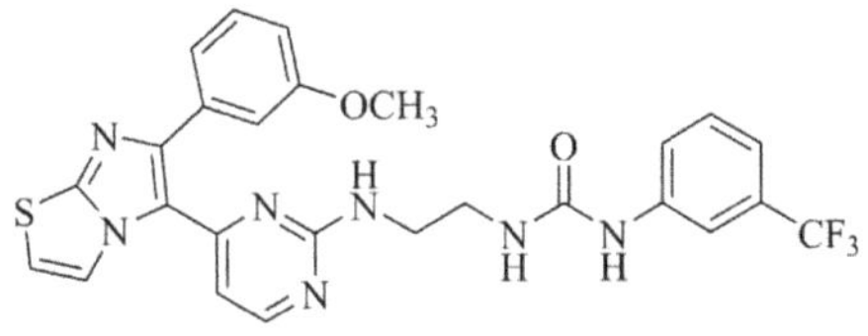

21

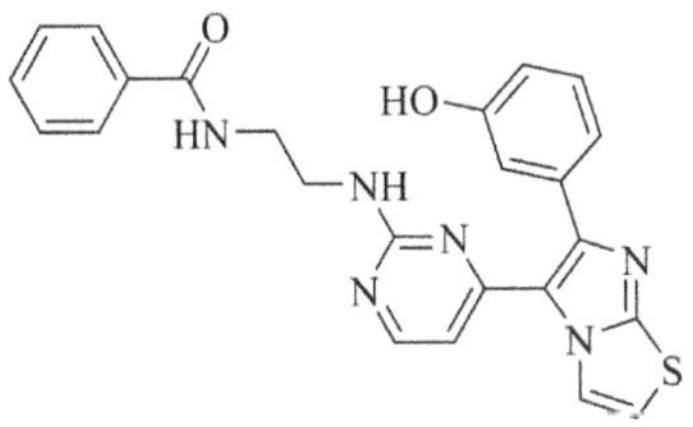

22

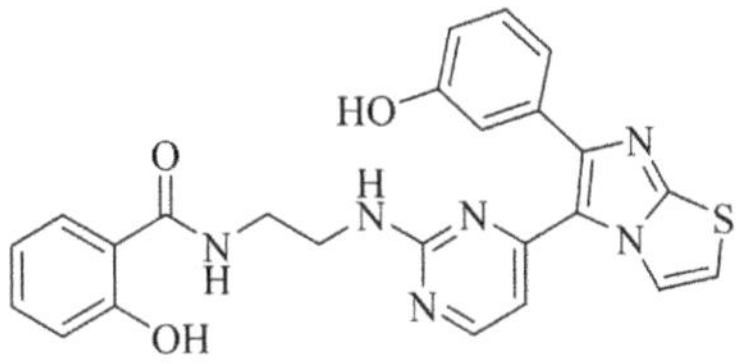

23

A series of Imidazothiazole compounds were reported by Tojo *et al.* in 2014 as potent inhibitors of the enzyme indoleamine 2,3-dioxygenase 1 (IDO1) which displayed an IC_{50} value of 1.9 μM (Compound **24**) against recombinant human IDO1 (rhIDO1), (Kamal *et al.*, 2012).

Figure 12. Imidazothiazole compounds were reported by Tojo et al.

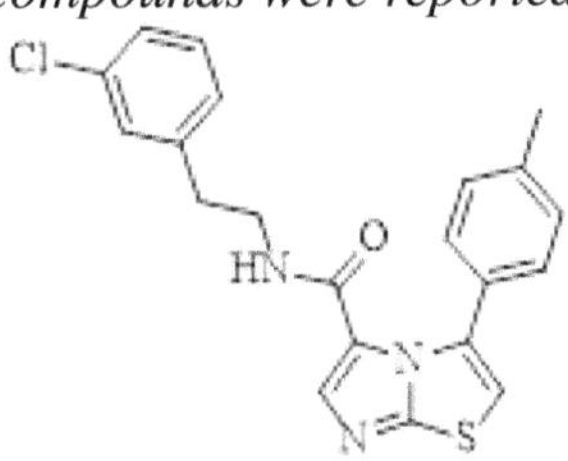

24

Gali *et al.* (2014) assessed the cytotoxic properties of a range of new coumarinylimidazo[2,1-b] thiazoles using the MTT assay on several cancer cell lines, such as MCF-7 (breast), HepG2 (liver), HeLa (cervical), and NCI-H460 (lung). The MTT test is used to measure the viability and proliferation of cells by converting the tetrazolium dye (MTT) into a formazan product. This product is then quantified colorimetrically to determine the number of viable cells. This method enables the evaluation of the potential cytotoxicity of the substances on several cancer cell lines. Compound **25c** had higher efficacy against MCF-7, HepG2, and HeLa cell lines, with IC_{50} values of 10.83±0.5 and 6.77±0.2 μM. Compound **27** exhibited efficacy against MCF-7 and HeLa cell lines. In addition, compounds **25a**, **25b**, and **26a** exhibited significant activity against HeLa cell lines, with IC50 values of 7.13±0.14, 14.21±0.6, and 17.87±0.5 μM, respectively. Compound **29b** exhibited significant efficacy against NCI-H460, as indicated by an IC_{50} value of 16.27±0.5 μM. Hence, it was shown that compounds **25a**, **25b**, **26a**, **26b**, **26c**, and **27** had elevated cytotoxicity.

Figure 13. Coumarinylimidazo[2,1-b] thiazoles structures synthesized by Gali et al. (2014)

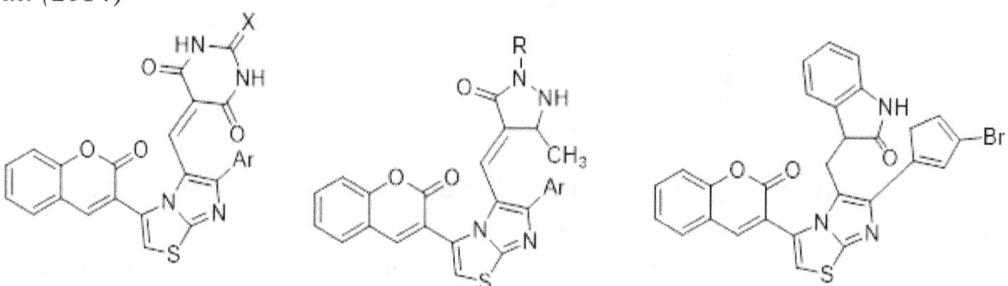

25a Ar = 4-BrC_6H_4, X = O
25b Ar = 4-BrC_6H_4, X = S
25c Ar = 4-ClC_6H_4, X = S

26a Ar = 4-BrC_6H_4, R = H
26b Ar = 4-BrC_6H_4, R = CHC_6H_5

27

Using the MTT method, Shaik *et al.* (2017) investigated novel 1,2,3-triazolo linked benzo[*d*]imidazo[2,1-*b*] thiazole conjugates (**28**) to evaluate their cytotoxic activity against different cell lines e.g. DU145, HeLa, MCF-7, HepG2, and A549 cell lines. Compounds **28a** and **28b** were identified as the most active among the tested compounds, with IC_{50} values of 0.607 and 0.781 μM, respectively, against the MCF-7 cell line. Compounds **28a** and **28b** were effective at arresting cells in the G2/M phase of the cell cycle. They also inhibited tubulin polymerization by 49.9% and 52.3%, with IC_{50} values of 1.23 and 1.65 μM, respectively. Molecular docking studies revealed that the active compounds **28a** and **28b** had enhanced binding and interaction with the colchicine binding site on tubulin. This site is situated at the interface between the α and β subunits of the tubulin protein. The effective binding of these compounds to this specific site suggests they may inhibit tubulin polymerization similarly to colchicine, potentially disrupting microtubule formation and contributing to their cytotoxic effects.

Figure 14. 1,2,3-triazole containg benzo[d]imidazo[2,1-b] thiazole derivatives

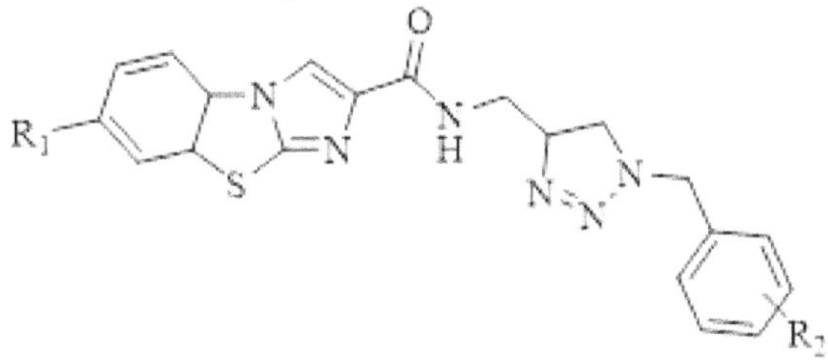

28

28a R_1= H, R_2= 4-F
28b R_1= Me, R_2= 4-F

Cardiovascular Activity

Some substituted imidazothiazoles have shown promise in the treatment of certain cardiovascular diseases due to their potential therapeutic effects. These compounds can act as inhibitors of various enzymes and receptors involved in cardiovascular function. 1,4-dihydropyridines (1,4-DHPs) like Nifedipine (**29**) are indeed L-type calcium channel (LTCC) blockers. These compounds work by inhibiting the influx of calcium ions through L-type calcium channels, which are crucial for the contraction of cardiac and smooth muscle but there is a significant challenge in the study of brain pathologies induced by high doses of nifedipine-like drugs. These drugs, which are used to block LTCCs, can cause brain pathologies, but it remains unclear whether these effects are due to LTCC inhibition or the influence of other mechanisms. This uncertainty is exacerbated by the lack of selective LTCC modulators for the brain. Its interaction with different subtypes of L-type calcium channels (LTCCs) is crucial for unravelling its effects on the brain and cardiovascular system, (Fascio, Errea, & D'accorso *et al.*, 2015).

Figure 15. nifedipine (29) and new Imidazo[2,1-b] thiazole derivatives (30,31, and 32)

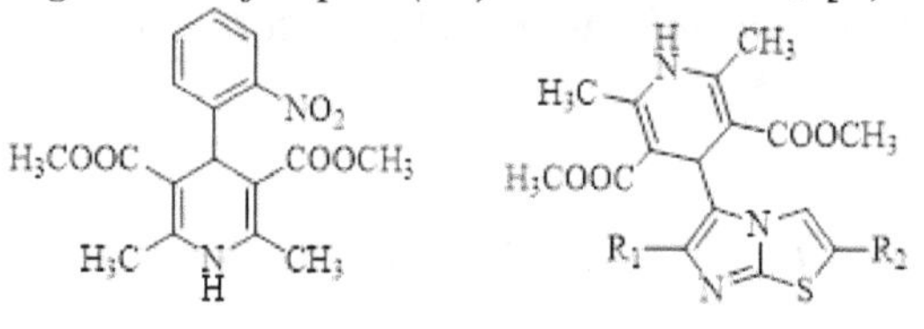

29

30.R_1= Methyl, R_2= 4 nitro 2.5 dimetroxyphenyle
31.R_1= Methyl, R_2= Cl
32.R_1= H, R_2= 2 -(trifluromethoxy phenyl)

Anti-Inflammatory Activity

The 6-arylidenimidazothiazolones synthesized by Saliyeva *et al.* (2021) were investigated for their antimicrobial and antiexudative properties. These compounds are interesting due to their potential therapeutic effects. The compound (*Z*)-6-(4-hydroxy-3-methoxy-5-nitrobenzylidene)-2-methyl-2,3-dihydroimidazo[2,1-*b*] [1,3] thiazol-5(6*H*)-one demonstrated antibacterial activity against *Candida albicans* ATCC 885/653 with a Minimum Inhibitory Concentration (MIC) of 15.62 µg/mL. In the study of novel 6-arylidene-2-methyl-2,3-dihydroimidazo[2,1-b] [1,3] thiazole-5(6H)-ones for anti-inflammatory activity, compounds **33** and **34** showed notable effectiveness. Specifically, compound **33** exhibited a 40.3% inhibition of inflammation, while compound **34** demonstrated a 38.8% inhibition. These percentages indicate that compounds **33** and **34** were among the most effective in reducing inflammation compared to the other 12 compounds tested. When comparing the anti-inflammatory activity of compound **33** to Diclofenac, a widely used anti-inflammatory medication with a percentage inhibition of 48.2%, compound **33** showed a percentage inhibition of 40.3%. This indicates that compound **33** has nearly the same impact as Diclofenac in terms of anti-exudative activity.

Figure 16. Structure of 6-arylidene-2-methyl-2,3-dihydroimidazo[2,1-b] [1,3] thiazol-5(6H)-ones

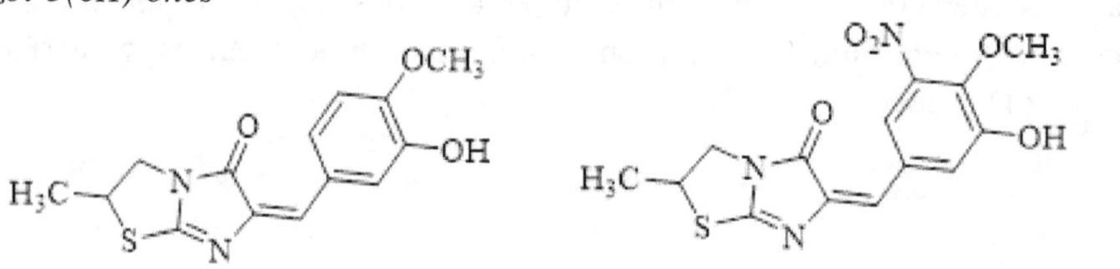

33 **34**

Shetty *et al.* (2010) conducted an in-depth study to assess the anti-inflammatory efficacy of two series of novel compounds 6-aryl-3-(3,4-dimethoxyphenyl)-2-phenylsulfonyl-imidazo[2,1-*b*] thiazole (**35**) and 6-aryl-2-benzenesulfonyl-3-(3,4-dimethoxyphenyl) imidazo[2,1-*b*] thiazole (**36**) In the study by Shetty et al., among the 6 compounds tested for anti-inflammatory efficacy, compound **35** exhibited the highest activity. The results were Showed a percentage inhibition of 40% at 2 hours Increased to 48% inhibition at 3 hours. The findings suggest that compound **36a**, with its significant anti-inflammatory activity, holds promise as an effective anti-inflammatory agent. The comparison between derivatives **36a-c** and **35a-c** highlights the superior efficacy of the 40 series, particularly compound 40a, which nearly matches the standard drug in its ability to inhibit inflammation. Further research and development are warranted to explore the potential clinical applications of these compounds.

Figure 17. Imidazothiazole compounds active as anti-inflammatory

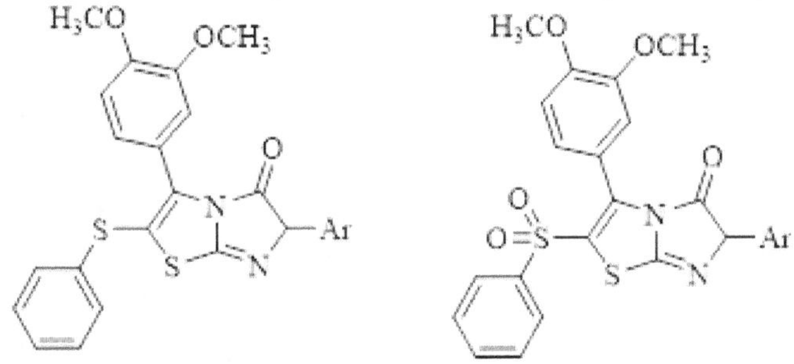

35

Ar 35a = p-Brph
Ar 35b = p-Clph
Ar 35c = p-NO_2ph

36

Ar 36a = p-Brph
Ar 36b = p-Clph
Ar 36c = p-NO_2ph

Antiviral Activity

Barradas *et al.* (2011) produced a range of imidazo[2,1-b] thiazole compounds that were altered at positions 3 and 5 with a carbohydrate component and various halobenzoyl groups. A virus yield inhibition assay was performed on Vero and A548 cells that were infected with the JUNV Strain IV4454. Rifampicin was used as the reference drug in the assay. The amount of virus produced outside of the cells was measured using a plaque test conducted on Vero cells that were infected with JUNV at a multiplicity of infection (MOI) of 0.1, without the presence of the chemical. Compounds **37a** and **37b** exhibited superior inhibitory efficacy compared to rifampicin, with EC50 values of around 1 μM and selectivity indices (SI) of 268.2 and 463.7 respectively. Nevertheless, compound **37a** exhibited efficacy solely during the process of intracellular expansion of JUNV. In addition, compound **37a** was assessed for its ability to hinder human haemorrhagic fever and was discovered to effectively impede JUNV replication in A549 human cells. The CC_{50} (MTT) and EC_{50} (JUNV inhibition assay) values were determined to be 274 μM and 2.8±0.5 μM, respectively. Hence, these compounds show great potential for further investigation and characterisation in order to determine their therapeutic efficacy against haemorrhagic fever viruses.

Figure 18. 3,5-disubstituted imidazo[2,1-b] thiazole

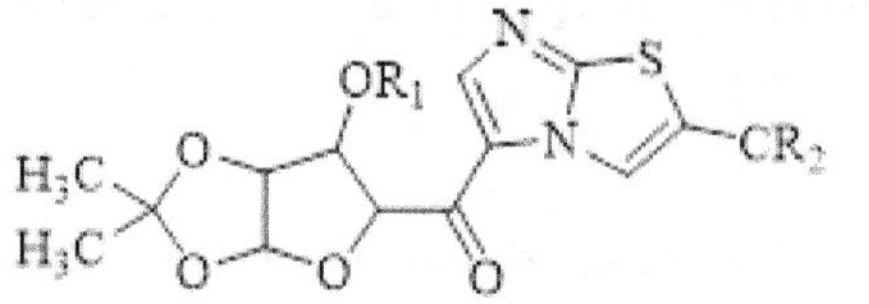

37a $R_1 = CH_3$; $R_2 = COC_6H_4Cl$-p
37b $R_2 = CH_3$; $R_2 = COC_6H_4F$-p

Guzeldemirci *et al.* (2018) synthesized a series of novel 4-thiazolidinone derivatives and evaluated their activity against various DNA and RNA viruses in mammalian cell cultures. Some of the synthesized compounds demonstrated moderate antiviral activity. For instance, 3-Propyl-2-[((6-(4-chlorophenyl)imidazo[2,1-b]thiazol-3-yl) acetyl)hydrazono]-5-methyl-4-thiazolidinone (compound 13) showed modest yet consistent activity against three strains of influenza A virus, including the 2009 pandemic virus A/H1N1 Virginia/ATCC3/2009. The cytotoxicity of compound **38** was greater than 100 μM. Additionally, compounds **38b** and **38c** displayed activity against vesicular stomatitis virus in HeLa cells, with compound **38b** having an antiviral EC_{50} value of 9 μM and a cytotoxicity of 100 μM, and compound **38c** having an antiviral EC50 value of 2 μM and a cytotoxicity of 20 μM.

Figure 19. Structures of series of new 4-thiazolidinone derivatives

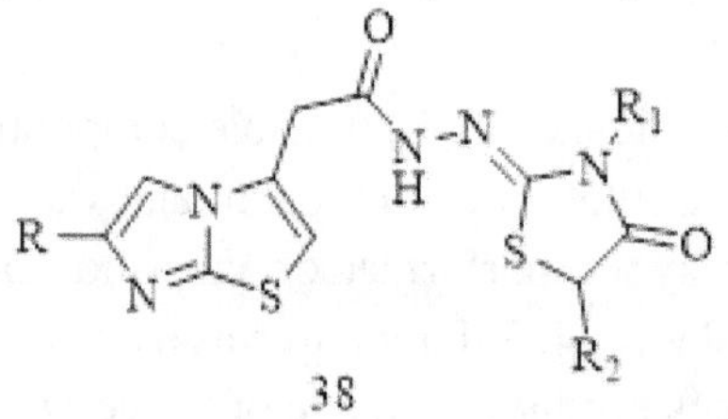

38

38a R= Cl, $R_1 = C_3H_7$, $R_2 = CH_3$
38b R= H, $R_1 = CH_2$=CH-CH_2, $R_2 = CH_3$
38c R= Cl, $R_1 = CH_3$, $R_2 = CH_3$

Antimicrobial Activity

Many researchers have designed and synthesized novel compounds containing imidazothiazoles and tested their antimicrobial properties, (Mahfouz & Elhabashy, 1990). In a recent published research by Vekariya *et al.* (2017) a series of imidazothiazole was synthesized. Those compounds were tested against both gram-positive and gram-negative strains of bacteria, and also against some fungal strains. Two of those compounds **39a** and **39b** were found to show better activity against gram-positive *Staphylococcus aureus,* with a MIC value of 50 μg/mL and 62.5 μg/mL which is lower than the ampicillin (250 μg/mL), taken as standard control. Compound **39a** was also showed better activity against *Streptococcus pyogenes,* bearing the MIC value of 62.5 μg/mL comparing with ampicillin (100 μg/mL). One compound (**39e**) from the series performed well against gram-negative *Pseudomonas aeruginosa* and *Escherichia coli* with a MIC value of 62.5 μg/mL.

Figure 20. General structure of imidazothiazole derivatives synthesized by Vekariya et al. (2017)

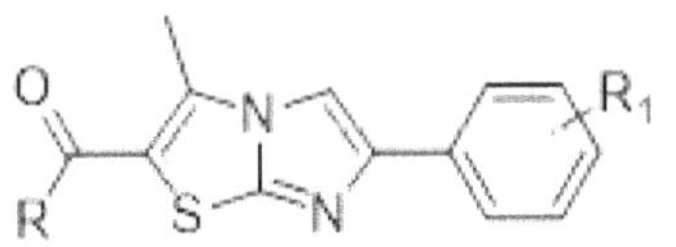

39a R= CH_3, R_1= 2-Cl-C_6H_4
39b R= OC_2H_5, R_1= 4-OCH_3-C_6H_4
39e R= OC_2H_5, R_1= 2-NO_2-C_6H_4

Two series of benzofuran linked derivatives of imidazothiazole were prepared by Santoshkumar *et al.* (2017) Antimicrobial study was performed against *S. aureus, P. aeruginosa, Enterococcus feltis, Klbesilla pneumona* and *E. coli* as the bacterial strains and *A. niger* and *Trichioderma virdae* as fungal strains. Out of the produced compounds, **40a** and **41a** showed excellent effectiveness against *S. aureus*, *P. aeruginosa*, and *E. feltis*, resulting in the largest zone of inhibition. When examining the antifungal activity, compounds **40a**, **40d**, and **41b** shown outstanding effectiveness against two fungus species that are harmful to humans, (Shankerrai, Bodke, & Santoshumar, 2017). These compounds exhibited a larger zone of inhibition compared to other compounds and were nearly as effective as the standard drug fluconazole.

Figure 21. Imidazothiazole derivatives synthesized by Santoshkumar et al. (2017)

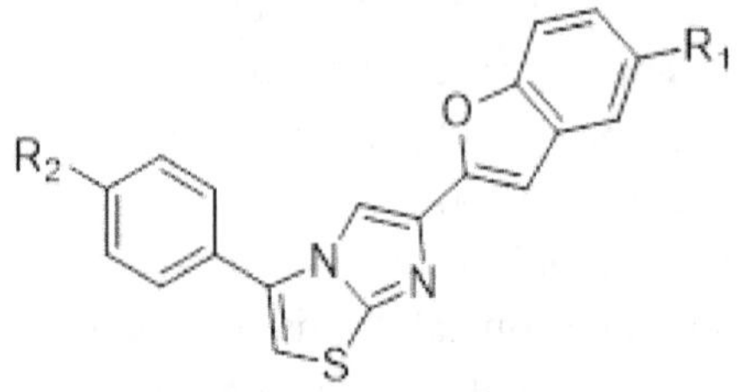

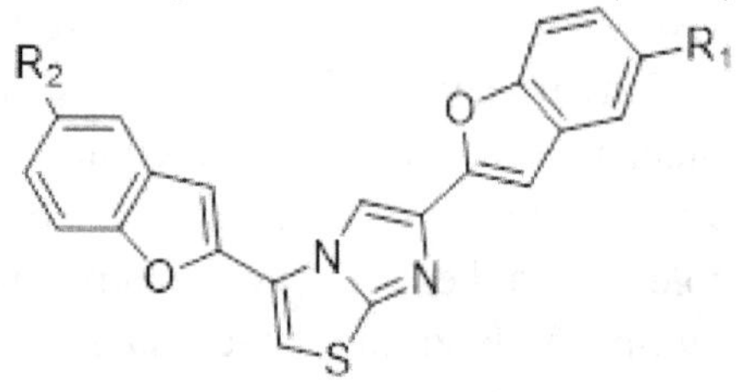

40a R_1 = Br, R_2= H
40b R_1 = H, R_2= F
40c R_1 = Br, R_2= F
40d r_1=Br, R_2=CH_3

41a R_1 = H, R_2= Br
41b R_1=Br, R_2= Br

Koudad *et al.* (2020) synthesized imidazothiazole derivative (**42**) using Claisen-Schmidt condensation and antimicrobial study was performed on various strains of bacteria like *P. aeruginosa, E. coli* and *S. aureus*. The substance exhibited significant inhibitory activity against Pseudomonas aeruginosa and Escherichia coli, with a minimum inhibitory concentration (MIC) of $\leq$ 0.2 mg/mL. For Staphylococcus aureus, the MIC was less than or equal to $\leq$ 0.5 mg/mL.

Figure 22. Imidazothiazole analogue synthesized by Koudad et al. (2020)

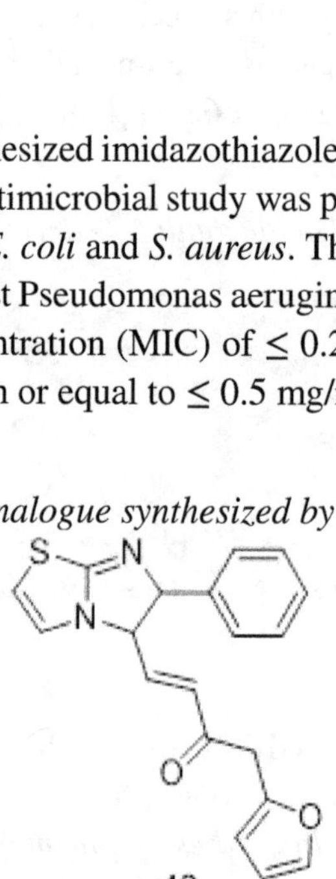

42

Antitubercular Activity

Tuberculosis is an infectious disease caused by *Mycobacterium tuberculosis* (*Mtb*), (Venugopala *et al.*, 2021). Long duration of treatment and development of resistance motivates research to develop novel agents for the treatment for tuberculosis. Some derivatives of imidazothiazole were also evaluated for treatment of tuberculosis. Samala *et al.* (2016) synthesized series of compounds bearing imidazolthiazole core. They have screened the synthesized compounds against *Mtb* H37Rv using MABA (microplate alamar blue assay) method. Two of the compounds **43** and **44** showed most potent activity among all the compounds tested with IC_{50} values of 0.52±0.24 and 0.53±0.13 μM respectively. Adult zebrafish model was used for the *in vivo* study which showed good result and less cytotoxity.

Figure 23. Imidazothiazole analogue synthesized by Samala et al. (2016)

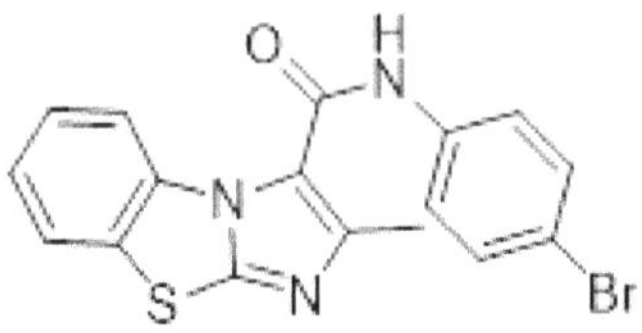

43

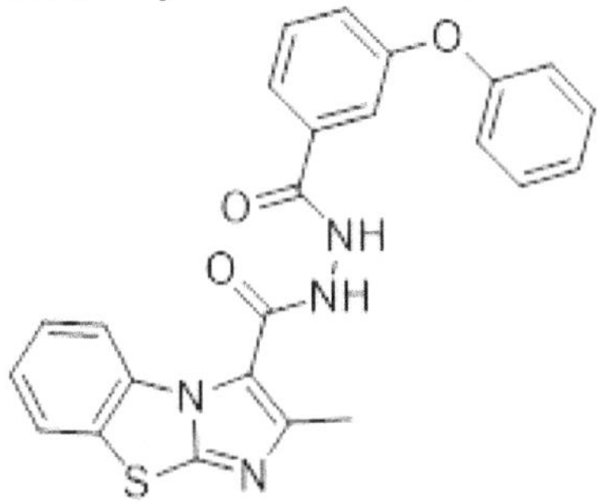

44

Vekariya et al. (2017) has synthesized a series of imidazothiazole derivatives under Microwave irradiation using polyethylene glycol-400 (PEG-400) as green chemistry approach. All the synthesized compounds showed good potency against *Mtb*. Among those compounds two compounds (**45a** and **45b**) showed the best activity in the series having MIC of 50 μg/mL.

Figure 24. Imidazothiazole analogue synthesized by Vekariya et al. (2017) showing antitubercular activity

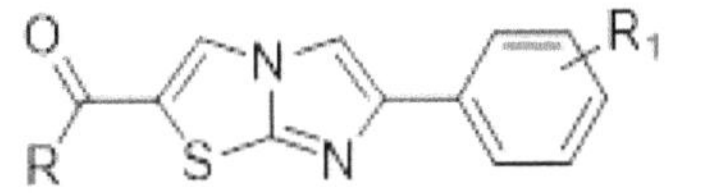

45a R=CH_3, R_1=$_4$-OCH_3-C_6H_4
45b R=CH_3, R_1=$_4$-NO_2-C_6H_4

Ulusoy and Gürsoy (2016) synthesized a series of imidazo[2,1-*b*]thiazoles and evaluated those compounds against *Mtb* H37Rv employing the BACTEC 460 radiometric system. Primary *in vitro* screening was performed and the most potent compound (**46**) was found to show MIC of 6.25 μg/mL.

Figure 25. Active molecule in the series synthesized by Nuray et al. (n.d.)

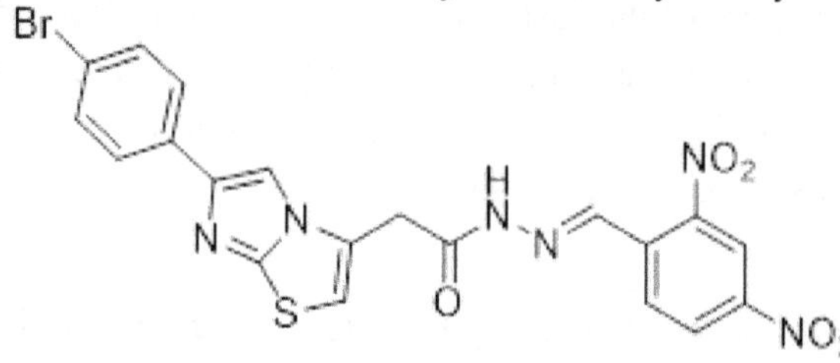

46

Inhibition of COX-2

Imidazo[2,1-*b*]thiazole analogues have also proved to have COX-2 inhibitory activity. In 2018 Shahrasbi *et al.* (2018) have synthesized a series of new Imidazo[2,1-*b*]thiazole analoguesand tested those compound as COX-1 and COX-2 inhibitor. But the compounds showed selective COX-2 inhibition with IC_{50} value of 0.08-0.16 µM range. In vitro inhibition assay was performed by the Cayman chemical chemiluminescent COX inhibitor screening assay which utilizes the heme-catalyzed hydroperoxidase activity of ovine cyclooxygenases to generate luminescence when a cyclic naphthalene hydrazide and the substrate arachidonic acid are present. Among all the synthesized compounds in the series compound **47** showed comparable potency (IC_{50}= 0.08 µM) with the standard agent Celecoxib (IC_{50}= 0.06 µM).

Figure 26. Active molecule in the series synthesized by Shahrasbi et al. (2018) having COX-2 inhibitory activity

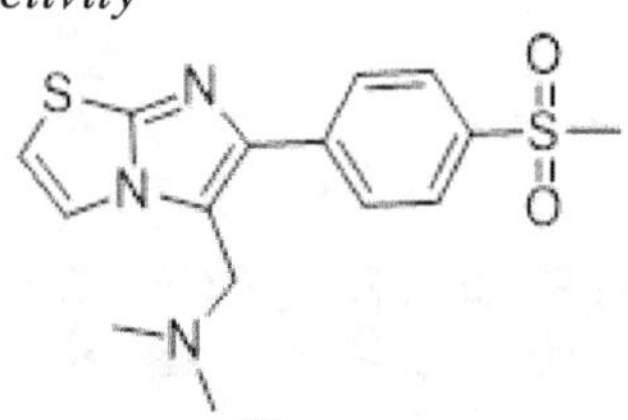

47

Anti-Oxidant Activity

Dincel *et al.* (2020) synthesized and assessed the antioxidant activity of a range of new oxo-hydrazone and spirocondensed-thiazolidine derivatives of imidazo[2,1-*b*]thiazole. The evaluation was conducted using three distinct methods: suppression of lipid peroxidation in lecithin liposomes induced by the $FeCl_3$/ascorbate system (anti-LPO), scavenging activity against ABTS radical, and Ferric Reducing Anti-oxidant Power (FRAP) activity. Three compounds (**48a**, **49**, and **50**) in the series showed best anti-LPO and ABTS radical removal activity, and two compounds (**48b** and **48c**) were identified to show best results in FRAP analysis. α-Tocopherol was used as the reference standard.

Figure 27. Imidazothiazole analogue synthesized by Dincel et al. (2020) showing antioxidant activity

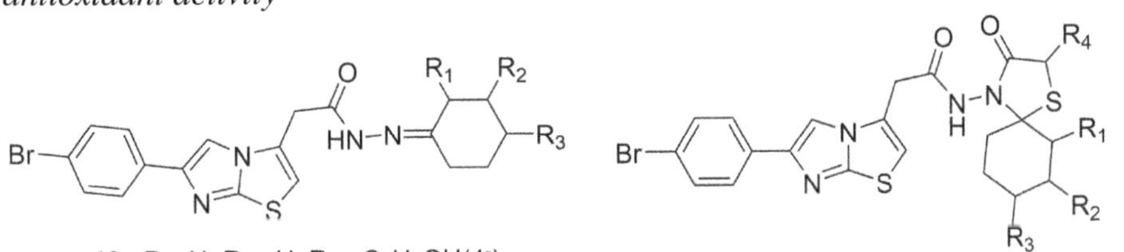

48a R_1=H, R_2= H, R_3= $C_6H_4OH(4^-)$
48b R_1=H, R_2= H, R_3= H
48c R_1=CH_3, R_2= H, R_3= H

49 R_1= H, R_2=H, R_3= C_6H_4OH (4-), R_4=H
50 R_1= H, R_2=H, R_3= C_6H_4OH (4-), R_4=CH_3

Antimalarial Activity

A series of imidazo[2,1-*b*]thiazole was synthesized under Microwave condition and was evaluated for its antimalarial activity against strain of *Plasmodium falciparum* using quinine and chloroquine as reference standard. Two of the compounds (**51a** and **52b**) were found to be active with IC_{50} of 0.2 μg/mL and 0.22 μg/mL comparing with quinine (IC_{50} = 0.268), (Vekariya *et al.,* 2017).

Figure 28. Imidazothiazole analogue showing antimalarial activity

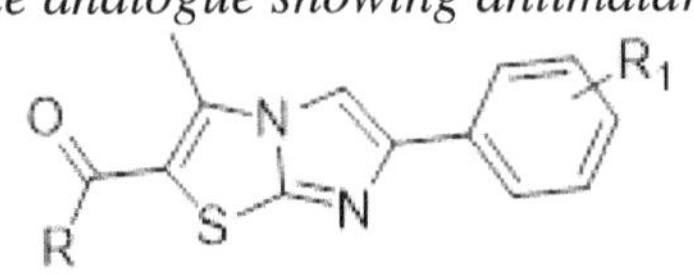

51a R= CH_3, R_1= 2-OCH_3-C_6H_4
51b R=OC_2H_5, R_1=2-OCH_3-C_6H_4

CONCLUSION

Imidazothiazole is an important class of heterocycle in medicinal chemistry. This book chapter is mainly focused on the various biological application of imidazothiazole derivatives are summarised. Imidazo[2,1-b]thiazole and its derivatives contain different biological activity such as anticancer, antiviral, anthelmintic, antitubercular, antimicrobial, anti-inflammatory, antimalarial etc. Many compounds of imidazothaizole series have been reported to show potent activity. Further modification of the scaffold may produce more potent molecule.

REFERENCES

Afsina Abdulla, C., Neetha, M., Aneeja, T., & Anilkumar, G. (2020). Synthesis and applications of imidazothiazoles: An Overview. *ChemistrySelect*, 5(33), 10374–10386. DOI: 10.1002/slct.202002842

Al-Mulla, A. (2017). A review: Biological importance of heterocyclic compounds. *Der Pharma Chemica*, 9(13), 141–147.

Amery, W. K., & Bruynseels, J. P. (1992). Levamisole, the story and the lessons. *International Journal of Immunopharmacology*, 14(3), 481–486. DOI: 10.1016/0192-0561(92)90179-O PMID: 1618599

Andreani, A., Granaiola, M., Locatelli, A., Morigi, R., Rambaldi, M., Varoli, L., Calonghi, N., Cappadone, C., Farruggia, G., Stefanelli, C., Masotti, L., Nguyen, T. L., Hamel, E., & Shoemaker, R. H. (2012). Substituted 3-(5-imidazo [2, 1-b] thiazolylmethylene)-2-indolinones and analogues: Synthesis, cytotoxic activity, and study of the mechanism of action. *Journal of Medicinal Chemistry*, 55(5), 2078–2088. DOI: 10.1021/jm2012694 PMID: 22283430

Barradas, J. S., Errea, M. I., D'Accorso, N. B., Sepúlveda, C. S., & Damonte, E. B. (2011). Imidazo [2, 1-b] thiazole carbohydrate derivatives: Synthesis and antiviral activity against Junin virus, agent of Argentine hemorrhagic fever. *European Journal of Medicinal Chemistry*, 46(1), 259–264. DOI: 10.1016/j.ejmech.2010.11.012 PMID: 21115214

Bhongade, B. A., Talath, S., Gadad, R. A., & Gadad, A. K. (2016). Biological activities of imidazo [2, 1 b][1, 3, 4] thiadiazole derivatives: A review. *Journal of Saudi Chemical Society*, 20, S463–S475. DOI: 10.1016/j.jscs.2013.01.010

Buttar, D., Edge, M., Emery, S. C., Fitzek, M., Forder, C., Griffen, A., Hayter, B., Hayward, C. F., Hopcroft, P. J., Luke, R. W., Page, K., Stawpert, J., & Wright, A. (2008). Discovery of imidazole vinyl pyrimidines as a novel class of kinase inhibitors which inhibit Tie-2 and are orally bioavailable. *Bioorganic & Medicinal Chemistry Letters*, 18(16), 4723–4726. DOI: 10.1016/j.bmcl.2008.06.106 PMID: 18676144

Cabaj, W., Stankiewicz, M., Jonas, W., & Moore, L. (1995). Levamisole and its influence on the immune response of lambs. *Veterinary Research Communications*, 19(1), 17–26. DOI: 10.1007/BF01839248 PMID: 7762136

Castanedo, G., Liu, Y., Crawford, J. J., & Braun, M.-G. (2016). Synthesis of fused imidazole-containing ring systems via dual oxidative amination of C (sp3)–H bonds. *The Journal of Organic Chemistry*, 81(18), 8617–8624. DOI: 10.1021/acs.joc.6b01517 PMID: 27529722

Chirra, N., Shanigarapu, V., Abburi, N. P., Sinegubova, E. O., Pedapati, R. K., Esaulkova, Y. L., Zarubaev, V. V., & Kantevari, S. (2024). Synthesis and antiviral activity of novel imidazo [2, 1-b] thiazoles coupled with morpholine and thiomorpholines. *Journal of Heterocyclic Chemistry*, 61(3), 496–505. DOI: 10.1002/jhet.4778

Da Pozzo, E., La Pietra, V., Cosimelli, B., Da Settimo, F., Giacomelli, C., Marinelli, L., Martini, C., Novellino, E., Taliani, S., & Greco, G. (2014). p53 functional inhibitors behaving like pifithrin-β counteract the Alzheimer peptide non-β-amyloid component effects in human SH-SY5Y cells. *ACS Chemical Neuroscience*, 5(5), 390–399. DOI: 10.1021/cn4002208 PMID: 24646317

Dincel, E. D., Gürsoy, E., Yilmaz-Ozden, T., & Ulusoy-Güzeldemirci, N. (2020). Antioxidant activity of novel imidazo [2, 1-b] thiazole derivatives: Design, synthesis, biological evaluation, molecular docking study and in silico ADME prediction. *Bioorganic Chemistry*, 103, 104220. DOI: 10.1016/j.bioorg.2020.104220 PMID: 32896742

Dincel, E. D., Hasbal-Celikok, G., Yilmaz-Ozden, T., & Ulusoy-Güzeldemirci, N. (2022). Design, synthesis, biological evaluation, molecular docking, and dynamic simulation study of novel imidazo [2, 1-b] thiazole derivatives as potent antioxidant agents. *Journal of Molecular Structure*, 1258, 132673. DOI: 10.1016/j.molstruc.2022.132673

Dorai, S. T., Lakshmikanth, K., Tiwari, P., Saini, S. M., & Chandrashekharappa, S. (2023). One-pot construction of novel trifluoromethyl dihydro-imidazo [1, 2-a] pyridine: A greener approach. *Tetrahedron*, 148, 133691. DOI: 10.1016/j.tet.2023.133691

Ewida, M. A., Abou El Ella, D. A., Lasheen, D. S., Ewida, H. A., El-Gazzar, Y. I., & El-Subbagh, H. I. (2018). Imidazo [2′, 1′: 2, 3] thiazolo [4, 5-d] pyridazinone as a new scaffold of DHFR inhibitors: Synthesis, biological evaluation and molecular modeling study. *Bioorganic Chemistry*, 80, 11–23. DOI: 10.1016/j.bioorg.2018.05.025 PMID: 29864684

Fascio, M. L., Errea, M. I., & D'accorso, N. B. (2015). Imidazothiazole and related heterocyclic systems. Synthesis, chemical and biological properties. *European Journal of Medicinal Chemistry*, 90, 666–683. DOI: 10.1016/j.ejmech.2014.12.012 PMID: 25499987

Gali, R., Banothu, J., Porika, M., Velpula, R., Bavantula, R., & Abbagani, S. (2014). Synthesis and in vitro cytotoxic activity of novel coumarinylimidazo [2, 1-b] thiazole derivatives. *RSC Advances*, 4(96), 53812–53819. DOI: 10.1039/C4RA11428K

García-Valverde, M., & Torroba, T. (2005). Sulfur-nitrogen heterocycles. *MDPI*, 10, 318–320.

Guzeldemirci, N. U., Pehlivan, E., & Naesens, L. (2018). Synthesis and antiviral activity evaluation of new 4-thiazolidinones bearing an imidazo [2, 1-b] thiazole moiety. *Marmara Pharmaceutical Journal*, 22(2), 237–248. DOI: 10.12991/mpj.2018.61

Harrow, I. D., & Gration, K. A. (1985). Mode of action of the anthelmintics morantel, pyrantel and levamisole on muscle cell membrane of the nematode Ascaris suum. *Pesticide Science*, 16(6), 662–672. DOI: 10.1002/ps.2780160612

Kabir, E., & Uzzaman, M. (2022). A review on biological and medicinal impact of heterocyclic compounds. *Results in Chemistry*, 4, 100606. DOI: 10.1016/j.rechem.2022.100606

Kajita, Y., Sakai, M., Atsuta, S., & Kobayashi, M.KAJITA. (1990). Y.; SAKAI, M.; ATSUTA, S.; KOBAYASHI, M., The immunomodulatory effects of levamisole on rainbow trout, Oncorhynchus mykiss. *Fish Pathology*, 25(2), 93–98. DOI: 10.3147/jsfp.25.93

Kamal, A., Khanna, G. R., Krishnaji, T., & Ramu, R. (2005). A new facile chemoenzymatic synthesis of levamisole. *Bioorganic & Medicinal Chemistry Letters*, 15(3), 613–615. DOI: 10.1016/j.bmcl.2004.11.054 PMID: 15664823

Kamal, A., Sultana, F., Ramaiah, M. J., Srikanth, Y., Viswanath, A., Kishor, C., Sharma, P., Pushpavalli, S., Addlagatta, A., & Pal-Bhadra, M. (2012). 3-Substituted 2-Phenylimidazo [2, 1-b] benzothiazoles: Synthesis, Anticancer Activity, and Inhibition of Tubulin Polymerization. *ChemMedChem*, 7(2), 292–300. DOI: 10.1002/cmdc.201100511 PMID: 22241597

Kamboj, P., Mahore, A., Husain, A., Shaquiquzzaman, M., Alam, M. M., & Amir, M. (2023). A review on the synthesis and anticancer potentials of imidazothiazole derivatives. *Mini-Reviews in Medicinal Chemistry*, 23(17), 1733–1759. DOI: 10.2174/1389557523666230201113237 PMID: 36722483

Karaman, B., & Ulusoy Güzeldemirci, N. (2016). Synthesis and biological evaluation of new imidazo [2, 1-b] thiazole derivatives as anticancer agents. *Medicinal Chemistry Research*, 25(11), 2471–2484. DOI: 10.1007/s00044-016-1684-x

Koudad, M., El Hamouti, C., Elaatiaoui, A., Dadou, S., Oussaid, A., Abrigach, F., Pilet, G., Benchat, N., & Allali, M. (2020). Synthesis, crystal structure, antimicrobial activity and docking studies of new imidazothiazole derivatives. *Journal of the Indian Chemical Society*, 17, 297–306.

Kumar, R. S., Praveen, S., Shridharshini, K., Maruthamuthu, M., Mohanapriya, K., & Mythili, A. (2022). Biological applications of imidazothiazole scaffolds: A current review. *Journal of Advanced Chemical Sciences*, 756-769.

Liu, K. G., Robichaud, A. J., Bernotas, R. C., Yan, Y., Lo, J. R., Zhang, M.-Y., Hughes, Z. A., Huselton, C., Zhang, G. M., Zhang, J. Y., Kowal, D. M., Smith, D. L., Schechter, L. E., & Comery, T. A. (2010). 5-Piperazinyl-3-sulfonylindazoles as potent and selective 5-hydroxytryptamine-6 antagonists. *Journal of Medicinal Chemistry*, 53(21), 7639–7646. DOI: 10.1021/jm1007825 PMID: 20932009

Mahfouz, , Aziz, A. A., & Elhabashy, F. M. (1990). Mahfouz; Aziz, A. A.; Elhabashy, F., New synthesis of 2-substituted imidazo [2, 1-b] thiazoles and their antimicrobial activities. *Archives of Pharmacal Research*, 13(1), 9–13. DOI: 10.1007/BF02857826

Mishra, A., Srivastava, M., Rai, P., Yadav, S., Tripathi, B. P., Singh, J., & Singh, J. (2016). Visible light triggered, catalyst free approach for the synthesis of thiazoles and imidazo [2, 1-b] thiazoles in EtOH: H 2 O green medium. *RSC Advances*, 6(54), 49164–49172. DOI: 10.1039/C6RA05385H

Nagesh, H., Padmashali, B., Sandeep, C., Yuvaraj, T., Siddesh, M., & Mallikarjuna, S. (2014). Synthesis and antimicrobial activity of benzothiophene substituted coumarins, pyrimidines and pyrazole as new scaffold. *International Journal of Pharmaceutical Sciences Review and Research*, 28(2), 6–10.

Ningegowda, R., Chandrashekharappa, S., Singh, V., Mohanlall, V., & Venugopala, K. N. (2020). Design, synthesis and characterization of novel 2-(2, 3-dichlorophenyl)-5-aryl-1, 3, 4-oxadiazole derivatives for their anti-tubercular activity against Mycobacterium tuberculosis. *Chemical Data Collections*, 28, 100431. DOI: 10.1016/j.cdc.2020.100431

Park, J.-H., El-Gamal, M. I., Lee, Y. S., & Oh, C.-H. (2011). New imidazo [2, 1-b] thiazole derivatives: Synthesis, in vitro anticancer evaluation, and in silico studies. *European Journal of Medicinal Chemistry*, 46(12), 5769–5777. DOI: 10.1016/j.ejmech.2011.08.024 PMID: 22033063

Saliyeva, L., Holota, S., Grozav, A., Yakovychuk, N., Litvinchuk, M., Slyvka, N., & Vovk, M. (2021). Synthesis and Evaluation of Antimicrobial and Anti-inflammatory Activity of 6-aryliden-2-methyl-2, 3-dihydroimidazo [2, 1-b][1, 3] thiazoles. *Appl. Chem*, 12, 292–303.

Saliyeva, L., Vas'kevich, R., Slyvka, N. Y., & Vovk, M. (2018). The Synthesis and Structural Functionalization of 6-Substituted 2, 3-dihydroimidazo [2, 1-b][1, 3] thiazol-5-ones. *J. Org. Pharm. Chem*, 16(2(62)), 31–41. DOI: 10.24959/ophcj.18.940

Saliyeva, L. N., Diachenko, I. V., Vas'kevich, R. I., Slyvka, N. Y., & Vovk, M. V. (2020). Vas' kevich, R. I.; Slyvka, N. Y.; Vovk, M. V., Imidazothiazoles and their hydrogenated analogs: Methods of synthesis and biomedical potential. *Chemistry of Heterocyclic Compounds*, 56(11), 1394–1407. DOI: 10.1007/s10593-020-02827-w

Samala, G., Devi, P. B., Saxena, S., Meda, N., Yogeeswari, P., & Sriram, D. (2016). Design, synthesis and biological evaluation of imidazo [2, 1-b] thiazole and benzo [d] imidazo [2, 1-b] thiazole derivatives as Mycobacterium tuberculosis pantothenate synthetase inhibitors. *Bioorganic & Medicinal Chemistry*, 24(6), 1298–1307. DOI: 10.1016/j.bmc.2016.01.059 PMID: 26867485

Sandeep, C. (2013). B. P., Rashmi S. Kulkarni, Synthesis of Isomeric Subtituted 6-acetyl-3-benzoylindolizine-1-carboxylate and 8-acetyl-3-benzoylindolizine-1-carboxylate from subtituteded 3-acetyl pyridinium bromides and their antimicrobial activity. *Journal of Applicable Chemistry*, 2(5), 1049.

Sandeep, C., Padmashali, B., Kulkarni, R. S., Mallikarjuna, S., Siddesh, M., Nagesh, H., & Thriveni, K. (2014). Synthesis of substituted 5-acetyl-3-benzoylindolizine-1-carboxylate from substituted 2-acetyl pyridinium bromides. *Heterocyclic Letters*, 4(03), 371–376.

Sandeep, C., Venugopala, K. N., Khedr, M. A., Padmashali, B., Kulkarni, R. S., Rashmi, V., & Odhav, B. (2017). Design and synthesis of novel indolizine analogues as COX-2 inhibitors: Computational perspective and in vitro screening. *Indian J. Pharm. Educ. Res*, 51, 452–460. DOI: 10.5530/ijper.51.3.73

Schafer, G., Ahmetovic, M., & Abele, S. (2017). Scalable synthesis of trifluoromethylated imidazo-fused N-heterocycles using TFAA and trifluoroacetamide as CF3-reagents. *Organic Letters*, 19(24), 6578–6581. DOI: 10.1021/acs.orglett.7b03291 PMID: 29161044

Schechter, L. E., Lin, Q., Smith, D. L., Zhang, G., Shan, Q., Platt, B., Brandt, M. R., Dawson, L. A., Cole, D., Bernotas, R., Robichaud, A., Rosenzweig-Lipson, S., & Beyer, C. E. (2008). Neuropharmacological profile of novel and selective 5-HT6 receptor agonists: WAY-181187 and WAY-208466. *Neuropsychopharmacology*, 33(6), 1323–1335. DOI: 10.1038/sj.npp.1301503 PMID: 17625499

Shahin, A. I., Zaib, S., Zaraei, S.-O., Kedia, R. A., Anbar, H. S., Younas, M. T., Al-Tel, T. H., Khoder, G., & El-Gamal, M. I. (2023). Design and synthesis of novel anti-urease imidazothiazole derivatives with promising antibacterial activity against Helicobacter pylori. *PLoS One*, 18(6), e0286684. DOI: 10.1371/journal.pone.0286684 PMID: 37267378

Shahrasbi, M., Movahed, M. A., Dadras, O. G., Daraei, B., & Zarghi, A. (2018). Design, synthesis and biological evaluation of new Imidazo [2, 1-B] thiazole derivatives as selective COX-2 inhibitors. *Iranian Journal of Pharmaceutical Research: IJPR*, 17(4), 1288. PMID: 30568687

Shaik, S. P., Vishnuvardhan, M., Sultana, F., Rao, A. S., Bagul, C., Bhattacharjee, D., Kapure, J. S., Jain, N., & Kamal, A. (2017). Design and synthesis of 1, 2, 3-triazolo linked benzo [d] imidazo [2, 1-b] thiazole conjugates as tubulin polymerization inhibitors. *Bioorganic & Medicinal Chemistry*, 25(13), 3285–3297. DOI: 10.1016/j.bmc.2017.04.013 PMID: 28462842

Shankerrao, S., Bodke, Y. D., & Santoshkumar, S. (2017). Synthesis and antimicrobial activity of some imidazothiazole derivatives of benzofuran. *Arabian Journal of Chemistry*, 10, S554–S558. DOI: 10.1016/j.arabjc.2012.10.018

Shetty, N. S., Khazi, I. A. M., & Ahn, C.-J. (2010). Synthesis, anthelmintic and anti-inflammatory activities of some novel imidazothiazole sulfides and sulfones. *Bulletin of the Korean Chemical Society*, 31(8), 2337–2340. DOI: 10.5012/bkcs.2010.31.8.2337

Siddesh, M., Padmashali, B., Thriveni, K., & Sandeep, C. (2014). Synthesis of thiophene-linked pyrimidopyrimidines as pharmaceutical leads. *Journal of Chemical Sciences*, 126(3), 821–826. DOI: 10.1007/s12039-014-0614-z

Siddiqui, R., El-Gamal, M. I., Boghossian, A., Saeed, B. Q., Oh, C.-H., Abdel-Maksoud, M. S., Alharbi, A. M., Alfahemi, H., & Khan, N. A. (2022). Imidazothiazole derivatives exhibited potent effects against brain-eating amoebae. *Antibiotics (Basel, Switzerland)*, 11(11), 1515. DOI: 10.3390/antibiotics11111515 PMID: 36358170

Soyer Can, O., Ünlü, S., Ocak, H., Çöldür, E. Ç., Sipahi, H., & Bilgin Eran, B. (2022). Synthesis of new imidazothiazole derivatives and investigation of their anti-inflammatory and analgesic activities. *Journal of the Iranian Chemical Society*, 1-9.

Sumanth, G., Lakshmikanth, K., Saini, S. M., Mundhe, P., Shivaprasad, K., & Chandrashekharappa, S. (2023). Phenyl pyrrolo [1, 2-a] quinolines-finding of a key by-product during quinolinium salt preparation. *Journal of Molecular Structure*, 1273, 134350. DOI: 10.1016/j.molstruc.2022.134350

Tong, W., Li, W.-H., He, Y., Mo, Z.-Y., Tang, H.-T., Wang, H.-S., & Pan, Y.-M. (2018). Palladium-metalated porous organic polymers as recyclable catalysts for the chemioselective synthesis of thiazoles from thiobenzamides and isonitrilcs. *Organic Letters*, 20(8), 2494–2498. DOI: 10.1021/acs.orglett.8b00886 PMID: 29620903

Ulusoy, N., & Gürsoy, E. (2016). Synthesis and evaluation of new imidazo [2, 1-b] thiazoles as antituberculosis agents. *Marmara Pharmaceutical Journal*, 21(1), 102–109. DOI: 10.12991/marupj.259887

Uppar, V., Chandrashekharappa, S., Basarikatti, A. I., Banuprakash, G., Mohan, M. K., Chougala, M., ... & Padmashali, B. (2020). Synthesis, antibacterial, and antioxidant studies of 7-amino-3-(4-fluorobenzoyl) indolizine-1-carboxylate derivatives. *Journal of Applied Pharmaceutical Science, 10*(2), 077-085.

Uppar, V., Mudnakudu-Nagaraju, K. K., Basarikatti, A. I., Chougala, M., Chandrashekharappa, S., Mohan, M. K., & Padmashali, B. (2020). Microwave induced synthesis, and pharmacological properties of novel 1-benzoyl-4-bromopyrrolo [1, 2-a] quinoline-3-carboxylate analogues. *Chemical Data Collections*, 25, 100316.

Vekariya, R. H., Patel, K. D., Vekariya, M. K., Prajapati, N. P., Rajani, D. P., Rajani, S. D., & Patel, H. D. (2017). Microwave-assisted green synthesis of new imidazo [2, 1-b] thiazole derivatives and their antimicrobial, antimalarial, and antitubercular activities. *Research on Chemical Intermediates*, 43(11), 6207–6231. DOI: 10.1007/s11164-017-2985-5

Venugopala, K. N., Al-Attraqchi, O. H. A., Tratrat, C., Nayak, S. K., Morsy, M. A., Aldhubiab, B. E., Attimarad, M., Nair, A. B., Sreeharsha, N., Venugopala, R., Haroun, M., Girish, M. B., Chandrashekharappa, S., Alwassil, O. I., & Odhav, B. (2019). Novel Series of Methyl 3-(Substituted Benzoyl)-7-Substituted-2-Phenylindolizine-1-Carboxylates as Promising Anti-Inflammatory Agents: Molecular Modeling Studies. *Biomolecules*, 9(11), 661. DOI: 10.3390/biom9110661 PMID: 31661893

Venugopala, K. N., Chandrashekharappa, S., Bhandary, S., Chopra, D., Khedr, M. A., Aldhubiab, B. E., Attimarad, M., & Odhav, B. (2018). Efficient Synthesis and Characterization of Novel Substituted 3-Benzoylindolizine Analogues via the Cyclization of Aromatic Cycloimmoniumylides with Electrondeficient Alkenes. *Current Organic Synthesis*, 15(3), 388–395. DOI: 10.2174/1570179414666171024155051

Venugopala, K. N., Chandrashekharappa, S., Deb, P. K., Tratrat, C., Pillay, M., Chopra, D., Al-Shar'i, N. A., Hourani, W., Dahabiyeh, L. A., Borah, P., Nagdeve, R. D., Nayak, S. K., Padmashali, B., Morsy, M. A., Aldhubiab, B. E., Attimarad, M., Nair, A. B., Sreeharsha, N., Haroun, M., & Mailavaram, R. (2021). Anti-tubercular activity and molecular docking studies of indolizine derivatives targeting mycobacterial InhA enzyme. *Journal of Enzyme Inhibition and Medicinal Chemistry*, 36(1), 1471–1486. DOI: 10.1080/14756366.2021.1919889 PMID: 34210233

Venugopala, K. N., Tratrat, C., Pillay, M., Chandrashekharappa, S., Al-Attraqchi, O. H. A., Aldhubiab, B. E., Attimarad, M., Alwassil, O. I., Nair, A. B., Sreeharsha, N., Venugopala, R., Morsy, M. A., Haroun, M., Kumalo, H. M., Odhav, B., & Mlisana, K. (2020). In silico design and synthesis of tetrahydropyrimidinones and tetrahydropyrimidinethiones as potential thymidylate kinase inhibitors exerting anti-TB activity against Mycobacterium tuberculosis. *Drug Design, Development and Therapy*, 14, 1027–1039. DOI: 10.2147/DDDT.S228381 PMID: 32214795

Chapter 2
Biological Relevance and Applications of Triazolothiadiazole Exploring Interactions, Synthesis, and Mechanisms:
Structural Study of Triazolothiadiazoles

Biki Hazarika
Mizoram University, India

Lal Ropuia
Mizoram University, India

Ved Prakash Singh
Mizoram University, India

ABSTRACT

Fused sulfur heterocycles, such as triazolothiadiazoles, contain four nitrogen atoms and one sulfur atom as heteroatoms in their structure. These elements impart electron-donating and electron-withdrawing properties, influencing the reactivity and chemical behaviour of the compounds. Such effects contribute to various chemical interactions, including coordination with metals and hydrogen bonding. For example, in drug discovery, these interactions play a role in binding with proteins

DOI: 10.4018/979-8-3693-7520-4.ch002

through non-covalent interactions. Additionally, the electronic properties of this heterocycle can interact with plant pathogens, showing potential for applications in the agricultural sector. This chapter discusses the structural significance of triazolothiadiazoles and how they affect their reactivity, as well as their biological importance in medicinal and agricultural contexts. The triazolothiadiazole heterocycle shows significant promise for medicinal as well as agricultural applications, warranting further exploration.

BACKGROUND

Heterocyclic compounds are a type of homocyclic compound in which the cyclic ring structure is composed of carbon atoms and one or more heteroatoms (non-carbon atoms). On the other hand, cyclic compounds that consist of more than one hetero-atom are inorganic heterocycles, like 1,3,5,2,4,6-triazaborinane and silazene (Ram et al., 2019). Sulfur-containing heterocycles are organic compounds with S being part of their cyclic structure. They have applications in pharmaceuticals and agrochemicals, materials science, and organic synthesis.

Sulfur heterocycles play a crucial role in both natural and synthetic chemistry, forming essential building blocks. The presence of sulfur in these heterocycles often imparts distinct reactivity and stability, making them valuable intermediates and active agents in numerous chemical reactions. Common sulfur-containing heterocycles include thiophene, isothiazole, thiopyran, thiazole, thiazolidine (shown in Figure 1), etc. which impart unique structural and bio-functional properties. For example, dopamine D2 inhibitor olanzapine which consists of the thiophene moiety in its chemical structure (Szasz et al., 2023). It is an antipsychotic drug commercially known as ZYPREXA and was approved by the FDA in 2004. Another drug ritonavir possesses a thiazole heterocycle which acts as an inhibitor for HIV protease (*Thiazoles | DrugBank Online*, 2024). S-heterocycles exist in different ring sizes, ranging from 3-membered to 8-membered.

Figure 1. Chemical structures of different S-heterocycles

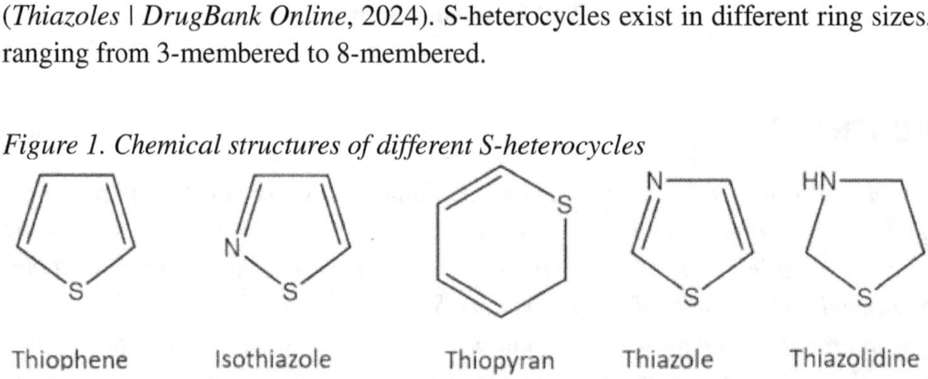

In a S-heterocycle, the electronic state of sulfur, and variation in ring size affect their reactivity and stability due to geometric ring strain and the degree of unsaturation in the molecules. These influence the overall physical, chemical, and bio characteristics of the heterocyclic system. Among the various S-heterocyclic systems, they form mono- to polycyclic systems, appearing either in fused or isolated configurations. These are of great biological significance, as they can be found in vitamins and coenzymes. For instance, essential biomolecules contain S-heterocycles like a thiazolium ring in thiamine or vitamin B1. This has a key role in cellular metabolism. Thiamine pyrophosphate acts as a coenzyme for several important enzymes involved in carbohydrate regulation. This includes pyruvate dehydrogenase, alpha-ketoglutarate dehydrogenase, and transketolase. This thiazolium ring's structure allows it to participate in crucial enzymatic reactions (Bettendorff, 2021). Moreover, fused systems also participate in biological systems that contain sulfur atoms fused with other ring systems.

Fused sulfur heterocycles provide structural diversity that can interact with various biological targets. This structural complexity allows them to fit into different biological fields and perform specialized functions. Such the nature of fused S-heterocycles imparts pharmaceutical significance. Antibiotics like penicillin and cephalosporins contain fused S-heterocycles. These structures are crucial for their antibacterial activity and ability to inhibit bacterial cell wall synthesis. In biological systems, they take part in enzyme inhibition which blocks the activity of enzymes making it critical for the survival of pathogens or cancer cells. Certain plant-derived alkaloids also contain fused sulfur heterocycles and have been used in traditional medicine for their therapeutic properties. Due to their ability to interact with various biological targets, fused sulfur heterocycles are valuable scaffolds in drug design and development. They are used to develop new therapies for diseases such as infections, cancer, and neurological disorders. Zhang et al. reported S-indole alkaloids obtained from the roots of *I. tinctoria*, which showed anti-inflammatory activity in RAW264.7 cells (Zhang et al., 2020). Although not always classified strictly as alkaloids, benzothiazoles are a class of compounds that include a fused benzene and thiazole ring. Some natural products and synthetic derivatives exhibit significant biological activities, including antimicrobial, antitumor, and anti-inflammatory effects (Yadav et al., 2023). Some multifunctional benzothiazole derivatives were reported by et al., that showed UVA & UVB protection, and antioxidant, antibacterial, and antiproliferative activity against several cell lines (Djuidje et al., 2020). Despite their medicinal potency benzothiazoles are less explored when compared to other heterocycles which can be due to their synthesis complexity and toxicity concerns (El Alami et al., 2024; Yin et al., 2024).

Fused sulfur compounds as purine nucleoside phosphorylase (PNP) inhibitors represent an interesting area of research due to their potential therapeutic applications. While traditional PNP inhibitors like forodesine and ulodesine have shown promise, exploring fused sulfur heterocycles could provide new avenues for drug development with potentially improved efficacy and specificity (Mohamed et al., 2017)

In the sulfur heterocyclic family, polymeric fused sulfur heterocycles are an intriguing class of compounds with potential applications in medicinal chemistry. These compounds consist of polymer chains that include fused sulfur-containing heterocyclic units. Their unique structural features enhance the thermal and chemical stability of these polymers, making them suitable for various biomedical applications. S-containing polyamides were reported with thiazole rings and sulfur linkages that showed anticancer and antimicrobial activity. New organo-sulfur polymers were also reported whose flexibility characteristics were utilized as polymeric biomaterials essentially capable of sustained drug release (Islam & Zeng, 2024). Investigating this field can aid in creating new drug delivery techniques that leverage the distinctive characteristics of organosulfur-based polymers, significantly influencing the pharmaceutical industry and benefiting patients globally.

Sulfur heterocycles play a significant role in agriculture due to their diverse biological activities, which can be harnessed for pest control, plant growth regulation, and disease management. Sulfur is among the important macronutrients essential for plant growth along with others like N, P, K, etc. Sulfur is a component of protein-building amino acids like methionine and cysteine, as well as glutathione, vitamins (biotin and thiamine), phytochelatins, chlorophyll, coenzyme A, and S-adenosylmethionine (Dawar et al., 2023; Riffat et al., 2023). Recently, thiazole derivatives were reported by Yang Shua et al., which result in insecticidal properties against *Plutellaxylostella* (Shuai et al., 2024). Another novel thioacetamide derivative showed algaecide characteristics that can disrupt cyanobacteria (Huang et al., 2024). The compound activity against *M. aeruginosa* & *Synechocystis* sp bacterial strains. Sulfur inhibits oxidative effects through glutathione derivatives and forms disulfide bonds with proteins and enzymes, preventing redox reactions. In soil, it exists as organic matter, making it challenging for plants to absorb. The most common form available to plants is the anion SO_4^{2-}. However, its water solubility can lead to leaching from the soil. Due to its crucial role in plant growth and development, sulfur remains a significant area of interest for researchers.

In Chemical Industries S-heterocyclesare used in the synthesis of dyes and pigments due to their ability to absorb light and produce vibrant colors. S-heterocycles can act as efficient pi electron donor in dyes via. intramolecular charge transfer (Rachi et al., 2018). The presence of sulfur atoms can influence the electronic distribution within the molecule, resulting in vivid and stable colors. They have good thermal and photostability, which is essential for dyes and pigments that need to withstand exposure to light, heat, and other environmental factors. Moreover, sulfur heterocycles can be designed to be soluble in various solvents, making them versatile for use in different dyeing processes. They can also be made compatible with different substrates, including natural and synthetic fibers. Sulfur-containing heterocycles are used in reactive dyes that form covalent bonds with fibers, ensuring strong attachment and high washfastness. These dyes are used for dyeing cotton, wool, and other fabrics. S-heterocycles have applications in pigments, optical brighteners, and fluorescent and phosphorescent dyes like bioimaging, authenticated printing, and biosensors. S-heterocycles are also needed in the creation of specialized polymers and materials with unique properties, including high thermal stability and conductivity. For instance, polythiophenes and polythiazoles. These are among the most well-known sulfur-containing conductive polymers. They exhibit excellent electrical conductivity and are used in organic solar cells, light-emitting diodes (LEDs), and field-effect transistors.

Sulfur-containing heterocycles have significant potential in pollution control due to their unique chemical properties, including high reactivity and the ability to form stable complexes with various pollutants. These compounds can be used in a variety of applications to mitigate environmental contamination, including the removal of heavy metals, degradation of organic pollutants, and emission control. Some sulfur heterocycles are used in processes that help remove pollutants from the environment, such as sulfur-containing catalysts in desulfurization reactions (Mirhoseini & Taghdiri, 2016).

Understanding the chemistry of sulfur heterocycles is essential for the development of innovative solutions in drug design, materials development, and industrial processes. Their study not only enhances our knowledge of organic chemistry but also drives advancements across multiple scientific and technological domains.

Historical Context of Sulfur-Containing Heterocycles

Heterocyclic systems have occupied the paramount interest of researchers around the globe. The history of sulfur-containing heterocycles is rich and spans over a century of scientific exploration and discovery. These compounds have played a significant role in the development of organic chemistry, pharmaceuticals, agriculture, and materials science. The first sulfur-containing heterocycle, thiophene, was discovered by Viktor Meyer in 1883 (Ashe et al., 2000). Meyer isolated thiophene from coal tar and noted its similarity to benzene in terms of chemical reactivity.

Shortly after, thiophene derivatives such as benzothiophene and dibenzothiophene were also identified (Gronowitz, 1986, 1991; Katritzky, 1995).

The structural elucidation of these compounds was a significant achievement of early organic chemistry. Understanding the arrangement of sulfur within the heterocyclic ring paved the way for further exploration of their chemical properties. The early to mid-20th century saw a growing interest in the medicinal properties of sulfur-containing heterocycles (Pathania et al., 2019). Thiazoles and benzothiazoles were investigated for their antimicrobial and anti-inflammatory properties. Sulfa drugs, such as sulfanilamide, became the first widely used antibiotics before the discovery of penicillin. Later, the development of synthetic methods for sulfur-containing heterocycles advanced significantly. New techniques allowed for the preparation of a wide range of thiophenes, thiazoles, and other sulfur heterocycles. Notable reactions include the Hantzschthiazole synthesis (Bouherrou et al., 2017)and the Gewald reaction (Kavitha et al., 2019), which allowed for the efficient production of these compounds. After adequate research, their functional potential was known which encouraged industrial applications. Sulfur-containing heterocycles found applications in the dye and pigment industries. Thiophene and its derivatives were used to produce vibrant and stable dyes (Mason & Nord, 1951).These compounds were also utilized as vulcanization accelerators in the rubber industry, improving the properties of rubber products (Porter, 1977). The latter half of the 20th century saw the rise of conductive polymers(Armour et al., 1967). Polythiophene, a polymer of thiophene, became a cornerstone of this new class of materials, finding applications in electronics and optoelectronics (Yamamoto et al., 1980).

In the modern Era (21st Century) advanced pharmaceuticals came to light. The drug discovery has continued to explore sulfur-containing heterocycles for their therapeutic potential. Many drugs, including antiviral, anticancer, and antimicrobial agents, are based on these structures (Bergman & Janosik, 2002). Notable examples include the HIV protease inhibitor darunavir, which contains a thiazole moiety. Green Chemistry and environmental applications of S-heterocycles have driven the development of more sustainable synthetic methods for sulfur-containing heterocycles. These compounds have been investigated for their roles in pollution control, including heavy metal chelation and the degradation of organic pollutants. Sulfur-containing heterocycles are integral to the development of advanced materials, including organic solar cells, field-effect transistors, and light-emitting diodes (LEDs). Research in nanotechnology has led to the creation of novel materials with unique electronic and optical properties based on thiophene and its derivatives. These heterocycles are important in agricultural chemistry, where they are used to develop pesticides, herbicides, and fungicides. Compounds like benzothiadiazoles have been employed to protect crops from pests and diseases. The most commonly explored are thiophene, thiazole, benzothiazole, and thiadiazole.

The history of sulfur-containing heterocycles reflects their versatility and importance across multiple fields. From their initial discovery in the 19th century to their current applications in advanced technology and medicine, these compounds continue to be a focal point of scientific research and industrial application. Advances in synthetic methods, an understanding of their chemical properties, and exploration of new applications will ensure that sulfur-containing heterocycles remain a vital area of study in the future.

STRUCTURAL PROPERTIES OF TRIAZOLOTHIADIAZOLE

Triazolothiadiazole refers to a class of heterocyclic compounds that incorporate both triazole and thiadiazole rings fused (Figure 2). The triazole ring is a five-membered ring with three nitrogen atoms, which is commonly denoted as 1,2,3-triazole or 1,2,4-triazole, depending on the positions of the nitrogen atoms. On the other end, the thiadiazole ring is a five-membered ring with two nitrogen atoms and one sulfur atom. It is known as 1,3,4-thiadiazole when the sulfur is at position 1, and the nitrogens are at positions 3 and 4. The rings are fused by sharing two adjacent carbon atoms which creates a bicyclic system. The fused ring system creates a planar or nearly planar structure due to the sp2 hybridization of the carbon and nitrogen atoms in the rings. The nitrogen atoms introduce electron-donating and electron-withdrawing effects, which can influence the reactivity and chemical properties of the molecule. The sulfur atom adds to the heterocyclic nature and can participate in various chemical interactions, including coordination with metals or hydrogen bonding. The nitrogen and sulfur atoms contribute to the electronic characteristics of the molecule, making it interesting for applications in electronics and optoelectronics (Waddar et al., 2024).

Figure 2. Structure of triazolothiadiazole

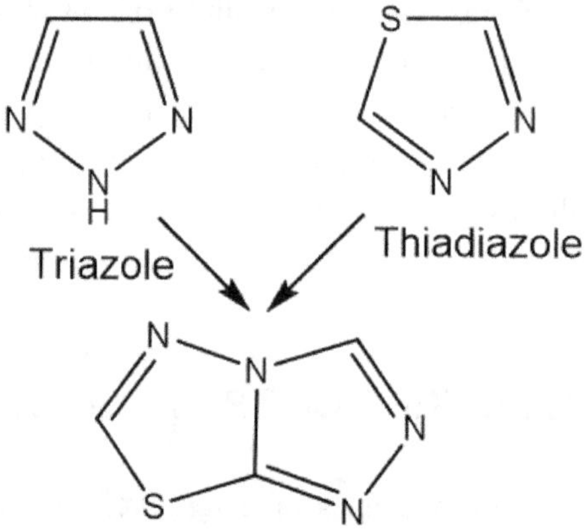

Triazolothiadiazole

In triazolothiadiazole, the presence of multiple heteroatoms (nitrogen and sulfur) makes these compounds versatile in chemical reactions. They can engage in multiple interactions and participate in various organic reactions. The heteroatoms in triazolothiadiazole can participate in both electrophilic and nucleophilic reactions. Nitrogen atoms can act as nucleophiles, attacking electrophilic centers in chemical reactions. Conversely, the presence of electron-rich nitrogen and sulfur atoms can attract electrophiles, leading to functionalization at these positions. Analysis of the molecular electrostatic potential (MEP) surface of triazolothiadiazole revealed areas of negative potential concentrated around the N1 and N2 atoms, highlighting their electron-rich nature (Kerassa et al., 2016) (Figure 3). The N5 atom also exhibited negative potential, however to a lesser extent compared to N1 and N2. The regions adjacent to the hydrogen atoms displayed the highest positive potential, indicated by dark blue coloring, while the areas near sulfur (S7) and N4 showed a lighter blue, signifying a lesser positive charge.

Figure 3. Three-dimensional MEP of triazolothiadiazole

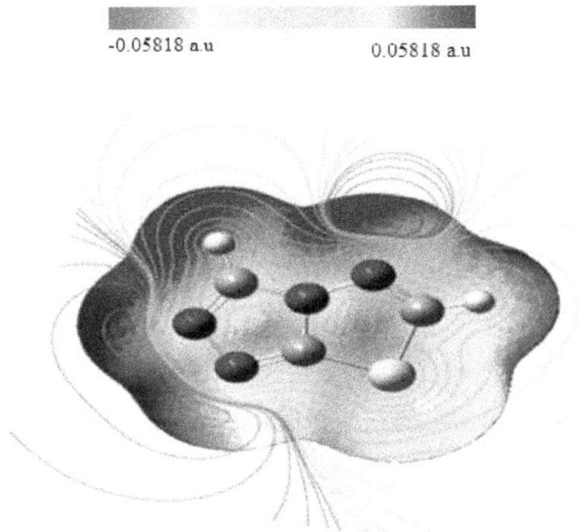

The triazolothiadiazole system can be considered a cyclic counterpart to two significant compounds, biguanide, and thiosemicarbazide, both of which frequently exhibit a wide range of biological activities (Fan et al., 2010). These compounds are characterized by their unique structural framework, which combines the properties of both triazole and thiadiazole rings, leading to diverse chemical reactivity and potential applications in various fields such as medicinal chemistry (Khalili Ghomi et al., 2023)and agriculture (Uddin et al., 2023). The specific arrangement of the atoms can vary, leading to different isomers of triazolothiadiazole.

Figure 4. H-bonding of triazolothiadiazole with MurB enzyme

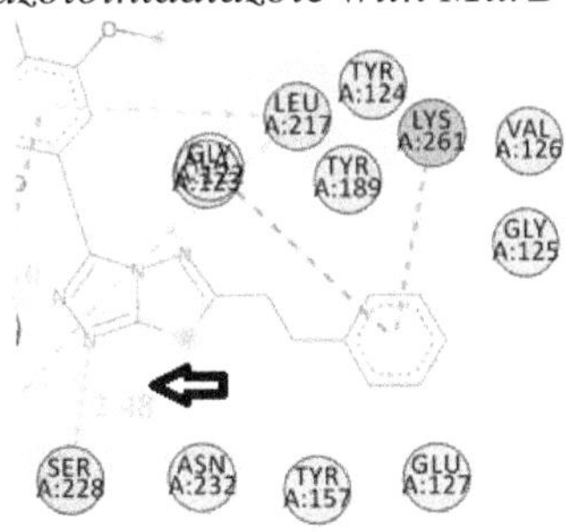

Nitrogen atoms in the triazole and thiadiazole rings act as hydrogen bond acceptors due to their lone pairs of electrons. This allows triazolothiadiazole to form hydrogen bonds with hydrogen bond donors such as water, alcohol, amines, etc. These H-bonding interactions are important in biological systems, influencing the binding of triazolothiadiazole to proteins, enzymes, and nucleic acids. For example, the H-bonding of N with serine228 residue of MurB enzyme within 2.48 Å (Kamoutsis et al., 2021)(Figure 4). Similarly, the sulfur atom can also act as a hydrogen bond acceptor due to its lone pairs of electrons (Altowyan et al., 2023).

The sulfur atom in the thiadiazole ring has a lone pair of electrons that can coordinate with metal ions. Thus, they are useful in coordination chemistry and metal chelation. Triazolothiadiazole can form stable complexes with transition metals, which are useful in catalysis, metal extraction, and designing metal-based drugs (Mostafa et al., 2024).

Figure 5. Involvement of triazolothiadiazoles in non-covalent interactions

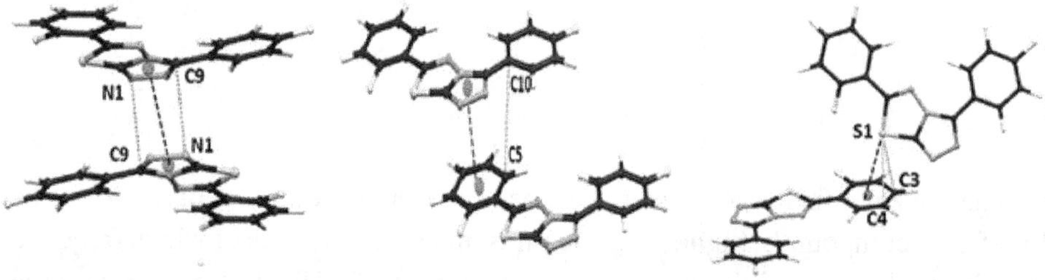

The fused ring system of triazolothiadiazole often displays aromaticity, contributing to π-π stacking interactions. For example, dimeric interactions between triazolothiadiazole derivatives which were stabilized by π-π and sulfur-π interactions (Al-Wahaibi et al., 2019) (Figure 5). These interactions occur between the aromatic rings of triazolothiadiazole and other aromatic systems, such as those found in DNA base pairs or aromatic amino acid residues in proteins. π-π interactions are crucial for the binding affinity and specificity of triazolothiadiazole in biological targets.

Triazolothiadiazoles consist of both triazole and thiadiazole, which can exhibit hydrophobic interactions. Their aromatic rings and sulfur atoms, contribute to their hydrophobic nature. Non-polar regions of these molecules interact favorably with other hydrophobic molecules or molecular regions. Such interactions were shown by 1,2,4-triazolo[3,4-b] 1,3,4-thiadiazole-based thieno[2,3-d]pyrimidine derivatives which interacted with multiple residues of C(30) carotenoid dehydrosqualene synthase (Settypalli et al., 2019). (Figure 6).

Figure 6. Hydrophobic interactions of triazolothiadiazole with 3ACX protein

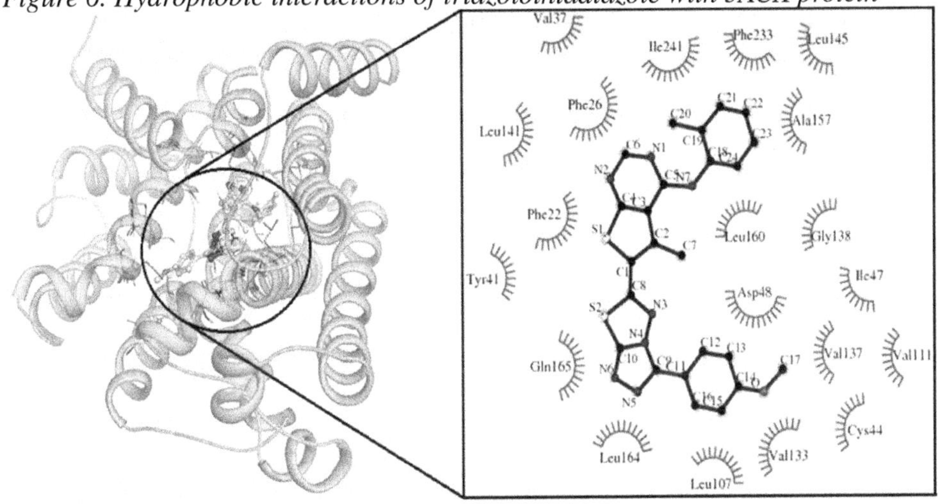

The presence of heteroatoms introduces polarity to the triazolothiadiazole molecules. Dipole-dipole interactions can occur between the dipoles of triazolothiadiazole and other polar molecules. These interactions are significant in the solubility, crystallization, and overall physicochemical properties of triazolothiadiazoles (Waddar et al., 2024).

Though weaker than hydrogen bonds or π-π interactions, Van der Waals forces contribute to the overall stability and interactions of triazolothiadiazoles. For instance, triazolo[3,4-*b*]thiadiazole derivatives with anti-tumor activity formed stable complexes with Akt1 and Akt2 proteins via van der Waals interactions (Trafalis et al., 2021). These forces are important in the packing of molecules in the solid state and their interaction with biological membranes.

Synthesis of Triazolothiadiazoles

Research on triazolothiadiazoles has explored their synthesis and application in various fields. For example, studies have utilized Density Functional Theory (DFT) to investigate the electronic properties and reactivity indices of these compounds(Kerassa et al., 2016). Such theoretical studies help in understanding how triazolothiadiazoles might behave in different chemical environments and how they can be optimized for specific uses. The synthesis of these heterocyclic compounds typically involves the formation of the triazole and thiadiazole rings, followed by their fusion to form the triazolothiadiazole core. Various methods have been developed to achieve this, involving different starting materials and reaction conditions. Common 1,3-dipolar cycloaddition reactions are performed by cyclization of hydrazine derivatives with thiocyanates. This leads to the formation of the thiadiazole ring, followed by subsequent cyclization to form the triazole ring (Scheme 1).

Figure 7. Scheme 1: Synthesis of triazolothiadiazole

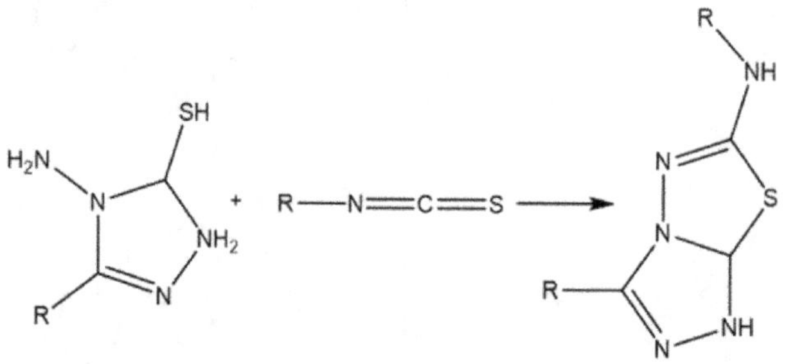

Synthesis of structurally diverse triazolothiadiazole derivatives typically involves a combination of addition, cyclization, and condensation reactions. Researchers often start with molecules containing different acidic COOH groups, which are then converted to various hydrazine derivatives. The addition of carbon disulfide in the presence of ethanolic KOH produces ionic potassium thiocarbamate salts. These salts undergo cyclocondensation with an excess of hydrazine hydrate, forming thiols. Subsequent reactions of these thiols with benzoyl chlorides(Hamdy et al., 2021)or acids (Seelolla & Ponneri, 2016; Suresh Kumar et al., 2010) in the presence of phosphorus oxychloride yield triazolothiadiazoles (Scheme 2). Acids are sometimes reacted with hydrazide derivatives in the presence of methanesulfonic acid and sulfolane to synthesize thiols (H. Yuan et al., 2018).

Figure 8. Scheme 2: Synthetic routes for the synthesis of triazolothiadiazoles

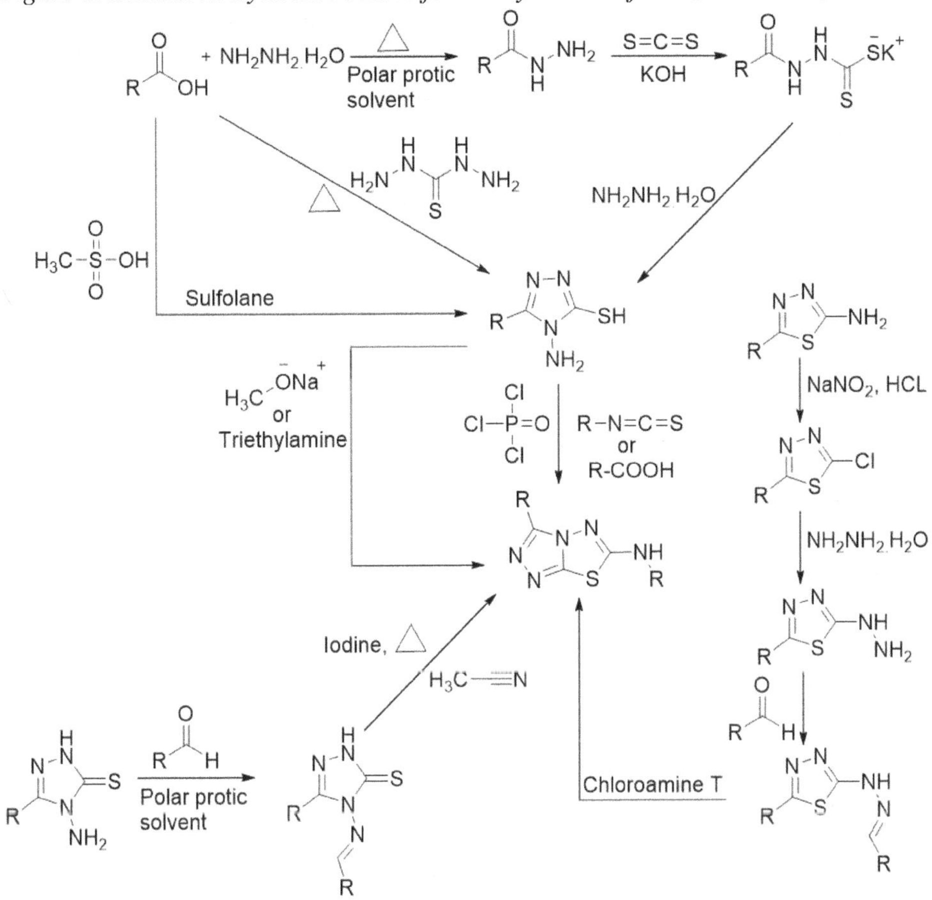

Triazolothiadiazoles can be synthesized through various one-step reactions, which involve straightforward procedures often yielding these heterocycles directly from simpler precursors. Methods include direct cyclization, condensation reactions, oxidative cyclization, and one-pot multicomponent reactions, each providing a valuable approach for the rapid synthesis of triazolothiadiazoles. For instance, acid derivatives undergo a one-step condensation reaction with thiocarbohydrazides under thermal conditions to give thiols in a very short time. This subsequent reaction with substitute isothiocyanates gives triazolothiadiazoles as the product (Badr& Barwa, 2011). To achieve rapid synthesis of these heterocycles, microwave irradiation can be used as a catalyst in the presence of sodium methoxide or triethylamine (El-Sayed et al., 2020).

Oxidative cyclization is a key method for synthesizing triazolothiadiazoles. This process involves the transformation of precursor compounds through the introduction of an oxidizing agent, leading to the formation of the triazolothiadiazole ring. Sulfur-containing triazoles can be reacted with aldehydes to synthesize an aryl diene compound. Iodine can catalyze oxidative cyclization for a diene cyclization which forms a thiadiazole ring (Al-Wahaibi et al., 2018). Similarly, bromine (Gomha & Riyadh, 2011), phenacyl bromide, or ethyl 2-bromoacetate(Zhao et al., 2012) can also participate in such catalysis. Chloramine-T is a widely used oxidizing agent in organic synthesis. It can facilitate the chlorination of various substrates, including aromatic compounds and heterocycles. Its mild oxidizing properties make it suitable for selective transformations of thiadiazolylhydrazones to triazolothiadiazoles (Himaja et al., 2012).

Structure-Activity Relationship and Bioactivity of Triazolothiadiazole

Understanding the structure-activity relationship and structural determinants of triazolothiadiazoles is crucial for the rational design of compounds with enhanced bioactivity. By systematically modifying the core structure, substituents, and other structural features, researchers can optimize these compounds for specific biological targets and therapeutic applications. Triazolothiadiazoles are a class of heterocyclic compounds known for their diverse biological activities. Substituents like alkyl, alkoxy, and amino groups generally enhance biological activity by increasing electron density on the triazole ring. Whereas, groups such as nitro, cyano, and halogens can either enhance or reduce activity depending on their position and the type of biological target. For instance, triazolothiadiazole derivatives showed antiurease activity with OCH3-bearing substituents at the triazole ring and ortho-OH group in the thiadiazole group. When the ortho-OH group was altered with para-OH from the thiadiazole ring, a significant increase in antiurease activity was observed when this p-OH group was replaced with an electronegative F, it further enhanced the antiurease activity (Uddin et al., 2023).

Similar to the triazole ring, both EDGs and EWGs on the thiadiazole ring influence the overall electron distribution, which in turn affects the compound's interaction with biological targets. Thiadiazole was reported by Mathew et al., in which EWGs like the NO_2 group increase Anti-tubercular activity. Whereas, EDGs like the methoxy group can decrease its activity (B. Mathew et al., 2013). An electron-rich pyridine group in the thiazole ring enhances anticancer characteristics than the pyrazine derivative (J. Wu et al., 2014). The heterocyclic nitrogen may change the electron density on the aromatic ring, therefore affecting hydrophobic interactions.

As discussed in Section 2, the positions of nitrogen and sulfur atoms vary in electronic distribution in the overall structure of the triazolothiadiazole heterocycle. This in turn produces reactivity differences and variation in biological interactions. Substituents at position 1 (N1) and position 2 (N2) on the triazole are crucial for activity. Though a wide range of synthetic routes were reported for triazolothiadiazoles, substitution in the N1, N2, and N5 positions was not reported so far. It is important to explore their SAR activities to discover their biological potential. For example, N-4 of triazoloquinazolines interacted via H-bonding with ASN1436 residue of histone PfGCN5 acetyltransferase (El-Shershaby et al., 2021).

BIOLOGICAL SIGNIFICANCE AND APPLICATIONS OF TRIAZOLOTHIADIAZOLE DERIVATIVES IN PHARMACEUTICALS

Nitrogen-containing heterocycles, particularly triazolothiadiazoles and triazolothiadiazines, serve as the core structural motifs in various compounds that exhibit significant activities in diverse fields, including pharmaceuticals and agrochemicals. The design, synthesis, and development of molecules with therapeutic value for humans remain key objectives in organic and medicinal chemistry. Over the past decade, combinatorial chemistry has facilitated access to chemical libraries founded on privileged structures, with particular emphasis on heterocyclic motifs due to their established efficacy in medicinal chemistry (Früchtel& Jung, 1996; Horton et al., 2003; Thompson & Ellman, 1996). The synthesis of nitrogen-containing heterocyclic systems has garnered significant interest over the past decade due to their utility in a variety of applications. The fusion of triazole and thiadiazole rings results in a class of heterocyclic compounds with a broad spectrum of pharmacological properties, including antifungal (Zhan et al., 2024), antibacterial (S. Wu et al., 2021), antiviral (Khramchikhin et al., 2022), anti-inflammatory (Al-Wahaibi et al., 2022; El Shehry et al., 2010), anticancer (Baburajeev et al., 2017; Trafalis et al., 2021), analgesic (Mathew et al., 2006, 2007), and antitubercular (Jaiswal et al., 2021) activities. These versatile compounds have garnered significant interest in both pharmaceutical and agricultural research (Holla et al., 2001; Shivarama Holla et al., 2002). Several key biological applications of triazolothiadiazole and its derivatives are highlighted below.

Antimicrobial

Antimicrobials kills microorganisms or inhibits their growth. These agents can be classified based on their target: antibacterial agents work against bacteria, antifungal agents against fungi, microbicides kill microbes, and microbiostatic agents inhibit microbial growth. Typically, antimicrobial agents are defined as disinfectants, antiseptics, antibiotics, bactericidal, or bacteriostatic agents. Antimicrobial therapy refers to a treatment approach that specifically targets the microorganisms causing a particular disease or condition (Mandal, 2016). Antimicrobial resistance has emerged and spread widely among various microbes, including both Gram-positive and Gram-negative bacteria (Dessen et al., 2001, 2001; Pfeltz & Wilkinson, 2004, 2004). Utilizing 1,2,4-triazole derivatives as precursor starting materials is particularly appealing in the synthesis of various biologically active heterocycles. These compounds represent a significant class of organic molecules with a wide range of biological activities (Reddy et al., 2010). Antimicrobial drugs primarily target the biosynthesis of proteins, RNA, DNA, cell walls, and folic acid. Numerous inhibitors for these critical biological processes have been successfully identified and developed (Bax et al., 2010; Phillips et al., 2011). Antimicrobial drugs were introduced for use in food animals with minimal controlled experimental studies, leading to frequent calls for a shift from reliance on wonder to a focus on science. Similar to human medicine, early dosing practices were largely empirical and based on inadequately controlled small-scale trials (Prescott, 2019). As the significant and diverse capabilities of triazole rings fused with thiadiazole moieties have been highlighted, there is considerable global research interest in their antimicrobial activity (Hitendra S. Joshi et al., 2011; Kamoutsis et al., 2021; Karegoudar et al., 2008a; Zhan et al., 2024). Various studies and findings related to this specific biological activity are mentioned below.

Antimicrobial application of fused heterocyclic [1,2,4]triazolo nucleus with [1,3,4]thiadiazole derivatives was reported (Purohit et al., 2011) that inhibited bacterial and fungal growth. When compared to standard drugs like amoxicillin, ciprofloxacin, erythromycin, benzylpenicillin, Greseofulvin, triazolothiadiazole derivative (Figure 9. Compound 1), containing the SO_2CH_3 group at position-4) good antimicrobial activities against *Bacillus megaterium*, *Staphylococcus aureus*, *Escherichia coli*, and *Pseudomonas aeruginosa*, as well as antifungal activity against *Aspergillus niger*. Another derivative containing trichlorophenyl moiety (Figure 9. Compound 2) revealed the antimicrobial activity against *Escherichia coli*, *Staphylococcus aureus*, *Pseudomonas aeruginosa* and *Klebsiella pneumoniae* strains. Additionally, it was observed that the thiadiazole derivatives are found to be more active than thiadiazine derivatives (Karegoudar et al., 2008). Another triazolo-thiadiazole derivatives were reported by Kamoutsis et al., which showed antibacterial activity with MIC ranging from 0.002-0.150 mg/mL and MBC between 0.005-0.200 mg/mL (Kamoutsis et al., 2021). The compounds (Figure 9. Compound 3) demonstrated higher antibacterial activity when compared to ampicillin and streptomycin (write MIC and MBC for ampicillin and streptomycin). In addition to this, they obtained biofilm formation inhibition, surpassing the efficacy of the reference antibiotics. The antifungal activity of the tested triazolo-thiadiazole derivatives was also noteworthy, exceeding that of the standard antifungal agents ketoconazole and bifonazole.

Chalcone derivatives of triazolothiadiazoles are observed as antifungal agents (Zhan et al., 2024). *In-vitro* analysis against *Rhizoctonia solani* (RS), *Sclerotinia sclerotiorum* (SS), *Botryosphaeria dothidea* (BD), *Fusarium graminearum* (FG), *Colletotrichum capsici* (CC), *Phytophthora capsici* (PC), and *Phomopsis sp.* (PS) pathogens. Two of the compounds (Figure 9. Compounds 4 and 5) showed inhibition effect at EC_{50} values of 15.9 and 44.2 μg/mL which were found corresponding with the presence of an electron-donating group.

Figure 9. Synthesized compounds of triazole thiadiazole derivatives showing potential antimicrobial activity

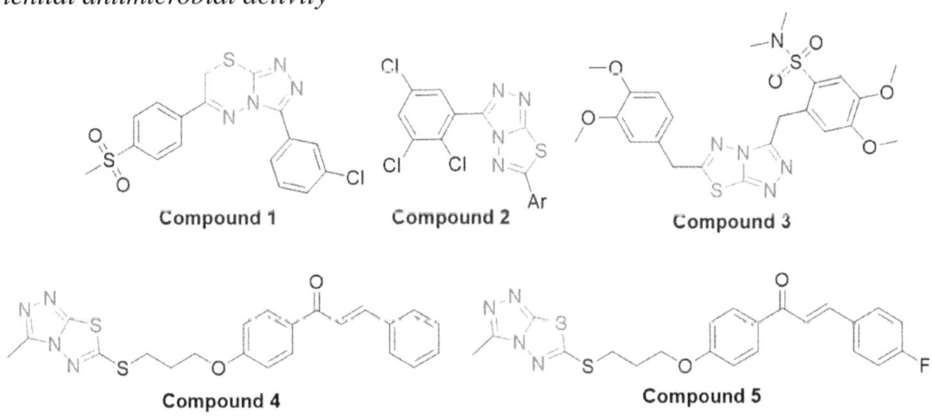

Anticancer

Cancer is an extremely dangerous disease characterized by the uncontrolled growth and rapid spread of abnormal cells that has been a major global health issue for decades. While the term "chemotherapy" is often associated specifically with anticancer drugs, it more broadly refers to the use of chemical compounds to treat diseases in general (Janet L. Stringer & Irvin S. Snyder, 2023). Many anticancer drugs face limitations due to toxicity to normal cells and potential drug resistance. Recently, various heterocyclic and fused heterocyclic compounds have been identified as promising anticancer agents through molecular biology, empirical screening, and rational drug design (Hartwell & Kastan, 1994; Jemal et al., 2011; Kamel & Megally Abdo, 2014; Kidwai et al., 2002). Numerous researchers have conducted extensive studies on various cancer cell lines, specifically investigating the triazolo-fused thiadiazole moiety (Bhat et al., 2004; Charitos et al., 2019; Husain et al., 2013; Ibrahim, 2009). Their research has demonstrated significant promise in the development of these compounds as effective anticancer agents (Baburajeev et al., 2017; Trafalis et al., 2021). In the following sections, their findings will be discuss, exploring the capability and potential of triazolo-fused thiadiazoles in targeting and inhibiting cancer cell proliferation. This discussion aims to shed light on the mechanisms, efficacy, and prospects of these compounds in the realm of cancer therapy.

Triazolo thiadiazole derivatives have shown anticancer effect among various cancer types. However, these fused heterocyclic compounds have not been utilised as an anticancer drug as real application. Their potential has been reported by various researchers demonstrating their clinical characteristics. Triazolothiadiazole have shown their anticancer activity in PC-3 (Prostate Cancer Cell Line), SKOV-3 (Ovarian Cancer Cell Line), HT-29 (Colorectal Cancer Cell Line), LoVo (Colorectal Cancer Cell Line) and SW403 (Colorectal Cancer Cell Line) (Trafalis et al., 2021). The biological evaluation of triazole-thiadiazole derivatives (Figure 10. Compounds 6,7 and 8) exhibit strong anticancer properties with cytotoxic effect.

Heparanase, an endoglucuronidase that cleaves heparan sulfate, plays a crucial role in regulating the structure and function of heparan sulfate proteoglycans, leading to the breakdown of the extracellular matrix. Its activity is linked to various regulatory processes, including increased bioavailability of growth factors and cytokines. By inhibiting heparanase, tumor growth, angiogenesis, and metastasis can be suppressed through the modulation of growth factor-mediated signaling pathways (Pisano et al., 2014). Triazolothiadiazole derivatives (Figure 10. Compound 9) are also heparanase inhibitors that exhibit significant antiproliferative effects, effectively suppressing the migration and invasion of cancer cells (Baburajeev et al., 2017). These sulfur-heterocycles can also inhibit DNA topoisomerase II nuclear enzymes (Larsen et al., 2003). Sofia and her team studied the triazolo fused thiadiazoles on DLD-1, HT-29, and LoVo cancer cell lines. Their findings revealed one highly potent compound (Figure 10. Compound 10) which can essentially induce both early and late apoptosis and cause cell cycle arrest in the S phase. Furthermore, they observed significantly reduced topIIα phosphorylation ($P<0.001$) upon treatment (Sagredou et al., 2020).

Figure 10. Synthesized compounds of triazolo thiadiazole derivatives showing potential anticancer activity

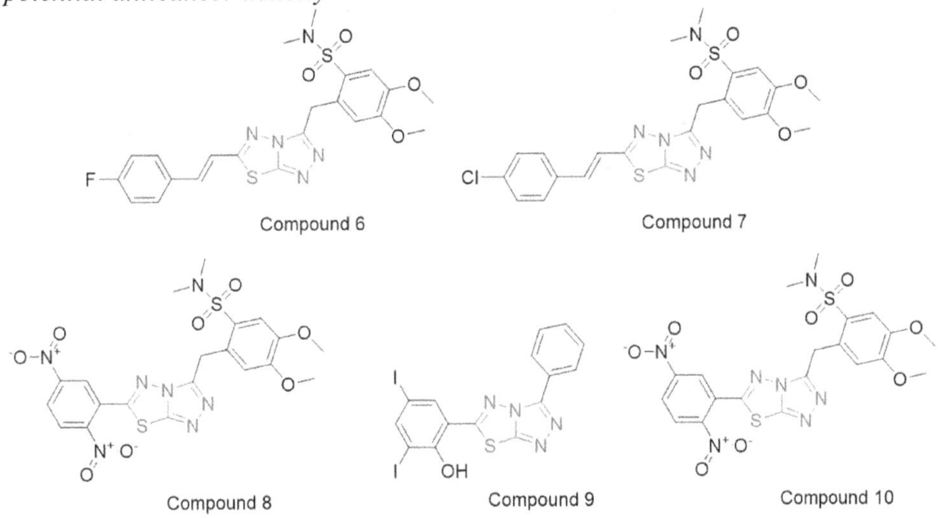

Anti-Inflammatory

Inflammation is triggered by the release of chemicals from tissues and migrating cells. Key contributors to this process include prostaglandins (PGs), leukotrienes (LTs), histamine, bradykinin, platelet-activating factor (PAF), and interleukin-1. The role of these substances has been substantiated through studies utilizing competitive antagonists for their receptors and inhibitors of their synthesis (Vane & Botting, 1987). Inflammation occurs during significant disruptions of homeostasis, such as tissue injury, infections, and exposure to contaminants. Consequently, extensive research has focused on elucidating the cellular and molecular mechanisms underlying the acute inflammatory response to infection (Medzhitov, 2008). Triazolo-thiadiazole derivatives have been evaluated for their anti-inflammatory properties (El Shehry et al., 2010; Karegoudar et al., 2008a; Malladi et al., 2012), showing significant protection against inflammation in experimental models. Compounds with specific substituents, such as adamantane, demonstrated high COX inhibitory activity, indicating their potential as anti-inflammatory drugs (Al-Wahaibi et al., 2022). These findings highlight the possibility of developing triazole derivatives into effective treatments for inflammatory conditions.

Triazolo-fused thiadiazole compounds are known to work against inflammation (El Shehry et al., 2010). Triazolo fused with thiadiazole derivatives introduced with adamantane and phenyl substituents showed anti-inflammatory properties against

COX-1 & COX-2 enzymes. The inhibition concentration (IC_{50}) values of the control drugs, celecoxib (a selective COX-2 inhibitor) and diclofenac (a non-selective COX inhibitor), were compared to evaluate the activity of Synthesized compounds (Figure 11. Compounds 11 and 12). *In vitro* data indicates that compound 12 exhibits nearly three times higher binding affinity for the COX-1 enzyme compared to compound 11. Conversely, compound 11 shows approximately twice the binding affinity for the COX-2 enzyme compared to compound 12. Assay results further reveal that compound 11, which contains an adamantane moiety, demonstrates significant COX inhibitory activity with a ten-fold selectivity for COX-2 (SI = 9.82) over compound 12, which features a phenyl ring instead of the adamantane structure. These findings highlight the impact of the adamantane and phenyl substituents on COX inhibition activity(Al-Wahaibi et al., 2022). This study identified that terminal groups (di-substituted and phenyl/adamantane moieties) are important in establishing noncovalent interactions with the active site residues of the COX-1/2 enzymes. In the future, these groups could be altered to improve the affinity and selectivity against these enzymes.

Figure 11. Synthesized compounds of triazolo thiadiazole derivatives showing potential anti-inflammatory activity

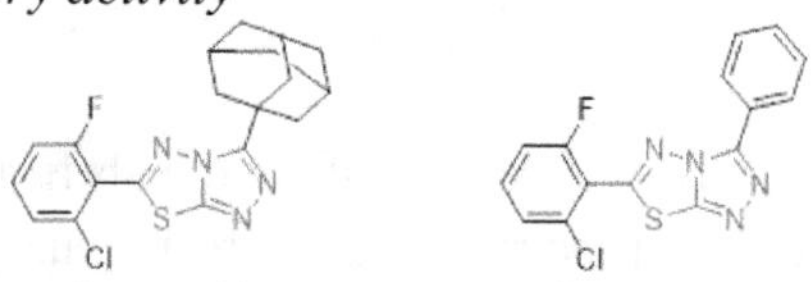

Compound 11 Compound 12

Anticonvulsant

Antiepileptic drugs (AEDs), also known as anticonvulsants, are the primary treatment for most individuals with epilepsy, aiming to prevent seizures (Tolou Ghamari et al., 2014). These drugs function through various mechanisms: (i) modulation of voltage-gated ion channels and membrane cation conductance (Na+, Ca2+, or K+); (ii) enhancement of gamma-aminobutyric acid (GABA)-mediated inhibitory neurotransmission; (iii) reduction of glutamate-mediated excitatory neurotransmission; and (iv) stabilization of thalamic neurons by inhibiting T-type calcium channels (Lasoń et al., 2013) (Polatin, 2004). A series of triazolothiadiazole-fused compounds (Compound 15) were synthesized and evaluated for anticonvulsant activity using maximal electroshock seizure (MES) and subcutaneous pentylenetetrazol (scPTZ) methods. The results showed that these triazolothiadiazole derivatives exhibited promising anticonvulsant effects and favourable safety profiles. However, the efficacy of these compounds did not surpass that of standard antiepileptic drugs like phenytoin and carbamazepine. The promising initial results suggest that modifications to their chemical structure could lead to more effective antiepileptic agents, making continued investigation into these triazolothiadiazole compounds crucial for developing new and improved treatments for epilepsy (Sarafroz et al., 2021).

Epileptical seizures are critical medical situations that creates electrical imbalance in brain. Several drugs that contain sulfur and without sulfur are reported for these cases (Grover et al., 2020; Hrncic et al., 2018; Kamiński et al., 2015). However, triazolo-thiadiazole-fused heterocycles did not achieve pharmaceutical applications. Druggable properties were reported for these heterocycles using the Maximal Electroshock Seizure (MES) method (Asif Hussain et al., 2009). These evaluations indicate the potential of these compounds as effective anticonvulsant agents. Notably, Compound 13, depicted in Figure 10, demonstrated significant anticonvulsant activity, comparable to established antiepileptic drugs such as phenytoin and carbamazepine. The promising results of this study suggest that the synthesized triazolo-thiadiazole derivatives could serve as a strong foundation for developing new anticonvulsant drugs. The observed efficacy highlights the potential of these compounds in managing seizures and other related disorders. To further enhance their therapeutic potential, additional research and optimization are necessary. This should focus particularly on refining their structural features and improving their pharmacokinetic properties. The exploration and optimization of triazolo-thiadiazole compounds hold considerable promise for the future of epilepsy treatment. Continued research in this area could lead to the discovery of highly effective and safe treatments for epilepsy and other seizure disorders, ultimately improving patient outcomes and quality of life.

Figure 12. Synthesized compounds of triazolo thiadiazole derivatives showing potential anticonvulsant activity

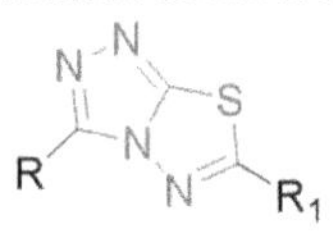

R=C_6H5CH_2, $C_6H_5OCH_2$,
R_1=$C_6H_5CONHCH_2$, $2BrC_6H_4$, $3BrC_6H_4$, $2C_6H_5COC_6H_4$, $C_{10}H_7CH_2$, $C_8H_6NCH_2$, $C_6H_5COCH_2CH_2$

Compound 13

Analgesic

Pain is the most common reason people seek medical care (Melnikova, 2010). Current pharmacotherapy primarily relies on established drug classes such as narcotic analgesics and nonsteroidal anti-inflammatory agents (Daniel A. Queremel Milan & Donald D. Davis, 2024). Additionally, cyclooxygenase-2 (COX-2) selective inhibitors, anticonvulsants, and antidepressants are widely used for pain treatment. Despite the availability of these treatments, not all patients achieve significant pain relief, highlighting a substantial unmet need for new, safe, and effective therapies for both acute and chronic pain. The global market for pain therapeutics is significant, estimated at $50 billion in 2009, underscoring the importance and attractiveness of this therapeutic area to the pharmaceutical industry. Addressing the limitations of current pain management options remains a critical focus for ongoing research and development (Borsook et al., 2014). Researchers have been particularly interested in the potential of these derivatives due to their unique chemical structure and multifaceted mechanism of action. By targeting specific pain pathways and receptors, triazolothiadiazole derivatives could provide effective relief for various types of pain, from acute to chronic conditions. The exploration of these compounds could lead to the development of novel analgesic medications that offer better efficacy and fewer side effects compared to existing treatments. As research progresses, the role of triazolothiadiazole derivatives in pain management may become increasingly significant, paving the way for advanced therapies in the field of analgesia.

Substituted 1,2,4-triazolo[3,4-b]-1,3,4-thiadiazoles (Figure 13. Compound 14) revealed analgesic activities upto 31.9% increase in reaction time within 90 minutes, compared to the standard drug diclofenac sodium, which demonstrated a 52.3% increase in the same timeframe. The study also found that introducing electron-donating groups such as methyl and methoxy at position 5 of the indole ring led to a decrease in analgesic activity. This suggests that the electronic nature of substituents significantly influences the biological activity of these derivatives (V. Mathew et al., 2006). Continuing their research later that year, Matthew and his team focused on the analgesic properties of 3,6-disubstituted 1,2,4-triazolo[3,4-b]-1,3,4-thiadiazoles and their dihydro analogs. They discovered that some compounds, particularly Compound 15 depicted in Figure 13, exhibited superior analgesic properties, with a 35.33% increase in reaction time within 90 minutes, compared to diclofenac sodium, which showed a 43.25% increase in the same period. The research conducted by Matthew and his team underscores the potential of triazolo-thiadiazole derivatives as analgesic agents. Their findings suggest that with further studies and optimization, these compounds could be developed into effective pain management therapies, highlighting the importance of structural modifications in enhancing their analgesic efficacy (V. Mathew et al., 2007). The promising analgesic properties of triazolo fused thiadiazole derivatives underscore their potential as effective pain management agents. Exploring the introduction of various substituents and their electronic effects could lead to the discovery of more potent derivatives.

Figure 13. Synthesized compounds of triazolo thiadiazole derivatives showing potential analgesic properties

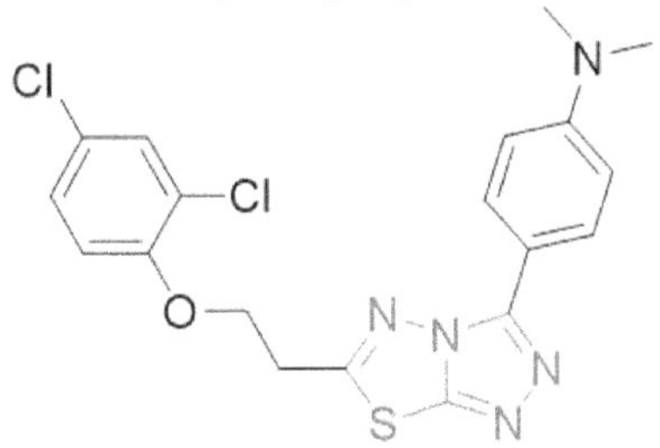

Compound 14

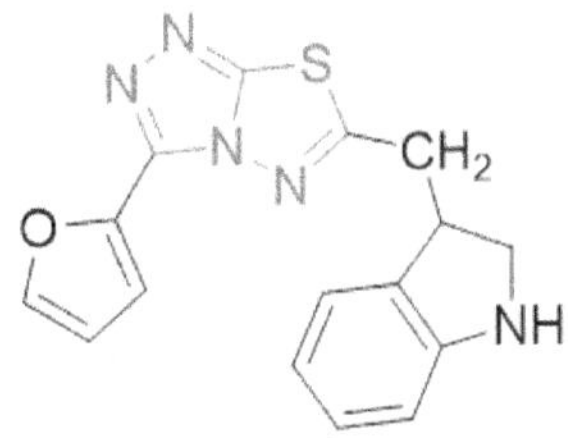

Compound 15

Antitubercular

Tuberculosis has been a leading cause of death worldwide, prompting scientists and researchers to explore all possible avenues to combat this persistent threat. The causative agent of tuberculosis (TB), the *Mycobacterium tuberculosis* complex, can cause disease in various parts of the body (extra-pulmonary TB), though it primarily affects the lungs (pulmonary TB). A significant portion of the human population remains infected with latent TB infection (LTBI), which is asymptomatic and non-transmissible. While LTBI remains controlled in most individuals, some progress to active TB disease (Pai et al., 2016; Zumla et al., 2013). In-silico studies indicated that these compounds effectively inhibit the DprE1 enzyme, suggesting a novel mechanism of action that could help overcome drug resistance (Jaiswal et al., 2021).

The research on triazolothiadiazole derivatives has revealed their promising potential across various therapeutic domains due to their diverse biological activities. The diverse biological activity underscores the versatility and efficacy of these compounds. The synthesis and evaluation of these compounds have underscored their effectiveness as antimicrobial, anticancer, anti-inflammatory, anticonvulsant, analgesic, and antitubercular agents, providing a robust foundation for future research and drug development. This promising class of compounds represents a valuable addition to the arsenal of drugs aimed at addressing some of the most pressing health challenges today.

A series of benzothiazinone derivatives containing a triazolo-thiadiazole moiety and evaluated for their antitubercular activity using the Löwenstein-Jensen (L.J.) slope method. This method was employed to determine the minimal inhibitory concentrations (MIC) of the synthesized compounds against *Mycobacterium tuberculosis* H37Rv, comparing their efficacy with standard drugs like Isoniazid and Rifampicin. The study involved preparing various dilutions of the compounds and assessing their effectiveness in inhibiting bacterial growth. The results showed that certain compounds, specifically Compounds 16 and 17 depicted in Figure 14, exhibited significant antitubercular activity. Compound 16 demonstrated an MIC of 0.28 μg/ml compared to Isoniazid's MIC of 0.39 μg/ml, while Compound 17 showed an MIC of 0.37 μg/ml compared to Rifampicin's MIC of 0.48 μg/ml. Furthermore, molecular docking studies were conducted for the synthesized compounds against the DprE1 enzyme, yielding promising results. The effective inhibition of the DprE1 enzyme by these compounds suggests a novel mechanism of action, which could be advantageous in overcoming drug resistance in tuberculosis treatment. This highlights the potential of triazolo-thiadiazole derivatives as promising candidates for developing new antitubercular agents (Jaiswal et al., 2021).

Figure 14. Synthesized compounds of triazolo thiadiazole derivatives showing potential antitubercular properties

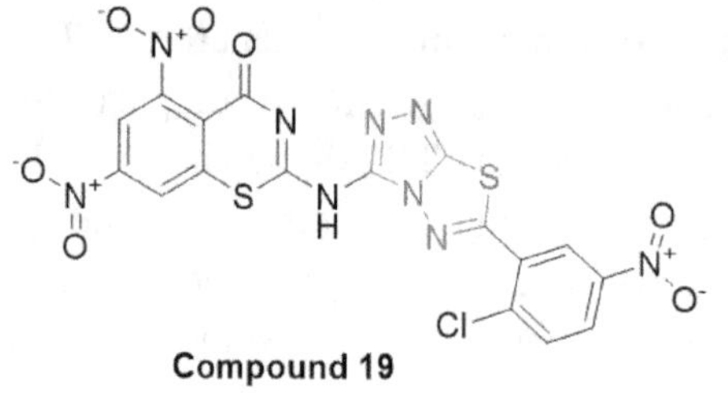

Compound 19

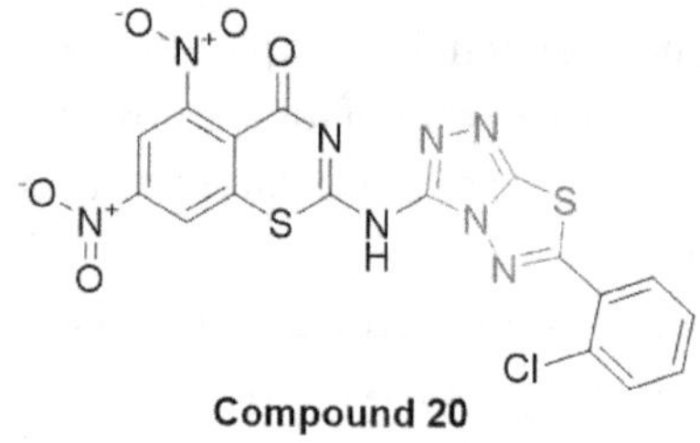

Compound 20

APPLICATIONS OF TRIAZOLOTHIADIAZOLE ANALOGS IN AGRICULTURE

Agriculture encounters substantial challenges from pests and microorganisms that can drastically affect crop yields. Triazolothiadiazole derivatives have emerged as promising solutions to these issues due to their wide-ranging efficacy and capacity to target diverse biological mechanisms. By interfering with crucial enzymes and pathways, these compounds offer valuable protection for crops and contribute to enhanced agricultural productivity.

Plant pathogens are responsible for numerous serious and often fatal plant diseases, significantly jeopardizing the safety, yield, and stability of crop production (C. Li et al., 2021; Silva et al., 2019). Moreover, various bacterial infections can cause significant harm to plants, leading to diseases like bacterial leaf streak (BLS) and bacterial leaf blight (BLF). These infections are particularly concerning as they can drastically reduce crop yields, with potential losses ranging from 20% to 50% (Jiang et al., 2020; Sundaram et al., 2014; T. Yuan et al., 2011). Such reductions in yield not only impact agricultural productivity but also affect the economic stability of farmers who rely on these crops for their livelihood. Consequently, understanding and managing these fungal and bacterial diseases is crucial for maintaining healthy crop production and ensuring food security. Utilizing chemical antifungals and antibacterials is a crucial strategy for preventing and managing crop diseases and pests, which in turn helps secure agricultural productivity and income (P. Li et al., 2018).

As discussed in previous sections, the investigation into the potential use of triazolothiadiazole derivatives has garnered significant attention in recent research. These compounds have shown promising antifungal and antibacterial properties, which are particularly valuable in the agricultural sector. The application of triazolothiadiazole derivatives is being explored to enhance plant growth and increase crop yields, addressing the challenges posed by various plant pathogens. Several studies have demonstrated the efficacy of these compounds in reducing the incidence of fungal infections (Joshi et al., 2021; Kamoutsis et al., 2021; N. Wu et al., 2023; Zhan et al., 2024) and bacterial diseases (El-Sayed et al., 2020; Joshi et al., 2021; Kamoutsis et al., 2021; N. Wu et al., 2023; S. Wu et al., 2021) that adversely affect crops. By mitigating these threats, triazolothiadiazole derivatives could play a crucial role in ensuring.

Future Perspectives

Triazolothiadiazole derivatives are promising candidates in the search for new drugs due to their diverse biological activities. Future research could focus on optimizing these compounds to enhance their efficacy and reduce toxicity. These efforts could lead to new treatments for infectious diseases, cancer, and neurological disorders. Advances in molecular docking and computational chemistry can help identify specific targets for triazolothiadiazole compounds, enabling the development of targeted therapies that minimize side effects and maximize therapeutic benefits (Khramchikhin et al., 2022; Z. Li et al., 2015). Given their biological activity, triazolothiadiazole compounds can be explored for use as pesticides and herbicides. Research into their mode of action could lead to the development of more effective and environmentally friendly agricultural chemicals (Khramchikhin et al., 2022). The most promising triazolothiadiazole derivatives should progress to clinical trials to evaluate their safety and efficacy in humans. Success in clinical trials could pave the way for these compounds to become part of the therapeutic arsenal against various diseases. Detailed studies on the pharmacokinetics and pharmacodynamics of triazolothiadiazole derivatives will be crucial. Understanding how these compounds are absorbed, distributed, metabolized, and excreted will help optimize their use in clinical settings (Khramchikhin et al., 2022; Z. Li et al., 2015). Triazolothiadiazole compounds have not been studied much with coordination complexes so far. Their chelation with metals can unveil biological interactions useful for drug development and catalyzing effects (Daisylet et al., 2024; Pervaiz et al., 2020; Tiwari et al., 2024). Crystal structures of triazolothiadiazoles need further exploration to help identify the binding sites of small molecules on their target proteins. This is crucial for rational drug design as it allows for the optimization of interactions between the drug and its target. By visualizing how a drug interacts with its target at the atomic level, researchers can understand the mechanisms of action, which can aid in improving efficacy and reducing side effects (Khramchikhin et al., 2022; Z. Li et al., 2015). These interactions are useful for supramolecular chemistry. Moreover, the triazolothiadiazole molecule consists of multiple heteroatoms that have the potential for dipole-dipole interactions that need attention for application in biological and material sciences.

Heterocycles such as triazoles and thiadiazoles exhibit structure-activity relationships (SARs) detailed in section 6. The triazolothiadiazole compounds, with potential interaction sites at N1, N2, N5, and S atoms, show promise for protein interactions and bioactivity. However, these interactions have not been thoroughly investigated for their SARs. Furthermore, triazolothiadiazoles require evaluation in material science, as their structurally similar counterparts, triazolothiadiazines have demonstrated anticorrosive properties (Arshad et al., 2017).

The future of triazolothiadiazole derivatives in agriculture looks highly promising, particularly for addressing the challenges of pest resistance and crop diseases. These compounds have proven effective against fungal infections and bacterial diseases that threaten crop yields, making them essential for pest and disease management. Future research will likely expand the range of pathogens these derivatives can target, including those emerging due to climate change. Advancements in formulations, such as controlled-release and nanotechnology-based delivery systems, will enhance their stability and bioavailability, ensuring effective pest control with minimal environmental impact. Integrating triazolothiadiazole derivatives into sustainable agricultural practices is crucial. Combining them with biological controls and organic farming techniques will reduce reliance on traditional pesticides, promoting eco-friendly pest management and long-term sustainability (Muneret et al., 2018). Future research should optimize these compounds for different climates, ensuring consistent crop protection and yield stability. Collaborative efforts between researchers, agricultural companies, and farmers will advance the understanding and application of triazolothiadiazole derivatives, identifying effective use cases and optimizing application methods. In conclusion, triazolothiadiazole derivatives have a bright future in agriculture, with significant potential to enhance crop protection and productivity. By addressing pest resistance, disease management, and sustainability, these compounds can play a pivotal role in modern agriculture, contributing to global food security and sustainable practices.

CONCLUSION AND DISCUSSION

Triazolo[3,4-b][1,3,4]thiadiazoles represent a significant class of fused heterocyclic systems comprising two five-membered rings: 1,2,4-triazole and 1,3,4-thiadiazole. The fusion of these rings into a single molecular scaffold can yield synergistic effects and/or an expanded pharmacological profile. This hetero-system can be viewed as a cyclic analog of thiosemicarbazide and biguanide, both known for their diverse biological activities. Consequently, the 1,2,4-triazolo[3,4-b][1,3,4]thiadiazole core has been incorporated into a variety of promising drug candidates with therapeutic potential, including antitumor, antimicrobial, antifungal, antioxidant, antiviral, anti-inflammatory, and analgesic agents. However, there is limited research available for this heterocycle which may be due to structural and chemical effects discussed in previous sections.

Agriculture faces significant challenges from pests and microorganisms that threaten crop yields. Triazolothiadiazole derivatives have emerged as promising solutions due to their broad-spectrum efficacy and ability to target various biological mechanisms. These compounds interfere with crucial enzymes and pathways, providing valuable protection for crops and enhancing agricultural productivity. Plant pathogens cause serious and often fatal diseases, jeopardizing crop yield and stability. Bacterial infections, such as bacterial leaf streak (BLS) and bacterial leaf blight (BLF), can significantly reduce yields by 20% to 50%, affecting both agricultural productivity and the economic stability of farmers. Managing these fungal and bacterial diseases is crucial for maintaining healthy crop production and ensuring food security. Chemical antifungals and antibacterials play a vital role in preventing and managing crop diseases and pests, securing agricultural productivity and income. Recent research into triazolothiadiazole derivatives has shown promising antifungal and antibacterial properties, particularly valuable in agriculture. These compounds have been demonstrated to enhance plant growth and increase crop yields by effectively reducing fungal infections and bacterial diseases. By mitigating the threats posed by various plant pathogens, triazolothiadiazole derivatives could play a crucial role in ensuring better productivity and stability in agricultural production, addressing the challenges faced by modern agriculture. The triazolothiadiazole heterocycle holds significant promise for both medicinal and agricultural applications, necessitating further exploration.

REFERENCES

Al-Wahaibi, L. H., Abu-Melha, H. M., & Ibrahim, D. A. (2018). Synthesis of Novel 1,2,4-Triazolyl Coumarin Derivatives as Potential Anticancer Agents. *Journal of Chemistry*, 2018, 1–8. DOI: 10.1155/2018/5201374

Al-Wahaibi, L. H., Akilandeswari, G., Anusha, R., Al-Shaalan, N. H., Alkmali, O. M., El-Emam, A. A., Percino, J. M., & Thamotharan, S. (2019). Insights into the nature of weak noncovalent interactions in 3-(4-fluorophenyl)-6-(2-fluorophenyl)-1,2,4-triazolo[3,4-b][1,3,4]thiadiazole, a potential bioactive agent: X-ray, QTAIM and molecular docking analysis. *Journal of Molecular Structure*, 1183, 331–341. DOI: 10.1016/j.molstruc.2019.01.106

Al-Wahaibi, L. H., Karthikeyan, S., Blacque, O., El-Masry, A. A., Hassan, H. M., Percino, M. J., El-Emam, A. A., & Thamotharan, S. (2022). Structural and Energetic Properties of Weak Noncovalent Interactions in Two Closely Related 3,6-Disubstituted-[1,2,4]triazolo[3,4- *b*][1,3,4]thiadiazole Derivatives: In Vitro Cyclooxygenase Activity, Crystallography, and Computational Investigations. *ACS Omega*, 7(38), 34506–34520. DOI: 10.1021/acsomega.2c04252 PMID: 36188268

Altowyan, M. S., Haukka, M., Soliman, S. M., Barakat, A., Alaswad, S. O., Boraei, A. T. A., Gad, E. M., & Youssef, M. F. (2023). Synthesis, Characterization and Single Crystal X-ray Diffraction Analysis of Fused Triazolo/Thiadiazole Clubbed with Indole Scaffold. *Crystals*, 13(3), 423. DOI: 10.3390/cryst13030423

Armour, M., Davies, A. G., Upadhyay, J., & Wassermann, A. (1967). Colored electrically conducting polymers from furan, pyrrole, and thiophene. *Journal of Polymer Science. Part A-1, Polymer Chemistry*, 5(7), 1527–1538. DOI: 10.1002/pol.1967.150050704

Ashe, A. J., Fang, X., Schiesher, M., Richardson, A. D., & Hedberg, K. (2000). Aromatic Heterocycles Containing Arsenic and Sulfur: The Synthesis of 1,2- and 1,3-Thiaarsole and the Gas-Phase Molecular Structure of 1,2-Thiaarsole. *Journal of the American Chemical Society*, 122(29), 7012–7016. DOI: 10.1021/ja993924z

Baburajeev, C. P., Mohan, C. D., Rangappa, S., Mason, D. J., Fuchs, J. E., Bender, A., Barash, U., Vlodavsky, I., Basappa, , & Rangappa, K. S. (2017). Identification of Novel Class of Triazolo-Thiadiazoles as Potent Inhibitors of Human Heparanase and their Anticancer Activity. *BMC Cancer*, 17(1), 235. DOI: 10.1186/s12885-017-3214-8 PMID: 28359266

Badr, S. M. I., & Barwa, R. M. (2011). Synthesis of some new [1,2,4]triazolo[3,4-b][1,3,4]thiadiazines and [1,2,4]triazolo[3,4-b][1,3,4] thiadiazoles starting from 5-nitro-2-furoic acid and evaluation of their antimicrobial activity. *Bioorganic & Medicinal Chemistry*, 19(15), 4506–4512. DOI: 10.1016/j.bmc.2011.06.024 PMID: 21723735

Bergman, J., & Janosik, T. (2002). Recent progress in the chemistry of sulfur-containing indoles. In *Progress in Heterocyclic Chemistry* (Vol. 14, pp. 1–18). Elsevier., DOI: 10.1016/S0959-6380(02)80003-4

Bettendorff, L. (2021). Update on Thiamine Triphosphorylated Derivatives and Metabolizing Enzymatic Complexes. *Biomolecules*, 11(11), 1645. DOI: 10.3390/biom11111645 PMID: 34827643

Bouherrou, H., Saidoun, A., Abderrahmani, A., Abdellaziz, L., Rachedi, Y., Dumas, F., & Demenceau, A. (2017). Synthesis and Biological Evaluation of New Substituted Hantzsch Thiazole Derivatives from Environmentally Benign One-Pot Synthesis Using Silica Supported Tungstosilisic Acid as Reusable Catalyst. *Molecules (Basel, Switzerland)*, 22(5), 757. DOI: 10.3390/molecules22050757 PMID: 28481282

Dawar, R., Karan, S., Bhardwaj, S., Meena, D. K., Padhan, S. R., Reddy, K. S., & Bana, R. S. (2023). Role of Sulphur Fertilization in Legume Crops: A Comprehensive Review. *International Journal of Plant and Soil Science*, 35(21), 718–727. DOI: 10.9734/ijpss/2023/v35i214033

Djuidje, E. N., Sciabica, S., Buzzi, R., Dissette, V., Balzarini, J., Liekens, S., Serra, E., Andreotti, E., Manfredini, S., Vertuani, S., & Baldisserotto, A. (2020). Design, synthesis and evaluation of benzothiazole derivatives as multifunctional agents. *Bioorganic Chemistry*, 101, 103960. DOI: 10.1016/j.bioorg.2020.103960 PMID: 32559579

El Alami, A., El Maraghi, A., & Sdassi, H. (2024). Review of synthesis process of benzoxazole and benzothiazole derivatives. *Synthetic Communications*, 54(10), 769–801. DOI: 10.1080/00397911.2024.2337089

El-Sayed, H. A., Assy, M. G., & Mohamed, S., A. (. (2020). An efficient synthesis and antimicrobial activity of *N*-bridged triazolo[3,4- *b*]thiadiazine and triazolo[3,4- *b*]thiadiazole derivatives under microwave irradiation. *Synthetic Communications*, 50(7), 997–1007. DOI: 10.1080/00397911.2020.1726397

El Shehry, M. F., Abu-Hashem, A. A., & El-Telbani, E. M. (2010). Synthesis of 3-((2,4-dichlorophenoxy)methyl)-1,2,4-triazolo(thiadiazoles and thiadiazines) as anti-inflammatory and molluscicidal agents. *European Journal of Medicinal Chemistry*, 45(5), 1906–1911. DOI: 10.1016/j.ejmech.2010.01.030 PMID: 20153090

El-Shershaby, M. H., Ghiaty, A., Bayoumi, A. H., Ahmed, H. E. A., El-Zoghbi, M. S., El-Adl, K., & Abulkhair, H. S. (2021). 1,2,4-Triazolo[4,3- *c*]quinazolines: A bioisosterism-guided approach towards the development of novel PCAF inhibitors with potential anticancer activity. *New Journal of Chemistry*, 45(25), 11136–11152. DOI: 10.1039/D1NJ00710F

Fan, Z., Yang, Z., Zhang, H., Mi, N., Wang, H., Cai, F., Zuo, X., Zheng, Q., & Song, H. (2010). Synthesis, Crystal Structure, and Biological Activity of 4-Methyl-1,2,3-thiadiazole-Containing 1,2,4-Triazolo[3,4- *b*][1,3,4]thiadiazoles. *Journal of Agricultural and Food Chemistry*, 58(5), 2630–2636. DOI: 10.1021/jf9029628 PMID: 20014761

Gomha, S. M., & Riyadh, S. M. (2011). Synthesis under Microwave Irradiation of [1,2,4]Triazolo[3,4-b] [1,3,4]thiadiazoles and Other Diazoles Bearing Indole Moieties and Their Antimicrobial Evaluation. *Molecules (Basel, Switzerland)*, 16(10), 8244–8256. DOI: 10.3390/molecules16108244 PMID: 21959297

Gronowitz, S. (1986). *Thiophene and its derivatives*. J. Wiley & sons.

Gronowitz, S. (Ed.). (1991). *Chemistry of Heterocyclic Compounds: Thiophene and Its Derivatives, Part Four* (1st ed., Vol. 44). Wiley., DOI: 10.1002/9780470187265

Grover, G., Nath, R., Bhatia, R., & Akhtar, M. J. (2020). Synthetic and therapeutic perspectives of nitrogen containing heterocycles as anti-convulsants. *Bioorganic & Medicinal Chemistry*, 28(15), 115585. DOI: 10.1016/j.bmc.2020.115585 PMID: 32631563

Hamdy, R., Jones, A. T., El-Sadek, M., Hamoda, A. M., Shakartalla, S. B., Al Shareef, Z. M., Soliman, S. S. M., & Westwell, A. D. (2021). New Bioactive Fused Triazolothiadiazoles as Bcl-2-Targeted Anticancer Agents. *International Journal of Molecular Sciences*, 22(22), 12272. DOI: 10.3390/ijms222212272 PMID: 34830153

Himaja, M., Prathap, K. J., & Mali, S. V. (2012). Facile Synthesis of 3,6-Disubstituted-1,2,4-triazolo-[3,4- *b*]-1,3,4-thiadiazoles via Oxidative Cyclization of *n* -Heteroaryl-Substituted Hydrazones and Their Biological Activity. *Journal of Heterocyclic Chemistry*, 49(4), 823–828. DOI: 10.1002/jhet.870

Hrncic, D., Rasic-Markovic, A., Macut, D., Mladenovic, D., Susic, V., Djuric, D., & Stanojlovic, O. (2018). Sulfur – Containing Amino Acids in Seizures: Current State of the Art. *Current Medicinal Chemistry*, 25(3), 378–390. DOI: 10.2174/0929867324666170609090613 PMID: 28595553

Huang, S., Zuo, L., Zhang, L., Guo, X., Cheng, C., He, Y., Cheng, G., Yu, J., Liu, Y., Chen, R., Tang, G., Fan, Y., & Feng, L. (2024). Design, Synthesis, and Mode of Action of Thioacetamide Derivatives as the Algicide Candidate Based on Active Substructure Splicing Strategy. *Journal of Agricultural and Food Chemistry*, 72(13), 7021–7032. DOI: 10.1021/acs.jafc.4c00912 PMID: 38501582

Hussain, Asif, MOHAMMAD A. NASEER, & MOHAMMAD SARAFROZ. (2009). SYNTHESIS AND ANTICONVULSANT ACTIVITY OF SOME NOVEL FUSED HETEROCYCLIC 1,2,4-TRIAZOLO-[3,4-b]-1,3,4-THIADIAZOLE DERIVATIVES. *ACTA POLONIAE PHARMACEUTICA –. Drug Research*, 66(2), 135–140.

Islam, F., & Zeng, Q. (2024). Advances in Organosulfur-Based Polymers for Drug Delivery Systems. *Polymers*, 16(9), 1207. DOI: 10.3390/polym16091207 PMID: 38732676

Jaiswal, P., Kawathekar, N., Jain, G., & Jain, S. (2021). *In Silico Design, Synthesis, Anti-Microbial and Anti-Tubercular Evaluation of New Series of Benzothiazinone Derivatives Containing Triazolo Thiadiazole Moiety*. DOI: 10.21203/rs.3.rs-688359/v1

Jiang, N., Yan, J., Liang, Y., Shi, Y., He, Z., Wu, Y., Zeng, Q., Liu, X., & Peng, J. (2020). Resistance Genes and their Interactions with Bacterial Blight/Leaf Streak Pathogens (Xanthomonas oryzae) in Rice (Oryza sativa L.)—An Updated Review. *Rice (New York, N.Y.)*, 13(1), 3. DOI: 10.1186/s12284-019-0358-y PMID: 31915945

Joshi, K. A., Dhalani, J. M., Bhatt, H. B., & Kapadiya, K. M. (2021). Synthesis of some new 6-aryl-3-(4-isopropylphenyl)[1,2,4]triazolo[3,4-b][1,3,4] thiadiazoles and its anti-microbial study. *Current Chemistry Letters*, 10(4), 479–488. DOI: 10.5267/j.ccl.2021.4.003

Kamiński, K., Zagaja, M., Łuszczki, J. J., Rapacz, A., Andres-Mach, M., Latacz, G., & Kieć-Kononowicz, K. (2015). Design, Synthesis, and Anticonvulsant Activity of New Hybrid Compounds Derived from 2-(2,5-Dioxopyrrolidin-1-yl)propanamides and 2-(2,5-Dioxopyrrolidin-1-yl)butanamides. *Journal of Medicinal Chemistry*, 58(13), 5274–5286. DOI: 10.1021/acs.jmedchem.5b00578 PMID: 26052884

Kamoutsis, C., Fesatidou, M., Petrou, A., Geronikaki, A., Poroikov, V., Ivanov, M., Soković, M., Ćirić, A., Carazo, A., & Mladěnka, P. (2021). Triazolo Based-Thiadiazole Derivatives. Synthesis, Biological Evaluation and Molecular Docking Studies. *Antibiotics (Basel, Switzerland)*, 10(7), 804. DOI: 10.3390/antibiotics10070804 PMID: 34356726

Karegoudar, P., Prasad, D. J., Ashok, M., Mahalinga, M., Poojary, B., & Holla, B. S. (2008). Synthesis, antimicrobial and anti-inflammatory activities of some 1,2,4-triazolo[3,4-b][1,3,4]thiadiazoles and 1,2,4-triazolo[3,4-b][1,3,4]thiadiazines bearing trichlorophenyl moiety. *European Journal of Medicinal Chemistry*, 43(4), 808–815. DOI: 10.1016/j.ejmech.2007.06.026 PMID: 17804121

Katritzky, A. R. (Ed.). (1995). *Comprehensive organic functional group transformations: 7-volume-set*. Pergamon Press.

Kavitha, K., Srikrishna, D., Dubey, P. K., & Aparna, P. (2019). An efficient one-pot four-component Gewald reaction: Synthesis of substituted 2-aminothiophenes with coumarin–thiazole scaffolds under environmentally benign conditions. *Journal of Sulfur Chemistry*, 40(2), 195–208. DOI: 10.1080/17415993.2018.1556275

Kerassa, A., Belaidi, S., Harkati, D., Lanez, T., Prasad, O., & Sinha, L. (2016). Investigations on Molecular Structure, Electronic Properties, NLO Properties and Comparison of Drug-Likeness of Triazolothiadiazole Derivatives by Quantum Methods and QSAR Analysis. *Reviews in Theoretical Science*, 4(1), 85–96. DOI: 10.1166/rits.2016.1050

Khalili Ghomi, M., Noori, M., Nazari Montazer, M., Zomorodian, K., Dastyafteh, N., Yazdanpanah, S., Sayahi, M. H., Javanshir, S., Nouri, A., Asadi, M., Badali, H., Larijani, B., Irajie, C., Iraji, A., & Mahdavi, M. (2023). [1,2,4]triazolo[3,4-b][1,3,4]thiadiazole derivatives as new therapeutic candidates against urease positive microorganisms: Design, synthesis, pharmacological evaluations, and in silico studies. *Scientific Reports*, 13(1), 10136. DOI: 10.1038/s41598-023-37203-z PMID: 37349372

Larsen, A. K., Escargueil, A. E., & Skladanowski, A. (2003). Catalytic topoisomerase II inhibitors in cancer therapy. *Pharmacology & Therapeutics*, 99(2), 167–181. DOI: 10.1016/S0163-7258(03)00058-5 PMID: 12888111

Li, C., Zhang, J., Ren, Z., Xie, R., Yin, C., Ma, W., Zhou, F., Chen, H., & Lin, Y. (2021). Development of 'multiresistance rice' by an assembly of herbicide, insect and disease resistance genes with a transgene stacking system. *Pest Management Science*, 77(3), 1536–1547. DOI: 10.1002/ps.6178 PMID: 33201594

Li, P., Tian, P., Chen, Y., Song, X., Xue, W., Jin, L., Hu, D., Yang, S., & Song, B. (2018). Novel bisthioether derivatives containing a 1,3,4-oxadiazole moiety: Design, synthesis, antibacterial and nematocidal activities. *Pest Management Science*, 74(4), 844–852. DOI: 10.1002/ps.4762 PMID: 29024290

Mason, C. D., & Nord, F. F. (1951). STUDIES ON THE CHEMISTRY OF HETEROCYCLICS. XIII. TRIARYLMETHANE DYES CONTAINING A THIOPHENE RING [1]. *The Journal of Organic Chemistry*, 16(5), 722–727. DOI: 10.1021/jo01145a011

Mathew, B., Mathew, G. E., Sonia, G., Kumar, A., Charles, N. P., & Kumar, P. (2013). Design of 1-(furan-2-yl)-N-(5-substituted phenyl-1, 3, 4-thiadiazol-2-yl) methanimine derivatives as Enoyl-ACP reductase inhibitors: Synthesis, molecular docking studies and anti-tubercular activity. *Bangladesh Journal of Pharmacology*, 8(3), 242–248. DOI: 10.3329/bjp.v8i3.14778

Mathew, V., Keshavayya, J., & Vaidya, V. P. (2006). Heterocyclic system containing bridgehead nitrogen atom: Synthesis and pharmacological activities of some substituted 1,2,4-triazolo[3,4-b]-1,3,4-thiadiazoles. *European Journal of Medicinal Chemistry*, 41(9), 1048–1058. DOI: 10.1016/j.ejmech.2006.03.018 PMID: 16822595

Mathew, V., Keshavayya, J., Vaidya, V. P., & Giles, D. (2007). Studies on synthesis and pharmacological activities of 3,6-disubstituted-1,2,4-triazolo[3,4-b]-1,3,4-thiadiazoles and their dihydro analogues. *European Journal of Medicinal Chemistry*, 42(6), 823–840. DOI: 10.1016/j.ejmech.2006.12.010 PMID: 17331622

Mirhoseini, H., & Taghdiri, M. (2016). Extractive oxidation desulfurization of sulfur-containing model fuel using hexamine–phosphotungstate hybrid as effective heterogeneous catalyst. *Fuel*, 167, 60–67. DOI: 10.1016/j.fuel.2015.11.042

Mohamed, K. O., Nissan, Y. M., El-Malah, A. A., Ahmed, W. A., Ibrahim, D. M., Sakr, T. M., & Motaleb, M. A. (2017). Design, synthesis and biological evaluation of some novel sulfonamide derivatives as apoptosis inducers. *European Journal of Medicinal Chemistry*, 135, 424–433. DOI: 10.1016/j.ejmech.2017.04.069 PMID: 28463785

Mostafa, M. S., Radini, I. A. M., El-Rahman, N. M. A., & Khidre, R. E. (2024). Synthetic Methods and Pharmacological Potentials of Triazolothiadiazines: A Review. *Molecules (Basel, Switzerland)*, 29(6), 1326. DOI: 10.3390/molecules29061326 PMID: 38542962

Pathania, S., Narang, R. K., & Rawal, R. K. (2019). Role of sulphur-heterocycles in medicinal chemistry: An update. *European Journal of Medicinal Chemistry*, 180, 486–508. DOI: 10.1016/j.ejmech.2019.07.043 PMID: 31330449

Pisano, C., Vlodavsky, I., Ilan, N., & Zunino, F. (2014). The potential of heparanase as a therapeutic target in cancer. *Biochemical Pharmacology*, 89(1), 12–19. DOI: 10.1016/j.bcp.2014.02.010 PMID: 24565907

Porter, M. (1977). Vulcanization of Rubber. In Oae, S. (Ed.), *Organic Chemistry of Sulfur* (pp. 71–118). Springer US., DOI: 10.1007/978-1-4684-2049-4_3

Purohit, D. H., Dodiya, B. L., Ghetiya, R. M., Vekariya, P. B., & Joshi, H. S. (2011). Synthesis and antimicrobial activity of some new 1,3,4-thiadiazoles and 1,3,4-thiadiazines containing 1,2,4-Triazolo nucleus. *Acta Chimica Slovenica*, 58(1), 53–59. PMID: 24061943

Rachi, Y., Yamakado, R., & Okada, S. (2018). Synthesis of dyes with a sulfur-containing heterocyclic donor and a pyrroline-type acceptor. *Dyes and Pigments*, 159, 345–351. DOI: 10.1016/j.dyepig.2018.06.032

Ram, V. J., Sethi, A., Nath, M., & Pratap, R. (2019). *The chemistry of heterocycles: Nomenclature and chemistry of three- to- five- membered heterocycles*. Elsevier.

Riffat, A., Ahmad, N., Sajid Aqeel Ahmad, M., & Alvi, A. K. (2023). Sulfur: A macronutrient having potential to improve salinity tolerance in plants. *Pakistan Journal of Botany*, 55(4). Advance online publication. DOI: 10.30848/PJB2023-4(33)

Sagredou, S., Dalezis, P., Nikoleousakos, N., Nikolaou, M., Voura, M., Almpanakis, K., Panayiotidis, M. I., Sarli, V., & Trafalis, D. T. (2020). 3,6-Disubstituted 1,2,4-Triazolo[3,4-b]Thiadiazoles with Anticancer Activity Targeting Topoisomerase II Alpha. *OncoTargets and Therapy*, 13, 7369–7386. DOI: 10.2147/OTT.S254856 PMID: 32801761

Seelolla, G., & Ponneri, V. (2016). Synthesis, Antimicrobial, and Antioxidant Activities of Some Fused Heterocyclic [1,2,4]Triazolo[3,4-b][1,3,4]thiadiazole Derivatives. *Journal of Heterocyclic Chemistry*, 53(3), 929–936. DOI: 10.1002/jhet.2348

Settypalli, T., Chunduri, V. R., Kerru, N., Nallapaneni, H. K., Chintha, V. R., Daggupati, T., Yeguvapalli, S., & Wudayagiri, R. (2019). Design, Synthesis, Neuroprotective and Antibacterial Activities of 1,2,4-Triazolo[3,4-b]1,3,4-thiadiazole Linked Thieno[2,3-d]pyrimidine Derivatives and In Silico Docking Studies. *ChemistrySelect*, 4(5), 1627–1634. DOI: 10.1002/slct.201803917

Shuai, Y., Haoyi, T., Chen, Z., & Hanhong, X. (2024). Design, synthesis and evaluation of insecticidal activity of phenylthiazole derivative containing primary amide and cyanogroup. *Journal of South China Agricultural University*, 45(3), 354–363. DOI: 10.7671/j.issn.1001-411X.202308001

Silva, R. N., Monteiro, V. N., Steindorff, A. S., Gomes, E. V., Noronha, E. F., & Ulhoa, C. J. (2019). Trichoderma/pathogen/plant interaction in pre-harvest food security. *Fungal Biology*, 123(8), 565–583. DOI: 10.1016/j.funbio.2019.06.010 PMID: 31345411

Sundaram, R. M., Chatterjee, S., Oliva, R., Laha, G. S., Cruz, C. V., Leach, J. E., & Sonti, R. V. (2014). Update on Bacterial Blight of Rice: Fourth International Conference on Bacterial Blight. *Rice, 7*(1), 12, s12284-014-0012–0017. DOI: 10.1186/s12284-014-0012-7

Suresh Kumar, G. V., Rajendraprasad, Y., Mallikarjuna, B. P., Chandrashekar, S. M., & Kistayya, C. (2010). Synthesis of some novel 2-substituted-5-[isopropylthiazole] clubbed 1,2,4-triazole and 1,3,4-oxadiazoles as potential antimicrobial and antitubercular agents. *European Journal of Medicinal Chemistry*, 45(5), 2063–2074. DOI: 10.1016/j.ejmech.2010.01.045 PMID: 20149496

Szasz, M., Casoni, D., & Cimpoiu, C. (2023). Study on binding affinity profile, bioactivity, and lipophilicity of selected antipsychotic drugs. *Journal of Liquid Chromatography & Related Technologies*, •••, 1–6. DOI: 10.1080/10826076.2023.2216780

Thiazoles | DrugBank Online. (n.d.). Retrieved July 28, 2024, from https://go.drugbank.com/categories/DBCAT000348

Trafalis, D. T., Sagredou, S., Dalezis, P., Voura, M., Fountoulaki, S., Nikoleousakos, N., Almpanakis, K., Deligiorgi, M. V., & Sarli, V. (2021). Anticancer Activity of Triazolo-Thiadiazole Derivatives and Inhibition of AKT1 and AKT2 Activation. *Pharmaceutics*, 13(4), 493. DOI: 10.3390/pharmaceutics13040493 PMID: 33916378

Uddin, J., Ullah, S., Halim, S. A., Waqas, M., Ibrar, A., Khan, I., Bin Muhsinah, A., Khan, A., & Al-Harrasi, A. (2023). Triazolothiadiazoles and Triazolothiadiazines as New and Potent Urease Inhibitors: Insights from In Vitro Assay, Kinetics Data, and In Silico Assessment. *ACS Omega*, 8(35), 31890–31898. DOI: 10.1021/acsomega.3c03546 PMID: 37692208

Waddar, B., Parne, S. R., Gandi, S., Prasanth, G. R., Yaseen, M., & Kariduraganavar, M. Y. (2024). The second-order nonlinear optical properties of novel triazolo[3,4-b][1, 3, 4] thiadiazole derivative chromophores using DFT calculations. *Structural Chemistry*, 35(1), 253–264. DOI: 10.1007/s11224-023-02178-0

Wu, J., Li, J., Xu, M.-H., & Liu, D. (2014). Structure–activity relationship and interaction studies of new SIRT1 inhibitors with the scaffold of 3-(furan-2-yl)-[1,2,4]triazolo[3,4-b][1,3,4]thiadiazole. *Bioorganic & Medicinal Chemistry Letters*, 24(14), 3050–3056. DOI: 10.1016/j.bmcl.2014.05.028 PMID: 24880902

Wu, N., Yang, Y., Tian, G., An, L., Liu, S., Yan, T., Yi, M., & Bao, X. (2023). Synthesis, X-ray Crystal Structure, and Antimicrobial Studies of New Quinazolin-4(3*H*)-one Derivatives Containing the 1,2,4-Triazolo[3,4-*b*][1,3,4]thiadiazole Moiety and 4-Piperidinyl Linker. *Journal of Agricultural and Food Chemistry*, 71(49), 19277–19287. DOI: 10.1021/acs.jafc.3c03670 PMID: 38038681

Wu, S., Shi, J., Chen, J., Hu, D., Zang, L., & Song, B. (2021). Synthesis, Antibacterial Activity, and Mechanisms of Novel 6-Sulfonyl-1,2,4-triazolo[3,4- *b*][1,3,4]thiadiazole Derivatives. *Journal of Agricultural and Food Chemistry*, 69(16), 4645–4654. DOI: 10.1021/acs.jafc.1c01204 PMID: 33871992

Yadav, K. P., Rahman, M. A., Nishad, S., Maurya, S. K., Anas, M., & Mujahid, M. (2023). Synthesis and biological activities of benzothiazole derivatives: A review. *Intelligent Pharmacy*, 1(3), 122–132. DOI: 10.1016/j.ipha.2023.06.001

Yamamoto, T., Sanechika, K., & Yamamoto, A. (1980). Preparation of thermostable and electric-conducting poly(2,5-thienylene). *Journal of Polymer Science. Polymer Letters Edition*, 18(1), 9–12. DOI: 10.1002/pol.1980.130180103

Yin, X., Wang, L., & Mao, L. (2024). Comparing the Developmental Toxicity Delay and Neurotoxicity of Benzothiazole and Its Derivatives (BTHs) in Juvenile Zebrafish. *Toxics*, 12(5), 341. DOI: 10.3390/toxics12050341 PMID: 38787120

Yuan, H., Liu, Q., Zhang, L., Hu, S., Chen, T., Li, H., Chen, Y., Xu, Y., & Lu, T. (2018). Discovery, optimization and biological evaluation for novel c-Met kinase inhibitors. *European Journal of Medicinal Chemistry*, 143, 491–502. DOI: 10.1016/j.ejmech.2017.11.073 PMID: 29202410

Yuan, T., Li, X., Xiao, J., & Wang, S. (2011). Characterization of Xanthomonas oryzae-Responsive cis-Acting Element in the Promoter of Rice Race-Specific Susceptibility Gene Xa13. *Molecular Plant*, 4(2), 300–309. DOI: 10.1093/mp/ssq076 PMID: 21208999

Zhan, W., Zhou, R., Mao, P., Yuan, C., Zhang, T., Liu, Y., Tian, J., Wang, H., & Xue, W. (2024). Synthesis, antifungal activity and mechanism of action of novel chalcone derivatives containing 1,2,4-triazolo-[3,4-b]-1,3,4-thiadiazole. *Molecular Diversity*, 28(2), 461–474. DOI: 10.1007/s11030-022-10593-4 PMID: 36964852

Zhang, D., Ruan, D., Li, J., Chen, Z., Zhu, W., Guo, F., Chen, K., Li, Y., & Wang, R. (2020). Four undescribed sulfur-containing indole alkaloids with nitric oxide inhibitory activities from Isatis tinctoria L. roots. *Phytochemistry*, 174, 112337. DOI: 10.1016/j.phytochem.2020.112337 PMID: 32163787

Zhao, P.-L., Duan, A.-N., Zou, M., Yang, H.-K., You, W.-W., & Wu, S.-G. (2012). Synthesis and cytotoxicity of 3,4-disubstituted-5-(3,4,5-trimethoxyphenyl)-4H-1,2,4-triazoles and novel 5,6-dihydro-[1,2,4]triazolo[3,4-b][1,3,4]thiadiazole derivatives bearing 3,4,5-trimethoxyphenyl moiety. *Bioorganic & Medicinal Chemistry Letters*, 22(13), 4471–4474. DOI: 10.1016/j.bmcl.2012.03.023 PMID: 22647723

Chapter 3
Imidazothiazole:
Different Synthetic Approach and their Anticancer Activity

Shalini Jaiswal
https://orcid.org/0000-0003-0137-7734
Amity University Greater Noida, India

ABSTRACT

Recent research on cancer therapy has revealed that the emergence of malignant cells is caused by several enzyme alterations. such include kinase inhibitors of ErB4 (HER4), B-Raf kinase, and indoleamine 2,3-dioxygenase 1 (IDO1). Researchers discovered that drugs based on imidazothiazoles, and their analogues exhibit more robust anticancer action by inhibiting these enzymatic receptors. In recent times, imidazothiazole scaffold is broadly explored for its anticancer activity, which acts through various mechanisms such as EGFR, B-RAF, DHFR kinase inhibition and tubulin polymerization inhibition and other molecular mechanisms of action. Due to their feasible synthetic accessibility and promising pharmacological profile, it has attracted various medicinal chemists to explore and develop imidazothiazole derivatives as potent and safe anticancer agents. In the present book chapter, we have reviewed various potent imidazothiazole scaffold-based derivatives reported as anticancer agents, their synthetic strategies, with their future perspective.

1. INTRODUCTION

Organic molecules known as heterocyclic compounds have at least one carbon atom and at least one extra heteroatom, such as N, O, or S. All living cells' metabolism depends on heterocyclic molecules, and most of these nitrogen-containing, five-membered fused heterocyclic molecules have biological function (Jaiswal,

DOI: 10.4018/979-8-3693-7520-4.ch003

2019, pp. 36-39). The fused heterocyclic compounds are one of the most important systems in medicine because of their wide range of biological activity. In the field of medicine, imidazothiazole is a crucial molecule. The fused heterocyclic molecule with a bridgehead nitrogen and sulfur atom that is physiologically active. It is made up of a five-membered thiazole ring and an imidazole ring. The imidazole is found in many synthetic and natural materials. (Shaik et al.,2017)

Imidazole is a planar five-membered ring **(Figure 1, Compd. 1)** (Guo et al.,2023) with two meta-nitrogen atoms at the first and third positions. Furthermore, since the unshared electron pair of the 1-position nitrogen atom in the structure participates in the cyclic conjugation, the electron's density decreases. A wide range of pharmacological properties is often exhibited by compounds containing imidazole rings, including antibacterial (Jaiswal & Dwivedi, 2017; Salahb et al.,2018; Senkardes et al.,2020), antifungal (Slassi et al.,2020), anticonvulsant (Jakovleva et al.,2020), antioxidant (Abdelhamid et al.,2020), anticancer (Bae et al.,2018), and antiparasitic (Adeyemi et al.,2020) properties. It has been shown that only substituents at the N- of the imidazole ring are tolerated without loss of activity in terms of druggability for the investigation of various N-substituted imidazole inhibitors of p38 mitogen-activated protein kinases, and appropriate substituents (Margutti & Laufer,2007). Furthermore, it has been documented that the compound's free nitrogen within the imidazole ring may initiate water-mediated associations with both the Arg85 side chain and the OH-4 of Globo H's GlcNAc moiety (Lal et al.,2021). Thiazole is also the five-membered, heterocyclic compound that contains nitrogen having smell like pyridine **(Figure 1, Compd. 2)**. Its chemical composition consists of one N-atom and one S-atom.

Figure 1. Imidazole and thiazole

Imidazole **1** Thiazole **2**

Thiazole, meantime, serves as a crucial building block for a variety of synthetic molecules. Thiazole molecules have a wide range of biological functions and may adhere to different targets in the microbe, such as enzymes or receptors. Thiazole and bis-thiazole derivatives have been shown to have a wide range of biological actions (Borcea et al.,2021), including analgesic (Thore et al.,2013), antioxidant (Liu et al.,2021), anticancer (Wan et al.,2021; Gomha, 2017), anti-inflammatory (Pricopie et al.,2020), antibacterial (Liaras et al.,2011), and cardioprotective (Drapak et al.,2019). The thiazole ring of these compounds is widely regarded as one of the key structural elements responsible for their pharmacological activity. It is also a structural element necessary for the compounds to be druggable. The imidazothiazole ring is a crucial place for activity and a structure that can increase the stability of the molecules, and it can also combine the biological features of both heterocycles (Shehata et al.,2023). Imidazothiazoles exhibit several fusion modes depending on where heteroatoms and unsaturated bonds are located **(Figure 2, Compd. 3-7)** (Saliyeva et al.,2021). Because of its varied pharmacological characteristics, the [2,1-b] structure has received the most attention. The chemical molecule binds to the target group is by the formation of a nitrogen–iron bond between the haemo group and the core of the imidazothiazole ring in one research (Serafini et al.,2019). Imidazo[2,1-b] thiazoles, imidazo[1,2-c] thiazoles, and their derivatives are nitrogen and sulphur containing heteroaromatic substances that are well-known for their biological activity and importance in medicinal chemistry. These drugs could be used to successfully treat a wide range of illnesses. **Figure 2** depicts the imidazothiazole core's chemical structure. Known medicinal properties of imidazothiazole core include anticancer (Andreani et al.,2008; Park et al.,2011), Anthelminthic (Robert et al.,1990), antifungal (Dangi, Hussain & Talesara, 2011), cardiotonic (Andreani et al.,1992), anti-hypertensive (Budriesi et al.,2008), anti-allergic (Jadhav et al.,2008), bactericidal (Mohan & Kiran,1991), antiviral in nature (Scribner et al.,2008), an antioxidant (Andreani et al.,2013), and so on.

Figure 2. Imidazothiazole core chemical structure

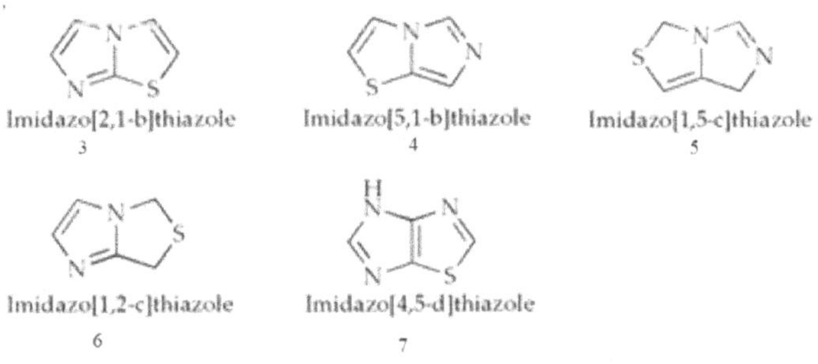

2. SYNTHETIC STRATEGIES OF IMIDAZOTHIAZOLES AND THEIR DERIVATIVES

Numerous techniques, including one-pot three-component (Jaiswal, S. & Siddiqui, 2014) and four-component microwave-assisted reaction (Jaiswal, S., & Sharma, 2014;Etivand et al.,2019; Jana et al.,2019) and without catalyst microwave procedure (Mukku &Maiti,2020), and others like catalyst-free visible light-triggered method, A^3 coupling utilizing CNTS includes nickel-ferrite magnetic nanoparticles assisted multicomponent reaction(Geedkar et al.,2020), FeCl3/ZnCl2 catalysed aerobic oxidation cyclization, (Mishra et al.,2014) copper-catalysed A3 coupling, (Rassokhina et al.,2017) Oxone mediated direct aryl selenation, (Rodrigues et al.,2018) etc. can be used to create imidazo[2,1-b]thiazoles and their analogues. Numerous reactions involving multicomponent reactions (MCRs) (Kishore et al.,2016) are used to synthesize imidazothiazoles and their derivatives. These reactions include the biosynthesis of staphyloxanthin, which is generated by staphylococcus aurease in the presence of the enzyme dehydrosqualene synthase (Shareef et al.,2019),Transition metal catalyzed reactions (Rassokhina,2017), visible light synthesis (Mishra et al.,2016), regioselective reaction (Serykh et al.,2018), Groebke–Blackburn–Bienayme–Reaction mechanism (Shivhare2018), Claisen–Schmidt condensation (Koudad et al.,2019), and ring-opening and ring-closing reconstruction reactions (Shi et al.,2016). Based on the easily accessible starting materials and the synthetic efficiency, especially in heterocyclic chemistry, green synthesis is quick, safe, selective, and highly yielding (Jaiswal et al., 2017; Jaiswal, S., & Mojahid,2020). The summary various methods of synthesis of imidazothiazoles by using as-aminothiazoles, imidazole's and miscellaneous reactions demonstrated various disconnection in **Figure 3(Compd. 8-17)** (Afsina et al.,2020).

Figure 3. Method for Synthesis of Imidazothiazole

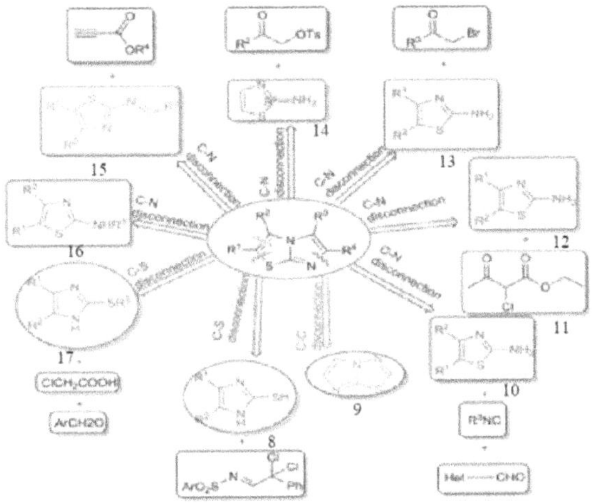

2.1 Synthesis From Aminothiazoles

Vilsmeier-Hack and Knoevenagel condensation processes, 3-(2-aminothiazol-4-yl)-2H-chromen-2-one and phenacyl bromides were converted into new coumarinyl imidazo[2,1-b] thiazole analogues. At ambient temperature and with catalytic NaF, this process produced yields upto 80 percent. (Gali *et al.*, 2014) **(Scheme 1, Compd. 18).**

Figure 4. Scheme 1

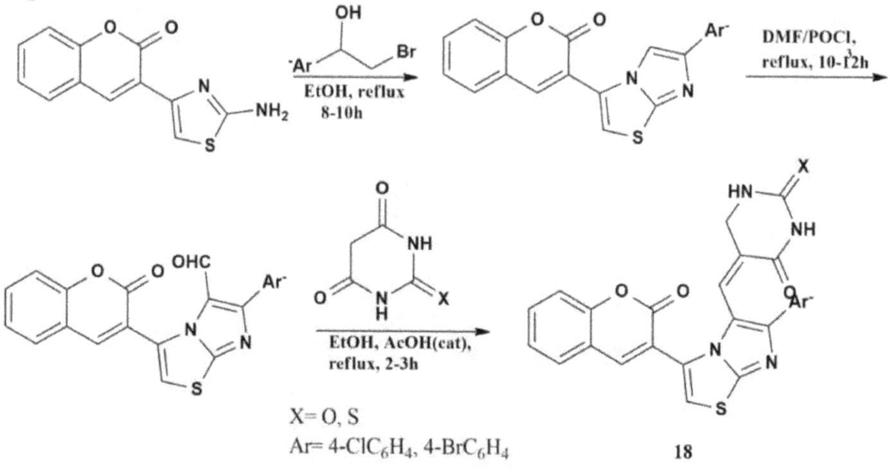

18

Vikas and associates devised a process using a mixture of $C_6H_5COCH_3$, hydroxyl(tosyloxy) iodobenzene (HTIB), and acetonitrile (CH_3CN) to produce 6-aryl imidazo[2,1-b] thiazole in a single pot. After two hours of condensing, α-tosyloxyketone was produced, which interacted with the ethanol-based solution of 2-aminothiazole (Sonawane,2015) **(Scheme 2, Compd. 19)**.

Figure 5. Scheme 2

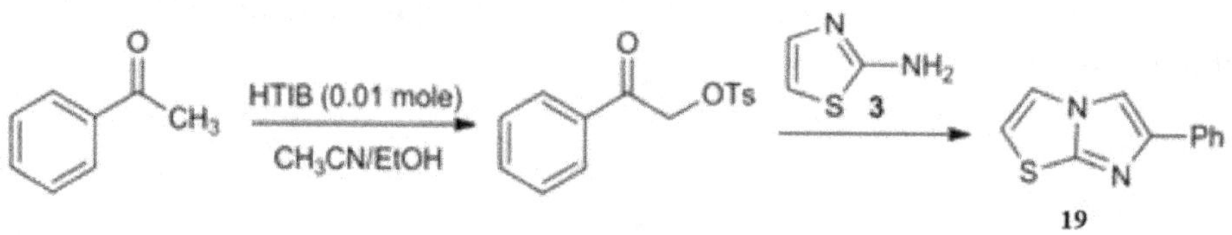

19

Imidazothiazole analogues were synthesized by Koudad and colleagues using the Claisen-Schmidt condensation of aldehyde and various methyl ketones. The substrates that were selected for reaction in the presence of ethanol were the substituted acetophenone and 2-aminothiazole. Imidazothiazole-chalcones were produced using the Vilsmeier-Haack reaction with product in dimethylformamide (DMF) and phosphoryl chloride ($POCl_3$) (Baig et al.,2018) **(Scheme 3, Compd. 20)**.

Figure 6. Scheme 3

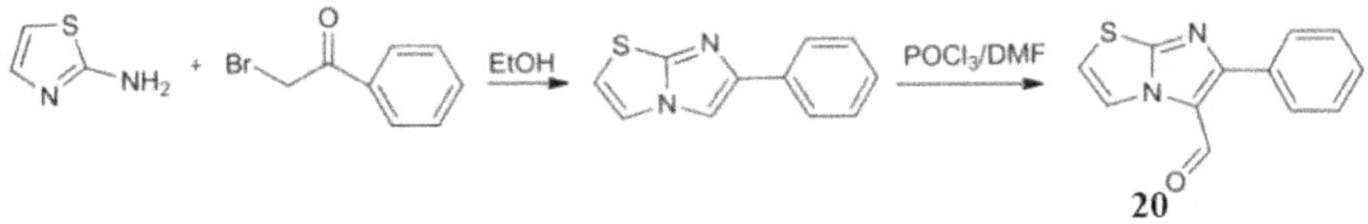

20

Substituted imidazo[2,1-b]thiazoles were created by combining N'-(5-arylthiazol-2-yl)-N,N-dimethylformamide with either 6-Bromo-6-deoxy-1,2-O-isopropylidene-3-O-methyl—D-Xylo- hexofuranos-5-ulose or 6-Bromo-6-deoxy-1,2-O-isopropylidene-3-O-acetyl—α-D-Xylo- hexafuranos-5-ulose.(Barradas et al.,2011) **(Scheme 4, Compd. 21).**

Figure 7. Scheme 4

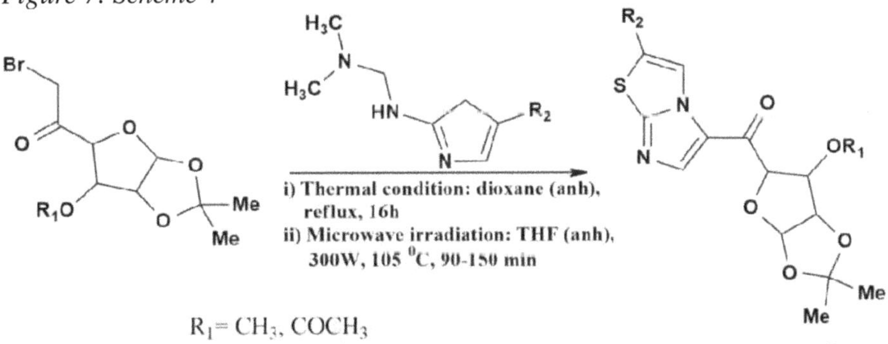

R_1= CH_3, $COCH_3$

R_2– COC_6H_5, COC_6H_4Br-p, COC_6H_4Cl-p, COC_6H_4F-p

21

Catalyst-free, visible light triggered method for synthesis of imidazo[2,1-b] were developed by Mishra et. al. They are utilizing 2-aminothiazole and substituted phenacyl bromide as reactants in green medium (EtOH: H_2O), with product, yields a 55% in 5 hours. The capacity to dissolve of reactants was enhanced by the benefits of EtOH and H_2O. They also attempted to raise the product yield by experimenting with varied ratios of the mixed solvent system. The optimal outcomes were obtained with EtOH: H_2O (4:1) (Mishra et al., 2016) **(Scheme 5, Compd. 22)**.

Figure 8. Scheme 5

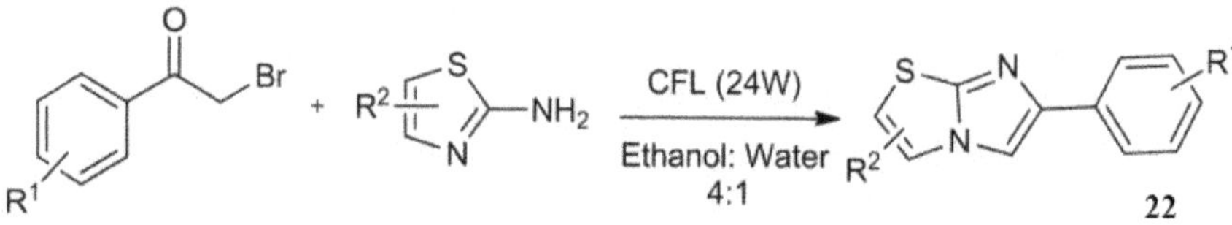

22

Vekariya et al. reported a unique sustainable synthesis using microwave assistance to produce imidazo[2,1-b] thiazole derivatives through the reaction of 1-(2-amino-4-methylthiazol-5- yl) ethenone and ethyl 2-amino-4-methyl thiazole-5-carboxylate with phenacyl bromide. Both thermal heating at 90°C and the presence of PEG-400 were used in this reaction. (Vekariya et al.,2017) **(Scheme 6, Compd. 23).**

Figure 9. Scheme 6

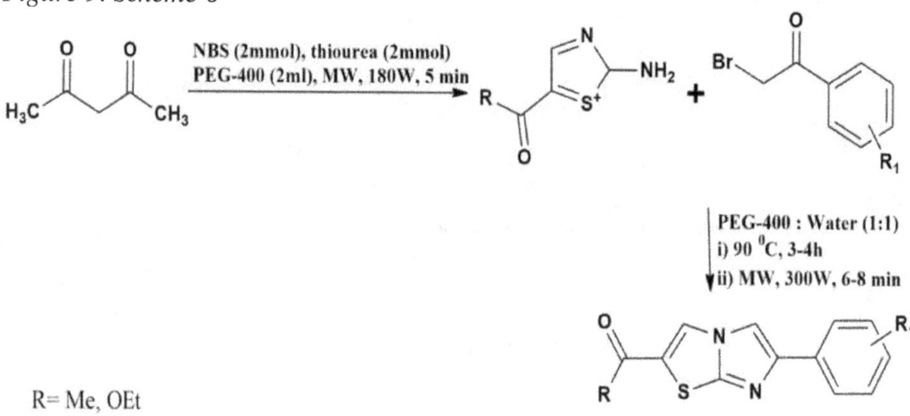

R= Me, OEt

R_1= H, 2-Cl, 4-Cl, 4-Br, 4-Me, 4-OMe, 4-NO_2, 4-F, 2-OMe

23

By reacting ethyl 2-aminothiazole-4-carboxylate with a range of 2-bromo-1-(4-substituted-phenyl)-ethanone derivatives in a solution of acetone at 60 °C for 24 hours, Ewida et al. created a process to synthesize ethyl 5-(4-substituted-phenyl) imidazo[2,1-b] thiazole-3-carboxylate (Ewida et al.,2018) **(Scheme 7, Compd. 24).**

Figure 10. Scheme 7

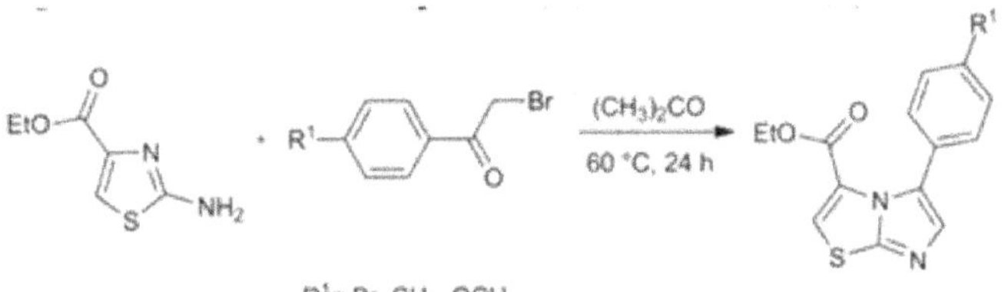

R¹= Br, CH3, OCH3

24

The 2-aminothiazole reacts with methanol, to from 3'-nitroacetophenone is α-brominated with N-bromo succinamide in DMF. Subsequently, the chemical was generated through the coupling by Haack reaction in the presence of $Pd(OAc)_3$, Ph_3P, and K_2CO_3. Potassiumperoxy monosulfate was used to oxidize the SCH_3 of the producing chemical to produce the intermediate. A specific N-Side chain reacted with the intermediate to produce imidazo[2,1-b] thiazole containing an m-nitrophenyl group. (Ammar et al.,2020) **(Scheme 8, Compd. 25).**

Figure 11. Scheme 8

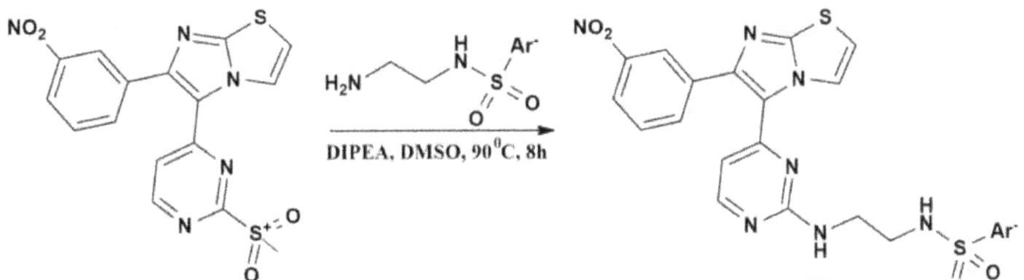

25

n= 1,2

Ar= 3-FPh, 4-CH3Ph, 4-CF3Ph, 4-OCH3Ph, 4-ClPh, 4-BrPh, 4-FPh, 1-NaPh

Imidazo[2,1-b] thiazole analogues were synthesized by Abdel-Maksoud and colleagues in 2019. In this case, the base material that is used is 2-bromo-1-(3-fluorophenyl)-ethan-1-one to yield 6-(3-fluorophenyl) imidazo[2,1-b] thiazole rather than α-bromo-4-fluoroacetophenone (Abdel-Maksoud, Ammar & Oh, 2019) **(Scheme 9, Compd. 26)**.

Figure 12. Scheme 9

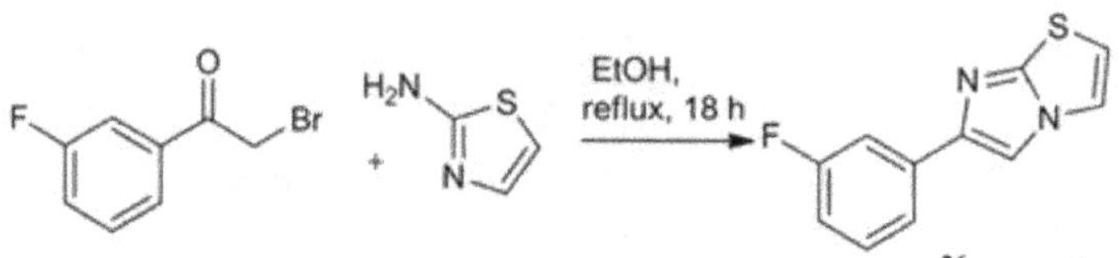

Using a metal-free synthesis method, imidazothiazoles tethered with sulfoximine were produced. A yield of 78–90% was obtained by synthesizing the compounds by treating the sulfoximine tethered imidazole solution with TBAF at 78^{0}C, followed by quenching with saturated ammonium chloride solution. (Battula etal.,2017) **(Scheme 10, Compd. 27).**

Figure 13. Scheme 10

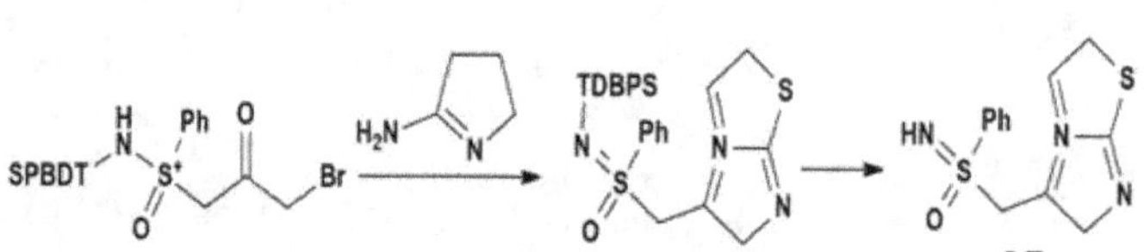

A variety of imidazo[2,1-b] thiazole-sulfonyl piperazine analogues were produced by condensation of ethyl-2-aminothiazole-4-carboxylate with ethyl bromopyruvate and thiourea under reflux conditions for 4 hours in ethanol. The desired product was obtained with good yields by cyclizing ethyl-2-aminothiazole-4-carboxylate in the presence of LiOH: H_2O during ester hydrolysis. The generated intermediate was combined with the ester hydrolysed molecule using EDCI and HOBt in DMF. (Manasa et al.,2020) **(Scheme 11, Compd. 28).**

Figure 14. Scheme 11

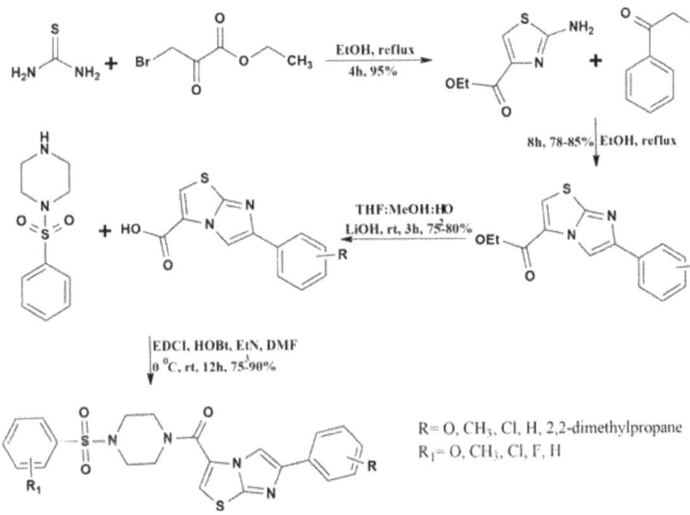

28

By combining of substituted 2-amino benzothiazoles with ethyl bromopyruvate, reduction with lithium aluminium hydride, and subsequent oxidation with DMP, benzo[d]imidazo[2,1-b] thiazole-chalcone derivatives, the benzo[d]imidazo[2,1-b] thiazole-2-carbaldehyde, were created. By using substituted acetophenones in the presence of 10% NaOH during the Claisen-Schmidt condensation of the obtained chemical, the chalcone-linked benzo[d]imidazo[2,1-b] thiazole derivatives were created. (Sultana et al.,2018) **(Scheme 12, Compd. 29).**

Figure 15. Scheme 12

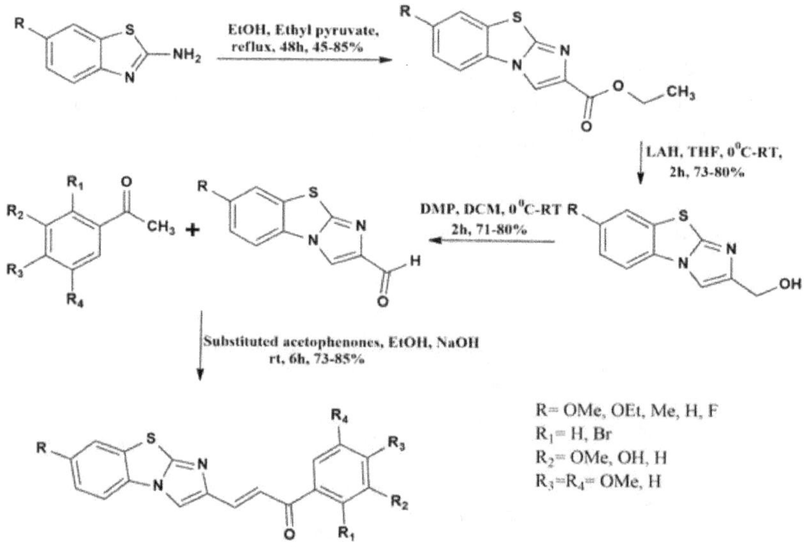

29

Imidazo[2,1-b] thiazoles were produced utilizing the one-pot visible light-activated Groebke Blackburn-Bienayme reaction (GBBR), which used aldehydes, isocyanides, and 2-aminoheterocycles in catalyst- and solvent-free conditions. The synthesis was carried out with visible light, and the stable and environmentally acceptable products were discovered (Shivhare et al.,2018) **(Scheme 13, Compd. 30)**. The compound produced 6-(1,2,3-triazol-4-yl)-5-(2-(thiazol-2-yl) hydrazonomethyl) imidazo[2,1-b] thiazoles by combining thiosemicarbazole in an equimolar proportion with phenacylbromides as well as hydrazonoyl chloride. (Abdel-Wahab etal.,2017) **(Scheme 14, Compd. 31).**

Figure 16. Scheme 13

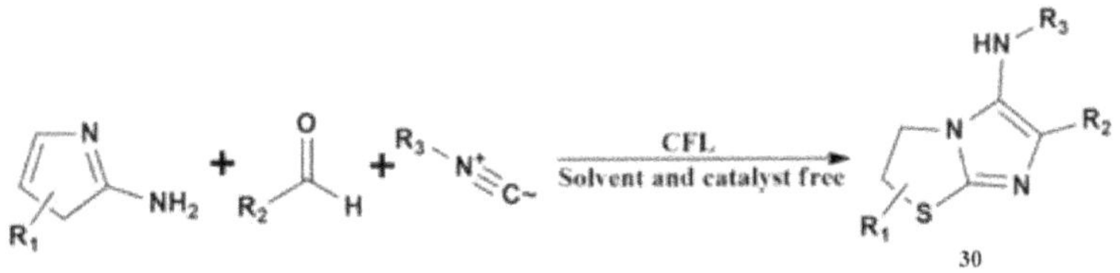

30

Figure 17. Scheme 14

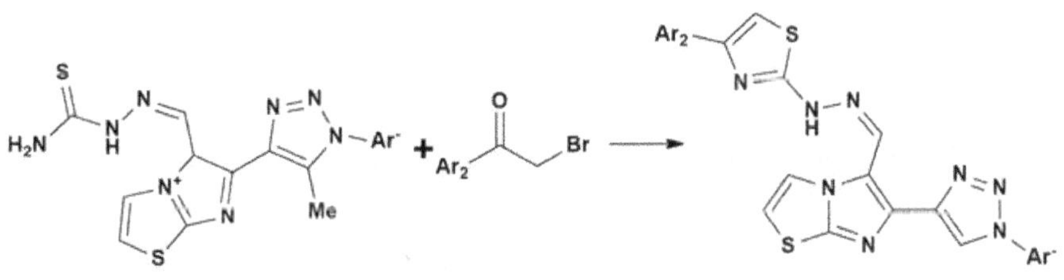

Ar= 4-tolyl
Ar_2= 4-ClC_6H_4, 4-BrC_6H_4, 2-benzofuryl, 3-coumarinyl

31

2.2 Synthesis From Imidazole and Other Compounds

Imidazo[2,1-b] thiazole is created by structurally altering 2-methyl-2,3-dihydroimidazo[2,1-b] [1,3] thiazol-5(6H)-one. Vanillin analogues were created by Knoevenagel condensation of the thiazole ring's sixth position. (Saliyeva et al.,2021) **(Scheme-15, Compd. 32).** A methodology for the divergent multicomponent tandem Palladium-catalyzed amino-carbonylation-cyclization approach towards functionalized imidazothiazoles has been devised by Veltri and colleagues (Peng etal.,2019) **(Scheme 16, Compd. 33).**

Figure 18. Scheme 15

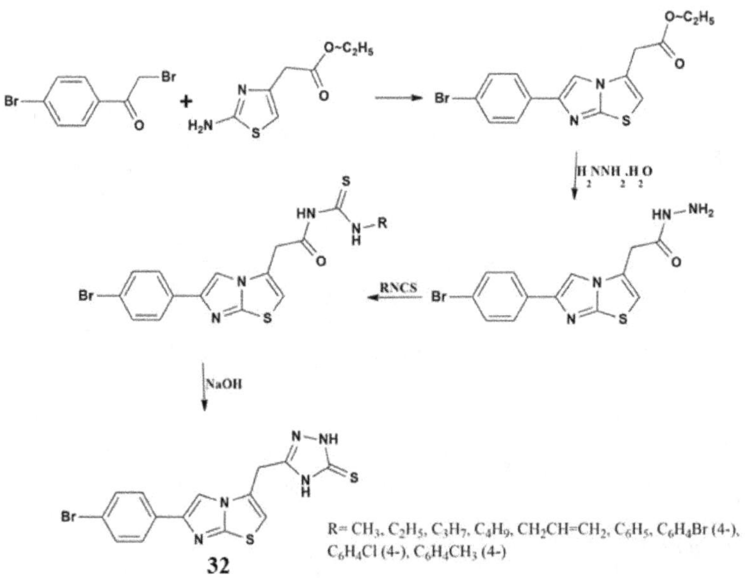

32

The 4-thiazolidinones were synthesized by the heating of ethyl(6-(phenyl/4-chlorophenyl)imidazo[2,1-b]thiazol-3-yl) acetate hydromides and hydrazine hydrate in ethanol, 2-(6-(phenyl/4- chlorophenyl)imidazo[2,1-b]thiazo-3-yl)acetohydrazide was obtained, followed by the heating with alkyl/aryl isothiocyanates in ethanol yielded 4-alkyl/aryl-1-(6-(phenyl/4-chlorophenyl)imidazo[2,1-b]thiazol-3-yl)acetyl)-3-thiosemicarbazides. The 3-alkyl/aryl-2-(6- (phenyl/4-chlorophenyl) imidazo[2,1-b] thiazol-3-yl) acetyl) hydrazono)-5-non-substituted/ methyl -4-thiazolidinones were produced by reacting ethyl/ethyl2- bromopropionate with anhydrous sodium acetate. (Güzeldemirci, Pehlivan & Naesens, 2018) **(Scheme 17, Compd. 34)**

Figure 19. Scheme 16

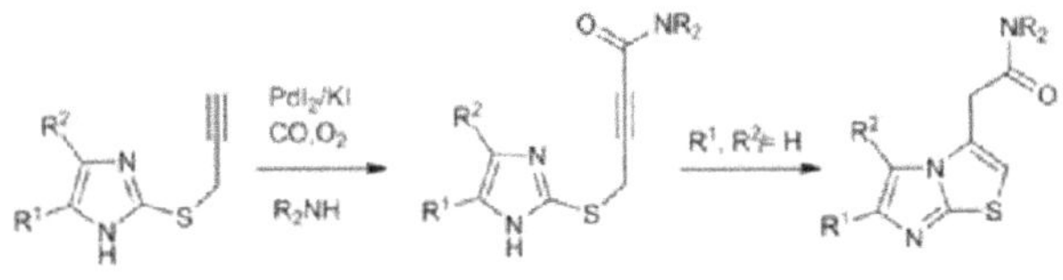

33

Deng and colleagues created compounds of benzo [4,5] imidazo[2,1-b] thiazoles. The 1H-benzo[d]imidazole-2-thiol was converted to the intermediate using ethyl-4-chloro-oxobutanoate. After that, another intermediate is created by cyclization in the presence of strong sulfuric acid at 50 ^{0}C. This intermediate was then immediately hydrolyzed and acidified to produce yet another intermediate. The benzo [4,5] imidazo[2,1-b] thiazole derivatives were produced by reacting aniline or its derivatives with the final intermediate. (Deng et al.,2019) **(Scheme 18, Compd. 35).**

Figure 20. Scheme 17

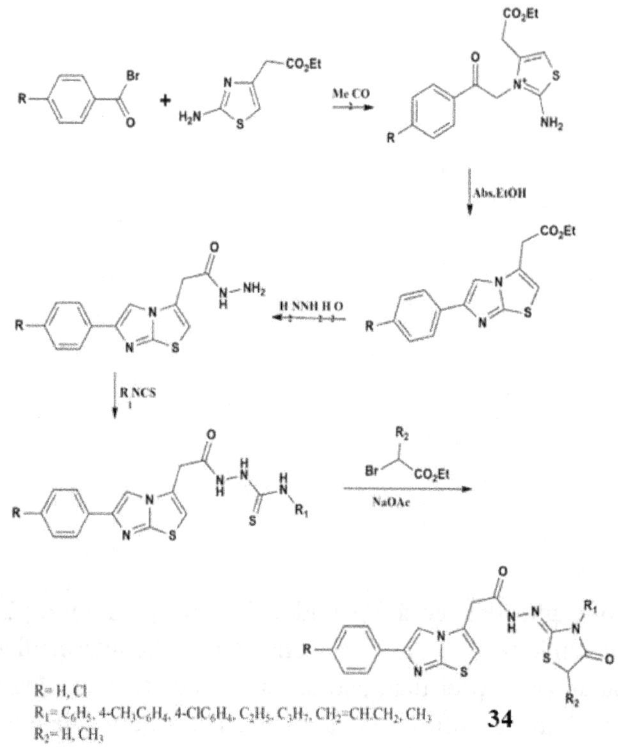

34

Biag and associates synthesized a series of imidazo[2,1-b] thiazole-linked benzimidazole derivatives. These drugs were obtained by oxidatively cyclizing imidazo[2,1-b] thiazole-5-carboxaldehyde and substituted O-phenylenediamine in ethanol alcohol with sodium metabisulphite. (Baig et al.,2018) **(Scheme 19, Compd. 36).**

Figure 21. Scheme 18

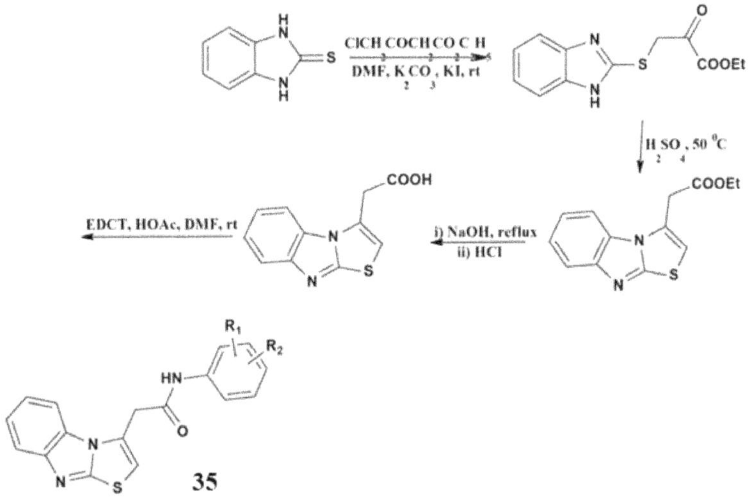

35

R_1= H, 3-trifluoromethyl, 3-fluoro, 3-bromo, 2-chloro
R_2= H, 4-hydroxy, 3-hydroxy, 2-hydroxy, 4-allyloxy, 3-nitro, 4-nitro, 2,3-benzo, 4-chloro, 4-bromo, 3-benzyloxy, 4-fluoro, 2-methyl, 5-fluoro, 3-trifluoromethyl

By reconstructing the imidazo[2,1-b] thiazole ring-opening and ring-closing with phenylacetylene, Shi and colleagues created a procedure for imidazo[2,1-b] thiazole derivatives. The reaction was conducted with potassium tert-butoxide (t-BuOK) at extremely mild conditions. Through a ring-opening and ring-closing reaction started by t-BuOK, imidazo[2,1-b] [1,3,4] thiadiazoles and arylacetylene reacted to produce imidazo[2,1-b] thiazole (Shi et al.,2016) **(Scheme20, Compd. 37)**.

Figure 22. Scheme 19

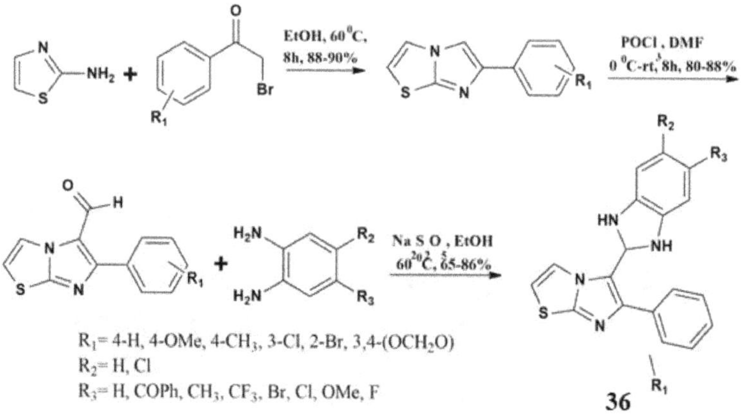

3. ANTICANCER ACTIVITY OF IMIDAZOTHIAZOLE AND THEIR DERIVATIVES

Melanoma is the most aggressive and fatal type of cancer of the skin, that develops from melanocytes (Davis & Shalin,2019). ARAF, BRAF, CRAF and RAF kinases are essential for the MAPK signaling pathway's stimulation (Yang et al., 2010) Treatment for melanoma could focus on the RAS-RAF-MEK-ERK pathway. BRAF is the primary stimulator of the MAPK signaling pathway among the subtypes of RAF kinase (Poudel et al., 2022) Certain people with melanoma have overexpressed and over-activated V600E mutant B-RRAF (Holderfield et al., 2014). It has been observed that RAF1 (C-RAF) and V600E-BRAF kinase inhibitors act as antiproliferative agents against melanoma (Qin et al., 2012; Rahman et al., 2015).

Figure 23. Scheme 20

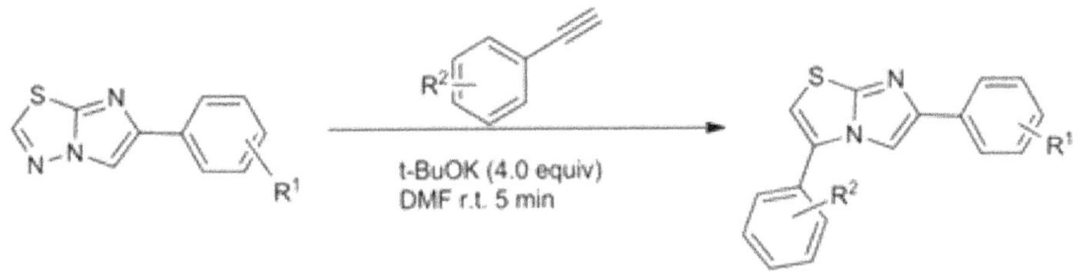

37

Medicinal chemists are interested in the fused heterocycle imidazo[2,1-b] thiazole because of its wide spectrum of biological actions. In the field of medicine, levamisole also called 2,3,5,6-tetrahydro-6-phenylimidazo[2,1-b] thiazole **(Figure 24, Compd. 38)** is a derivative of imidazothiazole that has been the subject of much research. Levamisole, a chemical of this structure, was initially employed as an immunomodulator and anthelmintic core unit. substances that induce apoptosis (Shareef et al., 2020). These scaffolds have been used in recent years to treat a variety of illnesses and disorders, including those that are anticancer(Nagireddy, et al.,2019), antiviral(Barradas et al.,2011), antitubercular((Vekariya et al.,2017; Moraski et al.,2020), antibacterial (Güzeldemirci et al., 2020), antioxidant(Andreani et al., 2013) and more.RSK2 Kinase (Andreani et al., 2011), B-Raf 4600E (Ammar et al.,2020), Carbonic anhydrase (Manasa et al.,2020), specific ErbB4 kinase(Zaraei et al., 2021), inhibitors of IDO1 (Tojo et al., 2014), selective inhibitors of COX-2 (Shahrasbi et al., 2018). These scaffolding molecules inhibit novel SIRT1 activators (Vu et al., 2009) and other related proteins. It was once intended to treat illnesses caused by parasites (Karaman et al., 2016) In cancer, it has also been employed as an immunostimulant agent (Ali et al., 2014).

Figure 24. L-isomer

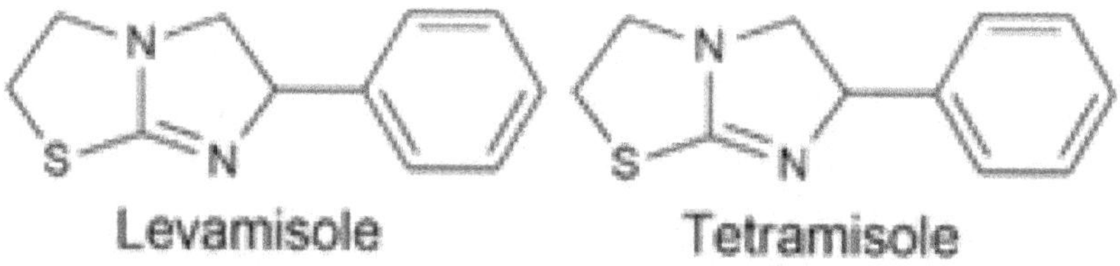

38 **39**

The primary aminothiazole analogue is tertamisole, a racemic combination of which the L-isomer is more potent and of which the D-isomer has no anthelmintic activity while levamisole is only the L-isomer. (Abongwa et al.,2017) (**Figure 24, Compd. 39**). Numerous imidazothiazole molecules, such as benzo[d]imidazo[2,1-b] thiazole conjugates, suppress tubulin polymerization. (Ammar et al.,2020). Numerous significant pharmaceutical compounds have the benzothiazole structural constituent and nitrogen bridgehead fused heterocycles. It was discovered that 3-substituted 2-phenylbenzo[d]imidazo[2,1-b] thiazoles and 2-aryl benzo[d]imidazo[2,1-b] thiazoles had tubulin polymerization inhibitory action to cause apoptosis (Shaik et al.,2017). A promising lead and p90 Ribosomal S6 Kinase 2 inhibitor with antiproliferative efficacy against a range of cell strains was made up of imidazo[2,1-b] thiazole–guanyl hydrazone. Nocodazole is an imidazothiazole-containing drug that inhibits the polymer formation of tubulin and slows down cell proliferation (Rashdan et al., 2020).

Anbar group created sequence of imidazothiazole with pyrimidine nucleus (**Figure 25, Compd. 40a & 40b**) in which Compd. 40a had the most promise for therapeutic use out of all the imidazothiazole compounds. The Compound 40a was the only one to achieve a sub-nano IC50 value against V600E-B-RAF and exhibited strong potency and relative selectivity towards the molecule. Furthermore, the chemical demonstrated submicron molar IC50 values for strong potency against various melanoma cell lines, and its selectivity for melanoma cells was higher than that of normal skin cells. In addition, it demonstrated the ability to cross the cell membrane and inhibit V600E-B-RAF kinase at an IC50 value of 0.19 μM, which was in line with its IC50 value of 0.18 μM against UACC-62 melanoma cells. The Compound 40b, has a strong anti-melanoma effect. According to the docking investigation and it is interacted molecularly with the amino acids Asn 580 and Phe 583. It also shared affinity groups with the active region of the BRAF kinase enzyme through H-bonding interaction (sulphonamides chemical moieties) and arene–arene interaction (imidazolyl ring). (Anbar et al.,2010)

Gali *et al.* (2014) examined the cytotoxicity of a range of new coumarinylimidazo[2,1-b] thiazoles (**Figure 25, Compd. 41-43**) against carcinoma cell lines MCF-7 (breast cancer.), HepG2 (liver cancer), HeLa (cervical cancer), and NCI-H460 (lung cancer). Compound 41c had greater activity against MCF-7, HepG2, and HeLa cell lines with IC50 values of 10.99±0.7, 13.92±0.2, and 50.18±0.0 μM, respectively. Compound 43 was active against MCF-7 and HeLa cell lines with IC50 values of 10.83±0.5 and 6.77±0.2 M, correspondingly. Additionally, it was shown that drugs 41a, 41b, and 42a were effective against HeLa cell strains with IC50 values of 7.13±0.14, 14.21±0.6, as well as 17.87±0.5 μM, respectively, and that compound 42b was significantly active against NCI-H460 with an IC50 value of 16.27±0.5 μM. (Gali *et al.,* 2014)

The compounds 44a-g (**Figure 25, Compd. 44**) were also tested for in-vitro EGFR kinase inhibitory activity using the solid phase enzyme linked immunosorbent assay (ELISA) method, with erlotinib used as the standard medication. Among the studied compounds, 4c, 4b, and 4a demonstrated substantial inhibitory activity with IC50 values of 18.35 μM, 24.34 μM, and 28.65 μM, respectively, and considerable binding activity to the EGFR kinase. Other compounds in the series demonstrated moderate activity, with IC50 values ranging from 29.13 to 75.41 μM. (Kamboji *et al.,* 2024)

A set of aryl hydrazones based on imidazo[2,1-b] thiazoles (**Figure 25, Compd. 45**) was evaluated by Shareef et al. against a variety of human cancer cell lines. Using the MTT assay, all the produced compounds were evaluated against the following cell lines: DU145 (prostate), MDA-MB-231 (breast), HeLa (cervical), A549 (lung), and HEK-293 (kidney). The most cytotoxic chemicals were 45a, 45b, and 45c, with IC50 values ranging from 1.12 to 5.2 μM. The MDA-MB-231 cell line was responsive to compounds 45b and 45c, with IC50 values of 1.65 and 1.12 μM, respectively. These putative chemicals were compatible with doxorubicin drug and causing the G0/G1 phase cell cycle to be arrested with 72 and 75.8% of deposition. (Shareef *et al.*, 2020).

Nagireddy et al. used the SRB test to assess the anticancer properties of several imidazothiazole-linked noscapinoids (**Figure 25, Compd. 46, 47**) against HeLa (cervical), MIA PaCa-2 (epithelial), SK-N-SH (neuroblastoma), and DU145 (prostate) cancer cell lines. The five-dose testing for each of the 31 synthetic compounds was done to ascertain their efficacy, and it was discovered that compounds 46, 47a, 47b, 47c, and 47d were potent against NCI- 60 panel cell lines. It was discovered that compounds 46 with p-Fluoro substitution and O-derived compound 49d were more effective among the derivatives. Compounds 47b and 47c also showed comparable effectiveness in suppressing the formation of colonies in pancreatic cells, as did compound 47d. These substances exhibited tubulin polymerization inhibitory effectiveness and stopped the G2/M cell cycle. It is the most efficient drug for G2/M phase cell cycle arresting, accumulating 76.94% of the compound through caspase-3 inhibition, apoptosis induction, and increased levels of CDK-1 and cyclin-B1. As a result, the imidazo[2,1-b] thiazole-linked noscapine derivatives were employed as drugs that induce apoptosis, and strong anticancer action. (Nagireddy et al., 2019).

Andreani et al. evaluated the effectiveness of imidazo[2,1-b] thiazoleguanyl-hydrazoles (**Figure 25, Compd. 48, 49**) as RSK2 inhibitors in p90 ribosomal s6 kinase 2. High-throughput kinase (enzyme) inhibition screening against RSK1, RSK2, CHK1, and CHK2 was performed after cell-based screening to ascertain their effectiveness. The antiproliferative activity of the synthesized compounds was tested using the following cell lines: HEK-293 (embryonic cell line), T47D (hypo-triploid human cancer), MCF-7 (breast), and MCF-10A (normal human breast cell line). When tested against kinase enzymes, compounds 48 and 49 were found to be active against RSK2, with IC50 values of 1.23 and 2.15 μM, respectively. The antiproliferative cell-based investigations were conducted against the human breast cancer cell line MCF-7 and the normal human breast cell line MCF-10A. The results showed that these compounds exhibited more effective kinase inhibitory activity to suppress tumor growth. Compound 49 inhibited the growth of cancer cells in both cell lines, while the other two compounds did not affect cancer cell growth. As a result, chemicals 48 and 49 selectively inhibited cancer cells and inhibited the GSK3 biomarker, indicating that they had an impact on RSK2 (Andreani et al., 2011)

Figure 25. Pyrimidine nucleus

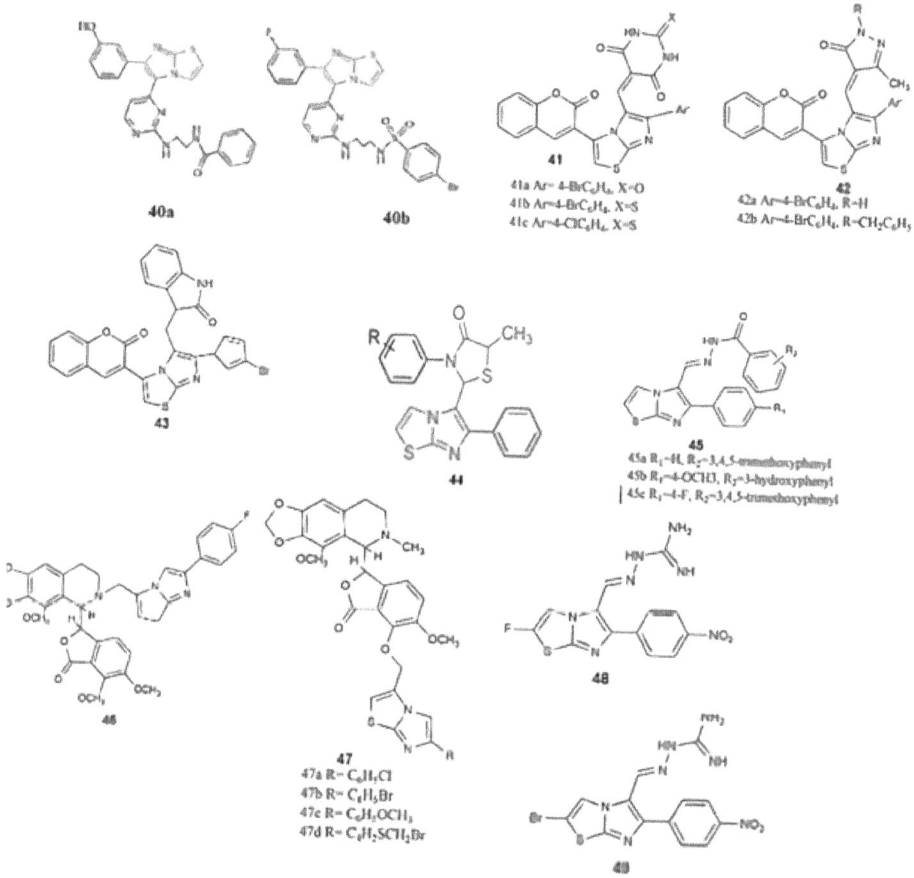

New imidazo[2,1-b] thiazole (**Figure 26, Compd. 50**) derivatives were assessed for cytotoxicity against NCI panel cell lines and examined their potential to block mutated B-Raf kinase. When tested against mutated B-Raf kinase, all the prepared formulations inhibited B-Raf V600E at 10 μM. It was discovered that compounds 50a, 50b, and 50c exhibited greater inhibition activity than the vemurafenib standard. One dose against NCI-60 tumor cell lines was used to estimate in vitro cytotoxicity. It was discovered that compounds 50a and 50b were highly potent, showing 160% and 90% suppression against leukaemia and prostate cancer. Compounds showed 100% suppression against the aneuploid BT-549, the breast MCF-7, the hypotriploid MDA-MB-468, and the aneuploid T47D cell lines. Molecular docking studies revealed that the additional NO_2 group in the terminal phenyl ring was effectively bound with B-Raf V600E kinase and was found to be an effective B-Raf V600E kinase inhibitor through H-bond interactions with amino acid residues. (Ammar et al.,2020).

To assess the potency of newer imidazothiazole derivatives against ErbR4 (HER4) inhibitors, Zaraei et al. tested (**Figure 26, Compd. 51 & 52**). Every manufactured compound's antiproliferative efficacy was evaluated using the NCI-60 panel. Following a 5-dose trial with sorafenib, a multi-kinase inhibitory anticancer drug, and ibrutinib, a pan-HER inhibitory anticancer drug, serving as reference standards, compounds 51 and 52 demonstrated enhanced antiproliferative activity. The compound 52 was found to be active against six cell lines, compound 51 was found to be more effective than sorafenib against seven different cell lines. Compound 51 demonstrated greater efficacy against six cell lines (including the hyperpentaploid SK-MEL-5) with an IC50 of 0.51 μM as compared to normal ibrutinib. The compound 52 demonstrated efficacy against five cell lines. After screening, it was discovered that the chosen compounds 51 and 52 had significant kinase inhibition against ErbB4 kinase, with percentage inhibitions of 96.10% and 60.21%, respectively. (Zaraei et al.,2021).

The efficacy of imidazo[2,1-b] thiazole scaffolds coupled thiadiazole derivatives (**Figure 26, Compd. 53)** against HepG2 (liver) cancer cell lines was examined by Rashdan et al. Compound 53 exhibits greater antiproliferative activity against HepG2 and normal cell lines, as evidenced by its IC50 values of 12.731.36 and 8.341.81 μg/mL, respectively, in contrast to doxorubicin's 3.560.46 and 3.210.32 μg/mL. Using the PyRx tool, in silico molecular docking of compound 53 against the GPC protein was carried out, and the binding affinity was determined to be -10.30 kcal/mol. At 6.42 and 6.49 Å, the compound 53 and the ARG197 form two pi-cation interactions. It therefore exhibited excellent non-covalent interactions and a high binding affinity with the target GPC-3 (Rashdan et al., 2020).

Karaman et al. tested various arylidene hydrazide (**Figure 26, Compd. 54**) derivatives on a panel of human cancer cell lines. After testing each of the prepared compounds on an NCI panel of cell lines, it was discovered that compounds 54a and 54b were active. Comparing compound 54a to compound 54b, the R-2-hydroxyphenyl substituted compound 54a demonstrated the most promising antiproliferative and inhibitory effect against malignant cell lines. The active ingredient of Compd. 54a was selectively active against the ovarian cancer cell line (OVACAR-3), as opposed to the standard medications' cisplatin and sorafenib. Additionally, this prevented the in vitro growth of ACHN, RXF-393, and UIO-31, whose log10GI50 values were, respectively, -5.75, -6.11, -5.57, and -6.30. (Karaman et al., 2016).

Several 5,6-diarylimidazo[2,1-b] thiazole (**Figure 26Compd. 55)** analogues were synthesized by Abdel-Maksoud et al. Baig, who tested them in vitro against NCI panel cell lines. The compounds 55e and 55f with propylene derivatives demonstrated as much activity than their ethylene derivatives, and the ethylene linked derivatives 55a, 55b, 55c, and 55d showed activity than their propylene derivatives based on one-dose in vitro antiproliferative testing.With IC50 values of 0.845 and 0.476 nM, respectively, the molecules 55d with the ethylene spacer and para-hydroxyphenyl terminal ring demonstrated superior antiproliferative action against the COLO205 colon Carcinoma cell line and the line. It was also discovered to be more potent than sorafenib based on the five-dose screening. The powerful antiproliferative compound 55d was tested against wild type B-RAF, V600E-B-RAF, and C-RAF kinases at a mono concentration (10 μM) to ascertain the suppression of kinase. It was reported that compound 55d had a higher IC50 value of 39.9 nM for inhibiting V600E-B-RAF kinase than sorafenib, and that it was more potent against overexpressed V600E-B-RAF in cell lines like COLO205, HT29 colon cancer, and SK-MEL-5 melanoma cancer cell. (Abdel-Maksoud et al., 2015)

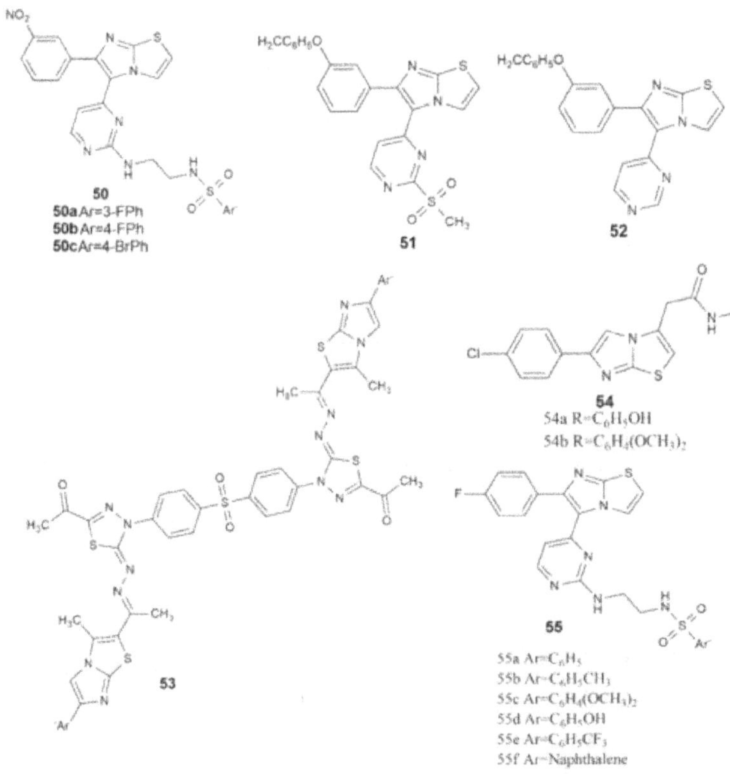

It was determined whether the imidazo[2,1-b] thiazole-benzimidazole (**Figure 27, Compd. 56**) derivatives were effective against the cancer cell lines HeLa (cervical), A549 (lung), MCF-7 (breast), and DU145 (prostate). With IC50 values of 1.08 and 1.13 μM, drug 56 was found to exhibit significant activity against the A549 cell line. Additionally, at doses of 1 and 2 μM, 56 was observed to have the ability to inhibit topoisomerase and arrest the G2/M cell cycle. The compound 56 was also tested for its ability to resist self-aggregation, and it was found to have a strong inhibition with an IC50 value of 1.68 μM. Molecular analysis indicates that the effective molecule has higher binding and interaction with the tubulin's colchicine site (Baig et al., 2018).

A sequence of derivative products of benzo[d]imidazo[2,1-b] thiazole-chalcone (**Figure 27, Compd. 57**) exhibits cytotoxic activity against the lung cancerous cell line A549, the breast line MDA-MB-231, the prostate line DU145, and the HT29. The cytotoxicity of compounds, 57a and 57b, against MDA-MB-231 were found to have IC50 values of 1.3 and 1.2 μM. (Sultana et al.,2018). The inhibition of tubulin polymerization, binding affinity, and apoptosis recognition were conducted to ascertain the cell cycle arrest capacity of compounds 57a and 57b'. Using the MDA-MB-231 cell line as the standard, flow cytometry was performed on the compounds 57a and 57b at 1 and 2μM concentration levels. The compound 57a showed 14.8% and 21.1% arresting in the S phase, while 57b showed 21.81% and 23.14% G2/M phase arresting. In the oligomerization inhibition study, the compounds 57a and 57b were inhibited by 59.16 and 61.38 percent, respectively. The results showed that the IC50 values were 1.93 and 1.88 μM, respectively, and they were equipotent concerning the standard drug nocodazole (1.97 μM). Additionally, molecular dynamics studies corroborated the compounds' binding at the tubulin's colchicine binding pocket. Flow cytometry analyses revealed that compounds 57a and 57b induced apoptosis with 17.79 and 24.27 percent at 1 μM and 23.61 and 33.67 percent at 2 μM, respectively.

Kamal et al. tested the potential of imidazothiazole-chalcone (**Figure 27, Compd. 58**) to cause cell death. With an IC50 value activity was evaluated at 0.26-4.74 μM. The compound 58a exhibited excellent cytotoxicity against leukemia, colon, prostate, and breast cells. The cytotoxic compounds were investigated against NCI-60 panel cell lines. It was discovered that compound 58b was effective against the SR and HC7-116 cell lines. Compounds 58a-d demonstrated greater cytotoxicity against MCF-7 at 4 μM concentration, with cell viability falling between 0.873±0.005 and 0.862±0. 006. Therefore, it was discovered that compound 58c had cytotoxic effects that were almost identical to doxorubicin (0.623±0.025). The compound 58c was potent as benchmark doxorubicin (93%), arresting the G0/G1 phase cell cycle with 91% inhibition. Moreover, it led to a decrease in the incorporation of BrdU into cellular DNA. The active chemicals caused phosphor-b to be down-regulated, which in turn caused G1/S-checkpoint associated tumor suppressor proteins to be up-regulated, leading to G1 phase cell cycle arrest. Consequently, the most effective and suitable compound for treating breast cancer was found to be compound 58b (Kamal et al., 2010).

Zhang et al. used the MTT assay with 5-fluorouracil as the standard to evaluate a series of steroidal[17,16-d] thiazolo[2,1-b] imidazole (**Figure 27, Compd. 59**) derivatives for their antiproliferative activity against EC109 (human oesophageal carcinoma), EC9706 (human esophageal carcinoma), and MGC-803 (human gastric carcinoma). Compounds 59a and 59b were effective against MGC-803 cell lines, and both showed greater levels of antiproliferative activity. These substances exhibited superior activity and offered the structure-activity relationship needed to create new antiproliferative drugs (Zhang et al., 2010).

Sorafenib was used by Park et al. to evaluate the in vitro antiproliferative activities of imidazo[2,1-b] thiazole (**Figure 27Compd. 60, 61**) derivatives against the A375P human melanoma cell line and the NCI panel. The 18 compounds were found to be active against the A375P cell line. The 61b and 61d compounds have greater potency, with IC50 values of 0.5 and 2.1 μM, respectively. Upon initial dose screening, the compounds 60a, 60c, 61b, and 61c demonstrated the highest antiproliferative activity against NCI-60 panel cell lines. Subsequently, the active constituents underwent five-dose screening, and the results indicated that compounds 60a and 60c exhibited superior efficacy against all tumour cell lines. Compounds 61c and 61d demonstrated good selectivity against melanoma, with selectivity ratios of 4.25 and 4.14. Additionally, the in-silico research showed that compounds 61c and 61d were effective and specific in their anti-melanoma properties (Park et al.,2011).

Using doxorubicin as a reference, a series of thiazolyl imidazothiazole (**Figure 27, Compd. 62–65**) derivatives were tested against the colon cancer cell line HCT-116 and the breast cancer cell line MCF-7 using the MTT assay. The MCF-7 cell line was found to be resistant to all the substances, while the HCT-116 cell line was more resistant to the substances 62, 63, 64a, 64b, 64c, 65a, 65b, 65c, 65d, and 65e, with IC50 values of 1.1, 5.6, 1.8, 1.1, 7.3, 8.0, 8.5, 8.9, and 6.2 μM, respectively. These substances were more potent than doxorubicin (IC50=9.4 μM). Compound 65's increased efficiency was brought about by the adhesion of amino and carbonyl groups with the tumor's DNA (Mahmoud et al., 2021).

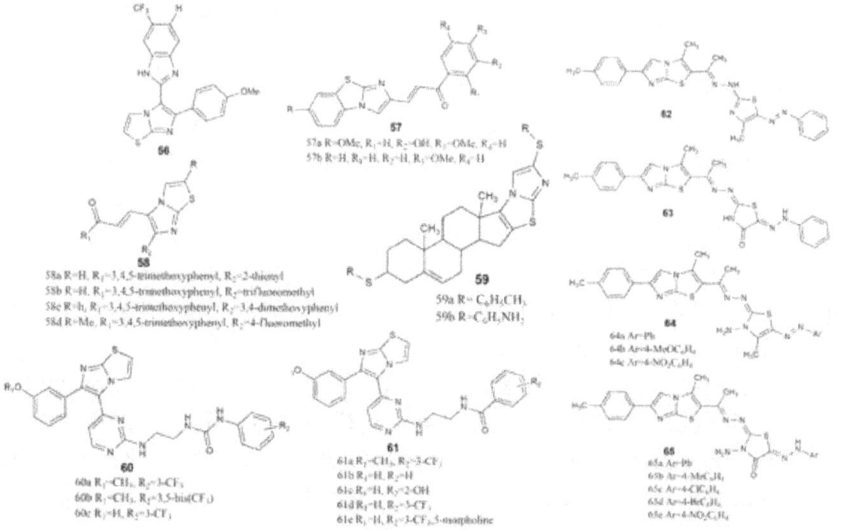

CONCLUSION

Imidazothiazoles are a significant group of heterocyclic substances. The imidazole ring is found in many synthetic and natural materials. The chapter focuses on several synthetic methods for creating imidazo[2,1-b] thiazole derivatives, including oxidative cyclization, Claisen-Schmidt condensation, multicomponent reactions (MCRs), ring-opening and ring-closing reconstruction reactions, visible light synthesis, GBBR mechanism, and microwave assisted green synthesis.

This chapter also provides an overview of the many biological activities of imidazo[2,1-b] thiazole derivatives, especially anticancer activities, and their docking investigations. It specifically inhibited protein kinases, triggered apoptosis, and functions as a tubulin targeting and polymerizing drug in the context of anticancer therapy. With increased IC50 values and inhibition rates, these scaffolds demonstrated improved activity against a range of illnesses.

The broad range of biological activities of the fused heterocycle imidazo[2,1-b] thiazole has drawn the attention of medicinal chemists. This structure's component, levamisole, was first used as an anthelmintic core unit and immunomodulator. In recent years, these scaffolds have been utilized to treat a variety of illnesses and disorders, including those that cause apoptosis and those that are antiviral, antitubercular, anticancer, antibacterial, antioxidant, and anti-inflammatory. Numerous enzymes, such as RSK2 Kinase, B-Raf 4600E, Selective ErbB4 Kinase, carbonic anhydrase, IDO1 inhibitors, Selective COX-2 inhibitors, Novel SIRT1 activators, and others are inhibited by these scaffolds.

REFERENCES

Abdel-Maksoud, M. S., Ammar, U. M., & Oh, C. H. (2019). Anticancer profile of newly synthesized BRAF inhibitors possess 5-(pyrimidin-4-yl) imidazo[2,1-b] thiazole scaffold. *Bioorganic & Medicinal Chemistry*, 2019(27), 2041–2051. DOI: 10.1016/j.bmc.2019.03.062 PMID: 30955995

Abdel-Maksoud, M. S., Kim, M. R., El-Gamal, M. I., Gamal El-Din, M. M., Tae, J., Choi, H. S., Lee, K.-T., Yoo, K. H., & Oh, C.-H. (2015, May 5). Gamal, El-Din, M.M., Tae, J., Choi, H.S., Lee, K.T., Yoo, K.H., Oh, C.H. (2015). Design, synthesis, in vitro antiproliferative evaluation, and kinase inhibitory effects of a new series of imidazo[2,1-b] thiazole derivatives. *European Journal of Medicinal Chemistry*, 95, 453–463. DOI: 10.1016/j.ejmech.2015.03.065 PMID: 25841200

Abdel-Wahab, B. F., Khidre, R. E., & Awad, G. E. (2017). Design and Synthesis of Novel 6-(5-Methyl-1 H-1, 2, 3-triazol-4-yl) -5-[(2-(thiazol-2-yl) hydrazono) methyl] imidazo [2, 1-b] thiazoles as Antimicrobial Agents. *Journal of Heterocyclic Chemistry*, 54(1), 489–494. DOI: 10.1002/jhet.2610

Abdelhamid, A. A., Salah, H. A., & Marzouk, A. A. (2020). Synthesis of imidazole derivatives: Ester and hydrazide compounds with antioxidant activity using ionic liquid as an efficient catalyst. *Journal of Heterocyclic Chemistry*, 57(2), 676–685. DOI: 10.1002/jhet.3808

Abongwa, M., Martin, R. J., & Robertson, A. P. (2017). A brief review on the mode of action of antinematodal drugs. *Acta Veterinaria*, 67(2), 137–152. DOI: 10.1515/acve-2017-0013 PMID: 29416226

Adeyemi, O. S., Eseola, A. O., Plass, W., Atolani, O., Sugi, T., Han, Y., Batiha, G. E. S., Kato, K., Awakan, O. J., Olaolu, T. D., Nwonuma, C. O., Alejolowo, O., Owolabi, A., Rotimi, D., & Kayode, O. T. (2020). Imidazole derivatives as anti-parasitic agents and use of molecular modelling to investigate the structure activity relationship. *Parasitology Research*, 119(6), 1925–1941. DOI: 10.1007/s00436-020-06668-6 PMID: 32279093

Afsina Abdulla, C. M., Neetha, M., Aneeja, T., & Anilkumar, G. (2020). Synthesis and Applications of Imidazothiazoles: An Overview. *ChemistrySelect*, 25(33), 10374–10386. DOI: 10.1002/slct.202002842

Ali, A. R., El-Bendary, E. R., Ghaly, M. A., & Shehata, I. A. (2014). Synthesis, in vitro anticancer evaluation and in silico studies of novel imidazo[2,1-b] thiazole derivatives bearing pyrazole moieties. *European Journal of Medicinal Chemistry*, 75, 492–500. DOI: 10.1016/j.ejmech.2013.12.010 PMID: 24576591

Ammar, U. M., Abdel-Maksoud, M. S., Ali, E. M. H., Mersal, K. I., Ho Yoo, K., & Oh, C. H. (2020). Structural optimization of imidazothiazole derivatives affords a new promising series as B-Raf V600E inhibitors, synthesis, in vitro assay and in silico screening. *Bioorganic Chemistry*, 100, 103967. DOI: 10.1016/j.bioorg.2020.103967 PMID: 32470760

Anbar, H. S., El-Gamal, M. I., Tarazi, H., Lee, B. S., Jeon, H. R., Kwon, D., & Oh, C. H. (2010). Imidazothiazole-based potent inhibitors of V600E-B-RAF kinase with promising anti-melanoma activity: Biological and computational studies. *Journal of Enzyme Inhibition and Medicinal Chemistry*, 35(1), 1712–1726. DOI: 10.1080/14756366.2020.1819260 PMID: 32962435

Andreani, A., Burnelli, S., Granaiola, M., Leoni, A., Locatelli, A., Morigi, R., Rambaldi, M., Varoli, L., Calonghi, N., Cappadone, C., Voltattorni, M., Zini, M., Stefanelli, C., Masotti, L., & Shoemaker, R. H. (2008). Antitumor activity of new substituted 3-(5-imidazo[2,1-b]thiazolylmethylene)-2-indolinones and 3-(5-imidazo[2,1-b] thiadiazolylmethylene)-2-indolinones: Selectivity against colon tumor cells and effect on cell cycle-related events. *Journal of Medicinal Chemistry*, 51, 809–816. DOI: 10.1021/jm701246g PMID: 19006285

Andreani, A., Granadilla, M., Leoni, A., Locatelli, A., Morigi, R., Rambaldi, M., Varoli, L., Lannigan, D., Smith, J., Scudiero, D., Kondapaka, S., & Shoemaker, R. H. (2011). Imidazo[2,1-b] thiazole guanylhydrazones as RSK2 inhibitors. *European Journal of Medicinal Chemistry*, 46(9), 4311–4323. DOI: 10.1016/j.ejmech.2011.07.001 PMID: 21794960

Andreani, A., Leoni, A., Locatelli, A., Morigi, R., Rambaldi, M., Cervellati, R., Greco, E., Kondratyuk, T. P., Park, E. J., Huang, K., VanBreemen, R. B., & Pezzuto, J. M. (2013). Chemopreventive and antioxidant activity of 6-substituted imidazo[2,1-b] thiazoles. *European Journal of Medicinal Chemistry*, 68, 412–421. DOI: 10.1016/j.ejmech.2013.07.052 PMID: 23994869

Andreani, A., Rambaldi, M., Locatelli, A., Bossa, R., Galatulas, I., & Salvatore, G. (1994). Antitumor and cardiotonic activity of imidazo[2,1-b] thiazole guanylhydrazones. *In Vivo (Athens, Greece)*, 8, 1031–1032. PMID: 7772733

Bae, S. H., Park, J. H., Choi, H. G., Kim, H., & Kim, S. H. (2018). Imidazole Antifungal Drugs Inhibit the Cell Proliferation and Invasion of Human Breast Cancer Cells. *Biomolecules & Therapeutics*, 26(5), 494–502. DOI: 10.4062/biomolther.2018.042 PMID: 30092625

Baig, M. F., Nayak, V. L., Budaganaboyina, P., Mullagiri, K., Sunkari, S., Gour, J., & Kamal, A. (2018). Synthesis and biological evaluation of imidazo [2, 1-b] thiazole-benzimidazole conjugates as microtubule-targeting agents. *Bioorganic Chemistry*, 77, 515–526.

Barradas, J. S., Errea, M. I., D'Accorso, N. B., Sepúlveda, C. S., & Damonte, E. B. (2010). Imidazo [2,1- b] thiazole carbohydrate derivatives: Synthesis and antiviral activity against Junin virus, agent of Argentine hemorrhagic fever. *European Journal of Medicinal Chemistry*, 46(1), 259–264. DOI: 10.1016/j.ejmech.2010.11.012 PMID: 21115214

Battula, S. R. K., Rama Kishore Putta, V. P., Subbareddy, G. V., Chakravarthy, I. E., & Saravanan, V. A. (2017). Divergent and metal free synthesis of sulfoximine tethered imidazoles, imidazopyridines, imidazothiazoles, imidazobenzothiazines, thiazoles and selenazoles. *Organic & Biomolecular Chemistry*, 15(17), 3742–3755. DOI: 10.1039/C7OB00601B PMID: 28406509

Borcea, A. M., Ionu , I., Cri an, O., & Oniga, O. (2021). An Overview of the Synthesis and Antimicrobial, Antiprotozoal, and Antitumor Activity of Thiazole and Bisthiazole Derivatives. *Molecules (Basel, Switzerland)*, 26(3), 624. DOI: 10.3390/molecules26030624 PMID: 33504100

Budriesi, R., Ioan, P., Locatelli, A., Cosconati, S., Leoni, A., Ugenti, M. P., Andreani, A., di Toro, R., Bedini, A., Spampinato, S., Marinelli, L., Novellino, E., & Chiarini, A. (2008). Imidazo[2,1-b] thiazole system: A scaffold endowing dihydropyridines with selective cardiodepressant activity. *Journal of Medicinal Chemistry*, 51(6), 1592–1600. DOI: 10.1021/jm070681+ PMID: 18303827

Dangi, R. R., Hussain, N., & Talesara, G. L. (2011). Synthesis characterization and biological evaluation of some alkoxyphthalimide derivatives of 3-(4-substituted phenyl)-6, 6-diphenyl-3,3a-dihydro-2H-imidazo[2,1-b] pyrazolo[3,4-d] [1,3] thiazol-7(6H)-one. *Medicinal Chemistry Research*, 20(9), 1490–1498. DOI: 10.1007/s00044-010-9392-4

Davis, L. E., Shalin, S. C., & Tackett, A. J. (2019). Current state of melanoma diagnosis and treatment. *Cancer Biology & Therapy*, 20(11), 1366–1379. DOI: 10.1080/15384047.2019.1640032 PMID: 31366280

Deng, X., Tan, X., An, T., Ma, Q., Jin, Z., Wang, C., Meng, Q., & Hu, C. (2019). Synthesis, Characterization, and Biological Activity of a Novel Series of Benzo [4,5] imidazo[2,1-b] thiazole Derivatives as Potential Epidermal Growth Factor Receptor Inhibitors. *Molecules (Basel, Switzerland)*, 24(4), 682. DOI: 10.3390/molecules24040682 PMID: 30769844

Drapak, I., Perekhoda, L., Demchenko, N., Suleiman, M., Rakhimova, M., Demchuk, I., Taran, S., Seredynska, N., & Gerashchenko, I. (2019). Cardioprotective Activity of Some 2-Arylimino-1,3-Thiazole Derivatives. *Scientia Pharmaceutica*, 87(1), 1–8. DOI: 10.3390/scipharm87010007

Etivand, N., Ahmadi Sabegh, M., & Khalafy, J. (2019). Synthesis of a new series of benzo[d] imidazo [2, 1-b] thiazole-1-ium hydroxides by a one-pot, three-component reaction in water. *Monatshefte für Chemie*, 150(2), 317–325. DOI: 10.1007/s00706-018-2315-7

Ewida, M. A., Abou El Ella, D. A., Lasheen, D. S., Ewida, H. A., El-Gazzar, Y. I., & El-Subbagh, H. I. (2018). Imidazo [2′,1′:2,3] thiazolo[4,5-d]pyridazinone as a new scaffold of DHFR inhibitors: Synthesis, biological evaluation and molecular modeling study. *Bioorganic Chemistry*, 80, 11–23. DOI: 10.1016/j.bioorg.2018.05.025 PMID: 29864684

Gali, R., Banothu, J., Porika, M., Velpula, R., Bavantula, R., Abbagani, S. (2014). Synthesis and in vitro cytotoxic activity of novel coumarinylimidazo [2, 1-b] thiazole derivatives. Royal Society of Chemistry advance,4, 53812-9.

Geedkar, D., Kumar, A., & Sharma, P. (2020). Multiwalled carbon nanotubes crowned with nickel-ferrite magnetic nanoparticles assisted heterogeneous catalytic strategy for the synthesis of benzo [d] imidazo [2, 1-b] thiazole scaffolds. *Journal of Heterocyclic Chemistry*, 57(12), 4331–4347. DOI: 10.1002/jhet.4140

Gomha, S. M., Kheder, N. A., Abdelaziz, M. R., Mabkhot, Y. N., & Alhajoj, A. M. (2017). A facile synthesis and anticancer activity of some novel thiazoles carrying 1,3,4-thiadiazole moiety. *Chemistry Central Journal*, 11(1), 25. DOI: 10.1186/s13065-017-0255-7 PMID: 29086817

Guo, M., Yu, X., Zhu, Y. Z., & Yu, Y. (2023). From Bench to Bedside: What Do We Know about Imidazothiazole Derivatives So Far? *Molecules (Basel, Switzerland)*, 28(13), 5052. DOI: 10.3390/molecules28135052 PMID: 37446714

Güzeldemirci, N. U., & Küçükbasmaci, O. (2010). Synthesis and antimicrobial activity evaluation of new 1,2,4-triazoles and 1,3,4-thiadiazoles bearing imidazo[2,1-b] thiazole moiety. *European Journal of Medicinal Chemistry*, 45(1), 63–68. DOI: 10.1016/j.ejmech.2009.09.024 PMID: 19939519

Güzeldemirci, N. U., Pehlivan, E., & Naesens, L. (2018). Synthesis and antiviral activity evaluation of new 4-thiazolidinones bearing an imidazo [2, 1-b] thiazole moiety. *Marmara Pharmaceutical Journal*, 22(2), 237–248. DOI: 10.12991/mpj.2018.61

Holderfield, M., Deuker, M., McCormick, F., & McMahon, M. (2014). Targeting RAF kinases for cancer therapy: BRAF-mutated melanoma and beyond. *Nature Reviews. Cancer*, 14(7), 455–467. DOI: 10.1038/nrc3760 PMID: 24957944

Jadhav, V. B., Kulkarni, M. V., Rasal, V. P., Biradar, S. S., & Vinay, M. D. (2008). Synthesis and anti-inflammatory evaluation of methylene bridged benzofuranyl imidazo[2,1-b][1,3,4]thiadiazoles. *European Journal of Medicinal Chemistry*, 3(8), 1721–1729. DOI: 10.1016/j.ejmech.2007.06.023 PMID: 17845827

Jaiswal, S. (2019). Five and Six Membered Heterocyclic Compound with Antimicrobial Activity. *Journal for Modern Trends in Science and Technology*, 5(06), 36–39.

Jaiswal, S., & Dwivedi, S. (2017). Microwave-assisted eco-friendly synthesis and antimicrobial evaluation of aryl-triazole-1, 3, 4-thiadiazols. *Journal of Medical Research and Innovation*, 1(3), 17–22. DOI: 10.15419/jmri.51

Jaiswal, S., & Dwivedi, S. (2017). Alumina Supported Solvent-free Synthesis and Structural Elucidation of New 3-aryl-5-benzylidene-2-thioxohydantoin Derivatives. *Indian Journal of Heterocyclic Chemistry*, 27(3), 265–268.

Jaiswal, S., & Mojahid, A. (2020). Innovation in Society through Green\Eco-Friendly Technology. *Journal for Modern Trends in Science and Technology*, 6(01), 37–43.

Jaiswal, S., & Sharma, A. (•••). Green Synthesis of Some Novel Substituted and Unsubstituted Benzimidazole Derivatives by Using Microwave Energy. *International Journal of Pharmaceutical Sciences Review and Research*, 27, 153–156.

Jaiswal, S. & Siddiqui, I. R. (2014). Montmorillonite K-10 Supported Solvent-Free Synthesis of Novel 2-thioxoimidazole-4-one N-nucleosides. *chemistry*, Int. J. Pharm. Sci. Rev. Res.,26(1),*16*, 18.

Jakovleva, E. E., Foksha, S. P., Brusina, M. A., Kubarskaja, L. G., Piotrovskij, L. B., Bychkov, E. R., & Shabanov, P. D. (2020). Studying the anticonvulsive activity of new ligands of NDMA-receptor complex—Imidazole-4,5-dicarbonic acid derivatives. *Rev. Clin. Pharm. Drug Ther.*, 18(2), 149–154. DOI: 10.17816/RCF182149-154

Jana, A., Bhaumick, P., Panday, A. K., Mishra, R., & Choudhury, L. H. (2019). I_2/DMSO mediated multicomponent reaction for the synthesis of 2-arylbenzo[d] imidazo[2,1-b] thiazole derivatives. *Organic & Biomolecular Chemistry*, 17(21), 5316–5330. DOI: 10.1039/C9OB00515C PMID: 31095156

Kamal, A., Dastagiri, D., Ramaiah, M. J., Reddy, J. S., Bharathi, E. V., Srinivas, C., Pushpavalli, S. N. C. V. L., Pal, D., & Pal-Bhadra, M. (2010). Synthesis of imidazothiazole–chalcone derivatives as anticancer and apoptosis inducing agents. *ChemMedChem*, 5(11), 1937–1947. DOI: 10.1002/cmdc.201000346 PMID: 20836120

Kamboj, P., Anjali, , Imtiyaz, K., Rizvi, M. A., Nath, V., Kumar, V., Husain, A., & Amir, M. (2024). Design, synthesis, biological assessment and molecular modelling studies of novel imidazothiazole-thiazolidinone hybrids as potential anticancer and anti-inflammatory agents. *Scientific Reports*, 14(1), 8457. DOI: 10.1038/s41598-024-59063-x PMID: 38605072

Karaman, B., & Ulusoy Güzeldemirci, N. (2016). Synthesis and biological evaluation of new imidazo [2, 1-b] thiazole derivatives as anticancer agents. *Medicinal Chemistry Research*, 25(11), 2471–2484. DOI: 10.1007/s00044-016-1684-x

Kishore, K. G., Jácome, A. I., Gómez, A. I., Conejo, A. S., Basavanag, U. M. V., Wrobel, K., & Montaño, R. G. (2016). Synthesis of unsymmetrical bis-heterocycles containing the imidazo[2,1-b] thiazole framework and their benzo[d]fused analogues by an acid-free Groebke–Blackburn–Bienaymé reaction. *Tetrahedron Letters*, 57(31), 3556 3560. DOI: 10.1016/j.tetlet.2016.06.120

Koudad, M. Hamouti, C. El, Elaatiaoui, A., Dadou, S., Oussaid, A., Abrigach, F., Pilet, G., Benchat, N., Allali, M. (2020). Synthesis, crystal structure, antimicrobial activity and docking studies of new imidazothiazole derivatives. *Journal of the Indian Chemical Society*, 17, 297–306.

Lal, K., Bermeo, R., Cramer, J., Vasile, F., Ernst, B., Imberty, A., Bernardi, A., Varrot, A., & Belvisi, L. (2021). Prediction and Validation of a Druggable Site on Virulence Factor of Drug Resistant Burkholderia cenocepacia. *Chemistry (Basel, Switzerland)*, 27(40), 10341–10348. DOI: 10.1002/chem.202100252 PMID: 33769626

Liaras, K., Geronikaki, A., Glamočlija, J., Ćirić, A., & Soković, M. (2011). Thiazole-based chalcones as potent antimicrobial agents. Synthesis and biological evaluation. *Bioorganic & Medicinal Chemistry*, 19(10), 3135–3140. DOI: 10.1016/j.bmc.2011.04.007 PMID: 21524583

Liu, J., Zhou, W., & Wu, F. (2021). Biological Evaluation and Synthesis of Thiazole Schiff Base Derivatives. *Heterocycles*, 102(7), 1337. DOI: 10.3987/COM-21-14467

Mahmoud, H. K., Gomha, S. M., Farghaly, T. A., & Awad, H. M. (2021). Synthesis of thiazole linked imidazo [2, 1-b] thiazoles as anticancer agents. *Polycyclic Aromatic Compounds*, 41(8), 1608–1622. DOI: 10.1080/10406638.2019.1689514

Manasa, K. L., Pujitha, S., Sethi, A., Mohammed, A., Alvala, M., Angeli, A., & Supuran, C. T. (2020). Synthesis and Biological Evaluation of Imidazo [2, 1-*b*] Thiazole based Sulfonyl Piperazines as Novel Carbonic Anhydrase II Inhibitors. *Metabolites*, 10(4), 136. DOI: 10.3390/metabo10040136 PMID: 32244413

Margutti, S., & Laufer, S. A. (2007). Are MAP Kinases Drug Targets? Yes, but Difficult Ones. *ChemMedChem*, 2(8), 1116–1140. DOI: 10.1002/cmdc.200600271 PMID: 17541990

Mishra, A., Srivastava, M., Rai, P., Yadav, S., Tripathi, B. P., Singh, J., & Singh, J. (2016). Visible light triggered, catalyst free approach for the synthesis of thiazoles and imidazo[2,1-b] thiazoles in EtOH : H_2O green medium. *RSC Advances*, 6(54), 49164–49172. DOI: 10.1039/C6RA05385H

Mohan, J., & Kiran, K. (1991). Novel bridgehead nitrogen heterocycles: Synthesis and antimicrobial activity of 2H-imidazo[2,1-a] pyrazolo[3,4-d] thiazoles. *Indian Journal of Chemistry*, 30B, 898–900.

Moraski, G. C., Deboosère, N., Marshall, K. L., Weaver, H. A., Vandeputte, A., Hastings, C., Woolhiser, L., Lenaerts, A. J., Brodin, P., & Miller, M. J. (2020). Intracellular and in vivo evaluation of imidazo[2,1-b] thiazole-5-carboxamide anti-tuberculosis compounds. *PLoS One*, 5(1), eo227224. Advance online publication. DOI: 10.1371/journal.pone.0227224 PMID: 31905374

Mukku, N., & Maiti, B. (2020). On water catalyst-free synthesis of benzo [d] imidazo [2, 1-b] thiazoles and novel N-alkylated 2-aminobenzo [d] oxazoles under microwave irradiation. *RSC Advances*, 10(2), 770–778. DOI: 10.1039/C9RA08929B PMID: 35494448

Nagireddy, P. K. R., Kommalapati, V. K., Siva Krishna, V., Sriram, D., Tangutur, A. D., & Kantevari, S. (2019). Imidazo[2,1-*b*] thiazole-Coupled Natural Noscapine Derivatives as Anticancer Agents. *ACS Omega*, 4(21), 19382–19388. DOI: 10.1021/acsomega.9b02789 PMID: 31763563

Park, J. H., El-Gamal, M. I., Lee, Y. S., & Oh, C. H. (2011). New imidazo [2, 1-b] thiazole derivatives: Synthesis, in vitro anticancer evaluation, and in silico studies. *European Journal of Medicinal Chemistry*, 46(12), 5769–5777. DOI: 10.1016/j.ejmech.2011.08.024 PMID: 22033063

Peng, X., Qin, F., Xu, M., Zhu, S., Pan, Y., Tang, H., Menga, X., & Wanga, H. (2019). Synthesis of imidazo[1,2-c] thiazoles through Pd-catalyzed bicyclization of tert-butyl isonitrile with thioamides. *Organic & Biomolecular Chemistry*, 17(36), 8403–8407. DOI: 10.1039/C9OB01664C PMID: 31482915

Poudel, M., Kim, G., Bhattarai, P. Y., Shin, S., Zaraei, S. O., Oh, C. H., & Choi, H. S. (2011). Potent Imidazothiazole-based Inhibitor of BRAF V600E Overcomes Acquired Resistance via Inhibition of RAF Dimerization in PLX4032-resistant Melanoma. *Anticancer Research*, 42(6), 2911–2921. DOI: 10.21873/anticanres.15773 PMID: 35641256

Pricopie, A. I., Foc an, M., Ionu , I., Marc, G., Vlase, L., Găină, L. I., Vodnar, D. C., Simon, E., Barta, G., Pîrnău, A., & Oniga, O. (2020). Novel 2,4-Disubstituted-1,3-Thiazole Derivatives: Synthesis, Anti-Candida Activity Evaluation and Interaction with Bovine Serum Albumine. *Molecules (Basel, Switzerland)*, 25s(5), 1079. DOI: 10.3390/molecules25051079 PMID: 32121062

Qin, J., Xie, P., Ventocilla, C., Zhou, G., Vultur, A., Chen, Q., Liu, Q., Herlyn, M., Winkler, J., & Marmorstein, R. (2012). Identification of a Novel Family of BRAFV600E Inhibitors. *Journal of Medicinal Chemistry*, 55(11), 5220–5230. DOI: 10.1021/jm3004416 PMID: 22537109

Rahman, M. A., Salajegheh, A., Smith, R. A., & Lam, A. K.-Y. (2015). Multiple proliferation-survival signalling pathways are simultaneously active in BRAF V600E mutated thyroid carcinomas. *Experimental and Molecular Pathology*, 99(3), 492–497. DOI: 10.1016/j.yexmp.2015.09.006 PMID: 26403329

Rashdan, H. R. M., Abdelmonsef, A. H., Shehadi, I. A., Gomha, S. M., Soliman, A. M. M., & Mahmoud, H. K. (2020). Synthesis, Molecular Docking Screening and Anti-Proliferative Potency Evaluation of Some New Imidazo[2,1-*b*] Thiazole Linked Thiadiazole Conjugates. *Molecules (Basel, Switzerland)*, 25(21), 4997. DOI: 10.3390/molecules25214997 PMID: 33126630

Rassokhina, I. V., Tikhonova, T. A., Kobylskoy, S. G., Babkin, I. Y., Shirinian, V. Z., Gevorgyan, V., Zavarzin, I. V., & Volkova, Y. A. (2017). Synthesis of imidazo [2, 1-b] thiazoles via copper-catalyzed A3-coupling in batch and continuous flow. *The Journal of Organic Chemistry*, 82(18), 9682–9692. DOI: 10.1021/acs.joc.7b01762 PMID: 28799762

Robert, J. F., Boukraa, S., Panouse, J. J., Loppinet, V., & Chaumont, J. P. (1990). Derivatives of imidazo[2,1-b]thiazoles X. Anthelmintic activityof imidazo[2,1-b]thiazoles. *European Journal of Medicinal Chemistry*, 25, 731–736. DOI: 10.1016/0223-5234(90)90192-6

Rodrigues, I., Barcellos, A. M., Belladona, A. L., Roehrs, J. A., Cargnelutti, R., Alves, D., Perin, G., & Schumacher, R. F. (2018). Oxone®-mediated direct arylselenylation of imidazo [2, 1-b] thiazoles, imidazo [1, 2-a] pyridines and 1H-pyrazoles. *Tetrahedron*, 74(31), 4242–4246. DOI: 10.1016/j.tet.2018.06.046

Salah, A. M. B., Fendri, L. B., Bataille, T., Herrera, R. P., & Naïli, H. (2018). Synthesis, structural determination and antimicrobial evaluation of two novel CoII and ZnII halogenometallates as efficient catalysts for the acetalization reaction of aldehydes. *Chemistry Central Journal*, 12(1), 24. DOI: 10.1186/s13065-018-0393-6 PMID: 29497955

Saliyeva, L., Diachenko, I., Vas'kevich, R., Slyvka, N., & Vovk, M. (2020). Imidazothiazoles and their Hydrogenated Analogs: Methods of Synthesis and Biomedical Potential. *Chemistry of Heterocyclic Compounds*, 56(11), 1394–1407. DOI: 10.1007/s10593-020-02827-w

Saliyeva, L., Holota, S., Grozav, A., Yakovychuk, N., Litvinchuk, M., Slyvka, N., & Vovk, M. (2021). Synthesis and Evaluation of Antimicrobial and Anti-inflammatory Activity of 6-aryliden- 2-methyl-2, 3-dihydroimidazo [2, 1-b] [1, 3] thiazoles. *Biointerface Research in Applied Chemistry*, 12(1), 292–303. DOI: 10.33263/BRIAC121.292303

Scribner, A., Meitz, S., Fisher, M., Wyvratt, M., Leavitt, P., Liberator, P., Gurnett, A. C., Brown, C., Mathew, J., Thompson, D., Schmatz, D., & Biftu, T. (2008). Synthesis and biological activity of anticoccidial agents: 5,6-Diarylimidazo[2,1-b][1,3] thiazoles. *Bioorganic & Medicinal Chemistry Letters*, 18(19), 5263–5267. DOI: 10.1016/j.bmcl.2008.08.063 PMID: 18789686

Şenkardeş, S., Kulabaş, N., Bíngöl Özakpinar, Ö., Kalayci, S., Şahín, F., Küçükgüzel, İ., & Küçükgüzel, Ş. G. (2020). Synthesis and Anticancer and Antimicrobial Evaluation of Novel Ether-linked Derivatives of Ornidazole. *Turkish Journal of Pharmaceutical Sciences*, 17(1), 81–93. DOI: 10.4274/tjps.galenos.2018.59389 PMID: 32454765

Serafini, M., Torre, E., Aprile, S., Massarotti, A., Fallarini, S., & Pirali, T. (2019). Synthesis, Docking and Biological Evaluation of a Novel Class of Imidazothiazoles as IDO1 Inhibitors. *Molecules (Basel, Switzerland)*, 24(10), 1874. DOI: 10.3390/molecules24101874 PMID: 31096672

Serykh, V. Y., Kaliev, A. R., Ushakov, I. A., Borodina, T. N., Smirnov, V. I., & Rozentsveiga, I. B. (2018). Regioselective reaction of imidazole-2-thiols with N-sulfonylphenyldichloroacetaldimines: En route to novel sulfonylamino-substituted imidazo[2,1-b] thiazoles and thiazolo[3,2-a] benzimidazoles. *ARKIVOC*, 3, 62–75.

Shahrasbi, M., AzamiMovahed, M., Ghorban Dadras, O., Daraei, B., Zarghi, A. (2018). Design, Synthesis and Biological Evaluation of New Imidazo[2,1-b] Thiazole Derivatives as Selective COX-2 Inhibitors. *Iranian Journal of Pharmaceutical Research*, 17, 1288–1296. PMID: 30568687

Shaik, S. P., Vishnuvardhan, M. V. P. S., Sultana, F., Subba Rao, A. V., Bagul, C., Bhattacharjee, D., Kapure, J. S., Jain, N., & Kamal, A. (2017). Design and synthesis of 1,2,3-triazolo linked benzo[d]imidazo[2,1-b] thiazole conjugates as tubulin polymerization inhibitors. *Bioorganic & Medicinal Chemistry*, 25(13), 3285–3297. DOI: 10.1016/j.bmc.2017.04.013 PMID: 28462842

Shareef, M. A., Devi, G. P., Rani Routhu, S., Kumar, C. G., Kamal, A., & Babu, B. N. (2020). New imidazo[2,1-*b*] thiazole-based aryl hydrazones: Unravelling their synthesis and antiproliferative and apoptosis-inducing potential. *RSC Medicinal Chemistry*, 11(10), 1178–1184. DOI: 10.1039/D0MD00188K PMID: 33479622

Shareef, M. A., Sirisha, K., Sayeed, I. B., Khan, I., Ganapathi, T., Akbar, S., Kumar, C. G., Kamal, A., & Babu, B. N. (2019). Synthesis of new triazole fused imidazo[2,1-b] thiazole hybrids with emphasis on Staphylococcus aureus virulence factors. *Bioorganic & Medicinal Chemistry Letters*, 29(19), 126621. DOI: 10.1016/j.bmcl.2019.08.025 PMID: 31431360

Shehata, M. K., Uzair, M., Zaraei, S. O., Shahin, A. I., Shah, S. J. A., Ullah, S., Iqbal, J., & El–Gamal, M. I. (2023). Synthesis, biological evaluation, and molecular modeling studies of a new series of imidazothiazole or imidazooxazole derivatives as inhibitors of ectonucleoside triphosphate diphosphohydrolases (NTPDases). *Medicinal Chemistry Research*, 32(2), 314–325. DOI: 10.1007/s00044-022-03000-y

Shi, B., Zhu, Z., Zhu, Y., Zhou, D., Wang, J., Zhou, P., & Jing, H. (2016). Approach for 2-(arylthio)imidazoles and imidazo[2,1-b]thiazoles from imidazo[2,1-b][1,3,4] thiadiazoles by ringopening and -reconstruction. *Organic & Biomolecular Chemistry*, 14(10), 2978–2984. DOI: 10.1039/C6OB00102E PMID: 26881829

Shivhare, K. N., Jaiswal, M. K., Srivastava, A., Tiwaria, S. K., Siddiqui, I. R. New (2018). light-activated C–C and C–N bond formation in the synthesis of imidazo [1,2- a] pyridines and imidazo [2,1-b] thiazoles under catalyst and solvent-free conditions. *New Journal of Chemistry, 42*, 16591.

Slassi, S., Zaki, H., Amine, A., Yamni, K., & Bouachrine, M. (2020). Quantum Chemical and Molecular Docking Studies of Imidazole and Its Derivatives as the Active Antifungal Components against C. Albicans. *Physical Chemistry Research.*, 8, 457–469. DOI: 10.22036/PCR.2020.211251.1706

Sultana, F., Reddy Bonam, S., Reddy, V. G., Nayak, V. L., Akunuri, R., Rani Routhu, S., Alarifi, A., Halmuthur, M. S. K., & Kamal, A. (2018). Synthesis of benzo[d] imidazo[2,1-b] thiazole-chalcone conjugates as microtubule targeting and apoptosis inducing agents. *Bioorganic Chemistry*, 76, 1–12. DOI: 10.1016/j.bioorg.2017.10.019 PMID: 29102724

Thore, S. N., Gupta, S. V., & Baheti, K. G. (2013). Docking, synthesis, and pharmacological investigation of novel substituted thiazole derivatives as non-carboxylic, anti-inflammatory, and analgesic agents. *Medicinal Chemistry Research*, 22(8), 3802–3811. DOI: 10.1007/s00044-012-0382-6

Tojo, S., Kohno, T., Tanaka, T., Kamioka, S., Ota, Y., Ishii, T., Kamimoto, K., Asano, S., & Isobe, Y. (2014). Crystal Structures and Structure-Activity Relationships of Imidazothiazole Derivatives as IDO1 Inhibitors. *ACS Medicinal Chemistry Letters*, 5(10), 1119–1123. DOI: 10.1021/ml500247w PMID: 25313323

Vekariya, R. H., Patel, K. D., Vekariya, M. K., Prajapati, N. P., Rajani, D. P., Rajani, S. D., & Patel, H. D. (2017). Microwave-assisted green synthesis of new imidazo[2,1-b] thiazole derivatives and their antimicrobial, antimalarial, and antitubercular activities. *Research on Chemical Intermediates*, 43(11), 6207–6231. DOI: 10.1007/s11164-017-2985-5

Vu, C. B., Bemis, J. E., Disch, J. S., Ng, P. Y., Nunes, J. J., Milne, J. C., Carney, D. P., Lynch, A. V., Smith, J. J., Lavu, S., Lambert, P. D., Gagne, D. J., Jirousek, M. R., Schenk, S., Olefsky, J. M., & Perni, R. B. (2009). Discovery of imidazo[1,2-b] thiazole derivatives as novel SIRT1 activators. *Journal of Medicinal Chemistry*, 52(5), 1275–1283. DOI: 10.1021/jm8012954 PMID: 19199480

Wan, Y., Long, J., Gao, H., & Tang, Z. (2021). 2-Aminothiazole: A privileged scaffold for the discovery of anti-cancer agents. *European Journal of Medicinal Chemistry*, 210, 112953. DOI: 10.1016/j.ejmech.2020.112953 PMID: 33148490

Yang, H., Higgins, B., Kolinsky, K., Packman, K., Go, Z., Iyer, R., Kolis, S., Zhao, S., Lee, R., Grippo, J., Schostack, K., Simcox, M. E., Heimbrook, D., Bollag, G., & Su, F. (2010). RG7204 (PLX4032), a Selective BRAF(V600E) Inhibitor, Displays Potent Antitumor Activity in Preclinical Melanoma Models. *Cancer Research*, 70(13), 5518–5527. DOI: 10.1158/0008-5472.CAN-10-0646 PMID: 20551065

Zaraei, S. O., Sbenati, R. M., Alach, N. N., Anbar, H. S., El-Gamal, R., Tarazi, H., Shehata, M. K., Abdel-Maksoud, M. S., Oh, C. H., & El-Gamal, M. I. (2021). Discovery of first-in-class imidazothiazole-based potent and selective ErbB4 (HER4) kinase inhibitors. *European Journal of Medicinal Chemistry*, 224, 113674. DOI: 10.1016/j.ejmech.2021.113674 PMID: 34237622

Zhang, A., Dastagiri, D., Ramaiah, M. J., Reddy, J. S., & Bharathi, E. V.. (2010). Synthesis of imidazothiazole–chalcone derivatives as anticancer and apoptosis inducing agents. *ChemMedChem*, 5(11), 1937–1947. DOI: 10.1002/cmdc.201000346 PMID: 20836120

Chapter 4

Thienopyrroles–Emerging Therapeutic Agents in Modern Medicine:
A Review of Medical Applications

Prashanth G. K.
https://orcid.org/0000-0001-6691-4030
Department of Chemistry, Sir M. Visvesvaraya Institute of Technology, Bengaluru, India

Lalithamba H. S.
Department of Chemistry, Siddaganga Institute of Technology, Tumkur, India

Srilatha Rao
Department of Chemistry, Nitte Meenakshi Institute of Technology, Bengaluru, India

Rashmi K. V.
Department of Biotechnology, Sir M. Visvesvaraya Institute of Technology, Bengaluru, India

Bhagya N. P.
Department of Chemistry, Sai Vidya Institute of Technology, Bengaluru, India

Manoj Gadewar
Department of Pharmacology, K.R.Mangalam University, Gurugram, India

ABSTRACT

Thienopyrroles, characterized by a thiophene ring fused to a pyrrole ring, have garnered significant interest in medicinal chemistry for their diverse therapeutic properties. This review explores the recent advancements in the medical applications of thienopyrroles, highlighting their roles as anticancer, antimicrobial, anti-inflammatory, antiviral, neuroprotective, analgesic, and antidiabetic agents. Through examining current research, we discuss the mechanisms by which thienopyrroles exert their effects, such as apoptosis induction, enzyme inhibition, cytokine suppression, and modulation of neurotrophic factors. The versatility and promise of

DOI: 10.4018/979-8-3693-7520-4.ch004

thienopyrroles in modern medicine underscore their potential as scaffolds for the development of novel therapeutics. However, further research and clinical trials are essential to fully realize their benefits and ensure their safety and efficacy.

INTRODUCTION

Heterocyclic compounds have become a cornerstone in chemical sciences, particularly in chemistry, due to their diverse structural possibilities that allow them to serve as the foundation for various applications. The incorporation of atoms such as nitrogen, sulfur, and oxygen into their unique ring structures makes these compounds highly versatile and valuable across various fields. In medicinal chemistry, heterocycles have proven especially important, with over 75% of FDA-approved drugs or drug candidates in clinical trials containing heterocyclic pharmacophores. This high prevalence is due to their ability to interact with biological targets in specific and effective ways, making them ideal for drug development. Beyond their pharmaceutical uses, heterocyclic compounds also find applications as theranostic agents, offering both therapeutic and diagnostic capabilities, and as key components in the creation of fluorescent materials or dyes. These compounds find utility in developing advanced materials, including copolymers, transistors, and thin films, where their distinctive chemical properties can be leveraged to enhance performance and functionality, (Huang & Yu, 2021). The broad applicability of heterocyclic compounds across these diverse areas underscores their critical role in both scientific research and technological innovation, (Wani et al., 2023) (Tomašič *et al.,* 2015) (Cai & Goodnow, 2008) (Barker, Greenhaff, & Timmons, 2006) (Barker, Faull, & Kettle, 2002).

Among heterocyclic compounds, those containing nitrogen and sulfur atoms have long been recognized as crucial pharmacophores in medicinal chemistry. These heteroatoms impart unique chemical properties to the compounds, enhancing their ability to interact with biological targets. Since ancient times, morphine and vincristine—two examples of pharmacologically active compounds—have been developed with the help of nitrogen and sulphur heterocycles, and these compounds have been essential in the treatment of a wide range of illnesses. Morphine, a nitrogen-containing alkaloid, has been widely used as a potent analgesic for pain management, showcasing the therapeutic potential of nitrogen heterocycles. Similarly, vincristine, a sulfur-containing natural product, has been instrumental in cancer therapy, particularly in the treatment of leukemia and lymphomas. The inclusion of these heteroatoms not only boosts the compounds' biological activity but also improves their metabolic stability and solubility, making them more effective as pharmaceuticals. The versatility and efficacy of nitrogen and sulfur heterocycles have made them indispensable in the pharmaceutical industry, driving ongoing research to discover new therapeutic agents based on these scaffolds. Their longstanding use and continued relevance highlight the importance of these heterocyclic structures in the development of modern medicine, (Zhang *et al.*, 2021a) (Zu *et al.,* 2020) (Zhang *et al.,* 2021b) (Eras, Galvez, & Garcta, 1984).

Thienopyrroles, a unique class of heterocyclic compounds characterized by a fused thiophene and pyrrole ring, have emerged as potent pharmacophores in medicinal chemistry. Over the past few decades, these compounds have been the focus of extensive research due to their wide range of therapeutic activities, including anticancer, antimicrobial, anti-inflammatory, antiviral, neuroprotective, analgesic, and antidiabetic effects, (Nandi *et al.,* 2024) (Patel *et al.,* 2023) (Yin *et al.*, 2012) (Zhong *et al.,* 2021) (Alamery *et al.,* 2023). Their chemical structure allows for significant functional diversity, making them versatile scaffolds for drug design and development.

The ability of thienopyrroles to modulate key biological targets, such as enzymes, receptors, and signaling pathways, has made them attractive candidates for treating various diseases. For instance, their role in inhibiting glycogen phosphorylase positions them as potential treatments for metabolic disorders, while their regulation of Lysine-specific demethylase 1 (LSD1) highlights their promise in oncology, (Bhatia & Kumar, 2023) (Choudhary, Kumar, & Thakur, 2024) (Kumar, Kumar, & Bhatia, 2023). Furthermore, their antiviral properties provide a basis for combating infections like hepatitis C and other viral diseases. Given the broad therapeutic potential of thienopyrroles, it is essential to explore their synthesis, structure-activity relationships (SAR), and mechanisms of action to fully harness their capabilities, (Cox, 2021). Despite the extensive research on heterocyclic compounds, their role in the development of anticancer drugs remains relatively underexplored. However, pharmacophores containing thieno[3,2-b]pyrrole have emerged as promising candidates in cancer treatment. These compounds demonstrate their anticancer potential by targeting a range of critical biological pathways and druggable targets, (Shirinian, Nikal, & Krayushk, 2003). Thieno[3,2-b]pyrrole derivatives have been developed as inhibitors of various cancer-related targets, including pyruvate kinase M2 (PKM2), a key enzyme involved in cancer cell metabolism and proliferation. Additionally, these compounds act as inhibitors of centromere-associated protein-E (CENP-E), which is crucial for mitotic spindle function and cell division. They also target RON receptor tyrosine kinase, involved in tumor growth and metastasis, and VEGFR-Akt pathways, which play significant roles in angiogenesis and cancer cell survival. Furthermore, thieno[3,2-b]pyrrole-based drugs are explored as inhibitors of Lysine-Specific Demethylase 1 (LSD1), an enzyme linked to gene expression and cancer progression, (Samsonenko *et al.,* 2022) (Lee, Kim, & Park, 2002) (Torosyan *et al.,* 2019). Beyond these traditional roles, these compounds are also utilized as photosensitizers in photodynamic therapy, where they help generate reactive oxygen species upon light activation to kill cancer cells. The diverse mechanisms through which thieno[3,2-b] pyrrole pharmacophores exert their anticancer effects highlight their potential as versatile tools in the development of innovative cancer therapies, (Le *et al.,* 2022) (Han *et al.,* 2024).

Thienopyrroles are a versatile class of heterocycles whose unique structures have led to their widespread application in diverse areas of research and development. Their synthesis and various structural types have been explored extensively; for instance, several synthetic methods have been developed for different thienopyrrole derivatives, including the thieno[2,3-b]-, -[3,2-b]-, and -[3,4-b]-pyrroles, based on modifications from thiophene or pyrrole starting materials, (Garcia, Francisco, & Gálvez, 1985).

Further, thienopyrrole-modified porphyrinoids have demonstrated unique aromatic properties, which are notably influenced by the substituents attached to their ring structures. This modification affects the aromatic or antiaromatic behavior of the macrocyclic systems, rendering them useful for electronic applications, (Valiev *et al.,* 2014). Advances in catalysis have also allowed efficient methods for constructing thienopyrrole frameworks; for example, a palladium-catalyzed intramolecular cyclization strategy has yielded high yields of aryl- or alkyl-substituted thienopyrroles, underscoring their potential in synthetic chemistry, (El-Atawy *et al.,* 2017).

Recent studies have further expanded the scope of thienopyrrole chemistry by synthesizing pyrido-fused thienopyrrole structures, which are of interest for their potential biological activities and material properties, (Stockmann & Fiksdahl, 2008). Furthermore, thienopyrrole-based S,N-heteroacenes, particularly those with dicyanomethylene end-capping, exhibit significant π-extension and red-shifted absorption characteristics, making them promising candidates for organic electronic devices, (Jiang *et al.,* 2017).

Emerging applications in medicinal chemistry have shown that thienopyrrole derivatives possess therapeutic potential, especially for autoimmune diseases and inflammatory conditions, (Sabnis, 2023). Additionally, improved synthetic routes for thienopyrroles and related heterocycles have been developed, allowing for more efficient production and enabling their use in various industrial and pharmaceutical applications, (Welch & Phillips, 1999).

This chapter aims to provide a comprehensive overview of the current advancements in thienopyrrole research, highlighting their medical applications and discussing future directions in this rapidly evolving field.

SYNTHESIS

Synthesis of Compound 2

In a 250mL round-bottom flask under an argon atmosphere, $Pd_2(dba)_3$ (0.1mmol) and ±2,2'-bis(diphenylphosphino)-1,1'-binaphthyl (0.1mmol) were mixed in toluene (20mL) and stirred for 5 minutes. Then, NaOtBu (3mmol) and dibromodibenzo[b] thiophene (**1**) (1mmol) were added. After adding 2-ethylhexylamine (1.2mmol) to the mixture, it was heated to reflux for a duration of six hours. Following the reaction, the liquid was allowed to cool to ambient temperature before being extracted using a brine and ethyl acetate mixture. After mixing the organic layers, anhydrous sodium sulphate was used to dry them, and then the mixture was filtered and concentrated under low pressure. Hexane was used as the eluent in column chromatography to purify the crude product that resulted, producing a white solid with a 70% yield of the desired chemical (**2**), (Balaji *et al.*, 2012).

Compound 3

To synthesize compound **3**, bromine (2mmol) was slowly added to a solution of compound **2** (2mmol) in chloroform (50mL) over 20 minutes. The reaction mixture was stirred at room temperature for 4 hours. Once the reaction was complete, the mixture was extracted twice using a dichloromethane-water mixture and then with a sodium thiosulfate solution. The combined organic layers were dried over anhydrous sodium sulfate, filtered, and concentrated under reduced pressure. The crude product was purified through column chromatography, resulting in a white solid with a 60% yield, (Balaji *et al.*, 2012).

Compound 4

Compound **2** (2mmol) was carefully dissolved in 50 mL of chloroform, then bromine (5mmol) was added to create compound **4**. For eight hours, the mixture was mixed at room temperature. Following the completion of the reaction, the mixture was twice extracted using a dichloromethane-water mixture and then again using a sodium thiosulphate solution. The organic layers that were mixed were dried with anhydrous sodium sulphate. Following the removal of the solvent under reduced pressure, column chromatography was used to purify the crude product, resulting in the dibromo compound **4** as a white solid with an 80% yield, (Balaji *et al.*, 2012).

Figure 1. Synthesis of thienopyrroles derivatives namely 2, 3, 5, 7, 8, 11 and 12

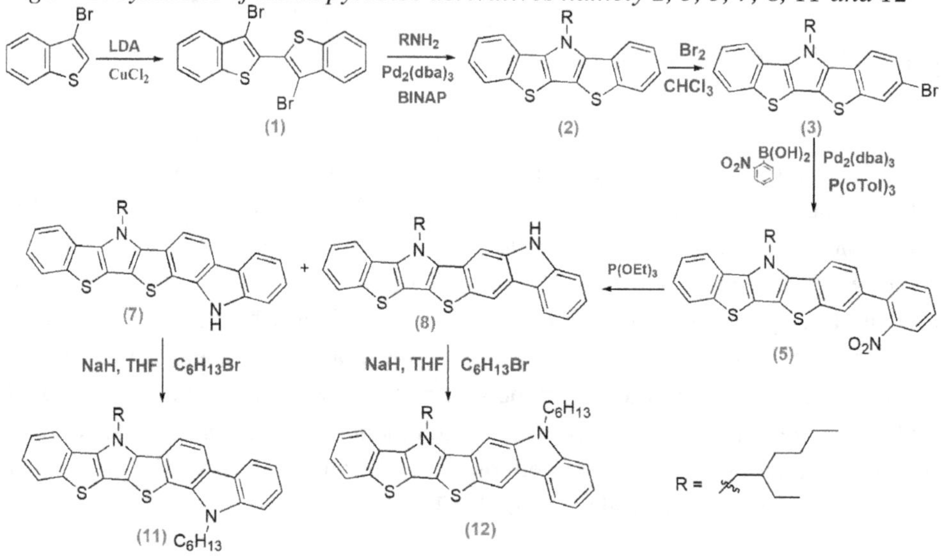

Compounds 5 and 6

To synthesize compounds **5** and **6**, a 250mL two-neck round-bottom flask was charged with THF (20mL), the monobromo or dibromo compound (**3 or 4**) (2.5mmol), and 2-nitrophenylboronic acid (3.0 mmol for compound **5** or 6.0mmol for compound **6**). Next, $Pd_2(dba)_3$ (0.25mmol), P(o-Tol)3 (0.25mmol), and 2M aqueous potassium carbonate (15mL) were added. The reaction mixture was heated to 50°C and stirred for 8 hours. After cooling to room temperature, the mixture was extracted with ethyl acetate and brine. The organic layer was dried with anhydrous sodium sulfate, filtered, and concentrated under reduced pressure. The crude product was purified using column chromatography with ethyl acetate-hexane as the eluent to obtain the pure compound. Compound **5** was obtained in a 60% yield, (Balaji *et al.,* 2012).

Compounds 7, 8, 9, and 10

To synthesize compounds **7, 8, 9,** and **10**, a 100mL two-neck round-bottom flask was set up under an argon atmosphere. The nitro compound (**5** or **6**) (0.5mmol) was added to the flask, followed by triethyl phosphite (5mL) and 1,2-dichlorobenzene (5mL). The reaction mixture was heated to reflux for 12 hours. After the reaction was completed, the mixture was cooled to room temperature and concentrated under reduced pressure. The crude product was then purified using column chromatography with ethyl acetate-hexane as the eluent to obtain the desired fused compounds, (Balaji *et al.*, 2012).

Figure 2. Synthesis of thienopyrroles derivatives namely 6, 9, 10 and 13

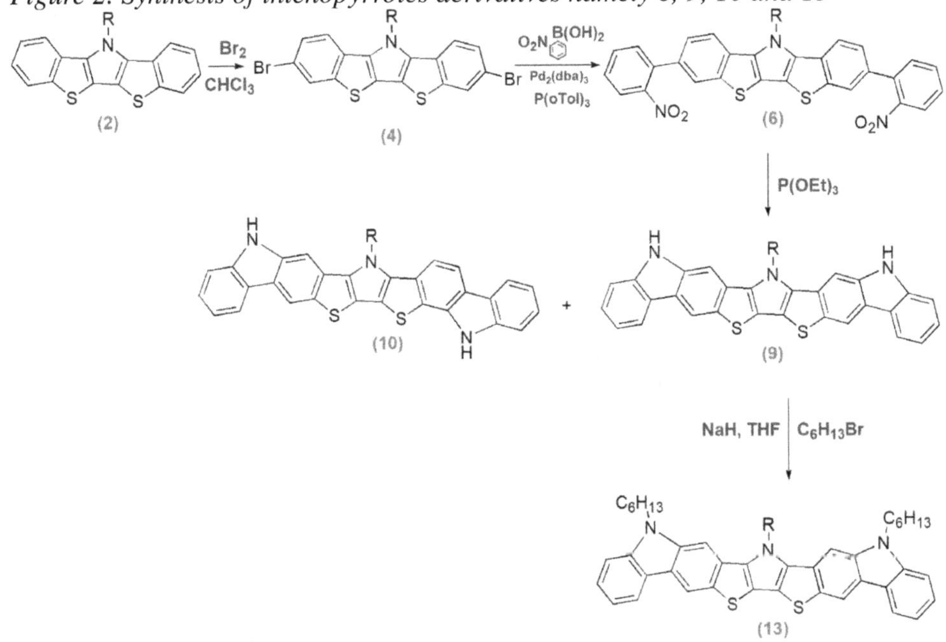

Compounds 11, 12, and 13

To synthesize compounds **11, 12,** and **13**, a 100mL round-bottom flask was set up under an argon atmosphere. The fused compound (**11, 12,** or **13**) (0.1mmol) was added to the flask along with dry THF (10mL) and alkyl bromide (0.3mmol). Sodium hydride (NaH, 60% in mineral oil, 0.25mmol) was then added to the reaction mixture, and it was heated to reflux for 6 hours. After the reaction was complete, it was quenched by carefully adding water and then extracted three times with ethyl acetate and water. The combined organic layers were dried over anhydrous sodium sulfate, filtered, and concentrated under reduced pressure. The crude product was purified using column chromatography to obtain the corresponding alkyl derivative, (Balaji *et al.*, 2012).

Synthesis of bis[1]benzothieno[3,2-b:2',3'-d]pyrroles (BBTPs) (14-16)

Various approaches have been employed to synthesize bis[1]benzothieno[3,2-b:2',3'-d]pyrroles (BBTPs), as shown in **Figure 3**. Initially, the synthesis involved cyclizing 3-nitro-2,2'-dibenzo[b]thiophene to produce the unfunctionalized BBTP **14**, which could then undergo N-functionalization. However, most current methods rely on using a 3,3'-dihalo-2,2'-bi(benzo[b]thiophene) (such as **17 or 18**) as the starting material. The central pyrrole ring is formed via catalytic amination. Intermediate **17** is the most commonly used pathway to synthesize BBTPs, and it is also a key route to other benzannulated compounds, including thieno[3,2-b:4,5-b']bis[1]benzothiophene and phospholo[3,2-b:4,5-b']bis[1]benzothiophene, (Jung *et al.*, 2015) (Rasmussen, 2016) (Qi *et al.*, 2008) (Gao *et al.*, 2007) (Gao *et al.*, 2010) (Jung *et al.*, 2016) (Dahlmann & Neidlein, 1997) (Balaji & Valiyaveettil, 2009) (Dienes *et al.*, 2008) (Balaji et al., 2010) (Balaji *et al.*, 2012b) (Wetzel *et al.*, 2014) (Gupta & Flynn, 2017) (Win *et al.*, 201) (Mitsudo *et al.*, 2012).

Figure 3. Synthesis of bis[1]benzothieno[3,2-b:2',3'-d]pyrroles (BBTPs) (14-16)

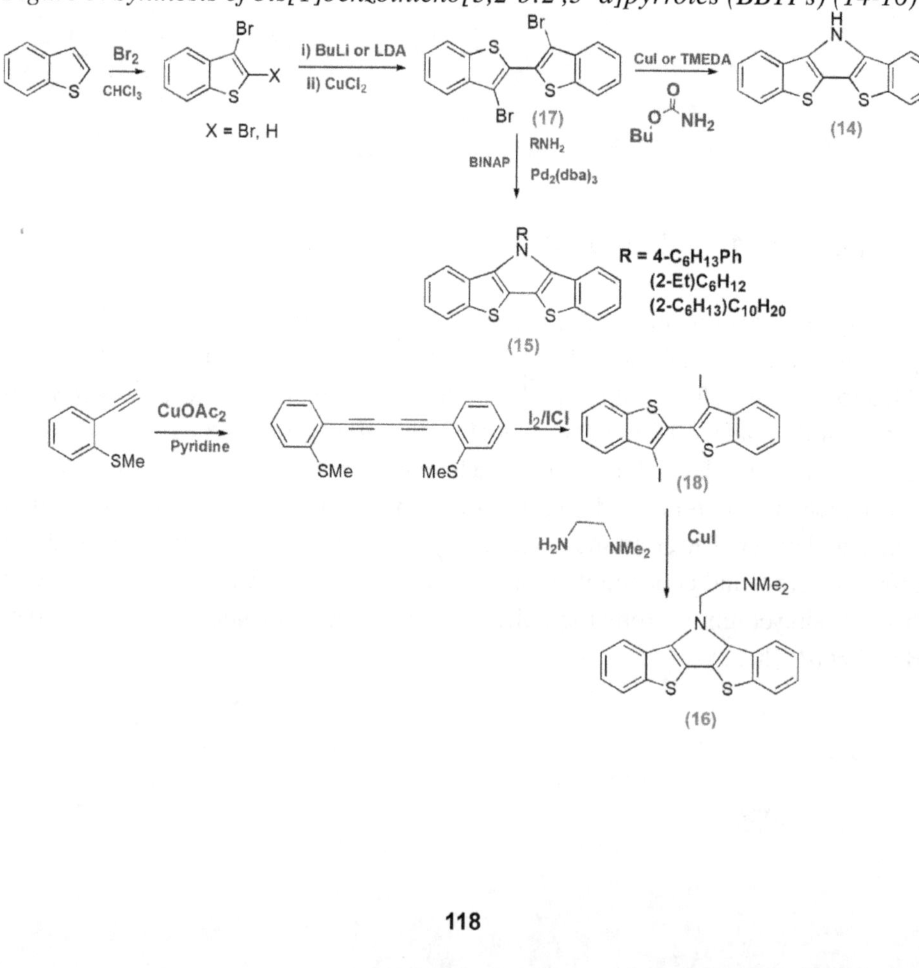

APPLICATIONS IN MEDICAL FILED

Thienopyrroles have shown significant promise in the field of oncology. Several studies have demonstrated their ability to inhibit the growth of cancer cells, induce apoptosis, and interfere with cell proliferation. For instance, research has indicated that specific thienopyrrole derivatives can target and disrupt critical signaling pathways involved in cancer progression. The potential of these compounds as chemotherapeutic agents is bolstered by their selective cytotoxicity towards cancer cells, minimizing damage to healthy tissues.

Castanheira et al. reported the synthesis of benzothieno[3,2-b]pyrrole (BTP, **19**), a fluorescent thienopyrrole derivative, with significant anticancer potential (**Figure 4**). In studies using human cancer cell lines, **BTP 55** showed GI50 values of 19.1 μM against MCF-7 (breast cancer), 38.7 μM against SF-268 (glioma), and 3.9 μM against NCI-H460 (lung cancer). Additionally, photophysical studies indicated that BTP **19** accumulates in the hydrophobic regions of lipid tails, suggesting its potential for developing targeted cancer therapies, (Castanheira *et al.*, 2009).

With AC50 values of 16nM and 11nM, respectively, compounds **20a** and **20b** showed significant anticancer action (**Figure 4**). While its removal resulted in a decrease in activity, the addition of a heteroatom at the second position of the core structure greatly increased potency. Studies on the structure-activity relationship (SAR) further demonstrated how crucial a methyl substituent is to preserving potency, since even little changes reduced efficacy. Both of the powerful compounds fit well into the pyruvate kinase target site (PDB ID: 5X1V), according to in silico study, mostly through hydrophobic Van der Waals interactions, (Jiang *et al.*, 2010).

Fang et al. (2021) explored the anticancer potential of 4H-thieno[3,2-b]pyrrole derivatives. After screening a library of compounds, four derivatives (**21-24**) were evaluated across various cancer cell lines (**Figure 4**). Following structural optimization, compound **25**, which contains a hydrophobic trifluoromethyl group, exhibited significant anticancer activity against colon cancer cells, showing 86% inhibition at a concentration of 20μM (IC50 = 10.93μM). A target fishing technique identified methionine aminopeptidase 2 as the primary target for compound **25**, leading to inhibition of cell growth. This was further confirmed through apoptotic and flow cytometry studies. Molecular docking studies revealed that compound **25** binds to the active site of methionine aminopeptidase 2 (PDB-ID:2OAZ) via two π-π interactions with the Hie-231 residue and hydrogen bonding with the CO-480 residue.

Figure 4. Some examples of thienopyrroles derivatives

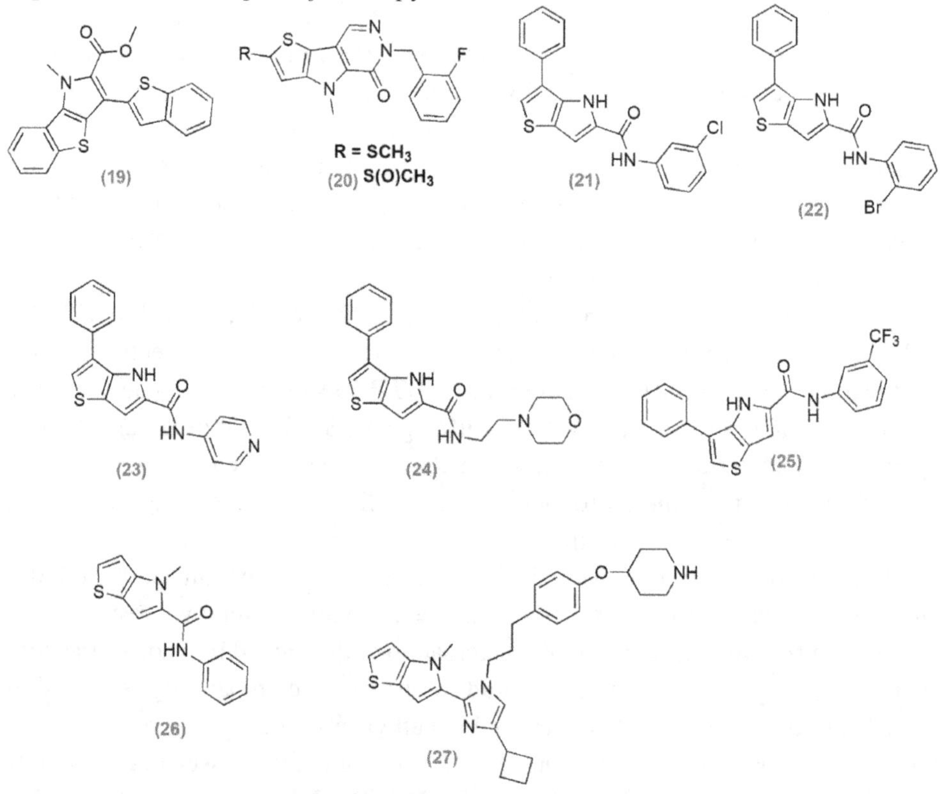

In models of acute myeloid leukaemia, Romussi et al. looked at reversible derivatives of a particular structural component. Using the previously found chemical **26** (IC50 = 25.7μM) as a starting point, the research team set out to alter the amide bridge (**Figure 4**). An imidazole group with different substituents was included, and a number of 5-imidazolylthieno[3,2-b]pyrrole derivatives were produced. Compound **27** stood out as the most effective among them, having an amazing IC50 against KDM1A of 0.10nM. Compound **27** suppressed colony formation in cellular experiments by 72% at 500nM concentration. Furthermore, at doses of 15 mgkg^{-1} and 30mgkg^{-1}, respectively, it increased lifespan by 29% and 70% in the MLLAF9 in vivo model. Additionally, the investigation showed that compound **27** binds to active sites in a U-shaped manner, which is in line with previous findings, (Romussi *et al.,* 2020).

Figure 5. Some examples of thienopyrroles derivatives

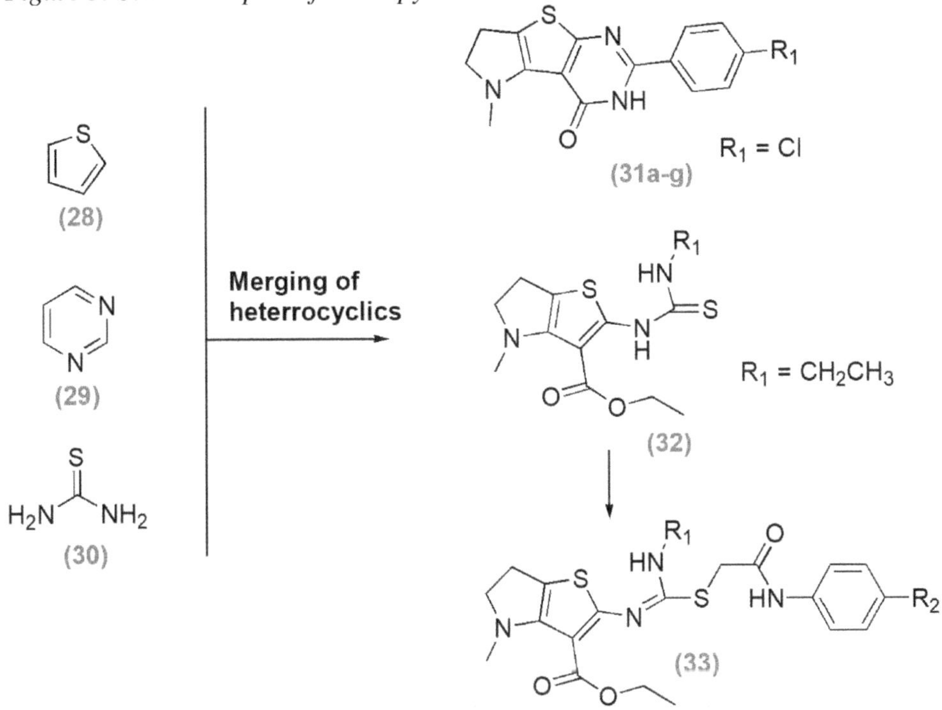

Three heterocyclic rings were combined by Abdelnaby et al. to create a novel class of fused thiophenes that have the potential to be dual inhibitors of VEGFR-2 and AKT with apoptotic action. Thiophene (**28**), pyrimidine (**29**), and thiourea (**30**) components were combined to form three distinct series (**31-33**) of derivatives of thieno[3,2-b]pyrroles (**Figure 5**). The compounds with the most promising anticancer characteristics were **31b** (IC50 = 3.1μM against HepG2 and 2.15μM against PC3 cells) and **31c** (IC50 =3.02μM against HepG2 and 3.12μM against PC3 cells). These substances spared healthy cells while specifically targeting cancer cells. In particular, compound **31b** showed IC50 values of 0.13μM and 6.96μM, whereas compound **32c** inhibited VEGFR-2 and AKT with IC50 values of 0.075μM and 4.60μM, respectively. These substances caused S-phase cell cycle arrest by caspase-3-mediated apoptosis, according to flow cytometry. The addition of an electron-withdrawing group at the para position of the phenyl ring increased activity against PC3 and HepG2 cells, according to SAR analysis of the **31a-g** series, (Abdelnaby *et al.*, 2022).

Figure 6. Few examples of thienopyrroles derivatives and uses in medicine

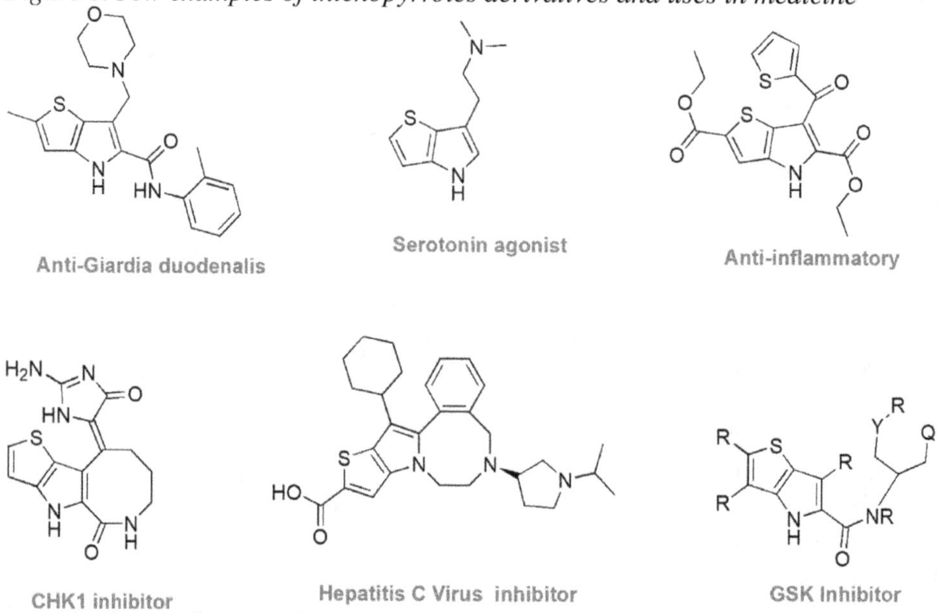

Thienopyrroles are an important class of compounds in medicinal chemistry due to their broad spectrum of pharmacological activities (**Figure 6**). These heterocyclic molecules have been employed in the treatment of serotonin agonists (Blair *et al.,* 1999), giardiasis (Hart *et al.,* 2023), CHK1 inhibitor (Parmentier *et al.,* 2009), anti-inflammatory (Kumar *et al.,* 2004), glycogen phosphorylase inhibitory activities, (Du Bois, 2004) (Bartlett *et al.,* 2003) (Du Bois, 2003), treatment of hepatitis C (Hernando *et al.,* 2009) antiviral (Attenni *et al.,* 2010) (Fu, Li, & Yu, 2021), and Lysine-specifc-demethylase-1-KDM1A (LSD1) regulators, (Attenni *et al.,* 2005). Thienopyrroles exhibit significant anti-inflammatory properties, making them useful in treating conditions where inflammation is a primary concern. One of their noteworthy applications is as CHK1 inhibitors, which are crucial in cancer therapy, as CHK1 plays a vital role in DNA damage repair, and its inhibition can enhance the effectiveness of chemotherapy.

Furthermore, thienopyrroles have shown promise in the treatment of hepatitis C, a viral infection that affects the liver. Their ability to inhibit glycogen phosphorylase also highlights their potential in managing metabolic disorders, as glycogen phosphorylase is an enzyme involved in glycogen breakdown. Additionally, these compounds regulate Lysine-specific demethylase 1 (KDM1A or LSD1), an enzyme implicated in gene expression and cancer progression. By modulating LSD1 activity, thienopyrroles could offer therapeutic benefits in oncology.

The antiviral properties of thienopyrroles further extend their therapeutic potential, making them valuable in combating various viral infections. Their diverse pharmacological attributes make thienopyrroles a promising scaffold for drug development, offering potential treatments for a range of diseases, from infectious conditions to cancer and metabolic disorders. This versatility underscores the importance of continued research into their drug delivery mechanisms and therapeutic applications.

Eco-Friendly Synthesis of Thienopyrroles

Eco-friendly synthesis methods for thienopyrroles have gained attention due to their potential for reducing environmental impact. For example, microwave-assisted synthesis enables rapid reaction times with reduced energy requirements and cleaner products, often utilizing non-toxic solvents such as water or ethanol, (Ghosh, Bedi, & Zade, 2015). Additionally, the use of biodegradable or non-metal catalysts, such as organocatalysts and biocatalysts, can eliminate the need for heavy metals or harsh reagents, thus reducing toxic waste, (Srebrenkoska *et al.,* 2013). Photocatalytic approaches, harnessing visible light or sunlight, offer another green alternative by lowering the need for high temperatures and pressures. The incorporation of ionic liquids, particularly biodegradable or non-toxic options, further minimizes environmental impact by replacing conventional, often harmful solvents with more sustainable choices, (Biehl, 2011). Together, these techniques contribute to more sustainable synthesis processes for thienopyrroles, aligning with the principles of green chemistry.

Future Work

While thienopyrroles have demonstrated significant potential in preclinical studies, there are several areas where further research is needed. One critical aspect is the optimization of their pharmacokinetic properties. Many thienopyrrole derivatives suffer from poor bioavailability, rapid metabolism, and limited water solubility, which can hinder their therapeutic effectiveness. Future work should focus on improving these properties through chemical modifications, prodrug strategies, or formulation approaches, such as nanoparticle delivery systems. Another important area for future research is the exploration of thienopyrroles in combination therapies. Given their ability to target multiple pathways, thienopyrroles could be used in conjunction with other drugs to enhance therapeutic outcomes, especially in cancer treatment. Investigating synergistic effects with existing chemotherapeutic agents could lead to the development of more effective treatment regimens. More in-depth studies on the toxicity and safety profiles of thienopyrroles are necessary. While these compounds have shown promising results in cell-based and animal models, their long-term effects and potential side effects in humans remain largely unknown. Conducting rigorous preclinical and clinical trials will be crucial in determining their safety and efficacy in a clinical setting. There is a growing interest in exploring the potential of thienopyrroles as neuroprotective agents. Initial studies have shown that these compounds can modulate neurotrophic factors and protect against neurodegeneration. Future research should delve deeper into their mechanisms of action in the nervous system and evaluate their potential in treating neurodegenerative diseases like Alzheimer's and Parkinson's.

Advantages and Disadvantages

Thienopyrroles exhibit a broad range of biological activities, making them valuable in treating various diseases, from cancer to viral infections. This versatility is a significant advantage as it allows for the development of multi-target drugs. The fused thiophene-pyrrole ring system offers a flexible scaffold for chemical modifications, enabling the design of derivatives with enhanced potency, selectivity, and pharmacokinetic properties. Thienopyrroles have demonstrated the ability to selectively target cancer cells, enzymes, and receptors, reducing the likelihood of off-target effects and minimizing damage to healthy tissues. The ability of thienopyrroles to interact with multiple biological pathways makes them ideal candidates for combination therapies, potentially enhancing treatment efficacy and overcoming drug resistance.

Many thienopyrrole derivatives suffer from issues such as poor bioavailability, rapid metabolism, and low solubility, which can limit their effectiveness as therapeutic agents. While thienopyrroles have shown promise in preclinical studies, there is a lack of comprehensive clinical data to support their widespread use in human medicine. More rigorous clinical trials are needed to establish their safety and efficacy. The potential toxicity of thienopyrroles, especially with long-term use, remains a significant concern. The development of safer derivatives with reduced toxicity profiles is essential for their successful clinical application. The synthesis of thienopyrrole derivatives can be complex and may require multiple steps, which can be a barrier to large-scale production and commercialization.

CONCLUSION

Thienopyrroles represent a promising class of compounds with significant potential in modern medicine. Their diverse pharmacological activities, ranging from anticancer and antimicrobial to antiviral and neuroprotective effects, make them versatile candidates for drug development. Despite their advantages, challenges such as poor pharmacokinetics, limited clinical data, and potential toxicity must be addressed to fully realize their therapeutic potential. Future research should focus on optimizing the pharmacological properties of thienopyrroles, exploring their use in combination therapies, and conducting more comprehensive toxicity studies. By overcoming these challenges, thienopyrroles could emerge as a valuable class of drugs in the treatment of a wide range of diseases, offering new hope for patients and advancing the field of medicinal chemistry. The continued exploration of thienopyrroles in various therapeutic areas, coupled with innovative approaches to drug design and delivery, will undoubtedly play a crucial role in shaping the future of these compounds in medicine.

REFERENCES

Abdelnaby, R. M., El-Malah, A. A., FakhrEldeen, R. R., Saeed, M. M., Nadeem, R. I., Younis, N. S., Abdel-Rahman, H. M., & El-Dydamony, N. M. (2022). In vitro anticancer activity screening of novel fused thiophene derivatives as VEGFR-2/AKT dual inhibitors and apoptosis inducers. *Pharmaceuticals (Basel, Switzerland)*, 15(6), 700. DOI: 10.3390/ph15060700 PMID: 35745619

Alamery, S., AlAjmi, A., Wani, T. A., & Zargar, S. (2023). In silico and in vitro exploration of poziotinib and olmutinib synergy in lung cancer: Role of hsa-miR-7-5p in regulating apoptotic pathway marker genes. *Medicina*, 59(11), 1923. DOI: 10.3390/medicina59111923 PMID: 38003971

Attenni, B., Hernando, J. I. M., Malancona, S., Narjes, F., Ontoria, J. M. O., & Rowley, M. (2005).

Attenni, B., Hernando, J. I. M., Malancona, S., Narjes, F., Ontoria, J. M. O., & Rowley, M. (2010). *U.S. Patent No. 7,781,431*. Washington, DC: U.S. Patent and Trademark Office.

Baker, D. J., Greenhaff, P. L., & Timmons, J. A. (2006). Glycogen phosphorylase inhibition as a therapeutic target: A review of the recent patent literature. *Expert Opinion on Therapeutic Patents*, 16(4), 459–466. DOI: 10.1517/13543776.16.4.459

Balaji, G., Della Pelle, A. M., Popere, B. C., Chandrasekaran, A., & Thayumanavan, S. (2012). Synthesis and properties of thienopyrrole based heteroacenes–indolodibenzothienopyrrole and dicarbazolodithienopyrrole. *Organic & Biomolecular Chemistry*, 10(17), 3455–3462. DOI: 10.1039/c2ob25087j PMID: 22421802

Balaji, G., Della Pelle, A. M., Popere, B. C., Chandrasekaran, A., & Thayumanavan, S. (2012). Synthesis and properties of thienopyrrole based heteroacenes–indolodibenzothienopyrrole and dicarbazolodithienopyrrole. *Organic & Biomolecular Chemistry*, 10(17), 3455–3462. DOI: 10.1039/c2ob25087j PMID: 22421802

Balaji, G., Phua, D. I., Shim, W. L., & Valiyaveettil, S. (2010). Synthesis and characterization of unsymmetric indolodithienopyrrole and extended diindolodithienopyrrole. *Organic Letters*, 12(2), 232–235. DOI: 10.1021/ol902528b PMID: 20000317

Balaji, G., & Valiyaveettil, S. (2009). Synthesis and properties of symmetric and unsymmetric dibenzothienopyrroles. *Organic Letters*, 11(15), 3358–3361. DOI: 10.1021/ol901133m PMID: 19572751

Barker, A. J., Faull, A. W., & Kettle, J. G. (2002). *U.S. Patent No. 6,479,527*. Washington, DC: U.S. Patent and Trademark Office.

Bartlett, J. B., Freeman, S., Kenny, P., Morley, A., & Whittamore, P. (2003).

Bhatia, R., & Kumar, B. (2023). p53 as a potential target for treatment of cancer: A perspective on recent advancements in small molecules with structural insights and SAR studies. *European Journal of Medicinal Chemistry*, 247, 115020. DOI: 10.1016/j.ejmech.2022.115020 PMID: 36543034

Biehl, E. R. (2011) "Five-Membered Ring Systems: Thiophenes and Se/Te Derivatives." In *Progress in Heterocyclic Chemistry*, vol. 22, pp. 109-141. Elsevier.

Blair, J. B., Marona-Lewicka, D., Kanthasamy, A., Lucaites, V. L., Nelson, D. L., & Nichols, D. E. (1999). Thieno [3, 2-b]-and thieno [2, 3-b] pyrrole bioisosteric analogues of the hallucinogen and serotonin agonist N, N-dimethyltryptamine. *Journal of Medicinal Chemistry*, 42(6), 1106–1111. DOI: 10.1021/jm980692q PMID: 10090793

Cai, J., & Goodnow, R. A., Jr. (2008). *U.S. Patent No. 7,396,940.* Washington, DC: U.S. Patent and Trademark Office.

Castanheira, E. M., Abreu, A. S., Queiroz, M.-J. R., Ferreira, P. M., Coutinho, P. J., Nazareth, N., & Nascimento, M. S.-J. (2009). Fluorescence properties of a potential antitumoral benzothieno [3, 2-b] pyrrole in solution and lipid membranes. *Journal of Photochemistry and Photobiology A Chemistry*, 206(2-3), 220–226. DOI: 10.1016/j.jphotochem.2009.07.007

Choudhary, D., Kumar, B., & Thakur, A. (2024). Cancer and insulin-like growth factor inhibitors: Recent advancements and SAR analysis. In *Current Molecular Targets of Heterocyclic Compounds for Cancer Therapy* (pp. 161–181). Academic Press. DOI: 10.1016/B978-0-323-96121-9.00008-5

Cox, T. R. (2021). The matrix in cancer. *Nature Reviews. Cancer*, 21(4), 217–238. DOI: 10.1038/s41568-020-00329-7 PMID: 33589810

Dahlmann, U., & Neidlein, R. (1997). The Diyne Reaction of 3, 3′-Bis (phenylethynyl)-2, 2′-bithiophene Derivatives via Rhodium Complexes: A novel approach to condensed benzo [2, 1-b: 3, 4-b′] dithiophenes. *Helvetica Chimica Acta*, 80(1), 111–120. DOI: 10.1002/hlca.19970800111

Dienes, Y., Eggenstein, M., Kárpáti, T., Sutherland, T. C., Nyulászi, L., & Baumgartner, T. (2008). Phosphorus-Based Heteropentacenes: Efficiently Tunable Materials for Organic n-Type Semiconductors. *Chemistry (Weinheim an der Bergstrasse, Germany)*, 14(32), 9878–9889. DOI: 10.1002/chem.200801549 PMID: 18844201

Du Bois, D. J. (2003). *U.S. Patent No. 6,576,653.* Washington, DC: U.S. Patent and Trademark Office.

Du Bois, D. J. (2004). *U.S. Patent No. 6,828,343*. Washington, DC: U.S. Patent and Trademark Office.

El-Atawy, M. A., Ferretti, F., & Ragaini, F. (2017). Palladium-Catalyzed Intramolecular Cyclization of Nitroalkenes: Synthesis of Thienopyrroles. *European Journal of Organic Chemistry*, 2017(14), 1902–1910. DOI: 10.1002/ejoc.201700165

Eras, J., Galvez, C., & Garcta, F. (1984). Reactivity of thienopyrroles. Synthesis of isomeric nitro and bromothienopyrroles. *Journal of Heterocyclic Chemistry*, 21(1), 215–217. DOI: 10.1002/jhet.5570210143

Fang, B., Hu, C., Ding, Y., Qin, H., Luo, Y., Xu, Z., Meng, J., & Chen, Z. (2021). Discovery of 4 H-thieno [3, 2-b] pyrrole derivatives as potential anticancer agents. *Journal of Heterocyclic Chemistry*, 58(8), 1610–1627. DOI: 10.1002/jhet.4285

Fu, D. J., Li, J., & Yu, B. (2021). Annual review of LSD1/KDM1A inhibitors in 2020. *European Journal of Medicinal Chemistry*, 214, 113254. DOI: 10.1016/j.ejmech.2021.113254 PMID: 33581557

Gao, J. H., Li, R. J., Li, L. Q., Meng, Q., Jiang, H., Li, H. X., & Hu, W. P. (2007). High-Performance Field-Effect Transistor Based on Dibenzo [d, d′] thieno [3, 2-b; 4, 5-b′] dithiophene, an Easily Synthesized Semiconductor with High Ionization Potential. *Advanced Materials*, 19(19), 3008–3011. DOI: 10.1002/adma.200701167

Gao, P., Cho, D., Yang, X., Enkelmann, V., Baumgarten, M., & Müllen, K. (2010). Heteroheptacenes with fused thiophene and pyrrole rings. *Chemistry (Weinheim an der Bergstrasse, Germany)*, 16(17), 5119–5128. DOI: 10.1002/chem.200903562 PMID: 20349476

Garcia, F., & Gálvez, C. A. R. M. E. N. (1985). The synthesis of hienopyrroles. *Synthesis*, 1985(02), 143–156. DOI: 10.1055/s-1985-31136

Ghosh, S., Bedi, A., & Zade, S. S. (2015). Thienopyrrole and selenophenopyrrole donor fused with benzotriazole acceptor: Microwave assisted synthesis and electrochemical polymerization. *RSC Advances*, 5(7), 5312–5320. DOI: 10.1039/C4RA14850A

Gupta, A., & Flynn, B. L. (2017). Linear and angular heteroacenes from double-electrophilic cyclization (DEC) and DEC-reductive elimination of diynes. *Organic Letters*, 19(8), 1939–1941. DOI: 10.1021/acs.orglett.7b00265 PMID: 28388059

Han, D., Lu, J., Fan, B., Lu, W., Xue, Y., Wang, M., Liu, T., Cui, S., Gao, Q., Duan, Y., & Xu, Y. (2024). Lysine-Specific Demethylase 1 Inhibitors: A Comprehensive Review Utilizing Computer-Aided Drug Design Technologies. *Molecules (Basel, Switzerland)*, 29(2), 550. DOI: 10.3390/molecules29020550 PMID: 38276629

Hart, C. J., Riches, A. G., Tiash, S., Abraham, R., Fayd'Herbe, K., Joch, E., Zulfiqar, B., Sykes, M. L., Avery, V. M., Šlapeta, J., Abraham, S., Ryan, J. H., & Skinner-Adams, T. S. (2023). Thieno [3, 2-b] pyrrole 5-carboxamides as potent and selective inhibitors of Giardia duodenalis. *International Journal for Parasitology. Drugs and Drug Resistance*, 23, 54–62. DOI: 10.1016/j.ijpddr.2023.09.002 PMID: 37776606

Huang, J., & Yu, G. (2021). Structural engineering in polymer semiconductors with aromatic N-heterocycles. *Chemistry of Materials*, 33(5), 1513–1539. DOI: 10.1021/acs.chemmater.0c03975

Jiang, H., Oniwa, K., Xu, Z., Bao, M., Yamamoto, Y., & Jin, T. (2017). Synthesis and properties of dicyanomethylene-endcapped thienopyrrole-based quinoidal S, N-heteroacenes. *Bulletin of the Chemical Society of Japan*, 90(7), 789–797. DOI: 10.1246/bcsj.20170083

Jiang, J. K., Boxer, M. B., Vander Heiden, M. G., Shen, M., Skoumbourdis, A. P., Southall, N., Veith, H., Leister, W., Austin, C. P., Park, H. W., Inglese, J., Cantley, L. C., Auld, D. S., & Thomas, C. J. (2010). Evaluation of thieno [3, 2-b] pyrrole [3, 2-d] pyridazinones as activators of the tumor cell specific M2 isoform of pyruvate kinase. *Bioorganic & Medicinal Chemistry Letters*, 20(11), 3387–3393. DOI: 10.1016/j.bmcl.2010.04.015 PMID: 20451379

Jung, I. H., Kim, J. H., Nam, S. Y., Lee, C., Hwang, D. H., & Yoon, S. C. (2015). Development of new photovoltaic conjugated polymers based on di (1-benzothieno) [3, 2-b: 2′, 3′-d] pyrrole: benzene ring extension strategy for improving open-circuit voltage. *Macromolecules*, 48(15), 5213–5221. DOI: 10.1021/acs.macromol.5b01129

Jung, I. H., Kim, J. H., Nam, S. Y., Lee, C., Hwang, D. H., & Yoon, S. C. (2016). A di (1-benzothieno)[3, 2-b: 2′, 3′-d] pyrrole and isoindigo-based electron donating conjugated polymer for efficient organic photovoltaics. *Journal of Materials Chemistry. C, Materials for Optical and Electronic Devices*, 4(4), 663–667. DOI: 10.1039/C5TC04220H

Kumar, A., Kumar, B., & Bhatia, R. (2023). Design, synthesis, molecular docking, and biological evaluation of isatin-based fused heterocycles as epidermal growth factor receptor inhibitors. *Assay and Drug Development Technologies*, 21(5), 222–233. DOI: 10.1089/adt.2022.120 PMID: 37439798

Kumar, P. R., Raju, S., Goud, P. S., Sailaja, M., Sarma, M. R., Reddy, G. O., Kumar, M. P., Reddy, V. K., Suresh, T., & Hegde, P. (2004). Synthesis and biological evaluation of thiophene [3, 2-b] pyrrole derivatives as potential anti-inflammatory agents. *Bioorganic & Medicinal Chemistry*, 12(5), 1221–1230. DOI: 10.1016/j.bmc.2003.11.003 PMID: 14980634

Le, D. T., Truong, N. T. T., Luu, T. H. T., Nguyen, L. T., Hoang, M. H., Huynh, H. P. K., Cu, S. T., Nguyen, Q. T., & Nguyen, H. T. (2022). Donor–acceptor and donor–donor alternating conjugated polymers based on dithieno [3, 2-b: 2′, 3′-d] pyrrole: synthesis, optical properties and organic solar cells applications. *Journal of Polymer Research*, 29(4), 111. DOI: 10.1007/s10965-022-02969-9

Lee, D. J., Kim, K., & Park, Y. J. (2002). Novel Synthesis of 5, 6-Dihydro-4 H-thieno [3, 2-b] pyrrol-5-ones via the Rhodium (II)-Mediated Wolff Rearrangement of 3-(Thieno-2-yl)-3-oxo-2-diazopropanoates. *Organic Letters*, 4(6), 873–876. DOI: 10.1021/ol016995n PMID: 11893174

Martin Hernando, J. I., Ontoria, J. M., Malancona, S., Attenni, B., Fiore, F., Bonelli, F., Koch, U., Di Marco, S., Colarusso, S., Ponzi, S., Gennari, N., Vignetti, S. E., del Rosario Rico Ferreira, M., Habermann, J., Rowley, M., & Narjes, F. (2009). Optimization of thienopyrrole-based finger-loop inhibitors of the hepatitis C virus NS5B polymerase. *ChemMedChem*, 4(10), 1695–1713. DOI: 10.1002/cmdc.200900184 PMID: 19672916

Mitsudo, K., Shimohara, S., Mizoguchi, J., Mandai, H., & Suga, S. (2012). Synthesis of nitrogen-bridged terthiophenes by tandem Buchwald–Hartwig coupling and their properties. *Organic Letters*, 14(11), 2702–2705. DOI: 10.1021/ol300887t PMID: 22594835

Nandi, N. K., Bhatia, R., Chauhan, G., Kalra, S., & Kumar, B. (2024). Cyclin-dependent kinase 4 and 6 in cancer: Exploration of CDK4/6 inhibitors as anticancer agents. In *Current Molecular Targets of Heterocyclic Compounds for Cancer Therapy* (pp. 115–137). Academic Press. DOI: 10.1016/B978-0-323-96121-9.00006-1

Parmentier, J. G., Portevin, B., Golsteyn, R. M., Pierré, A., Hickman, J., Gloanec, P., & De Nanteuil, G. (2009). Synthesis and CHK1 inhibitory potency of hymenialdisine analogues. *Bioorganic & Medicinal Chemistry Letters*, 19(3), 841–844. DOI: 10.1016/j.bmcl.2008.12.001 PMID: 19111462

Patel, S. A., Nilsson, M. B., Le, X., Cascone, T., Jain, R. K., & Heymach, J. V. (2023). Molecular mechanisms and future implications of VEGF/VEGFR in cancer therapy. *Clinical Cancer Research*, 29(1), 30–39. DOI: 10.1158/1078-0432.CCR-22-1366 PMID: 35969170

Qi, T., Guo, Y., Liu, Y., Xi, H., Zhang, H., Gao, X., Liu, Y., Lu, K., Du, C., Yu, G., & Zhu, D. (2008). Synthesis and properties of the anti and syn isomers of dibenzothieno [b, d] pyrrole. *Chemical Communications*, (46), 6227–6229. DOI: 10.1039/b813683a PMID: 19082128

Qin, P., Kast, H., Nazeeruddin, M. K., Zakeeruddin, S. M., Mishra, A., Bäuerle, P., & Grätzel, M. (2014). Low band gap S, N-heteroacene-based oligothiophenes as hole-transporting and light absorbing materials for efficient perovskite-based solar cells. *Energy & Environmental Science*, 7(9), 2981–2985. DOI: 10.1039/C4EE01220H

Rasmussen, S. C. (2016). The nomenclature of fused-ring arenes and heterocycles: A guide to an increasingly important dialect of organic chemistry. *Chemtexts*, 2(4), 1–13. DOI: 10.1007/s40828-016-0035-3

Romussi, A., Cappa, A., Vianello, P., Brambillasca, S., Cera, M. R., Dal Zuffo, R., Fagà, G., Fattori, R., Moretti, L., Trifirò, P., Villa, M., Vultaggio, S., Cecatiello, V., Pasqualato, S., Dondio, G., So, C. W. E., Minucci, S., Sartori, L., Varasi, M., & Mercurio, C. (2020). Discovery of reversible inhibitors of KDM1A efficacious in acute myeloid leukemia models. *ACS Medicinal Chemistry Letters*, 11(5), 754–759. DOI: 10.1021/acsmedchemlett.9b00604 PMID: 32435381

Sabnis, R. W. (2023). Novel Thienopyrrole Compounds for Treating Autoimmune Diseases and Inflammatory Conditions. *ACS Medicinal Chemistry Letters*, 14(5), 549–550. DOI: 10.1021/acsmedchemlett.3c00131 PMID: 37197455

Samsonenko, A. L., Kostyuchenko, A. S., Zheleznova, T. Y., Shuvalov, V. Y., Vlasov, I. S., & Fisyuk, A. S. (2022). Synthesis of new Fused 4H-Thieno [3, 2-b] pyrrole derivatives via decomposition of Methyl 4-Azido-5-arylthiophene-2-carboxylates. *Synthesis*, 54(14), 3227–3238. DOI: 10.1055/a-1799-9339

Shirinian, V. Z., Nikal, D. M., & Krayushk, M. M. (2003). Synthesis of 4-aryl-2, 6, 6-trimethyl-5-oxo-5, 6-dihydro-4 H-thieno [3, 2-b] pyrroles. *Russian Chemical Bulletin*, 52(8), 1873–1876. DOI: 10.1023/A:1026041728967

Srebrenkoska, V., Bogoeva-Gaceva, G., & Dimeski, D. (2013). Bio-based composites produced by conventional techniques using recycled polymer matrix.

Stockmann, V., & Fiksdahl, A. (2008). Preparation of new pyrido [3, 4-b] thienopyrroles and pyrido [4, 3-e] thienopyridazines. *Tetrahedron*, 64(32), 7626–7632. DOI: 10.1016/j.tet.2008.05.062

Tomašič, T., Nabergoj, D., Vrbek, S., Zidar, N., Jakopin, Ž., Žula, A., Hodnik, Ž., Jukič, M., Anderluh, M., Ilaš, J., Dolenc, M. S., Peluso, J., Ubeaud-Séquier, G., Muller, C. D., Mašič, L. P., & Kikelj, D. (2015). Analogues of the marine alkaloids oroidin, clathrodin, and hymenidin induce apoptosis in human HepG2 and THP-1 cancer cells. *MedChemComm*, 6(1), 105–110. DOI: 10.1039/C4MD00286E

Torosyan, S. A., Gimalova, F. A., Zagitov, V. V., Erastov, A. S., & Miftakhov, M. S. (2018). Synthesis of N-Substituted Methyl 4 H-Thieno [3, 2-b] pyrrole-5-carboxylates. *Russian Journal of Organic Chemistry*, 54(6), 912–917. DOI: 10.1134/S1070428018060131

Torosyan, S. A., Nuriakhmetova, Z. F., Gimalova, F. A., & Miftakhov, M. S. (2019). 4H-Thieno [3, 2-b] pyrrole-5-carboxylate Conjugates with Taurine and Its Tetrabutylammonium Salt. *Russian Journal of Organic Chemistry*, 55(12), 1902–1906. DOI: 10.1134/S1070428019120157

Valiev, R. R., Fliegl, H., & Sundholm, D. (2014). The aromatic character of thienopyrrole-modified 20π-electron porphyrinoids. *Physical Chemistry Chemical Physics*, 16(22), 11010–11016. DOI: 10.1039/c4cp00883a PMID: 24776774

Wani, T. A., Zargar, S., Alkahtani, H. M., Altwaijry, N., & Al-Rasheed, L. S. (2023). Anticancer potential of sulfonamide moieties via in-vitro and in-silico approaches: Comparative investigations for future drug development. *International Journal of Molecular Sciences*, 24(9), 7953. DOI: 10.3390/ijms24097953 PMID: 37175658

Welch, M., & Phillips, R. S. (1999). Improved Syntheses of [3, 2-b]-and [2, 3-b]-fused Selenolo-and Thienopyrroles, and of Furo [3, 2-b] pyrrole. *Heterocyclic Communications*, 5(4), 305–310. DOI: 10.1515/HC.1999.5.4.305

Wetzel, C., Mishra, A., Mena-Osteritz, E., Liess, A., Stolte, M., Würthner, F., & Bäuerle, P. (2014). Synthesis and structural analysis of thiophene-pyrrole-based S, N-heteroacenes. *Organic Letters*, 16(2), 362–365. DOI: 10.1021/ol403153z PMID: 24359234

Xu, Y., He, Z., Liu, H., Chen, Y., Gao, Y., Zhang, S., Wang, M., Lu, X., Wang, C., Zhao, Z., Liu, Y., Zhao, J., Yu, Y., & Yang, M. (2020). 3D-QSAR, molecular docking, and molecular dynamics simulation study of thieno [3, 2-b] pyrrole-5-carboxamide derivatives as LSD1 inhibitors. *RSC Advances*, 10(12), 6927–6943. DOI: 10.1039/C9RA10085G PMID: 35493862

Yin, S. Q., Wang, J. J., Zhang, C. M., & Liu, Z. P. (2012). The development of MetAP-2 inhibitors in cancer treatment. *Current Medicinal Chemistry*, 19(7), 1021–1035. DOI: 10.2174/092986712799320709 PMID: 22229417

Zhang, X., Yan, J., Wang, H., Wang, Y., Wang, J., & Zhao, D. (2021). Molecular docking, 3D-QSAR, and molecular dynamics simulations of thieno [3, 2-b] pyrrole derivatives against anticancer targets of KDM1A/LSD1. *Journal of Biomolecular Structure & Dynamics*, 39(4), 1189–1202. DOI: 10.1080/07391102.2020.1726819 PMID: 32036765

Zhang, X., Yan, J., Wang, H., Wang, Y., Wang, J., & Zhao, D. (2021). Molecular docking, 3D-QSAR, and molecular dynamics simulations of thieno [3, 2-b] pyrrole derivatives against anticancer targets of KDM1A/LSD1. *Journal of Biomolecular Structure & Dynamics*, 39(4), 1189–1202. DOI: 10.1080/07391102.2020.1726819 PMID: 32036765

Zhong, L., Li, Y., Xiong, L., Wang, W., Wu, M., Yuan, T., Yang, W., Tian, C., Miao, Z., Wang, T., & Yang, S. (2021). Small molecules in targeted cancer therapy: Advances, challenges, and future perspectives. *Signal Transduction and Targeted Therapy*, 6(1), 1–48. DOI: 10.1038/s41392-021-00572-w PMID: 34054126

Chapter 5
Examining Biological Relevence of Thiazolo Thiadiazole and Thiazolo Thiazole Fused Heterocyles

Vikrant Sudhakar Palekar
Gokhale Education Society's, Arts, Commerce, and Science College, India

Kulbhushan A. Sasane
Abasaheb Marathe Arts and New Commerce Science College, India

ABSTRACT

Aromatic Fused heterocycles containing Nitrogen (N) and Sulfur(S) has extraordinary bioactive behavior due to which they are attracting the interest of chemists world-wide & thus used as a potential material in medicinal/pharmaceutical industries. The nitrogen-sulphur fused heterocycles attracting attention of the researchers due to number of biological and pharmacological activities and observed to exhibit extensive applications in drug discovery and development. The efforts are made to understand biological activity based on position of heteroatom & the substituent present in 5-member aromatic heterocyles. These compounds reported to having anticancer, anticonvulsant, antimicrobial, and antitubercular activities. This chapter aims to provide comprehensive overview on their hidden potential and inspiring further exploration into their biological relevance with respect to thiazolo thiadi-azole & thiazolo thiazole derivatives are possessing considerable prospects in the field of drug development & material Science.

DOI: 10.4018/979-8-3693-7520-4.ch005

1. INTRODUCTION

Fused heterocycles with nitrogen-sulfur containing compounds are fascinating class in organic chemistry, where two or more heterocyclic rings share one or more common atoms. This fusion creates structures with unique chemical and physical properties, making them important in various fields such as medicinal, pharmaceutical, materials science, and organic synthesis. Among the fused five-membered bridgehead-nitrogen heterocycles, thiazolo thiadiazole & thiazolo thiazole with different substitution pattern form an important and potentially bioactive subclass due to their extensive use in medicinal chemistry as scaffolds as active agents (Kumar et al., 2010; Khan et al., 1988; Saundane et al. 2009).

Substituted thiazolothiazole heterocyles has gained attention due to its planar-rigid backbone having extended π-conjugation along with presence of fused five-membered ring containing nitrogen-sulphur (2N,2S). It shows biological activities like antimicrobial, anticancer, antitubercular, anti-inflammatory, analgesic, anticonvulsant, etc (Bevk et al., 2013; Reginato et al., 2016; Yadav et al., 1994; Hagiwara et. al., 1993; Gao et. al., 2013; Kumar et. al., 2017; Sarkar et. al., 2018). Thus, this scaffold explores and expand large potential for developing new therapeutic agents across a range of its biological applications.

Derivatives of thiazolo thiazole heterocycles are significant in medicinal chemistry, material science, and organic synthesis due to their diverse biological activities and chemical properties. In literature it has been reported that substituted thiazolo thiazole derivatives as potential biological applications such as anticancer, antimicrobial, anti-inflammatory, antioxidant activity (Kumar et. al., 2014; Zhang et. al., 2016; Koshelev et. al., 2023; Reddy et. al., 2018). As per the thiadiazole moiety [1,2,4-thiadiazoles or 1,3,4-thiadiazoles] arrangements of nitrogen & sulphur heteroatoms and the joining at bridgehead position of thiazole ring and thiadiazole ring the different structures are categorized **(Figure 1)**.

2. PROPERTIES

Fused heterocylces moieties with presence nitrogen and sulfur heteroatom has hydrogen bonding places & acts as electron donor system in certain context. If electronic properties are considered for thiazolo thiadiazole structures thiazole- thiadiazole rings are found to more electron-deficient due to the presence of heteroatoms, as they withdraw electron density from the ring. The electron-donor properties can be attributed due to conjugation effects within the heterocyclic rings or interactions with other electron-accepting species (Khan, 2017; Patel, 2016, Lee, 2015). The nitrogen and sulfur atoms in thiazolo thiadiazoles & thiazolo thiazole heterocycles shows electronic properties as they act as ligands & thus can form complexes with metal ions that has applications in catalysis and material science (Yuan, 2019).[18] This electron deficiency of these fused heterocycles can make these compounds reactive towards nucleophiles and electrophiles & exhibit range of chemical reactivities that can be harnessed for synthetic applications and in the development of new materials and pharmaceuticals (Li, 2019; Mojumdar, 2021).

Figure 1. Different structures of thiazolo thiadiazole & thiazolo thiazole fused heterocycles

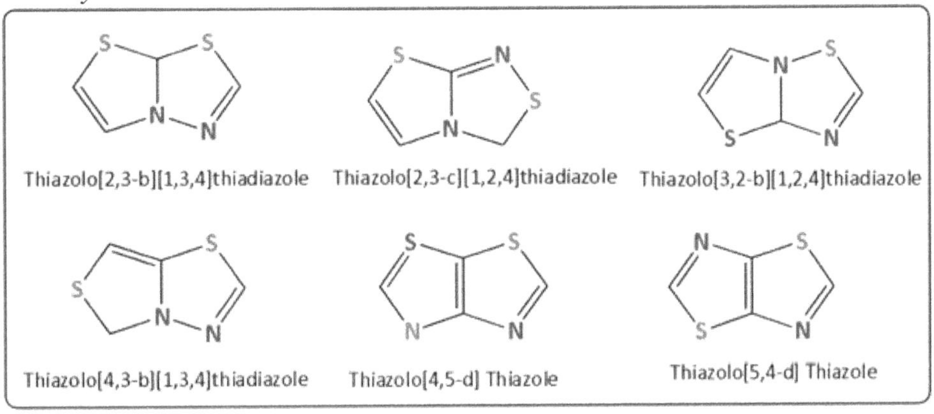

3. RECENT SYNTHESIS THIAZOLO THIADIAZOLE & THIAZOLO THIAZOLE HETEROCYLES

3.1 Synthesis of Substituted Thiazolo[4,3-b][1,3,4]-thiadiazol

When an equal amounts of an aromatic aldehyde **1**, thioglycolic acid & thiosemicarbazide mixture is refluxed by adding little amount of conc. H_2SO_4 the cyclized product 5-Aryl-[1,3]-thiazolo[4,3-b][1,3,4]-thiadiazole-2-amines are obtained with 72-91% yield **(Scheme 1)** (Saundane et. al., 2009; Ramprasad et. al., 2016; Khanet. al., 1988).

In an alternate method, the intramolecular cyclization aromatic aldehyde **1**, thioglycolic acid and methyl hydrazine carbodithioate was refluxed by adding little amount of conc. H_2SO_4 to produce product 5-aryl-2-(methyl sulfanyl)-[1,3]-thiazolo-[4,3-b][1,3,4]-thiadiazole (**7**) with 71-87% yield under similar conditions. **(Scheme 2)** (Shukurpov et. al. 1996; Bayoumy et. al., 1991; Bakr et. al., 2014).

Figure 2. Scheme 1

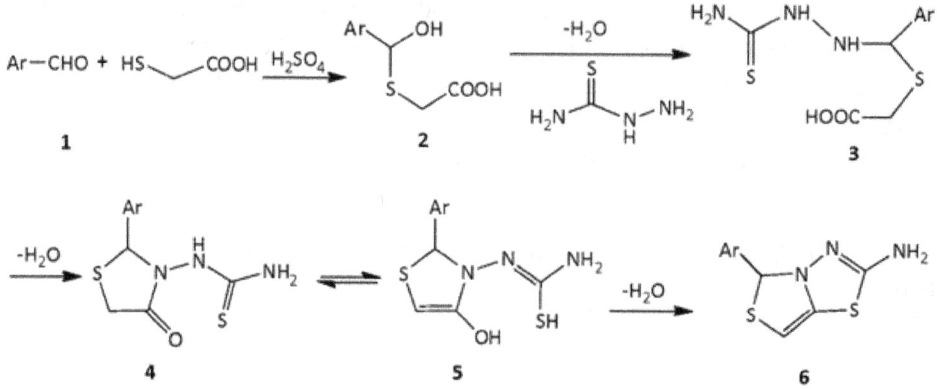

Figure 3. Scheme 2

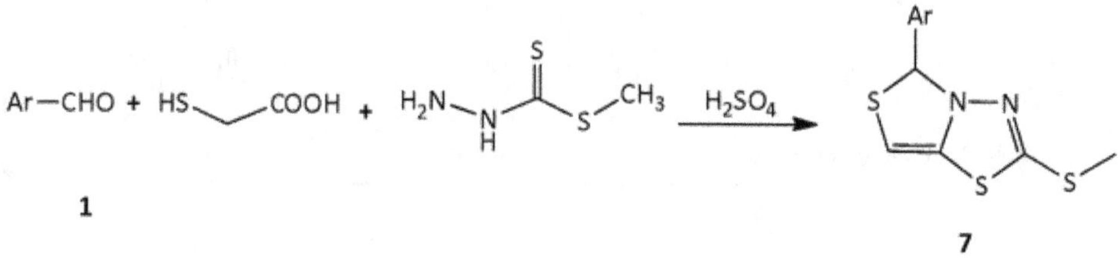

3.2 Synthesis of Substituted [1,3] thiazolo-[2,3-*b*][1,3,4]-thiadiazol

Kochhar et al. reported the synthesis in which reaction of the sulphanylthiadiazole **8** is refluxed with the phenacyl bromide in ethanol to give product [1,3]thiazolo-2,3-b][1,3,4]- thiadiazol **9** with yield upto 32% **(Scheme 3)** (Kochhar et. al., 1973).

Figure 4. Scheme 3

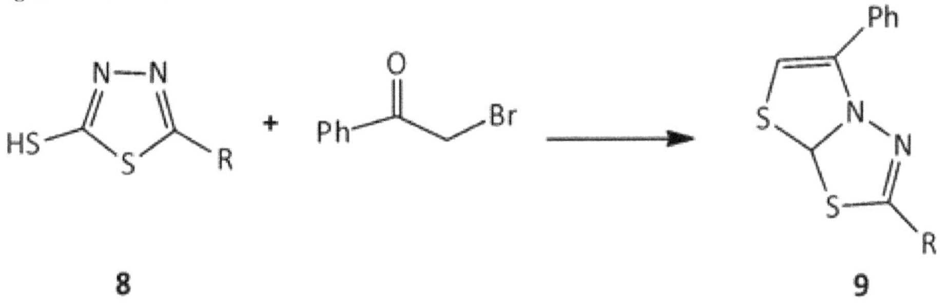

Molina et al. reported Aza-Wittig reaction by reacting iminophosphorane **10** with phenyl isothiocyanate by using solvent dry CH_2Cl_2 to obtained product [1,3]thiazolo-[2,3-*b*][1,3,4]-thiadiazol **11** with 48-56% yield **(Scheme 4)** (Molina et. al., 1992).

Figure 5. Scheme 4

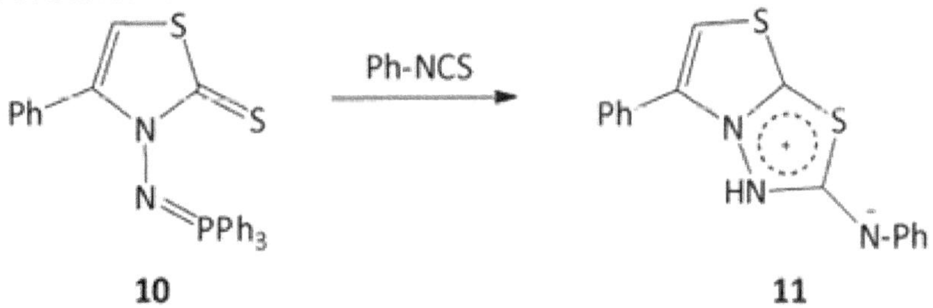

3.3 Synthesis of Substituted [1,3] Thiazolo[3,2-b][1,2,4]thiadiazoles

Labbe et al. reported addition of 2-benzyl-5-chloro-1,2,4-thiadiazol-3(2H)-one **12** with 4,5-dihydro-1,3-thiazol-2-amine **13** to obtained product substituted carbamoyllimino [1,3]Thiazolo[3,2-b][1,2,4]-thiadiazoles **14** compound (**Scheme 5**) (Labbe et. al., 1993).

Figure 6. Scheme 5

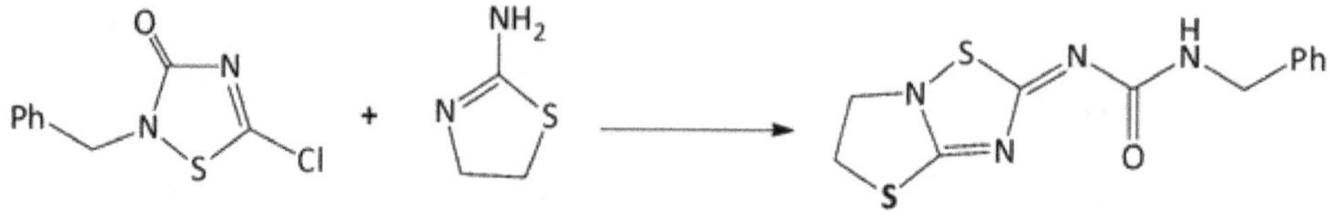

12 **13** **14**

Hagiwara et al. proposed ring closure of 3-(N-aryl-thiocarbamoyl)-2-iminothiazolidines **15** in presence of conc. H_2SO_4 to give the product 3-arylimino-thiazolo[2,3-c][1,2,4]-thiadiazoles **16** compound **(Scheme 6)** (Hagiwara et. al., 1993).

Figure 7. Scheme 6

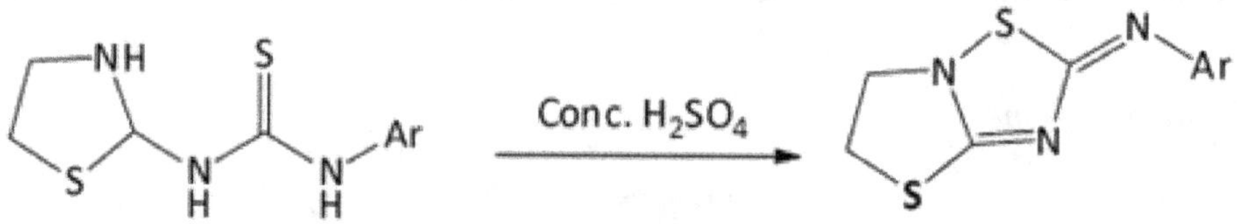

15 **16**

3.4 Synthesis of Substituted [1,3] thiazolo[2,3-c][1,2,4]-thiadiazoles

Hagiwara et. al. in their further study proposed the ring closure of 3-(N-aryl-thiocarbamoyl)-2-iminothiazolidines **17** to obtained the product 3-arylimino[1,3]-thiazolo-[2,3-c][1,2,4]-thiadiazoles **18** compound **(Scheme 7)** (Gardner et. al., 1979).

Figure 8. Scheme 7

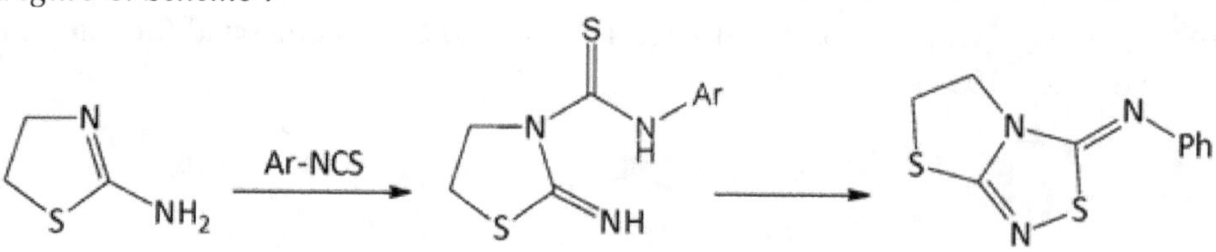

13 **17** **18**

On reacting compound [(trichloromethyl) sulphanyl] amino thiazoles **19** with the sunstituted anilines in chloroform by adding catalytic amount of triethylamine and sodium carbonate to obtained the product [1,3]-thiazolo[2,3-c][1,2,4]-thiadiazoles **20** with 10-95% yield **(Scheme 8)** (Potts et. al., 1975).

Figure 9. Scheme 8

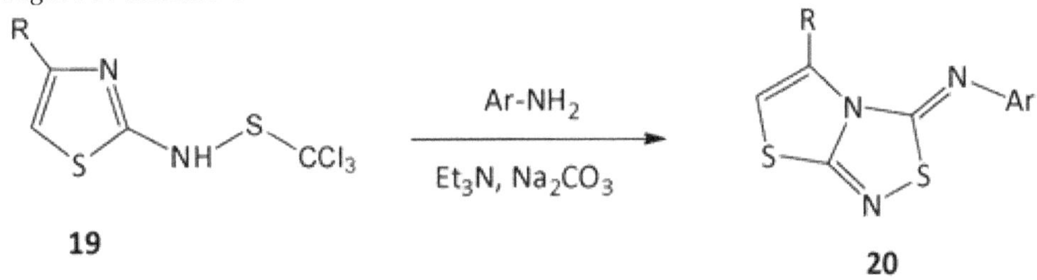

3.5 Synthesis of Substituted thiazolo[4,5-*d*]-thiazoles

As per the earlier report the Jacobsen cyclization was reported for the synthesis of thiazolo[4,5-*d*]-thiazoles which involves cyclisation of substituted thioamide **(21)** to the desired product 2,5-dimethyl-[1,3]thiazolo[4,5-d] [1,3]thiazole **(22)**.[33] The recent synthesis was reported by using potassium ferricyanide as an oxidant in alkaline of aldehydes & N-methylation to give conjugated products **(Scheme 9)** (Abo-Bakret. al., 2015).

Figure 10. Scheme 9

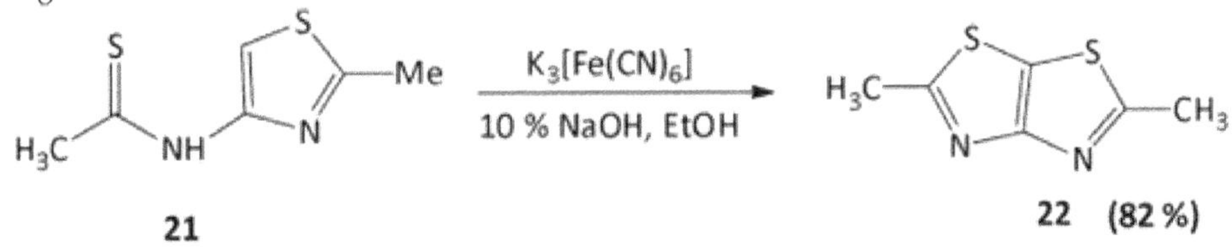

Demydchuk et. al. reacted N-(1-amino-2,2-dichloro ethyl)amides **(23)** with aryl isothiocyanates by adding Lawesson's reagent to give 2,5-disubstituted 3a,6a-dihydro[1,3]-thiazolo[4,5-*d*]-[1,3]-thiazoles compound **(24)** **(Scheme 10)** (Demydchuk et. al., 2011).

Figure 11. Scheme 10

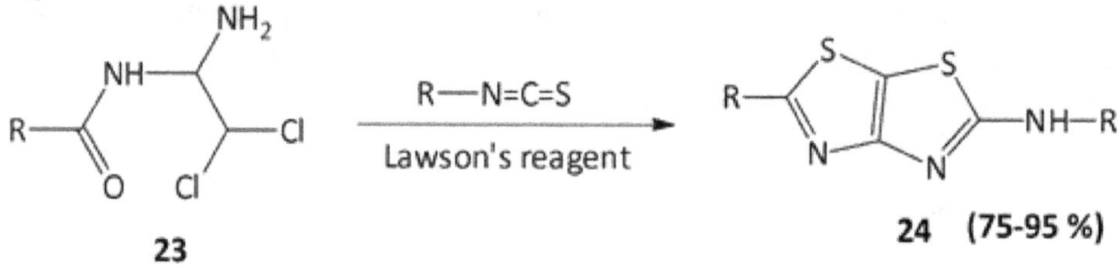

3.6 Synthesis of Substituted thiazolo[5,4-d]thiazoles

The condensation reaction was reported between dithiooxamide **(25)** and two moles of aryl aldehydes to give synthesis of bicyclic substituted thiazolo[5,4-d] thiazole compound **(26)** as a product **(Scheme 11)** (Johnson et. al., 1970).

Figure 12. Scheme 11

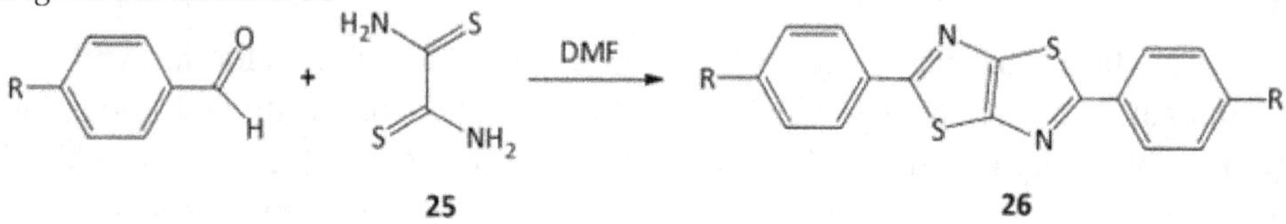

Dessi et al. reported an effective methodology for the synthesis of substituted thiazolo[5,4-d]-thiazole **(27)** by reacting of aldehydes with 1,4-Dithioxamide **(25)** in presence of various oxidants such as DDQ (Sayresmith et. al., 2019), chloranil (Dessi et. al., 2014), and SeO_2 (Papernaya et. al., 2017) under microwave irradiation under solvent free conditions & by using different solvents to obtained desired products in good yields. **(Scheme 12)**

Figure 13. Scheme 12

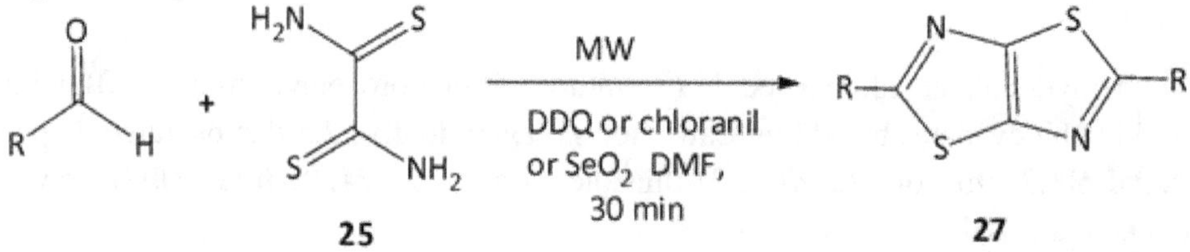

4 BIOLOGICAL APPLICATIONS

4.2 Biological Applications of Thiazolo Thiadiazole Derivatives

4.2.1 Antimicrobial Activity of thiazolo[4,3-b][1,3,4]-thiadiazole

The past research has indicated that thiazolo thiadiazole derivatives and its derivatives exhibit significant antimicrobial activity against a range of pathogens. The biological activity of thiazolo thiadiazole derivatives are associated with existence of the thiadiazole moiety. It has been reported that fused bridgehead heterocycles containing nitrogen & sulphur has different connecting position and the presence of substituents affects potency and the toxicity of these compounds when diverse functionality interacts with the present biological receptors mutually.

Ibrahim et. al. given the method for the preparation of 5-phenyl-5H-thiazolo-[4,3-b][1,3,4]-thiadiazol-2-amine (Mona et. al., 2017) as per the reported work acid catalyzed condensation with 2,4-dihydroxybenzaldehyde heating under reflux in ethanol yielded Schiff's base compounds **(28).** The synthesized Schiff's base then were reacted with thioglycolic acid & thiosemicarbazide in DMF to give substituted derivatives of thiazolo[4,3-*b*][1,3,4]-thiadiazol compounds **(29)** & **(30). (Figure 14)**

Figure 14. Thiazolo[4,3-b][1,3,4]thiadiazol derivatives having antimicrobial activity

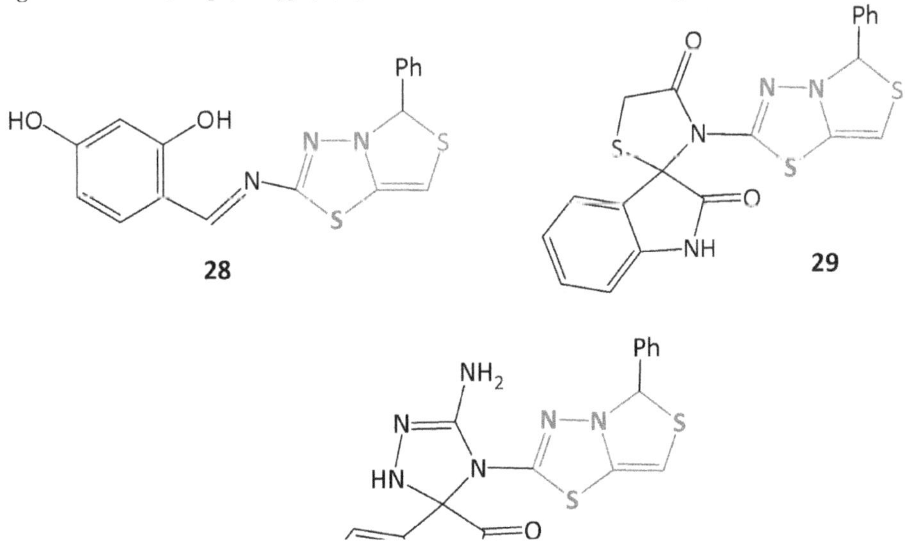

Hamama et.al. synthesized some novel 5-Phenyl-5H-thiazolo[4,3-b][1,3,4]thiadiazole (Wafaa et. al., 2017) systems by reacting 2-amino-5-phenyl-5H-thiazolo[4,3-*b*][1,3,4]-thiadiazoles with the ethyl cyanoacetate, ethyl bromoacetate, monochloroacetic acid respectively to give compounds **31-33 (Figure 15).** These compounds **31-33** are evaluated for their antimicrobial activities against the gram-positive bacteria *Bacillus subtilis* gram-negative bacteria *Escherichia coli* the obtained results evaluated for highest inhibitory activity against the standard samples tested.

Figure 15. Novel thiazolo[4,3-b][1,3,4]-thiadiazol derivatives having antimicrobial activity

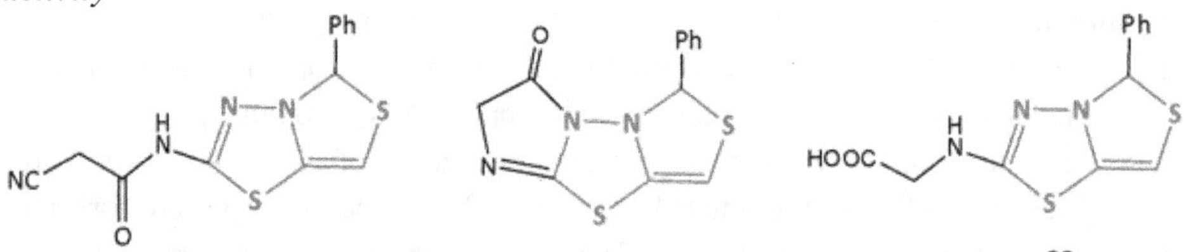

31 32 33

Hamama et.al. also reported synthesis of novel compounds **34-36** by reacting phenylbutanoic acid, malonic acid & acetyl chloride with 5-phenyl-5H-thiazolo[4,3-b][1,3,4]-thiadiazole respectively (Waffa et al., 2017). These compounds were evaluated for antimicrobial activity & evaluated for highest inhibitory activity against *Bacillus subtilis & E-coli strains.* **(Figure 16)**

Figure 16. Compounds showing antimicrobial activity

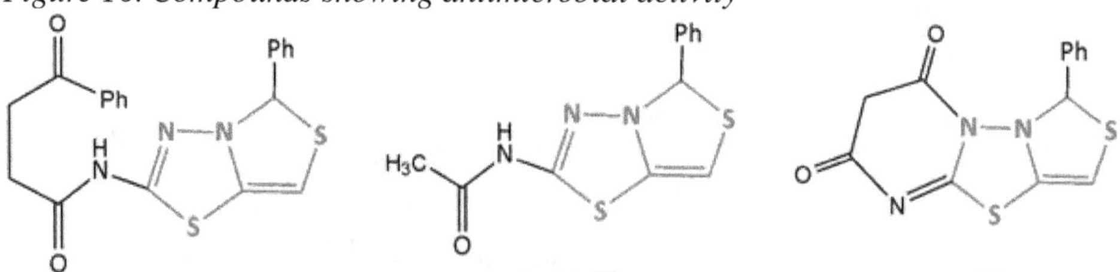

34 35 36

4.2.2 Thiazolo[4,3-b][1,3,4]thiadiazole as Anticancer Activity

Cancer results in the abnormal cell growth and results into spread of cancer cell into another part of body & hence cancer is recognized under agressive diseases. Due to increasing drug resistance and the high exclusion for cancer disease development, the research on anti-cancer drug is intense research area presently (Duwuri et. al., 2006). In recent scientific research studies, it has been observed that a combination of two different bioactive molecules overcome the pharmacokinetic disadvantage associated with the classical drugs (Yahya et. al., 2017).

Abo-salema et.al. reported synthesis of new series of thiazolo[4,3-b]-1,3,4-thiadiazoles attached to steroid moieties. In this work they have synthesized thiazolo[4,3-b]-1,3,4-thiadiazoles through one-pot reaction of N-substituted indole-3-carboxaldehydes with thioglycolic acid and thiosemicarbazide. The product obtained 2-amino-5-(N-substituted-1H-indol-3-yl)-5H-thiazolo[4,3-b][1,3,4]-thiadiazoles is further condensed with acetylated epiandrosterone and progestronee to afford the corresponding Schiff's bases **(37-38)**. The resultant compounds were tested against the lung cancer, breast cancer, colon cancer, prostate cancer & found to shown more activity towards cancer cells compared with the reference drug doxorubicin (Heba et. al., 2021) **(Figure 17).**

Figure 17. Compounds having fused heterocycles attached with steroid showing anticancer activity

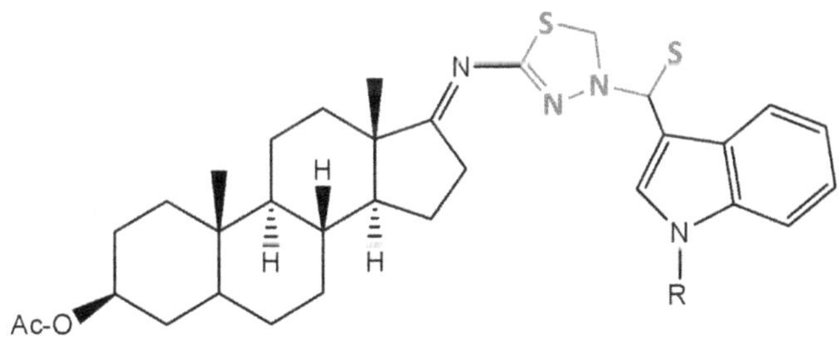

37

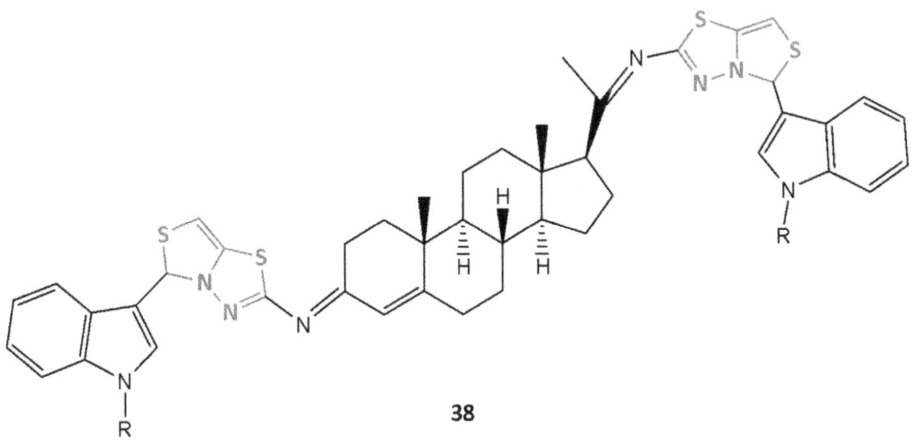

38

4.2.3 Thiazolo[4,3-b][1,3,4]thiadiazole as Antitubercular Activity

Tuberculosis is one of the deadly & infectious diseases that causes lung infection due to pathogenic bacterial species Mycobacterium tuberculosis. Around one-third of the world's population is infected with tuberculosis & declines are reported. The present treatment includes multiple drug doses for at least six months, but it is observed this drug doses are still results positive against pathogen & they could be eradicated completely from body. Thus, there is a serious need to develop novel anti-tuberculosis drugs possessing improved action against mycobactericidal organisms with less toxicity and an eradicating multiple drug-resistant against varied strains.

Malipeddi et.al. reported synthesis various Schiff bases **(39)** by reacting of 2-amino-5-aryl-5H-thiazolo[4,3-b]-1,3,4-thiadiazole with different aryl aldehydes. The prepared Schiff bases were evaluated against *M. tuberculosis* using the microplate alamar blue assay. Further he reacted these Schiff bases with thioglycollic acid in DMF to give substituted thiazolo[4,3-b][1,3,4]thiadiazol-2-yl)-1,3-thiazolidin-4-one compounds **(40) (Figure 18)** (Malipeddi et. al., 2012).

Figure 18. Compounds showing antitubercular activity

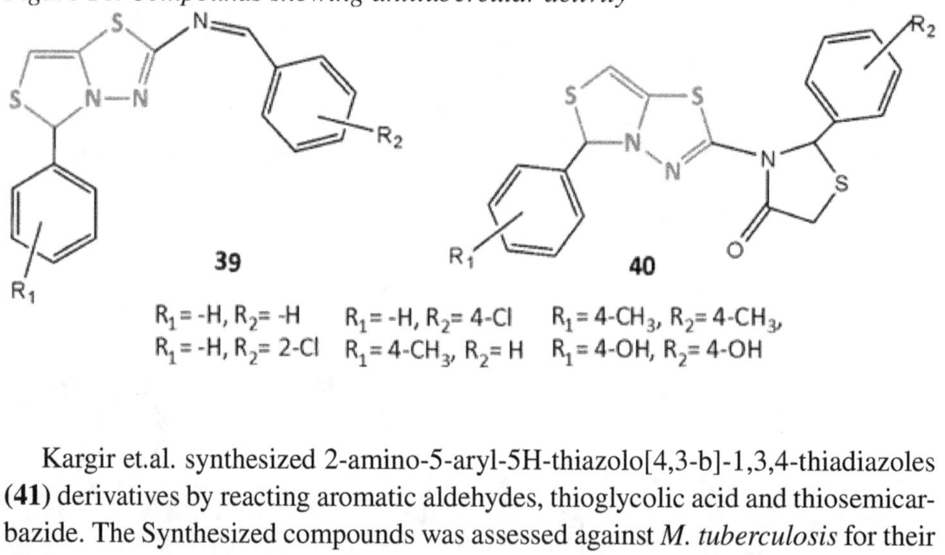

R_1 = -H, R_2 = -H R_1 = -H, R_2 = 4-Cl R_1 = 4-CH_3, R_2 = 4-CH_3,
R_1 = -H, R_2 = 2-Cl R_1 = 4-CH_3, R_2 = H R_1 = 4-OH, R_2 = 4-OH

Kargir et.al. synthesized 2-amino-5-aryl-5H-thiazolo[4,3-b]-1,3,4-thiadiazoles **(41)** derivatives by reacting aromatic aldehydes, thioglycolic acid and thiosemicarbazide. The Synthesized compounds was assessed against *M. tuberculosis* for their antitubercular activity & observed to show better in vitro antitubercular activity against standard used Streptomycin and Pyrazinamide samples **(Figure 19)** (Karigar et. al., 2011).

Figure 19. Compounds showing antitubercular activity

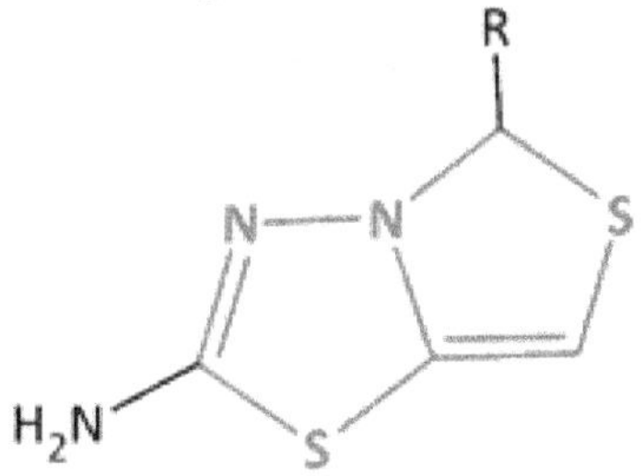

(41)

R= Ph **(a)**, 4-MeC_6H_4 **(b)**, 4-HOC_6H_4 **(c)**, 4-$O_2NC_6H_4$ **(d)**
4-$Me_2NC_6H_4$ **(e)**, 2-ClC_6H_4 **(f)**, 4-ClC_6H_4 **(g)**,
2,4-$Cl_2C_6H_3$ **(h)**, 2-$OMeC_6H_4$ **(i)**, 3-$OMeC_6H_4$

4.2.4 Thiazolo[4,3-b][1,3,4]thiadiazole as Herbicidal Activity

Kant et.al. reported methodology for the synthesis of 5-aryl-2-arylamino-5(H)-thiazolo-[4,3-b]-1,3,4-thiadiazoles & 1,5-diaryl-5(H)-2-mercaptothiazolo-[3,4-b]-1,2,4-triazoles compounds **(42)** & screened for their herbicidal potential against *Argemon maxicana* and *Cyperus rotundus* & compared with commercial herbicide 2,4-Dichlorophenoxyacetic acid (2,4-D) **(Figure 20)** (Kant et. al., 2015).)

Figure 20. Compounds with herbicidal activity

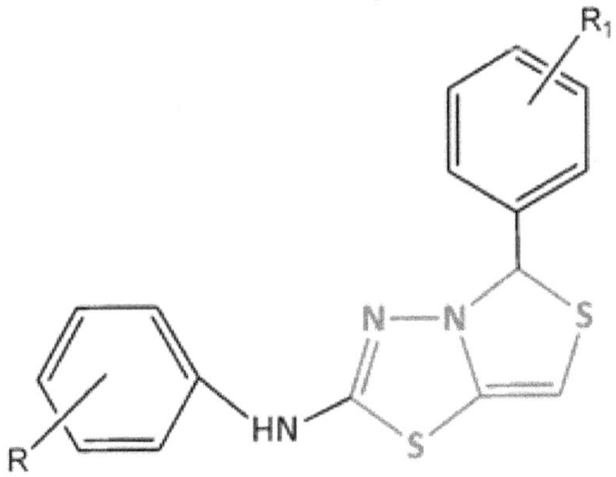

(42)

R = 4-Cl, $\mathbf{R_1}$= 2-Cl
R = 2,4-Cl_2, $\mathbf{R_1}$= 2-Cl

4.3 Biological Applications of Thiazolo Thiazole Derivatives

4.3.1 Thiazolo[5, 4-d]-thiazoles as Antibacterial Activity

Fikrat et.al. reported synthesis of bis-substituted thiazolo [5,4-d] thiazoles by condensation of pyridine-3-aldehyde and 2,2-dimethyl-4-cyanobutyraldehyde derivatives in reflux condition with pyridine. The synthesis was also carried out using quinoline-2-aldehyde with dithiooxamide & also dihydride of thiazolo[5,4-d] thiazole dicarboxylic acid were prepared by reacting corresponding carboxylic acid & hydrazine hydrate. They have also reported diisopropylidene hydrazide of thiazolo[5,4-d] thiazole dicarboxylic acid by reacting dihydrazide of thiazolothiazole dicarboxylic acid. The compounds **(43)** were tested against different strains & found to show moderate antibacterial activity **(Figure 21)** (Fikrat et. al., 1962).

Figure 21. Bis-2,5-disubstituted Thiazolo[5,4-d]-thiazoles Compounds

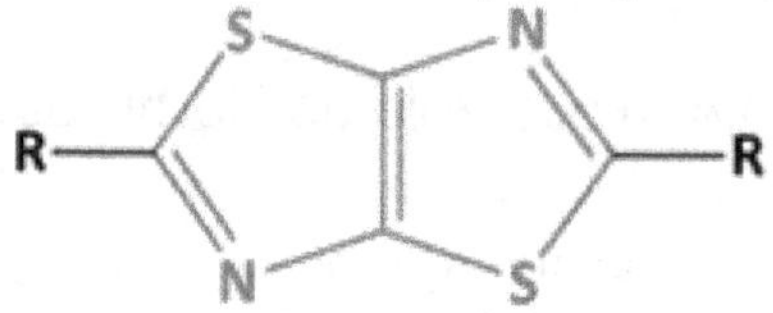

(43)

R = 2-Carhoxy phenyl, **R** = 2.4- Dimethoxy phenyl,
R = 3,4 Dimethoxy phenyl, **R** = 2-Hydroxy-l -naphthyl

Flores et.al. derived new ferrocenyl thiazolo[5,4-d]-thiazole catechol as inhibitor of human 5-lipoxygenase that could be used for the treatment inflammation associated with the diseases and as antibacterial activity against *Staphylococcus aureus.* They have also studied structural relationship and carried out computational studies & found to show antibacterial activity. First time they have reported synthesis of asymmetric organometallic thiazolo-[5,4-d]-thiazole by reacting ferrocene carboxaldehyde, dithiooxamide giving asymmetric ferrocenyl thiazolo-[5,4-d]-thiazole-catechol complex **(44)**. The study was also carried out by taking 3,4-dihydroxybenzaldehyde & dithiooxamide under reflux condition to give compound **(45).** The study was compared with ferrocenyl complex and the catechol precursor & it was found use of ferrocenyl complex enhances the antibacterial activity against gram-positive bacterial strain *Staphylocoocus Aureus***(Figure 22)** (Erick et. al. 2020).

Figure 22. Novel compounds with antibacterial activity

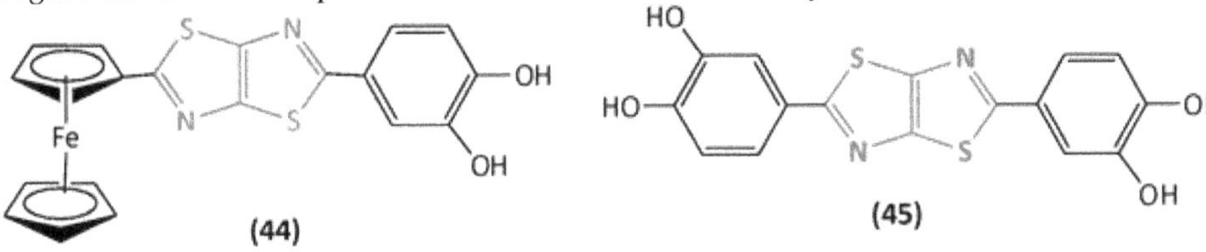

(44) (45)

4.2.2 Thiazolo[5,4-d]-thiazoles as Electron Transporting Properties

Researchers are working to produce renewable energy sources as it is difficult to depends on just fossil fuels as they have limited availability across the world & day by day there is increasing demand of energy sources with growing population. From few decades recent research demonstrated the transformation of the solar energy from traditional inorganic solar cells to organic solar cells due to their cost effective and difficult synthesis routes. Small molecules based organic solar cells with simple structures of donor-acceptor units are now the prime materials for organic solar cells.

The thiazolo-[5,4-d]-thiazole structure has a conjugated system which helps for electron delocalization this property enhances electrons transport. These fused heterocycles has high electron affinity thus accept electrons more feasibly as electron transport is critical in organic electronic devices. Thiazolo[5,4-d]-thiazoles can exhibit a smaller band which results better conductivity and easier electron transport, effective π-stacking, good overlapping of the orbital. These properties prove their usefulness in organic semiconductors and devices like organic light-emitting diodes & organic photovoltaic cells (Peng et. al., 2005; Ando et. al., 2004; Osaka et. al., 2007).

Recently there are increasing reports on synthesis of small molecule compounds and polymeric materials incorporating the thiazolo[5,4-d]-thiazole compounds that has application in various fields, including photovoltaics has been investigated. We herein try to include some of the molecules with applications in the organic, photovoltaics, luminescence etc. There are some reports that studied the application of thiazolothiazole containing compounds **(46-49)** as organic semiconductor materials and in solar cell devices **(Figure 23)** (Roman et. al., 2012; Yuanyuan et. al., 2014).

Figure 23. Thiazolothiazole containing compounds applications in organic semi-conductor materials and in solar cell devices

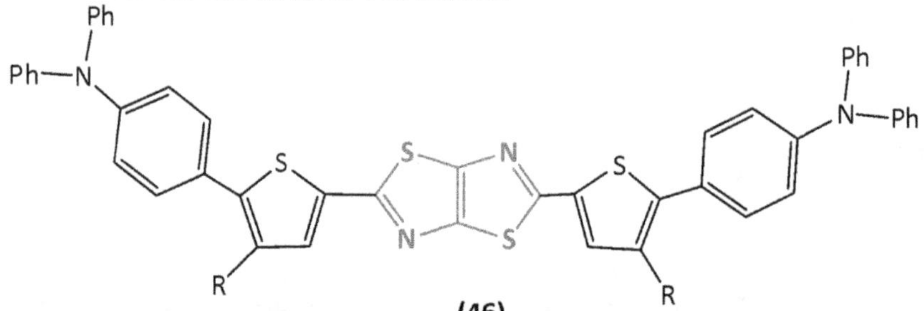

(46)

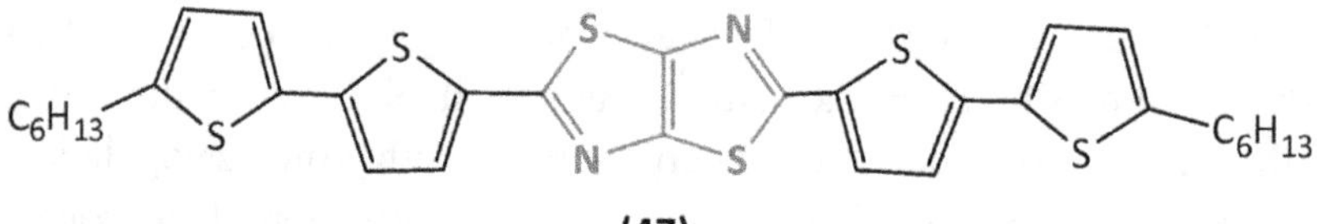

(47)

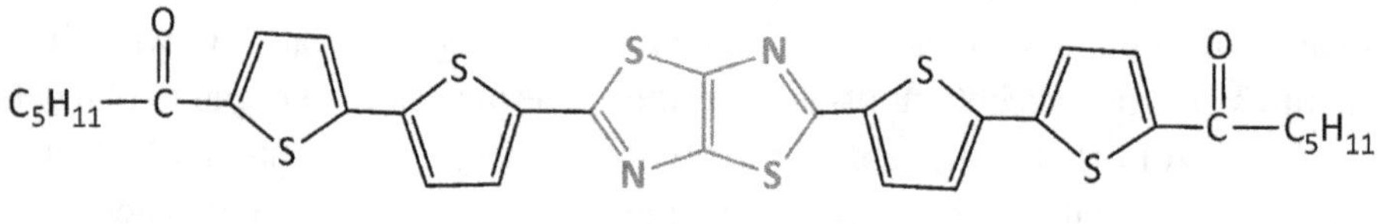

(48)

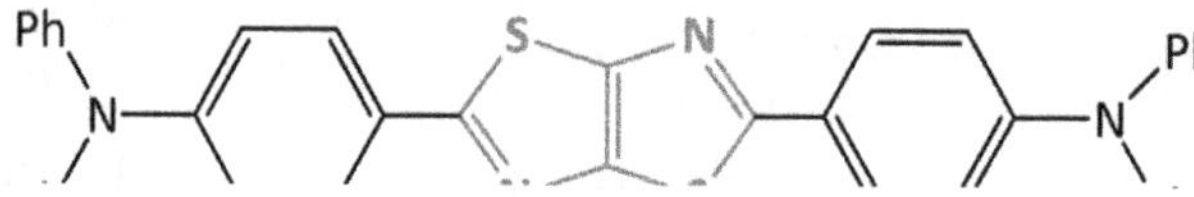

5 CONCLUSION

This chapter discusses the comprehensive overview of biological activities with respect to thiazolo thiadiazole & thiazolo thiazole derivatives that demonstrate them convincing prospects for medicinal chemistry and material science. The proven efficiency of these fused nitrogen & sulphur containing aromatic heterocyles for the treatment of several diseases & confirmed them as promising leaders for further studies as they exhibit range of chemical reactivities towards the development of new materials and pharmaceuticals.

REFERENCES

Abo-Bakr, A., & Abbas-Temirek, H. H. (2015)... *International Journal of Current Research*, 7, 14434.

Abo-salema, H. M., El-kadyb, D. S., Abd-Elhalimb, M. M., Solimanc, A. A. F., Ebaida, M. S., & El-Sawya, E. R. (2021)... *Egyptian Journal of Chemistry*, 64(2), 561–571.

Ando, S., Nishida, J. I., Inoue, Y., Tokito, S., & Yamashita, Y. (2004). Synthesis, physical properties, and field-effect transistors of novel thiophene/thiazolothiazole co-oligomers. *Journal of Materials Chemistry*, 14(12), 1787. DOI: 10.1039/b403699a

Bakr F Abdel-Wahab, S. S. (2014)..*Synthesis*, 46, 1709–1716. DOI: 10.1055/s-0033-1338627

Bayoumy, B. E. (1991). *Acta Poloniae Pharmaceutica*, 48, 13. PMID: 1669421

Bevk, D., Marin, L., Lutsen, L., & Vanderzande, D. (2013). Maes, W. *RSC Advances*, 3, 11418. DOI: 10.1039/c3ra40851e

(2014)... *Chemistry (Weinheim an der Bergstrasse, Germany)*, 3(2), 134–139.

Demydchuk, B. A., Sokolyuk, P. A., Vasilenko, O. M., Brovarets, V. S., & Drach, B. S. (2011)... *J. Org. Pharm. Chem.*, 1, 47.

Dessi, A., Calamante, M., Mordini, A., Zani, L., Taddei, M., & Reginato, G. (2014). Microwave-activated synthesis of thiazolo[5,4-d]thiazoles by a condensation/oxidation sequence. *RSC Advances*, 4(3), 1322–1328. DOI: 10.1039/C3RA45015E

Duwuri, M., Konkar, S., Hong, K. H., Blagg, B. S. J., & Krise, J. P. (2006). *ACS Chemical Biology*, 1(5), 309–315. DOI: 10.1021/cb6001202 PMID: 17163760

Gao, Y., Zhang, J., & Liu, Y. (2013).*European Journal of Medicinal Chemistry*, 66, 49–55.

Gardner, J. O., & Beard, C. C. (1979).Synthesis and Antimicrobial Activity of 3H-1,2,4-Thiadiazolo Fused Heterocycles. *Journal of Pharmaceutical Sciences*, 68(2), 182–185. DOI: 10.1002/jps.2600680216 PMID: 423088

Hagiwara, K., Saitoh, K., Iihama, T., Kawana, T., & Hosaka, H. (1993). Synthesis and Herbicidal Activity of Fused Δ²-1, 2, 4-Thiadiazolines. *Journal of Pesticide Science*, 18(4), 309–318. DOI: 10.1584/jpestics.18.4_309

Hagiwara, K., Saitoh, K., Lihama, T., Kawana, T., & Hosaka, H. (1993). Synthesis and Herbicidal Activity of Fused Δ²-1, 2, 4-Thiadiazolines. *Journal of Pesticide Science*, 18(4), 309–318. DOI: 10.1584/jpestics.18.4_309

Himaja, M., Karigar, A. A., Ramana, M. V., & Sikarwar, M. S. (2012). *Tropical Journal of Pharmaceutical Research*, 11(4), 611–620.

(2011). International Research. *Journal of Pharmaceutics*, 2(1), 153–158.

Johnson, J. R., & Ketcham, R. (1960). Thiazolothiazoles. I. The Reaction of Aromatic Aldehydes with Dithioöxamide. *Journal of the American Chemical Society*, 82(11), 2719–2724. DOI: 10.1021/ja01496a017

Johnson, J. R., Rotenberg, D. H., & Ketcham, R. (1970). Thiazolothiazoles. II. Parent heterocycle and its carboxylic and amino derivatives. *Journal of the American Chemical Society*, 92(13), 4046–4050. DOI: 10.1021/ja00716a036

Khan, M. N. (•••). . . *Journal of Electronic Materials*, 201746(5), 2772–2783.

Khan, R. H., & Rastogi, R. C. (1988). Indian J. Chem. *Sect. B: Org. Chem. Incl. Med. Chem*, 27, 501.

Khan, R. H., & Rastogi, R. C. (1988). Indian J. Chem. *Sect. B: Org. Chem. Incl. Med. Chem.*, 27, 501.

Kochhar, M. M., Salahi-Asbahi, M., & Williams, B. B. (1973). Bicyclic Thiadiazoles I: 2-(2-Thienyl)-5-phenylthiazolo[2,3-b]-1,3,4-thiadiazole. *Journal of Pharmaceutical Sciences*, 62(2), 336–337. DOI: 10.1002/jps.2600620239 PMID: 4686420

Koshelev, V. N., Primerova, O., Stepan, V. V., & Stupnikova Anna, S. (2023). Synthesis and Antioxidant Activity of Novel Thiazole and Thiazolidinone Derivatives with Phenolic Fragments. *Applied Sciences (Basel, Switzerland)*, 13(24), 13112. DOI: 10.3390/app132413112

Kumar, M. P., Ravi, T. K., Chawla, R., Bhuvana, S., Sonia, G., & Gopalakrishnan, S. (2010). Microwave assisted synthesis and biological activity of novel coumarinyltriazolothiadiazoles. *Indian Journal of Pharmaceutical Sciences*, 72(3), 357. DOI: 10.4103/0250-474X.70483 PMID: 21188046

Kumar, P., & Kumar, V. (2017). . . *Journal of Heterocyclic Chemistry*, 54(2), 465–471.

Kumar, R., & Pandey, M. (2014). A comprehensive review on synthetic approach for antimalarial agents. *European Journal of Medicinal Chemistry*, 85, 315–325. DOI: 10.1016/j.ejmech.2014.07.084

Labbe, G., Buelens, J., & Dehaen, W. A. (1993). A new approach to azapentalenes by an addition–rearrangement sequence: Synthesis of fused 1,2,4-thiadiazoles. *Journal of the Chemical Society, Perkin Transactions 1: Organic and Bio-Organic Chemistry*, 1(16), 1825–1826. DOI: 10.1039/P19930001825

Lee, J. H. (2015).. . *Organic Electronics*, 20, 108–115.

Li, H. (2019).. . *Chemical Reviews*, 119(10), 5421–5478.

(2020). Mascayano. *Organometallics*, 39(14), 2672–2681.

Mojumdar, S. (2021).. . *European Journal of Medicinal Chemistry*, 13, 113144.

Molina, P., Arques, A., & Alias, A. (1992). Aza Wittig-type reaction between the iminophosphorane derived from 3-amino-4-phenylthiazole-2(3H)-thione and iso(thio)cyanates: Preparation of mesoionic thiazolo[2,3-b]-1,3,4-thiadiazoles and N,N-bisheteroarylamines. *Tetrahedron*, 48(7), 1285–1298. DOI: 10.1016/S0040-4020(01)90791-4

Mona, E. (2017). Ibrahim, Wafaa S Hamama, Hadeer A Raoof, Hanafi H Zoorob. *Der Pharma Chemica*, 9(12), 28–33.

Osaka, I., Sauve, G., Zhang, R., Kowalewski, T., & McCullough, R. D. (2007). Novel Thiophene-Thiazolothiazole Copolymers for Organic Field-Effect Transistors. *Advanced Materials*, 19(23), 4160–4165. DOI: 10.1002/adma.200701058

Padam, K., Kumar, L. D., & Nisheeth, R. (2015).. . *Indian Journal of Heterocyclic Chemistry*, 25, 91–96.

Papernaya, L. K., Shatrova, A. A., Kletskov, A. V., Petkevich, S. K., Sterkhova, I. V., Klyba, L. V., & Levkovskaya, G. G. (2017). Microwave synthesis of new azolyl-substituted thiazolo[5,4-d]thiazoles. *Russian Journal of Organic Chemistry*, 53(4), 550–556. DOI: 10.1134/S107042801704008X

Patel, S. (2016).. . *Journal of Materials Chemistry*, 4(21), 4856–4866.

Peng, Q., Peng, J.-B., Kang, E. T., Neoh, K. G., & Cao, Y. (2005). Synthesis and Electroluminescent Properties of Copolymers Based on Fluorene and 2,5-Di(2-hexyloxyphenyl)thiazolothiazole. *Macromolecules*, 38(17), 7292–7298. DOI: 10.1021/ma050378n

Potts, K. T., & Kane, J. (1975). Bridgehead nitrogen heterocycles. IX. Fused-ring systems derived from fusion of the 1,2,4-thiadiazole system with the isoxazole, 1,3,4-oxadiazole, thiazole, 1,2,4-thiadiazole, and 1,3,4-thiadiazole systems. *The Journal of Organic Chemistry*, 40(18), 2600–2604. DOI: 10.1021/jo00906a006

Ramprasad J., Nayak N., Dalimba U., Yogeeswari P., Sriram D., *Med. Chem. Commun.* 2016, 7, 338 (2016).

Reddy, M., & Karthik, P. (2018).. . *Journal of Applied Pharmaceutical Science*, 8(3), 19–24.

Reginato, G., Mordini, A., Zani, L., Calamante, M., & Dessi, A. (2016). Photoactive Compounds Based on the Thiazolo[5,4- *d*]thiazole Core and Their Application in Organic and Hybrid Photovoltaics. *European Journal of Organic Chemistry*, 2016(2), 233–251. DOI: 10.1002/ejoc.201501237

Sarkar, R., Singh, S., & Dey, S. (2018).. . *Bioorganic & Medicinal Chemistry Letters*, 28(16), 2613–2618.

Saundane, A. R., Manjunatha, Y., & Walmik, P. (2009). Synthesis and antimicrobial activities of 5-(2′, 5′-disubstituted 1H- indol-3′-yl)-5h-thiazolo[4,3-b]1,3,4-oxadiazole,-1,3,4- thiadiazole-1,2,4-triazole and their derivatives. *Heterocyclic Communications*, 15(4), 303. DOI: 10.1515/HC.2009.15.4.303

Saundane, A. R., Manjunatha, Y., & Walmik, P. (2009). Synthesis and antimicrobial activities of 5-(2′, 5′-disubstituted 1H- indol-3′-yl)-5h-thiazolo[4,3-b]1,3,4-oxadiazole,-1,3,4- thiadiazole-1,2,4-triazole and their derivatives. *Heterocyclic Communications*, 15(4), 303. DOI: 10.1515/HC.2009.15.4.303

Sayresmith, N. A., Saminathan, A., Sailer, J. K., Patberg, S. M., Sandor, K., Krishnan, Y., & Walter, M. G. (2019). Photostable Voltage-Sensitive Dyes Based on Simple, Solvatofluorochromic, Asymmetric Thiazolothiazoles. *Journal of the American Chemical Society*, 141(47), 18780–18790. DOI: 10.1021/jacs.9b08959 PMID: 31660737

Hamama, W. S., Ibrahim, M. E., Raoof, H. A., & Zoorob, H. H. (2017). Synthesis and Antimicrobial Evaluation of Some Novel 5-Phenyl-5H-thiazolo [4, 3-b][1, 3, 4] thiadiazole Systems. *Journal of Heterocyclic Chemistry*, 54(4), 2360–2366.

Yadav, L. D. S., Vaish, A., & Sharma, S. (1994). New Fungitoxic Fused-Ring Synthetics Incorporating Azoles and Azines in Different Combinations. *Journal of Agricultural and Food Chemistry*, 42(3), 811–813. DOI: 10.1021/jf00039a043

Yahya, S. M. M., Abdelhamid, A. O., Abd-Elhalim, M. M., Elsayed, G. H., & Eskander, E. F. (2017). The effect of newly synthesized progesterone derivatives on apoptotic and angiogenic pathway in MCF-7 breast cancer cells. *Steroids*, 126, 15–23. DOI: 10.1016/j.steroids.2017.08.002 PMID: 28797724

Yuan, J. (2019).. . *Materials Science and Engineering R Reports*, 139, 100537.

Zahradnik, P., Magdolen, P., & Zahradnik, P. (2010). Thiazolo[4,5-d]thiazole—A new domain for potential optoelectronic application. *Tetrahedron Letters*, 51(44), 5819–5821. DOI: 10.1016/j.tetlet.2010.08.110

Zhang, X., Zhao, Y., & Wang, L. (2016).*Journal of Heterocyclic Chemistry*, 53(3), 1096–1102.

Chapter 6
Potential Impact on Science and Society: Key Findings of Thienothiophene

Hemchandra K. Chaudhari
Institute of Chemical Technology, Mumbai, India

Sujal Mondal
https://orcid.org/0009-0001-1778-4142
Institute of Chemical Technology, Mumbai, India

Sankalp Salunkhe
https://orcid.org/0009-0002-4153-7536
Institute of Chemical Technology, Mumbai, India

Mudassir Nagori
https://orcid.org/0009-0009-9164-3442
Institute of Chemical Technology, Mumbai, India

Kulbhushan A. Sasane
Abasaheb Marathe Arts and New Commerce Science College, India

ABSTRACT

Thienothiophene, among known fused ring structures, linearly symmetrical thieno[3,2-b]thiophene has been demonstrated as a good organic building block for organic semiconductors owing to its coplanar rigidity and longer π-conjugation, facilitating intermolecular π–π stacking in the solid state and hence providing high charge carrier mobility. The group of researchers Gulsen et al., who synthesized four novel organoboron copolymers consisting of thienothiophene and selenothiophene analogs with thiophene π-spacer and copolymers could be employed as promising

DOI: 10.4018/979-8-3693-7520-4.ch006

candidates for F– detection. Thienothiophene (two annulated thiophene rings) and their derivatives have been reported to have a many integral roles that show many applications comprising pharmacological such as anti-tumor, antiviral, antibiotic, antiglaucoma and as inhibitors of platelet aggregation and optoelectronic means light-harvesting, semiconducting, and electroluminescent properties.

INTRODUCTION

Thienothiophene (TTs), the simplest fused thiophene rings, are strong electron donating groups. There are four regioisomers of thienothiophene, based on the orientation of two sulphur atoms in both rings: Thieno [3,2-*b*] thiophene, Thieno[2,3-*b*] thiophene, Thieno[3,4-*b*]thiophene, and Thieno [3,4-*c*] thiophene. Thienothiophene and its derivatives have a diverse range of applications in various fields. Due to recent development of numerous derivatives of this class of compounds, the scope and field of applications has evolved considerably. Some of the applications of Thienothiophene derivatives include anti-cancer property (Hu *et al.*, 2022) (El-Remaily *et al.,* 2015) (Aboelez *et al.,* 2022) (Rosell *et al.,* 2012) (Di Noia *et al.,* 2021) (Sigismund, Avanzato, & Lanzetti, 2018) (Baniik, Bandyopadhyay, & Ganguloy, 2003) (Toda, 1995), antioxidant activity (Kidwai, Venkataramanan, & Dave, 2001) (Ravichandran & Karthikeyan, 2011) (Stuerga & Gaillard, 1996) (Feng *et al.,* 2016), antimicrobial activity (Mabkhoot, 2009), antiviral activity (Vince *et al.,* 2014), HIV protease inhibitory activity (Dorsey *et al.,* 1994), treatment of Central Nervous System (CNS) disorders (Fr 1979), analgesic activity (BRD, 1988), anti-inflammatory activity (BRD, 1977), thrombosis inhibitory activity (BRD, 1977), utilisation in Organic Light Emitting Diodes (OLEDs) (Rescep *et al.,* 2022)[(18)], utilisation in Organic Solar Cells (OSCs) (Zhang *et al.*, 2015) (Tamilavan *et al.,* 2017) (Lee et al., 2016) (Abbasriyaludeen *et al.,* 2021) (Fernandes et asl., 2017) (Zhu *et al.,* 2015) (Massin *et al.,* 2015) (Paek *et al.,* 2016) (Peng *et al.,* 2017) (Cao *et al.,* 2022)(Yang *et al.,* 2019) (Noh *et al.,* 2013), application in Organic Field Effective Transistors (OFETs) (Durso *et al.,* 2013) (Fei *et al.,* 2014) (Yue *et al.,* 2016) (Mu *et al.,* 2016) (Zhang *et al.,* 2022) (Rawcliffe *et al.,* 2008), application in Non Linear Optics (Wang *et al.,* 2019) (Marco *et al.,* 2018) (Marco *et al.,* 2013) (Jen *et al.,* 1994) (Blenkle *et al.,* 1996) (Mashraquiu *et al.,* 2007) (Mashraqui *et al.,* 2006) (Can, 1995) (BRD, 1994) (Tsivgoulis & Angew, 1995), application in Organic Thin Film Transistors (OTFTs) (Le *et al.,* 2016) (Redecker *et al.,* 1999), development of novel charge-transfer complexes and as ligands for metal ion coordination ((Fuller *et al.,* 1999), use in electro-optical switches and assessment devices as clathrate crystals (BRD, 1985) (BRD, 1996) (Tso *et al.,* 1998) (Haristoy *et al.,* 2000) (Bishop, 1996) and development of novel electro-photoactive multifunctional dyes (Oh-Kil *et al.,*

2000) (Kim *et al.,* 2000)(Braverma & Freund, 1990) (Luhman *et al.,* 1978). The above mentioned applications are discussed in detail as follows:

1. BIOLOGICAL ACTIVITY

A. Anti-Cancer Property

One popular therapy strategy for a number of malignancies is the use of EGFR (Epidermal Growth Factor Receptor) inhibitors to target EGFR-mediated signalling pathways (Hu *et al.,* 2022) (El-Remaily *et al.,* 2015) (Aboelez *et al.,* 2022). A series of Thienothiophene derivatives (Compounds 1-7) were evaluated for antiproliferative effect against cell lines MCF-7 and A549, *in vitro (Aboelez et al., 2022) (Rosell et al., 2012)* (Di Noia *et al.,* 2021). In this study, doxorubicin and erlotinib were utilised as controls in the MCF-7 and A549 cell lines, (Di Noia *et al.,* 2021) (Sigismund, Avanzato, & Lanzetti, 2018). Compound 2 (IC_{50} value = 4.92 ± 0.19 in MCF-7 and IC_{50} value = 4.69 ± 0.09 in A549 cell lines) was shown to be 4.42 and 4.12 times more effective than standard Erlotinib in MCF-7 (IC_{50} value = 21.76 ± 1.85) and A549 cells (IC_{50} value = 19.33 ± 1.85). Furthermore, compound 2 exhibited greater activity (IC_{50} value = 7.89 ± 0.55) against the MCF-7 cell line than doxorubicin.

Figure 1. Thienothiophene derivatives with anti-cancer and anti-oxidant property

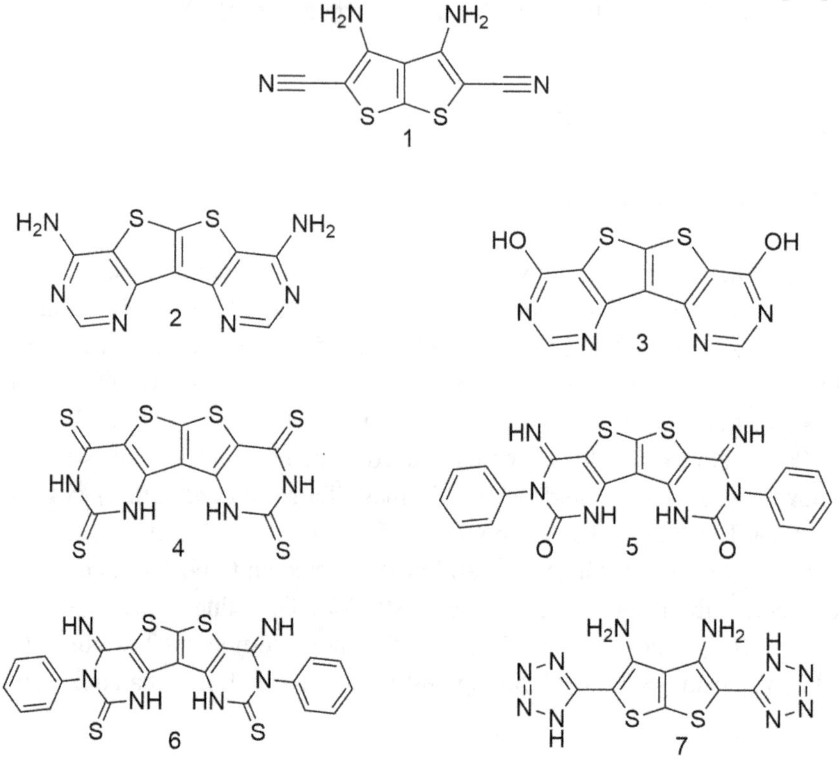

Using erlotinib as the control medication, the inhibitory effect of these compounds on the EGFRWT kinase enzyme was investigated using the homogeneous time-resolved fluorescence (HTRF) method, (Di Noia et al., 2021) (Banik, Bandyopadhyay, & Ganguly, 2003) (Todsa, 1995). Compounds 1, 2, and 3 (as illustrated in Figure 1) have been documented to exhibit greater efficacy than erlotinib, which has an IC_{50} value of 32 ± 0.05 μM against EGFRWT. The IC_{50} values for compounds 1, 2, and 3 are 0.29, 0.28, and 0.29 μM, respectively. The efficacy of compounds 1–7 (Figure 1) in inhibiting the mutant EGFRT790M was further investigated, particularly given their significant IC_{50} values against EGFRWT. Gefitinib served as the standard reference for comparison. The findings indicated that Compound 2, with an IC_{50} of 5.02 ± 0.19 μM, emerged as the most potent agent, exhibiting a potency 4.27 times greater than that of gefitinib, which has an IC_{50} of 21.44 ± 0.75 μM. Additionally, the most active derivatives identified were compounds 1, 3, and 4, demonstrating activities that were 2.71, 2.95, and 2.34 times that of gefitinib, respectively.

B. Antioxidant Activity

The antioxidant activity of compounds 1 through 7 (Figure 1) was assessed using the 2,2-azinobis(3-ethyl-benzothiazoline-6-sulfonic acid)diammonium salt (ABTS) method (Kidwai, Venkataramanan, & Dave, 2001) (Ravichandran & Karthikeyan, 2011) (Stuerga & Gaillard, 1996) (Feng *et al.*, 2016). The capability of antioxidants to scavenge the long-lived ABTS radical cation is taken into account by the ABTS index. The percent radical scavenging activity inhibited for Compounds 1-7 were evaluated. Compounds 2, 3, and 5 had the highest level of antioxidant activity, repressing radical scavenging activity to 78, 64, and 70%, respectively. In contrast, compounds 1, 4, 6, and 7 exhibited average antioxidant activity, ranging from 39 to 49%.

C. Antimicrobial Activity

In comparison to standard Amphotericin B, compound 8 (Figure 2), which is a biscyanopyrazole derivative of thienothiophene, has been shown to have more potent antimicrobial action against *Geotricum candidum*; more potent action against Gram negative bacteria, such as *Escherichia Coli* and *Pseudomonas aeruginosa*; and equipotent action against Gram-positive bacteria, such as *Staphylococcus aureus*. According to studies on the structure-activity relationship, maximal activity required a pyrazole nucleus with a chloro substituent in the phenyl ring (Mabkhoot, 2009).

Figure 2. Thienothiophene derivative with antimicrobial property

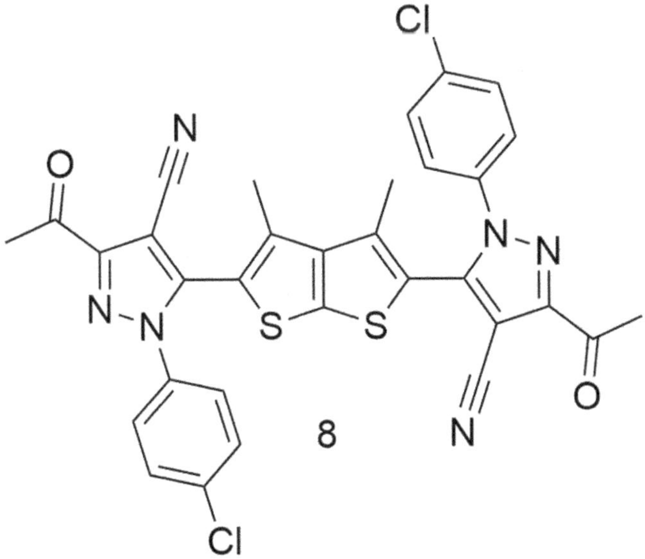

D. Antiviral Activity

Figure 3. Chemical structure of Samatasvir

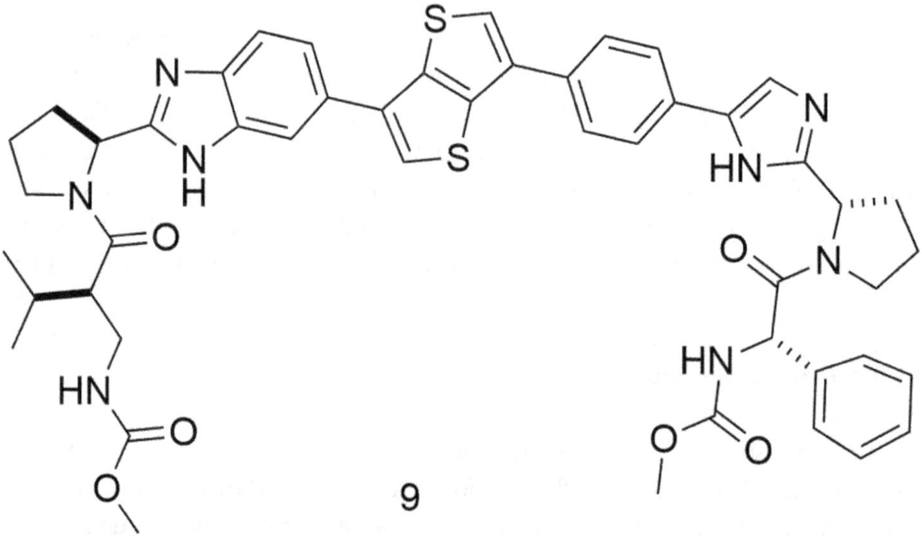

9

Compound 9 (Figure 3) is an inhibitor that targets the hepatitis C virus (HCV) non-structural protein 5A (NS5A) and is effective across different genotypes. This study investigated the therapeutic efficacy and safety of Compound 9 when used as monotherapy in patients new to treatment who were affected with genotypes 1 through 4. Participants included 34 individuals with genotype 1 and 30 infected with other 3 genotypes. They were randomly administered either a placebo or Compound 9 doses in range 25-100 milligram everyday for 3 days. Plasma specimens for analysing HCV RNA, pharmacokinetics, and sequence were taken till day 10. Compound 9 plasma levels were examined using a proven liquid chromatography-tandem mass spectrometry method. The chromatography involved use of a ZORBAX 300-SCX column (50 millimetre x 3 millimetre; 5 micrometre particles, Agilent Technologies, Santa Clara, CA) with isocratic elution at 1 millilitre per minute and a mobile phase of 4:1 (volume by volume) acetonitrile to ammonium formate (twenty-five milliMoles and pH 2.5). Mass spectrometry data were determined with an AB Sciex API 5500 triple quadrupole mass analyser (Framingham, MA), determining transitions of 443.3->659.2 m/z for Compound 9 and 446.3->659.2 m/z for 13C5-15N-Compound 9. HCV RNA levels were quantified using a proven concurrent PCR assay (COBAS AmpliPrep/COBAS Taqman HCV Test, Roche, Pleasanton, CA) with 25 IU/mL as lower limit for estimation. Results were compiled according to dose for each HCV genotype and sub-genotype. Additional examination using stepwise and logistic regression was done to determine pharmacokinetic and baseline factors considering characteristics such as sex, body-weight, Body Mass Index, race, baseline HCV RNA, HCV genotype, and IL28B genotype. Pharmacokinetic factors were summarized on the basis of dose in all HCV genotypes, while adverse events were classified by system organ class, favoured term and dosage. Safety information like vital signs, Electrocardiography and clinical laboratory results were compiled by dose for every slated recording.

Table 1. Demographics and baseline characteristics

	QD			**BD**	**Placebo**
	25 mg n=8	**50 mg n=8**	**100 mg n=18**	**50 mg n=18**	**n=12**
Mean age, yr (range)	45.1(4.2)	48.8(3.2)	41.9(2.1)	46.3(2.4)	44.7(3.2)
Male, n (%)	5(62.5)	7(87.5)	14(77.8)	12(66.7)	8(66.7)
Race, n (%)					
White	7(87.5)	7(87.5)	15(83.3)	13(72.2)	11(91.7)
African American	1(12.5)	1(12.5)	3(16.7)	4(22.2)	1(8.3)
Other	0	0	0	1(5.6)	0
Mean BMI, kg/m^3(range)	26.1(1.7)	26.9(1.1)	26.2(0.8)	25.6(0.7)	28.3(0.9)
Mean baseline HCV RNA, log_{10} IU/mL	6.6(0.18)	5.9(0.12)	6.5(0.13)	6.3(0.11)	6.3(0.18)
HCV genotypc					
1a	5	6	6	6	6
1b	3	2	0	0	0
2	0	0	1	1	1
2b	0	0	3	3	1
3a	0	0	4	4	2
4	0	0	4	4	2
IL28B genotype, n					
CC	3	3	6	4	4
CT	4	3	9	11	8
TT	1	2	3	3	0

(Vince *et al.*, 2014)

Table 2. Mean changes over time of plasma HCV RNA from baseline are shown in Fig. 2 for subjects with genotype 1, 3 or 4 HCV. Fig 3 depicts the individual and mean changes over time of plasma HCV RNA from baseline for subjects with genotype 2 HCV. The mean change in log10 HCV RNA from baseline to 24 h and

72 h, the mean maximum change from baseline and the corresponding time were summarized

Table 2.

	GT	Placebo	QD			BID
			25 mg	**50 mg**	**100 mg**	**50 mg**
Mean (min, max; n) change from baseline to 24h in HCV RNA, log_{10}	1a	0.1(-0.1,0.3;6)	3.1 (2.4, 3.5; 6)	3.2 (2.9,3.3; 6)	3.2 (2.6,3.8; 6)	2.7 (1.2,3.7; 6)
	1b	-	2.6 (2.0, 3.7; 3)	3.4 (3.3,3.5; 2)	-	-
Mean (min, max; n) change from baseline to 72h in HCV RNA, log_{10}	1a	0.1(-0.1,0.2)	3.1(2.5,3.6)	3.6(3.3,3.9)	3.1(2.2,4.2)	2.9(2.3,3.6)
	1b	-	2.8(1.6,4.1)	4.1(4.0,4.2)	-	-
Mean (min, max; n) maximum change from baseline in HCV RNA, log_{10}	1a	0.4(0.2-0.6)	3.3(2.9,3.7)	3.6(3.3,3.9)	3.5(2.6,4.3)	3.2(2.7,3.8)
	1b	-	3(2,4.3)	4.3(4.1,4.5)	-	-
Median (min, max) time to maximum change, day	1a	-	2(0.7,3)	2.3(2,5)	1.5(0.7,4)	3(1,4)
	1b	-	2(1,5)	4(4,4)	-	-

(Vince *et al.*, 2014)

Compound **9** showed significant antiviral ability among different HCV genotypes, achieving average maximum decrease in viral load from baseline between 3.0 and 4.3 log10 per millilitre, with no viral recoil seen during treatment period. In patients with HCV of genotype 2, Compound **9** was effective for those with the NS5A L31 polymorphism but had limited effect in patients with the M31 polymorphism. The drug's long plasma $t^{1/2}$ of about 20 hours allows one dose a day regimen. Compound 9 dosing had no serious adverse events or safety-related discontinuations reported. Common side effects were constipation, nausea, and headache, which were mild to moderate and were observed equally in both the active and placebo groups. There were no discernible patterns or dose-dependent effects in terms of adverse incidents, vital signs, laboratory findings or ECG readings.

In summary, Compound **9** was found to be safe and highly effective across all HCV genotypes at dosing ranging from 25–100 milligram per day for 3 days in treatment-naïve patients. Its pharmacokinetic and antiviral properties suggest it would be a valuable addition to all-oral direct-acting antiviral (DAA) combination therapies. Currently, a Phase II study is underway to evaluate the efficacy of one dose a day regimen of Compound 9 together with simeprevir in untreated patients with HCV of genotypes 1b or 4, (Vince *et al.*, 2014).

E. HIV Protease Inhibitors

Figure 4. HIV protease inhibiting thienothiophene derivative

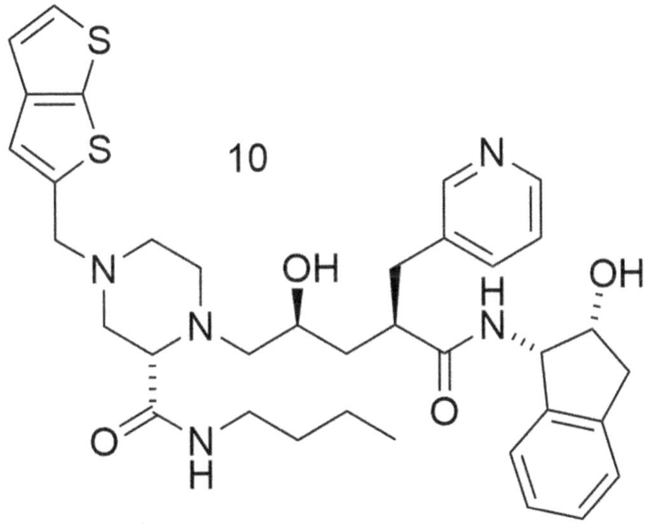

A study reported a series of thieno[2,3-*b*]thiophene-based HIV protease inhibitors. These inhibitors demonstrated significant efficacy against HIV at low concentrations, with some achieving satisfactory plasma levels in animals following peroral administration. Notably, compound **10** (Figure 4) emerged as the most promising candidate in this series, showing an IC_{50} of 0.12 nM and an IC95 of 6 ± 12 nM (Dorsey *et al.*, 1994).

F. CNS Disorders

Various patents have been issued for preparation of different thieno[2,3-*b*] thiophene derivatives and their applications. This include the preparation of 4(5)-(3-indolyl)imidazole **11** (Figure 5), which is useful for treating disorders related to central nervous system function, cerebrovascular dementia, senile dementia, and memory or learning impairments (Fr, 1979).

Figure 5. Thienothiophene derivative useful in treating CNS disorders

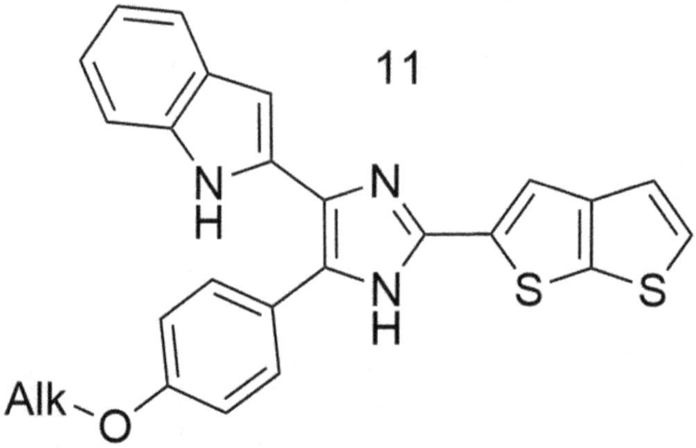

11

G. Analgesic, Anti-Inflammatory, and Thrombosis Inhibitor

Preparation of thieno[2,3-*b*]thienyl-containing carbonyl **12** (Figure 6) and its derivatives, is reported to be used in relieving pain and inhibit inflammation and thrombosis (BRD, 1977).

Figure 6. Thienothiophene with analgesic, anti-inflammatory and thrombosis inhibitor activity

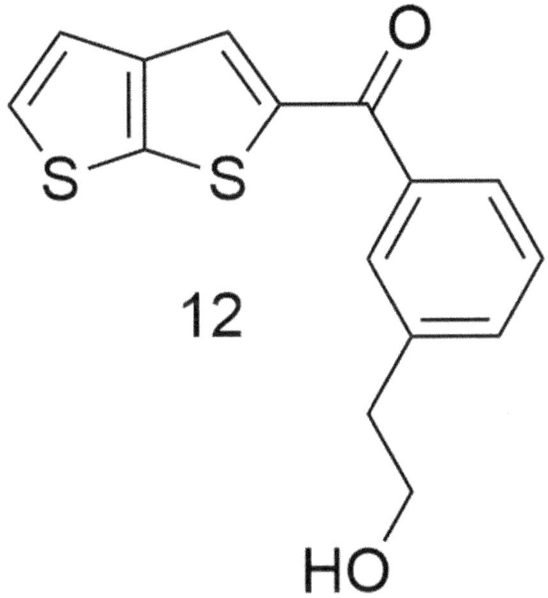

2. OTHER PRACTICAL APPLICATIONS

A. Organic Light Emitting Diodes

A group of thieno[3,2-*b*]thiophene moiety containing compounds was developed to be utilised as OLED emitters by Isci et. al. This group had 4-cyanophenyl moiety in common along with additional π-systems for fluorescence, such as triphenylamine or triphenylethylene group. Of them, Derivative 13 (Figure 7) was found to have highest brightness and current efficiency (Rescep *et al.,* 2022) (Table 3).

Figure 7. Thienothiophene derivatives developed to be used as OLED

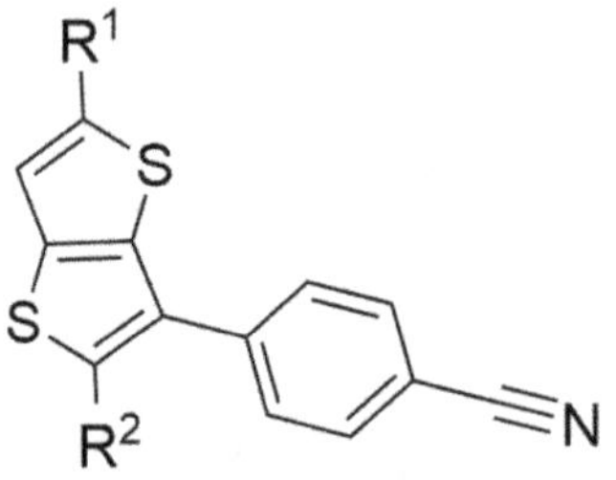

13, R1 = H, R2 = TPA
14, R1 = R2 = TPA
15, R1 = TPA, R2 = TPE
16, R1 = H, R2 = TPE
17, R1 = R2 = TPE

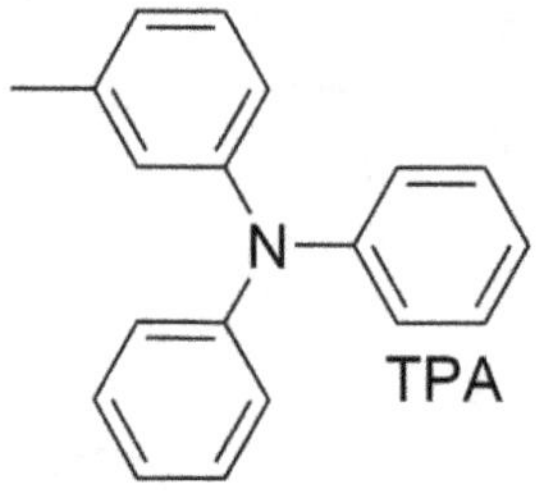

TPE

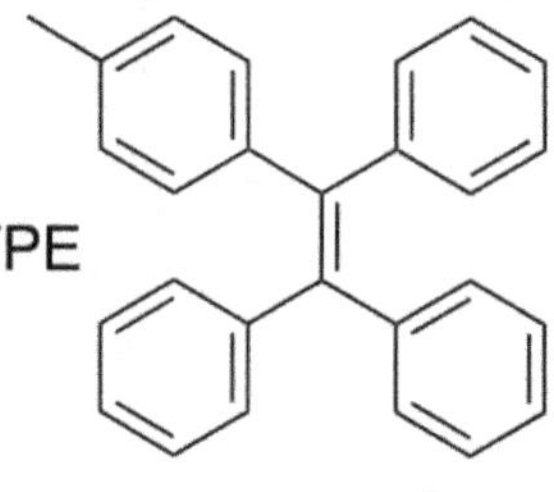

Table 3. Compound 13 absorbance and emission. [a] Measured in the THF solution; [b] Maximum Luminance; [c] HOMO=-($E_{onset(os)}$+4.40); [d] LUMO=HOMO+ΔE_{opt} ; [e] calculated from onset of the absorption spectra

Table 3

COMPOUND	λ_{max}(abs)[a] [nm]	λ_{max}(em)[a] [nm]	L[b] [cd m^{-2}]	η [cd A^{-1}]	E_{HOMO}[c][eV]	E_{LUMO}[d][eV]	ΔE_{opt}[e] [eV]
13	365	523	2790	4.70	-5.50	2.62	2.88

B. Organic Solar Cells (OSCs)

B.1. Bulk Heterojunction Solar Cells (BHJs)

Derivatives 18 (Zhang *et al.,* 2015) and 19 (Tamilavan *et al.,* 2017) (Figure 8) can be utilized as electron donor groups in BHJ's. Of this, 18 has A π-D πA arrangement with extra thienyl linkers and terminal acceptors at C_2 and C_5 of thieno[3,2-*b*] thiophene. 19 is polymeric with TT1 acting as alternating structural motif. Though BHJs constructed using both had identical hole mobility and fill factors (FF), BHJ made with derivative 18 had better power conversion efficiency (7.91 vs. 5.21%)

(Table 4). Derivative 20 (Lee *et al.*, 2016) (Figure 9) synthesised by combining TT2 with diketopyrrolopyrrole units showed PCE of 2.9% when incorporated in BHJ.

Figure 8. Thienothiophenes with application in BHJ's (compound 18 and 19)

18

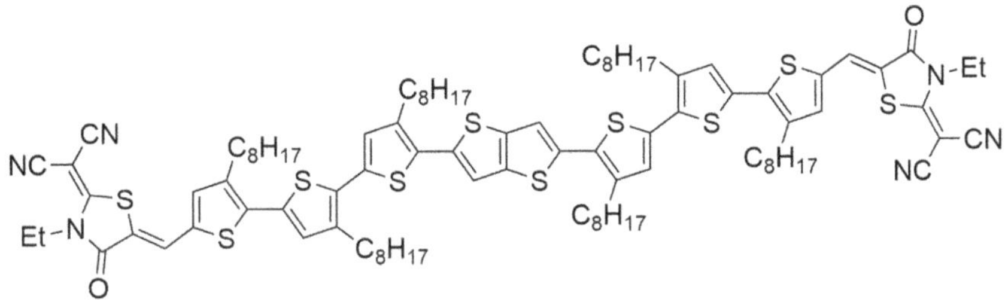

19

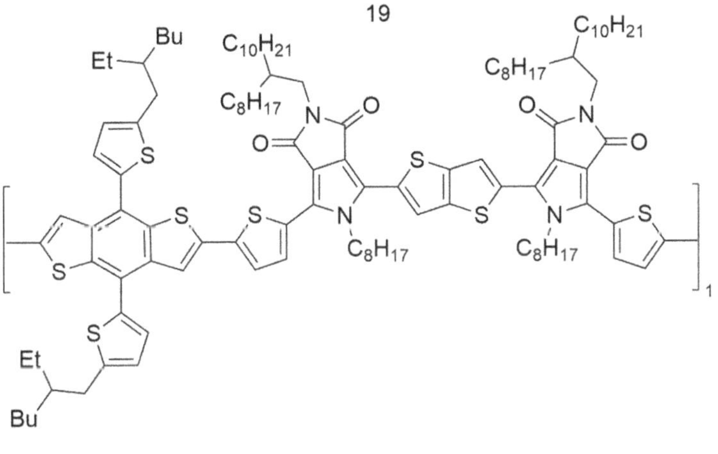

TT 1 has been utilised as a π-linker in the push–pull chromophores joining the tetracyanoquinodimethane (TCNQ) acceptor group and the diphenylamino donor group (Abbasriyaludeen *et al.*, 2021). The two chromophores 21 and 22 (Figure 9) are different in that the thiophene is attached at periphery. This structural feature does not affect their effectiveness in BHJs much. The push–pull arrangement of 21 and 22 provides with better absorption with the absorption maxima greater than 800 nanometre.

Table 4. [a]Measured in the thin film; [b] Measured by cyclic voltammetry in the DCM solution; [c] Measured by cyclic voltammetry in the acetonitrile solution; [d] LUMO = HOMO + ΔE; [e] Estimated from the onset wavelength of the optical absorption in the thin film; [f] Estimated from the organic field-effect transistor, [g] Average PCE values of five devices; [h] HOMO = LUMO – ΔE; [i] Measured by cyclic voltammetry in the chloroform solution; [j] Optical band gap determined from the onset wavelength of the optical absorption in the chloroform solution.

Table 4.

COMPOUND	λ_{max}(abs)[a] [nm]	E_{HOMO}[eV]	E_{LUMO}[eV]	ΔE [eV]	μ_h [$cm^2V^{-1}s^{-1}$]	PCE [%]	FF [%]
18	608	-5.05[b]	-346[b]	1.62	$6x10^{-4}$	7.91	65.5
19	505	-5.42[c]	3.36[d]	2.06[e]	$6.21x10^{-4}$[f]	5.21[g]	67

Figure 9. Thienothiophenes with application in BHJ's (compound 21 and 22)

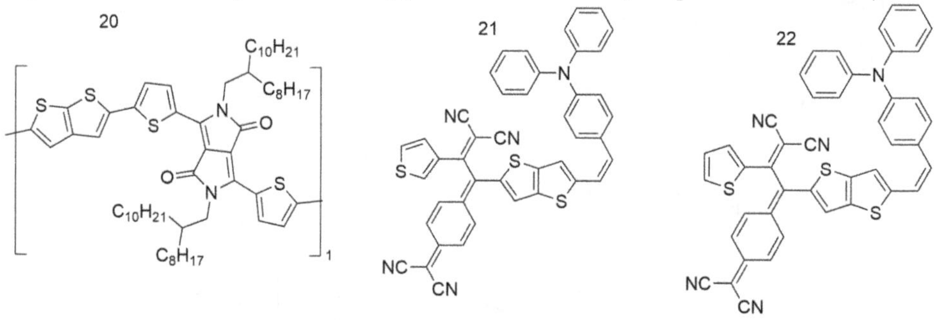

B.2. Dye Sensitized Solar Cells (DSSCs)

A series of TT1 derivatives with linear D πA arrangement were developed (23-26) (Figure 10). This arrangement had cyanoacrylic acid fragment as electron acceptor and anchor. Derivatives 23-25, differing only in peripheral donors showed similar performance in DSSC. Derivative 24 and 26 were developed with additional π-system made of carbazole and amino groups. Their PCE was slightly better than 23 (Fernandes *et al.,* 2017) and 25(Fernandes *et al.*, 2017) Compound 24 (Zhu *et al.*, 2015) had TT1 between the acceptor and donor groups while compound 26 (Massin *et al.*, 2015) carries 2 TT units as extra electron donors alongside carbazole moiety.

Figure 10. Thieno[3,2-b] thiophene derivatives with linear D πA arrangement (compound 23-26).

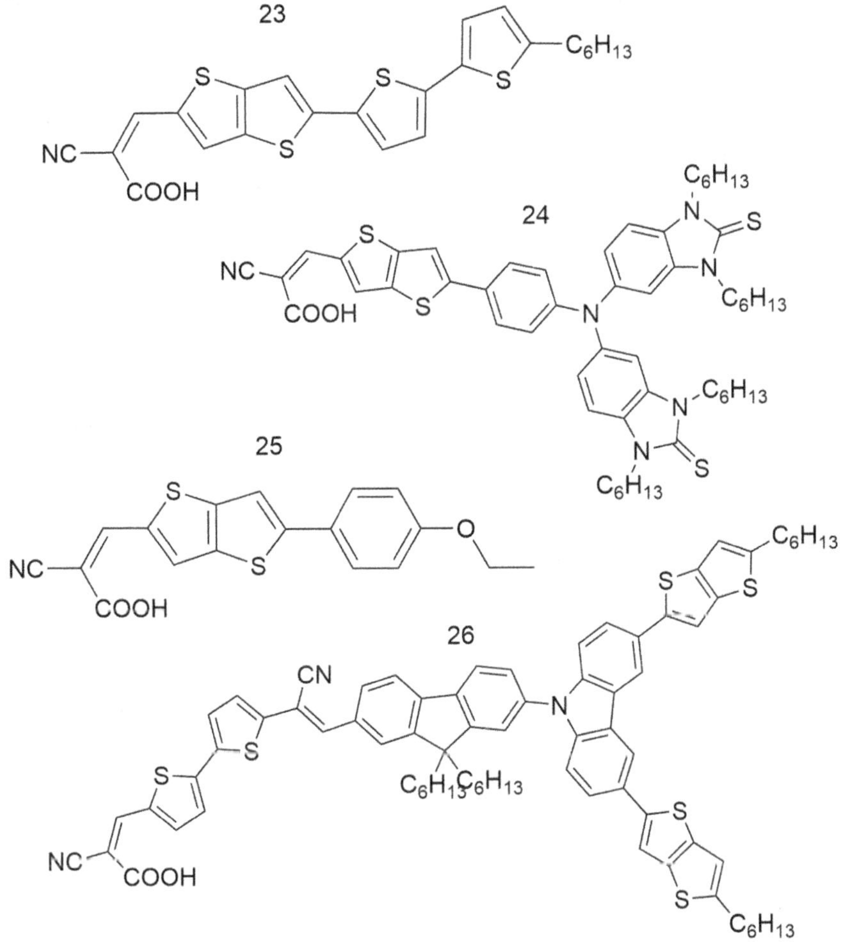

Table 5. [a] Measured in acetonitrile: DCM solution (1:1, volume ratio); [b] Measured by cyclic voltammetry in acetonitrile: DCM solution (3:1, volume ratio); [c] Measured in the DCM solution; [d] Determined by the TD-DFT calculations

Table 5.

COMPOUND	λ_{max} (abs) [nm]	E_{HOMO} [eV]	E_{LUMO} [eV]	ΔE [eV]	PCE [%]	FF [%]
24	488[a]	-4.95[b]	-3.75[b]	1.20	7.38	66
26	474[c]	-5.41[d]	-3.59[d]	1.82	6.50	74

B.3 Perovskite Solar Cells (PSCs)

Another series 27-30 (Figure 11) was developed to be utilized as hole transporting material (HTM) in PSCs. Among them except 27 (Paek *et al.*, 2016), other compounds were polymeric with incorporated TT units. Thienothiophene and diketopyrrolopyrrole units containing derivative 29 (Peng *et al.*, 2017) showed highest hole mobility value whereas derivative 30 (Cao *et al.*, 2022) showed highest PCE (Table 6). Derivative 30 had enlarged π system with 4 TT units. Another compound 31 (Yang *et al.*, 2019) (Figure 11) had significantly higher PCE and FF values (Table 6). It has central TT2 unit with four [bis(4-methoxyphenyl)amino] phenyl substituents. The latter group is also a common characteristic of Spiro-OMeTAD, a popular HTM used in PSCs (Noh *et al.*, 2013).

Table 6. [a] Measured by cyclic voltammetry; [b] Calculated from the optical band gap; [c] Optical band gap; [d] Measured by ultraviolet photoelectron spectroscopy in thin film

Table 6.

COMPOUND	E_{HOMO} [eV]	E_{LUMO} [eV]	ΔE [eV]	μh [cm2V−1s−1]	PCE [%]	FF [%]
30	-5.36[a]	-3.33[b]	2.03[c]	$1.24x10^{-4}$	15.80	66.6
31	-5.44[d]	-2.42[b]	3.02[c]	$3.76x10^{-4}$	18.78	76.2

Figure 11. Thienothiophene derivatives developed to be utilized as hole transporting material (HTM) in PSCs

27

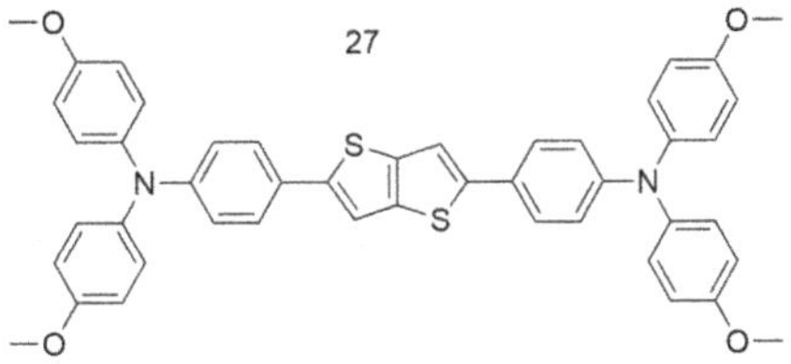

29

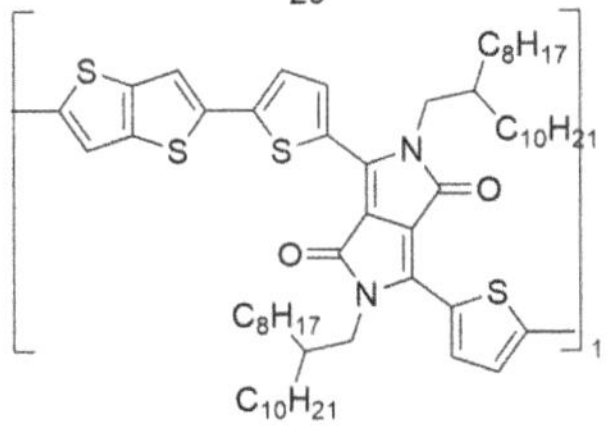

28

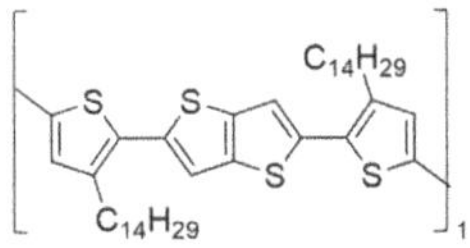

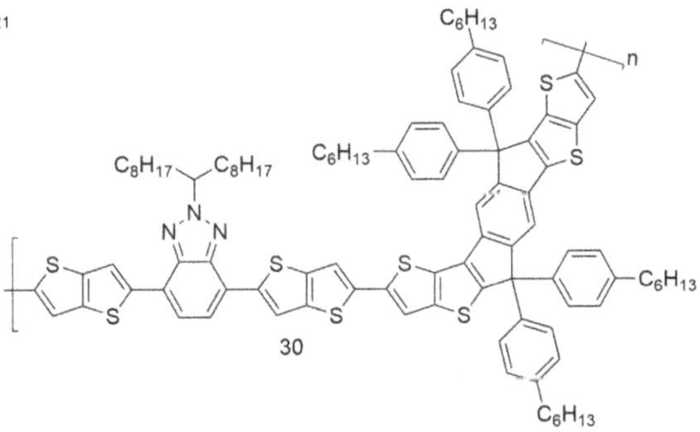

30

31

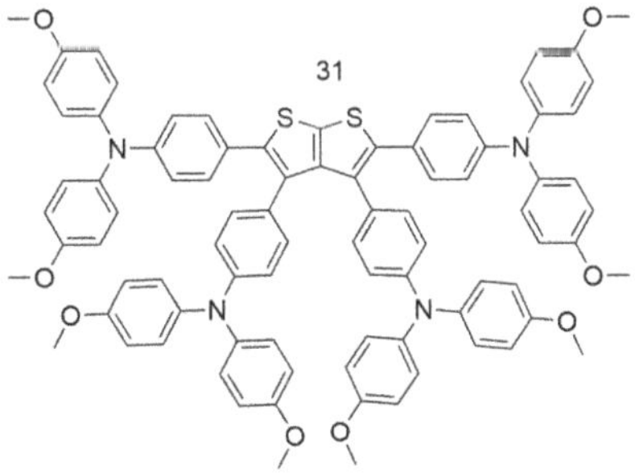

C. Organic Field Effective Transistors (OFETs)

When it came to OFETs, the semiconducting ability of these TT derivatives could be tapped. For example, the TT1 derivatives 32 (Durso *et al.*, 2013) and 33 (Fei *et al.*, 2014) (Figure 12) are representative n type organic semiconductors with somewhat similar absorption spectra and electron mobility to each other but different HOMO/ LUMO levels. Compounds 34 (Fei *et al.,* 2014) and 36 (Yue *et al.,* 2016) (Figure 12) are p type semiconducting materials with absorption shifts to red. Furthermore, ambipolar semi-conductivity with an electron-dominant mobility has been revealed in the isomer 35 (Mu *et al.*, 2016) (Figure 12). The symmetric derivative 37 (Figure 12) decorated with the TT1 core and diphenylamino donors combined to trifluoro-phenyl acceptors has also been disclosed to be a semiconductive molecule which possesses excellent photo-stability along exceptional mobilities (Zhang *et al.*, 2022) (Table 7). TT2 derivatives 20 (Lee *et al.*, 2016), 38 (Rawcliffe *et al.,* 2008) (Figure 12) for OFETs and in 2 examples, the bicyclic thienothiophene scaffold was used as a monomester within p-type semiconducting π-conjugated polymeric backbone. The highest hole mobility of 38 was located when copious with octadecyltrichlorosilane, and the by-product stands all over silane self-assembled monolayers.

Figure 12. Thienothiophene derivatives with application in OFETs

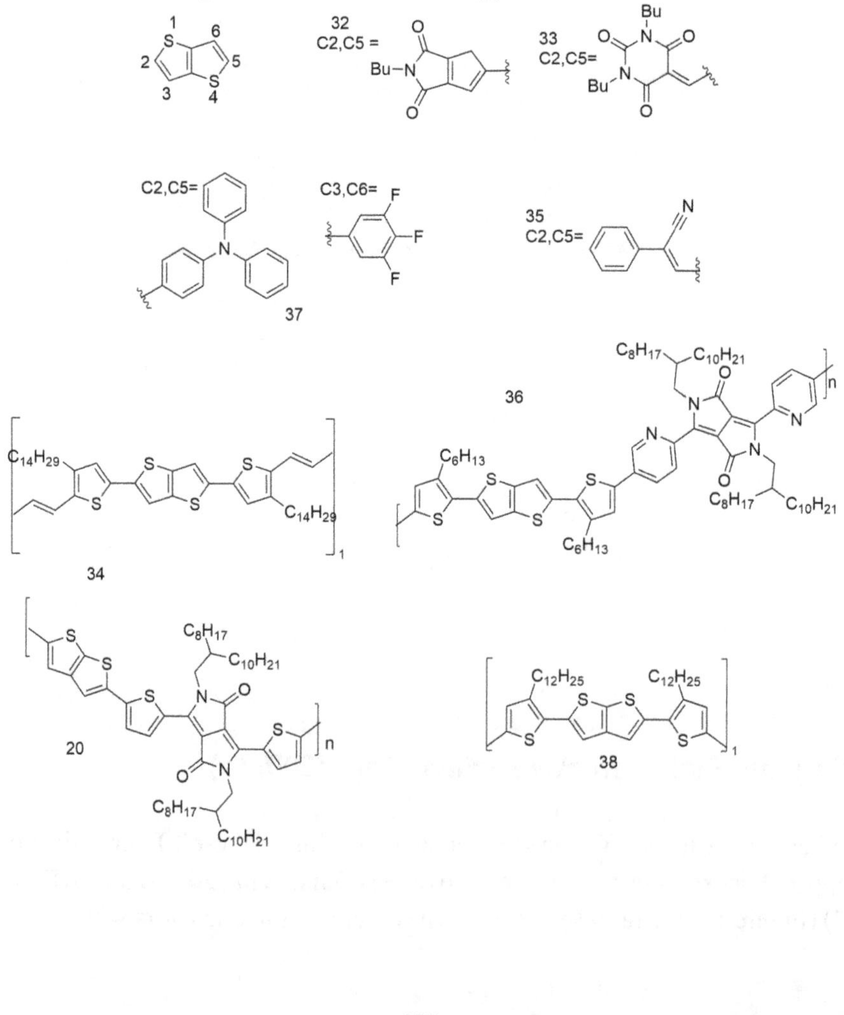

Table 7. [a] Measured in the chloroform solution; [b] Measured by cyclic voltammetry in the acetonitrile solution; [c] Measured in the DCM solution; [d] Measured by cyclic voltammetry in the DCM solution

Table 7.

COMPOUND	Λ_{max} (abs) [nm]	E_{HOMO} [eV]	E_{LUMO} [eV]	ΔE [eV]	μh [cm2V−1s−1]
20	706[a]	-5.40[b]	-3.74[b]	1.66	$1.60x10^{-1}$
37	389[c]	-4.92[d]	-2.30[d]	2.62	[N/S]$7.2x10^{-1}$

D. Non-Linear OPTICS (NLO)

Polarizable cloud of π electrons possessed by TTs makes them suitable for use as NLO medium. Graphene quantum dots doped derivative 39 (Figure 13) showed large amount of electron cloud polarization followed by electron transfer to the quantum dots (Wang *et al.*, 2019). Derivative 40 (Figure 13) was used for xerogel structure study and resulting micrographs were analyzed using SHG microscopy (Marco *et al.*, 2018). Derivatives 41-44 (Figure 13) were examined as 2° NLOphores by EFISH (Electric Field induced Second Harmonic generation). These derivatives had diphenyl pyran donors, thiobarbituric acid and tricyanofuran acceptors, (Marco *et al.*, 2013). Adjustments in π system were made to optimize optical nonlinearity with highest NLO response obtained for 44 (Jen *et al.*, 1994). Inclusion of methylthio and tricyanovinyl improved second order NLO response (47a and 48b) (Blenkle *et al.*, 1996) (Figure 13, Table 8). Mashraqui et. al. studied NLO propertics of a series of compounds having aromatic azo or electron withdrawing group at C_2(49a-d,50a-f) (Figure 13). Hyper Rayleigh Scattering determined NLO properties in range 4-17 x 10^{-30} esu (Mashraqui *et al.*, 2007). NLO properties of Compound 51 were reported in Mashraqui *et al.* (2006).

Table 8. [a] Measured in the chloroform solution. [b] Determined at 1320 nm; [c] Calculated using the two-level dispersion model

Table 8.

COMPOUND	X atom	λ_{max}[a] (abs) [nm]	B[b] [10^{-30} esu]	B[c] [10^{-30} esu]	μ [D]
47a	S	350	26	17	4.6
48b	S	516	124	43	9.5

Figure 13. Thienothiophene derivatives with application in NLOs (compounds 39-51)

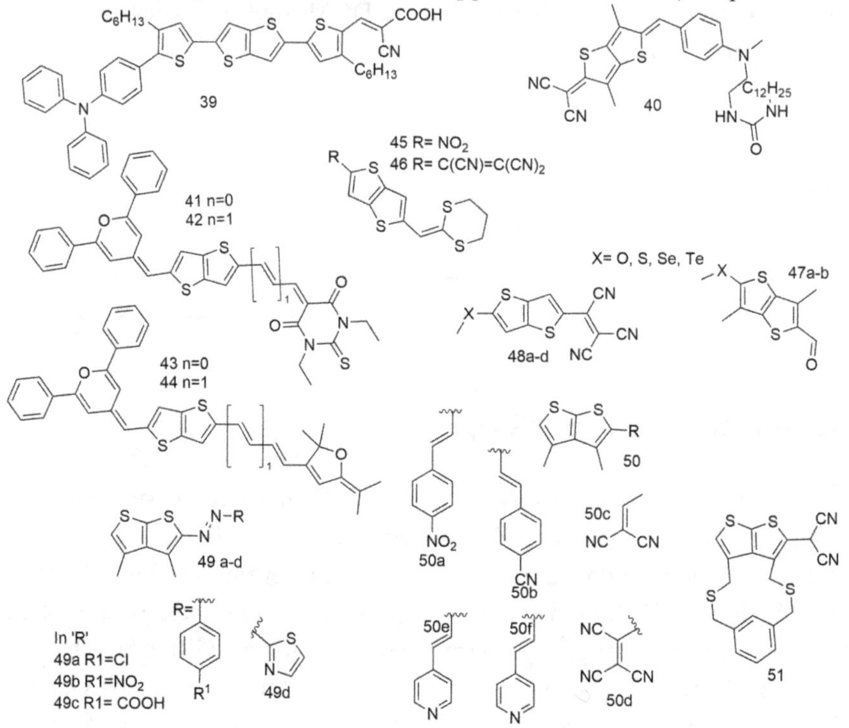

Recent discoveries regarding the physicochemical properties of thieno- and dithienothiophenes have opened up new possibilities for designing advanced nonlinear optical materials. 2-(1,3-Dithian-2-ylidenemethyl)-5-(tricyanovinyl)thieno[3,2-*b*] thiophene (52) (Figure14) has been utilized in the synthesis of nonlinear optical polymers. Substituted thioethers of thieno[3,2-*b*]thiophenes (53) (Figure14) exhibit similar optical characteristics. The preparation method for these compounds is detailed in a patent, (Can, 1995) (BRD, 1994).

Figure 14. Thienothiophene derivatives with application in NLOs (compounds 52 and 53)

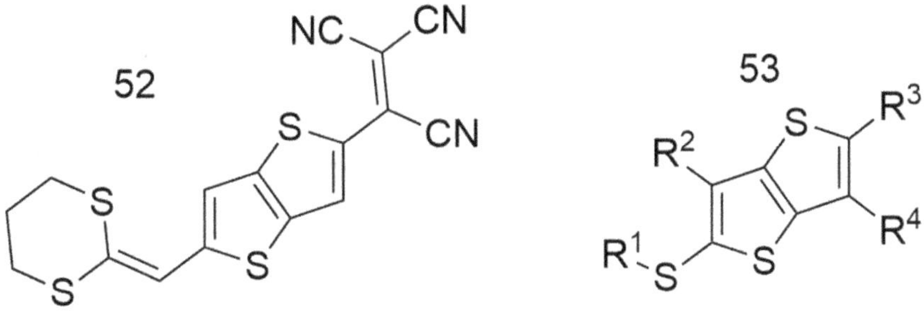

R^1=C_1-C_{22}-Alk,CH=C$(CN)_2$;R^2,R^3,R^4= H,Me;
R^2=H,R^3=NO_2,R^4=CN,NO_2

In the study of compounds 54a and 54b (Figure 15), which feature extensive conjugation between two fused dithienothiophene parts were developed. These compounds exhibit high heat stability, impressive exhaustion resistance, responsiveness to diode laser wavelengths, and almost full reversible interconversion. Notably, their ability to carry out non-destructive data evaluation makes them suitable for practical use as photo-switching elements, (Tsivgoulis *et al.,* 1995).

Figure 15. Thienothiophene derivatives with application in NLOs (compound 54 a-b)

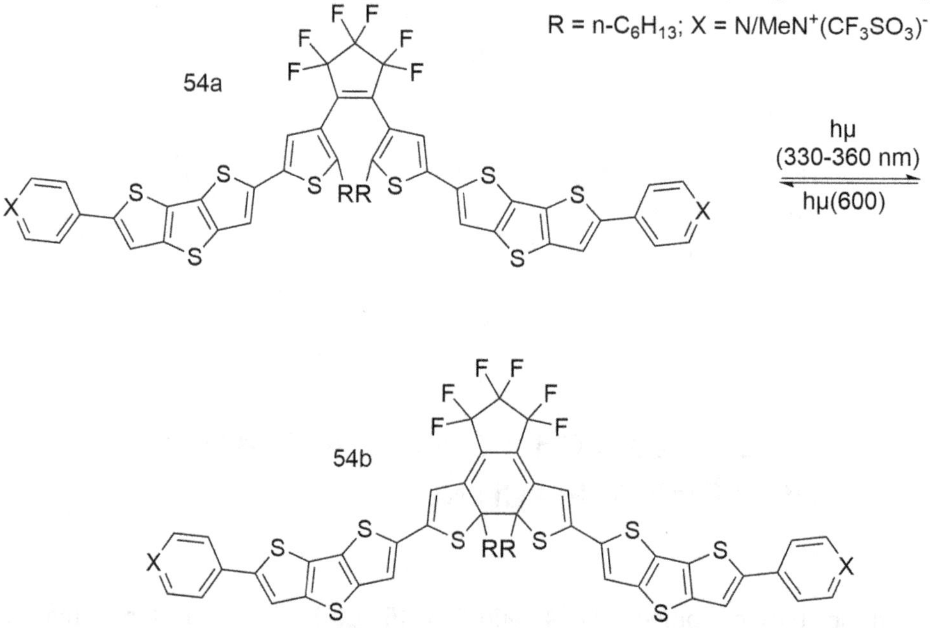

In general, heteroatom containing π-conjugated substances are recently garnering attention from both academic research and commercial interest. Devices like OLEDs, OFETs and some types of OSCs are operated by TT derivatives. Emissive and hole-injecting layers are a very successful application of TT derivatives in OLEDs. Due to their light-accumulating and electron-donating attributes, TT-derivatives have been used widely in BHJs and DSSCs, while their electron-hole transport properties have been used in PSCs. Thiothiophene compounds are also known for their semiconducting characteristics. Therefore, various OFETs have been fabricated with TT compounds that have n-type or p-type semiconductor attributes. The electron-transferring and electron-donating properties of TT have been used in nonlinear optics, especially SHG. The TT derivative in the push-pull configuration is mainly utilised in this field. On the basis of above results, thieno[3,2-*b*]thiophene 1(TT1) and thieno[2,3-*b*]thiophene 2 (TT2) can serve as small, easily prepared organic fundamental units, allowing for adjusting of the properties of many organic materials.

E. Organic Thin Film Transistors (OTFTs)

Figure 16. Thienothiophene derivatives with application in OTFTs (compounds F8TT and F8T2)

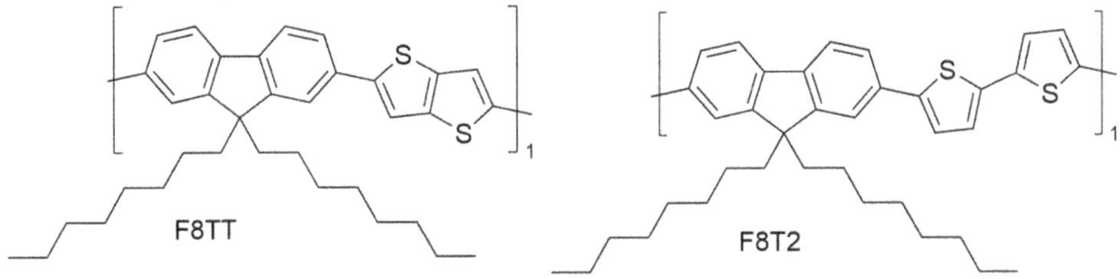

The polymers poly(9,9′-n-dioctylfluorene-alt-thieno[3,2-*b*] thiophene) (F8TT), poly(9,9′-n-dioctylfluorene-alt-bithiophene) (F8T2), and poly(9,9′-n-dioctylfluorene) (F8) were synthesized following established procedures as described in the literature, (Lee *et al.*, 2016)[(21)]Figure 16 illustrates the chemical structures of these polymers. The number-mean molecular weights (Mn) for F8TT, F8T2, and F8 are 23,000 (with a polydispersity index of 2.8), 31,000 (with a polydispersity index of 2.4), and 35,600 (with a polydispersity index of 2.8), respectively.

Figure 17. DSC thermograms of F8TT

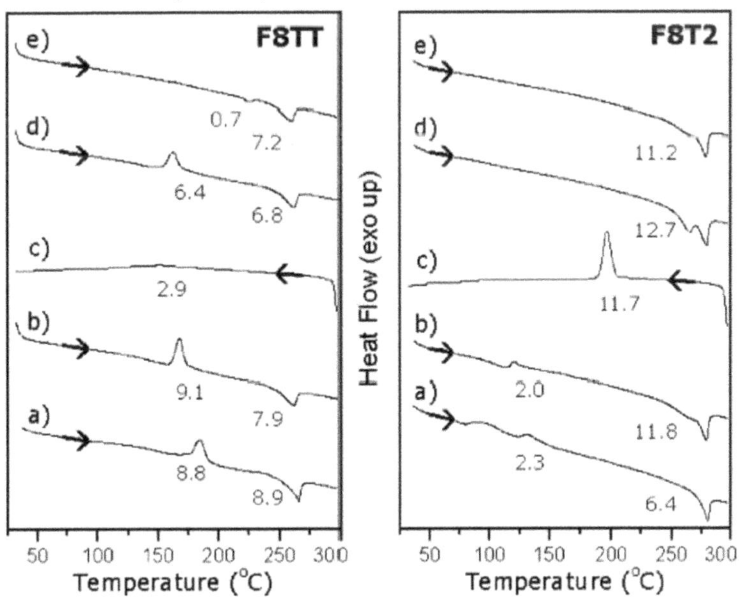

Figure 17. DSC thermograms of F8TT (left, sample weight: 4.9 mg) and F8T2 (right, sample weight: 5.2 mg) for the following thermal cycles: (a) heating at 10 °C/min of pristine sample; (b) heating at 10 °C/min after quenching from 300 °C; (c) sluggishly chilling at 10 °C/min after equilibration at 300 °C; (d) heating at 10 °C/min after sluggishly chilling from 300 °C; (e) heating at 10 °C /min after quenching from 220 °C. The enthalpic changes (J/g) are shown at each transition

According to Figure 17, in the case of F8TT, both heating scans (lines b and d) exhibited phase transitions between crystalline and liquid-crystalline states. However, the enthalpic changes associated with crystallization varied significantly: the quenched sample (line b) showed a higher crystallization enthalpy, approximately 2.5 J/g greater, compared to the sample that was cooled slowly (line d). This discrepancy aligns with the 2.9 J/g enthalpic change observed in a broad exothermic recrystallization peak during slow cooling (line c). This suggests that quenching nearly entirely suppresses the recrystallization of F8TT during the cooling process. Conversely, F8T2 displayed a distinct recrystallization peak at 195 °C in the slowly cooled scan (line c) and no exothermic phase transition in the heating scan following slow cooling from 300 °C (line d). This indicates that F8T2 undergoes complete recrystallization when cooled slowly at 10 °C/min. The enthalpic changes for F8T2 during cooling were influenced by the cooling rates: faster cooling resulted in smaller enthalpic changes. Additionally, the quenched F8T2 crystallized upon heating at 120 °C with an enthalpic change of approximately 2 J/g (line b), indicating that quenching partially suppresses recrystallization. It has been noted that recrystallization during cooling can decrease the mobility of polycrystalline films due to carrier trapping at grain boundaries. Therefore, it is crucial to achieve a liquid-crystalline glassy morphology in polymer films to maintain high mobility. This DSC study confirms that quenching effectively suppresses the recrystallization of F8TT, suggesting that F8TT films in a liquid-crystalline melt state can be obtained by quenching after heating above their melting transition.

Figure 18. Output characteristics of TFT devices

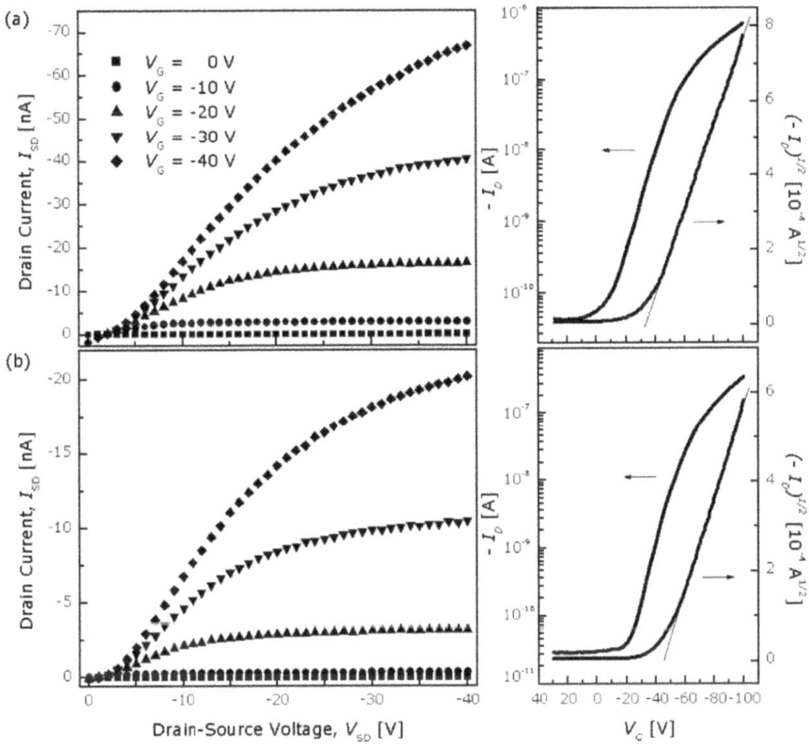

Figure 18. Output characteristics of the TFT devices at different gate biases (left) and plots of the transfer at constant VD) -100 V (right) (semilogarithmic plot of -ID vs VG (left axis) and plot of (-ID)1/2 vs VG (right axis)) for devices using (a) F8TT and (b) F8T2

The thin-film transistors made from F8TT and F8T2 exhibited typical p-channel Field Effect Transistor (FET) behaviour, as illustrated in Figure 18. Specifically, the drain-source current increased with the negative gate voltage when a negative bias was applied, due to a rise in the number of charge carriers (holes). Figure 17 also shows the transfer curves [ID vs. VG] at a constant drain voltage (VD) of -100 V. Field-effect mobilities were determined in the saturation regime at a drain voltage of VD = -40 V, while on/off ratios were measured at VD = -100 V. The films were annealed at 285 °C for 10 minutes and then quenched to room temperature to preserve the liquid-crystalline alignment of the polymer chains. For F8TT, the thin film exhibited a field-effect mobility of 1.1 × 10^-3 cm^2 V^-1 s^-1 and an on/off ratio of 1.4 × 10^4 without any alignment techniques, which is three times greater than the mobility of F8T2 (0.4 × 10^-3 cm^2 V^-1 s^-1) under the same conditions. The superior performance of F8TT suggests a well-defined molecular structure, which is supported by XRD analysis of its morphological characteristics. The transition from a comparatively flexible bithiophene to a more rigid fused thieno[3,2-*b*] thiophene enhances crystallinity and order within the material. This observation aligns with prior findings on linear compounds, which show that increased molecular rigidity typically leads to higher mobility, (Redecker *et al.,* 1999).

F. Charge Transfer Complexes

Isomeric thienothiophenes and its derivatives are noteworthy for their potential as electron donors or acceptors in development of novel charge-transfer complexes and as ligands for metal ion coordination. Notable examples include the use of 2,5-bis(dicyanomethylene)-2,5-dihydrothieno[3,2-*b*]thiophenes (55) and 2,6-dihydrothieno[3,2-*b*:2',3'-*d*]thiophenes (56) (Figure 19) as electron acceptors, which have resulted in charge-transfer complexes exhibiting very high conductivities, reaching levels comparable to metallic conductivity, (Fuller *et al.,* 1999).

Figure 19. Thienothiophene derivatives with application in charge transfer complexes

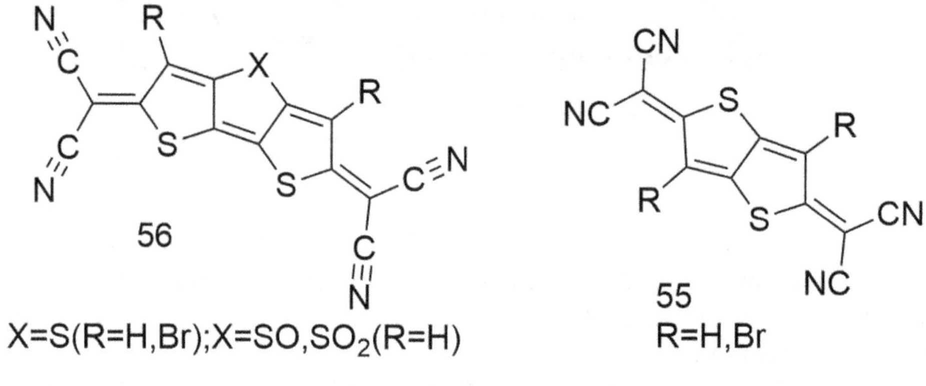

G. Liquid and Clathrate Crystals

Certain derivatives of thienothiophene can be integrated into liquid crystalline mixtures for use in electro-optical switches and assessment devices. Benzo[*b*]thieno[3,2-*b*]benzo[*b*]thiophene-2,7-dicarboxylate derivatives, in particular, demonstrate smectic liquid crystal properties, with their stability being influenced by the length and nature of their terminal chains. These liquid crystal samples exhibit notable photoconductive behaviour (BRD, 1985) (BRD, 1996) (Tso *et al.,* 1998) (Haristoy *et al.,* 2000). Additionally, a sequence of novel substance with host-type complexation properties, such as 2,5-bis(9-hydroxyfluoren-9-yl)thieno[3,2-*b*]thiophene (57c) (Figure 20), have been incorporated into crystals complexes. These compounds are referred to as "clathrate crystals," (Bishop, 1996).

Figure 20. Thienothiophene derivatives with application in electro optical switches and assessment devices and as clathrates.

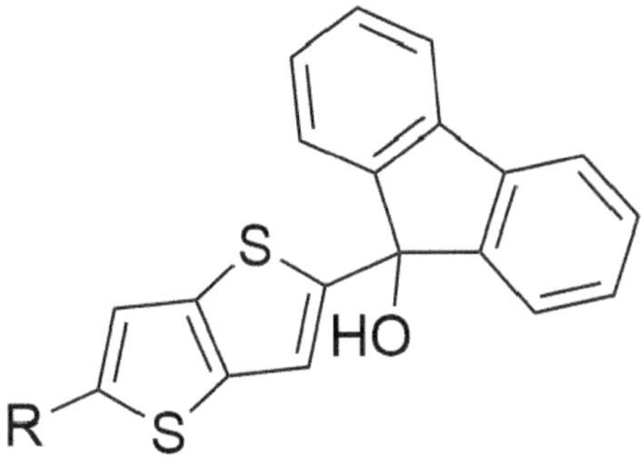

57a-c
R
a = Me
b = Cl
c = 9-hydroxyfluoren-9-yl

H. Dyes

Novel electro-photoactive multifunctional dyes were developed based on dithienothiophenes (Oh-Kil *et al.*, 2000) (Kim *et al.*, 2000). The pigment properties of 2,2,5,5-tetramethyl-2,3,5,6-tetrahydrothieno[3,2-*b*]thiophene (58) (Figure 21) were compared with those of related compounds comprising of thioindigoid part. It was found that thienothiophene 57 (Figure 20) is classified among indigoid dyes, (Braverman & Freund, 1990) (Luhman *et al.*, 1978).

Figure 21. Thienothiophene derivatives with application as Electro-photoactive multifunctional dyes

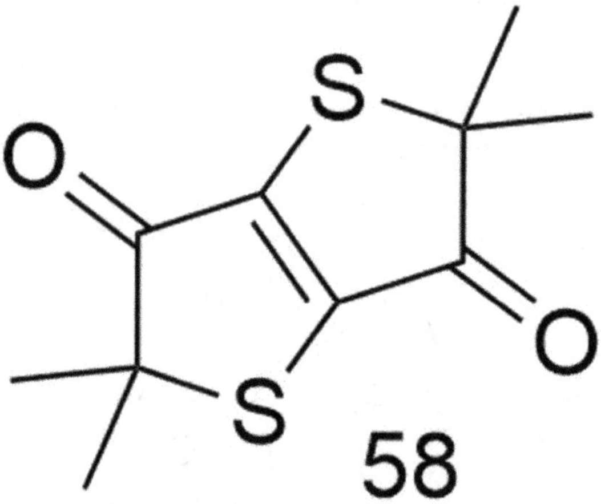

58

REFERENCES

Abbasriyaludeen, A. R., Chitra, K., Ramasamy, S., Murugan, P., & Chandrasekar, P. (2021). Molecular engineering of twisted dipolar chromophores for efficiency boosted BHJ solar cells. *Journal of Materials Chemistry. C, Materials for Optical and Electronic Devices*, 9(13), 4562–4575. DOI: 10.1039/D1TC00708D

Aboelez, M. O., Kamel, M. S., Belal, A., El Badry Abdel-Aziz, A., Abourehab, M. A., Abdel-Ghany, H., & El Hamd, A. (2022). M.; El-Remaily, M. A. E. A. A. Microwave-assisted synthesis, spectroscopic characterization, and biological evaluation of fused thieno [2, 3-*d*] pyrimidines as potential anti-cancer agents targeting EGFRWT and EGFRT790M. *Molecular Diversity*, •••, 1–17. PMID: 35780205

Banik, S., Bandyopadhyay, S., & Ganguly, S. (2003). Bioeffects of microwave–a brief review. *Bioresource Technology*, 87(2), 155–159. DOI: 10.1016/S0960-8524(02)00169-4 PMID: 12765354

Bishop, R. (1996). Designing new lattice inclusion hosts. *Chemical Society Reviews*, 25(5), 311. DOI: 10.1039/cs9962500311

Blenkle, M., Boldt, P., Bräuchle, C., Grahn, W., Ledoux, I., Nerenz, H., Stadler, S., Wichern, J., & Zyss, J. (1996). Chalcogens as electron donors for selected nonlinear optic phores. *Journal of the Chemical Society, Perkin Transactions 2: Physical Organic Chemistry*, 2(7), 1377–1384. DOI: 10.1039/P29960001377

Braverman, S., & Freund, M. (1990). Rearrangements of bridged diallenes. A facile synthesis of novel condensed heterocycles by tandem [3,3]-sigmatropic rearrangement and double intramolecular michael addition of diallenyl disulfides and diselenides. Isolation of stable diallenyl diselenides. *Tetrahedron*, 46(16), 5759–5776. DOI: 10.1016/S0040-4020(01)87773-5

Cao, J., Wang, L., You, G., Li, L., Yao, L., Zhou, Z., Yang, E., Cai, W., Ling, Q., & Zhen, H. (2022). Dopant-free, hole-transporting polymers containing benzotriazole acceptor unit for perovskite solar cells. *Dyes and Pigments*, 200, 110170. DOI: 10.1016/j.dyepig.2022.110170

Di Noia, V., D'Aveni, A., D'Argento, E., Rossi, S., Ghirardelli, P., Bortolotti, L., Vavassori, V., Bria, E., & Ceresoli, G. (2021). Treating disease progression with osimertinib in EGFR-mutated non-small-cell lung cancer: Novel targeted agents and combination strategies. *ESMO Open*, 6(6), 100280. DOI: 10.1016/j.esmoop.2021.100280 PMID: 34634633

Dorsey, F. B. D., McDaniel, S. L., Levin, R. B., Vacca, J. P., Darke, P. L., Zugay, J. A., Emini, E. A., Schleif, W. A., Lin, J. H., Chen, I. W., Holloway, M. K., Anderson, P. S., & Huff Bioorg, J. R. (1994). Synthesis and evaluation of pyridyl analogs of L-735,524: Potent HIV-1 protease inhibitors Med. *Bioorganic & Medicinal Chemistry Letters*, 4(23), 2769–2774. DOI: 10.1016/S0960-894X(01)80592-8

Durso, M., Gentili, D., Bettini, C., Zanelli, A., Cavallini, M., De Angelis, F., Grazia Lobello, M., Biondo, V., Muccini, M., Capelli, R., & Melucci, M. (2013). π-Core tailoring for new high performance thieno(bis)imide based n-type molecular semiconductors. *Chemical Communications (Cambridge)*, 49(39), 4298–4300. DOI: 10.1039/C2CC37053K PMID: 23165167

aEl-Remaily, M. A. E. A. A. A., Soliman, A. M., Khalifa, M. E., El-Metwaly, N. M., Alsoliemy, A., El-Dabea, T., & Abu-Dief, A. M. (2022). Rapidly, highly yielded and green synthesis of dihydrotetrazolo [1, 5-a] pyrimidine derivatives in aqueous media using recoverable Pd (II) thiazole catalyst accelerated by ultrasonic: Computational studies. *Applied Organometallic Chemistry*, 36(2, No. e6320), e6320. DOI: 10.1002/aoc.6320bEl Remaily, M. A. E. A. A. A., Naidu, V. R., Ni, S., & Franzén, J. (2015). Carbocation Catalysis: Oxa-Diels-Alder Reactions of Unactivated Aldehydes and Simple Dienes. *European Journal of Organic Chemistry*, 2015(30), 6610–6614. DOI: 10.1002/ejoc.201501112

Fei, Z., Pattanasattayavong, P., Han, Y., Schroeder, B. C., Yan, F., Kline, R. J., Anthopoulos, T. D., & Heeney, M. (2014). Influence of side-chain regiochemistry on the transistor performance of high-mobility, all-donor polymers. *Journal of the American Chemical Society*, 136(43), 15154–15157. DOI: 10.1021/ja508798s PMID: 25302474

Feng, M., Tang, B., Liang, S., & Jiang, X. (2016). Sulfur containing scaffolds in drugs: Synthesis and application in medicinal chemistry. *Current Topics in Medicinal Chemistry*, 16, 1200–1216. DOI: 10.2174/15680266156661509151117 41 PMID: 26369815

Fernandes, S. S. M., Castro, M. C. R., Mesquita, I., Andrade, L., Mendes, A., & Raposo, M. M. M. (2017). Synthesis and characterization of novel thieno [3,2-*b*] thiophene based metal-free organic dyes with different heteroaromatic donor moieties as sensitizers for dye-sensitized solar cells. *Dyes and Pigments*, 136, 46–53. DOI: 10.1016/j.dyepig.2016.08.020

Fuller, L. S., & Iddon, B. (1999). K A Smith J. Thienothiophenes. Part 3.1,2 On the ring-opening reactions of 3-lithiated and 3,6-dilithiated thieno[3,2-*b*]thiophenes; new routes to polyfunctionalized thiophenes and enediynes Chem. Soc. *Journal of the Chemical Society, Perkin Transactions 1: Organic and Bio-Organic Chemistry*, 1(10), 1273–1278. DOI: 10.1039/a901300h

Haristoy, D., Mery, S., Heinrich, B., Mager, L., Nicoud, J. F., & Guillon, D. (2000). Structure and photoconductive behaviour of a sanidic liquid crystal. *Liquid Crystals*, 27(3), 321–328. DOI: 10.1080/026782900202769

Hu, L., Fan, M., Shi, S., Song, X., Wang, F., He, H., & Qi, B. (2022). Dual target inhibitors based on EGFR: Promising anticancer agents for the treatment of cancers (2017-). *European Journal of Medicinal Chemistry*, 227, 113963. DOI: 10.1016/j.ejmech.2021.113963 PMID: 34749202

Isci, R., Rahimi Varzeghani, A., Kaya, K., Sütay, B., Tekin, E., & Ozturk, T. (2022). Triphenylamine/Tetraphenylethylene Substituted 4-Thieno[3,2-b]thiophen-3-ylbenzonitriles: Synthesis, Photophysical-Electronic Properties and Applications. *ACS Sustainable Chemistry & Engineering*, 10(4), 1605–1615. DOI: 10.1021/acssuschemeng.1c07240

Jen, A. K.-Y., Rao, V. P., Drost, K. J., Cai, Y., Mininni, R. M., Kenney, J. T., Binkley, E. S., & Dalton, L. R. (1994). The Development of Highly Active, Thermally Stable Chromophores and Polymers for Electrooptic Applications. *Proceedings of SPIE - the International Society for Optical Engineering*, 2285, 49–57. DOI: 10.1117/12.187500

Kidwai, M., Venkataramanan, R., & Dave, B. (2001). Solventless synthesis of thiohydantoins over K2CO3. *Green Chemistry*, 3(6), 278–279. DOI: 10.1039/b106034c

aKim, O. K., Woo, H. Y., Heuer, W. B., Kim, K. S., & Lee, K. S. (2000). Electronic/photonic property modulations of D-π-D (A) chromophores by DTT as π-center. *American Chemical Society, Polymer* Preprints, *Division of Polymer Chemistry, 41*(1), 795-796.

Lee, H. S., Lee, J. S., Jung, A. R., Cha, W., Kim, H., Son, H. J., Cho, J. H., & Kim, B. S. (2016). Processing temperature control of a diketopyrrolopyrrole-alt-thieno [2,3-*b*]thiophene polymer for high-mobility thin-film transistors and polymer solar cells with high open-circuit voltages. *Polymer*, 105, 79–8. DOI: 10.1016/j.polymer.2016.10.024

Luhman, U., Wentz, F. G., Knieriem, B., & Luttke, W. (1978). Theoretical and spectroscopical investigations of indigo dyes, xviii. properties of 2,2,5,5-tetramethylthieno(3,2-*b*)thiophene-3(2*H*),6(5*H*)-dione, a basic chromophoric system of thioindigo dyes with trans-s-trans-s-trans-configuration of the carbonyl groups. *Chemische Berichte*, 111, 3233. DOI: 10.1002/cber.19781110927

Mabkhoot, Y. N. (2009). Synthesis and analysis of some bis-heterocyclic compounds containing sulphur. *Molecules (Basel, Switzerland)*, 14(5), 1904–1914. DOI: 10.3390/molecules14051904 PMID: 19471210

Marco, A. B., Andreu, R., Franco, S., Garín, J., Orduna, J., Villacampa, B., & Alicante, R. (2013). Efficient second-order nonlinear optical chromophores based on dithienothiophene and thienothiophene bridges. *Tetrahedron*, 69(19), 3919–3926. DOI: 10.1016/j.tet.2013.03.027

Marco, A. B., Gindre, D., Iliopoulos, K., Franco, S., Andreu, R., Canevet, D., & Sallé, M. (2018). (Super)gelators derived from push-pull chromophores: Synthesis, gelling properties and second harmonic generation. *Organic & Biomolecular Chemistry*, 16(14), 2470–2478. DOI: 10.1039/C8OB00251G PMID: 29561017

Mashraqui, S. H., Ghadigaonkar, S., Ashraf, M., Sri Ranjini, A., Ghosh, S., & Das, P. K. (2007). Optically transparent and thermally stable nonlinear optic chromo. phores featuring a thieno [2,3-b]thiophene donor. *Tetrahedron*, 63(40), 10011–10017. DOI: 10.1016/j.tet.2007.07.059

Mashraqui, S. H., Sangvikar, Y. S., & Meetsma, A. (2006). Synthesis and structures of thieno [2,3-b]thiophene incorporated [3.3]dithiacyclophanes. Enhanced first hyperpolarizability in an unsymmetrically polarized cyclophane. *Tetrahedron Letters*, 47(31), 5599–5602. DOI: 10.1016/j.tetlet.2006.05.098

Massin, J., Ducasse, L., Abbas, M., Hirsch, L., Toupance, T., & Olivier, C. (2015). Molecular engineering of carbazole-fluorene sensitizers for high open-circuit voltage DSSCs: Synthesis and performance comparison with iodine and cobalt electrolytes. *Dyes and Pigments*, 118, 76–87. DOI: 10.1016/j.dyepig.2015.02.020

Mu, S., Oniwa, K., Jin, T., Asao, N., Yamashita, M., & Takaishi, S. (2016). A highly emissive distyrylthieno [3,2-b]thiophene based red luminescent organic single crystal: Aggregation induced emission, optical waveguide edge emission, and balanced ambipolar carrier transport. *Organic Electronics*, 34, 23–27. DOI: 10.1016/j.orgel.2016.04.001

Noh, J. H., Jeon, N. J., Choi, Y. C., Nazeeruddin, M. K., Grätzel, M., & Seok, S. (2013). Il Nanostructured TiO2/CH3NH3PbI3 heterojunction solar cells employing spiro-OMeTAD/Co-complex as hole-transporting material. *Journal of Materials Chemistry. A, Materials for Energy and Sustainability*, 1(38), 11842–11847. DOI: 10.1039/c3ta12681a

Kim, O. K., Woo, H., Kim, J. K., & Huang, Z. (2000, March). New class of light-emitting polymers/oligomers. In *Liquid Crystal Materials, Devices, and Flat Panel Displays* (Vol. 3955, pp. 134-140). SPIE.

Paek, S., Zimmermann, I., Gao, P., Gratia, P., Rakstys, K., Grancini, G., Nazeeruddin, M. K., Malik, A. R., Samia, A. K., & Khalid, A. A.. (2016). Donor-π-donor type hole transporting materials: Marked π-bridge effects on optoelectronic properties, solid-state structure, and perovskite solar cell efficiency. *Chemical Science (Cambridge)*, 7(9), 6068–6075. DOI: 10.1039/C6SC01478J PMID: 30034747

Peng, S. H., Huang, T. W., Gollavelli, G., & Hsu, C. S. (2017). Thiophene and diketopyrrolopyrrole based conjugated polymers as efficient alternatives to spiro-OMeTAD in perovskite solar cells as hole transporting layers. *Journal of Materials Chemistry. C, Materials for Optical and Electronic Devices*, 5(21), 5193–5198. DOI: 10.1039/C7TC00966F

Ravichandran, S., & Karthikeyan, E. (2011). Microwave synthesis-a potential tool for green chemistry. *International Journal of Chemtech Research*, 3, 466–470.

Rawcliffe, R., Shkunov, M., Heeney, M., Tierney, S., McCulloch, I., & Campbell, A. (2008). Organic field-effect transistors of poly(2,5-bis(3- dodecylthiophen-2-yl) thieno [2,3-*b*]thiophene) deposited on five different silane self-assembled monolayers. *Chemical Communications (Cambridge)*, (7), 871–873. DOI: 10.1039/B715536K

Redecker, M., Bradley, D. D. C., Inbasekaran, M., & Woo, E. P. (1999). Mobility enhancement through homogeneous nematic alignment of a liquid-crystalline polyfluorene. *Applied Physics Letters*, 74(10), 1400–1402. DOI: 10.1063/1.123563

Rosell, R., Carcereny, E., Gervais, R., Vergnenegre, A., Massuti, B., Felip, E., Palmero, R., Garcia-Gomez, R., Pallares, C., Sanchez, J. M., Porta, R., Cobo, M., Garrido, P., Longo, F., Moran, T., Insa, A., De Marinis, F., Corre, R., Bover, I., & Paz-Ares, L. (2012). Erlotinib versus standard chemotherapy as first-line treatment for European patients with advanced EGFR mutation-positive non-small-cell lung cancer (EURTAC): A multicentre, open-label, randomised phase 3 trial. *The Lancet. Oncology*, 13(3), 239–246. DOI: 10.1016/S1470-2045(11)70393-X PMID: 22285168

Sigismund, S., Avanzato, D., & Lanzetti, L. (2018). Emerging functions of the EGFR in cancer. *Molecular Oncology*, 12(1), 3–20. DOI: 10.1002/1878-0261.12155 PMID: 29124875

Stuerga, D., & Gaillard, P. (1996). Microwave heating as a new way to induce localized enhancements of reaction rate. Non-isothermal and heterogeneous kinetics. *Tetrahedron*, 52(15), 5505–5510. DOI: 10.1016/0040-4020(96)00241-4

Tamilavan, V., Kim, S., Sung, J., Lee, D. Y., Cho, S., Jin, Y., Jeong, J., Park, S. H., & Hyun, M. H. (2017). Enhanced photovoltaic performances of bis(pyrrolo [3,4-*c*]pyrrole-1,3-dione)-based wide band gap polymer via the incorporation of an appropriate spacer unit between pyrrolo [3,4-*c*]pyrrole-1,3-dione units. *Organic Electronics*, 42, 34–41. DOI: 10.1016/j.orgel.2016.12.012

Toda, F. (1995). Solid state organic chemistry: Efficient reactions, remarkable yields, and stereoselectivity. *Accounts of Chemical Research*, 28(12), 480–486. DOI: 10.1021/ar00060a003

Tsivgoulis, G. M., & Lehn Angew, J.-M. (1995). Photonic Molecular Devices: Reversibly Photoswitchable Fluorophores for Nondestructive Readout for Optical Memory Chem. *Angewandte Chemie International Edition in English*, 34(10), 1119–1122. DOI: 10.1002/anie.199511191

Tso, H.-H., Wang, J.-S., Wu, C.-Y., & Lin, H.-C. (1998). H-C Lin Novel Supramolecular Liquid- Crystalline Complexes Derived from 5-n-Decyl Substituted Thieno[3,2-*b*]thiophene-2- and Thiophene-2-Carboxylic Acids. *New Journal of Chemistry*, 22(8), 771–773. DOI: 10.1039/a803315c

Vince, B., Hill, J. M., Lawitz, E. J., O'Riordan, W., Webster, L. R., Gruener, D. M., Mofsen, R. S., Murillo, A., Donovan, E., Chen, J., McCarville, J. F., Sullivan-Bólyai, J. Z., Mayers, D., & Zhou, X. J. (2014). A randomized, double-blind, multiple-dose study of the pan-genotypic NS5A inhibitor Compound 9 in patients infected with hepatitis C virus genotype 1, 2, 3 or 4. *Journal of Hepatology*, 60(5), 920–927. DOI: 10.1016/j.jhep.2014.01.003 PMID: 24434503

Wang, X., Li, Y., Song, P., Ma, F., & Yang, Y. (2019). Second-Order Nonlinear Optical Switch Manipulation of Photosensitive Layer by an External Electric Field Coupled with Graphene Quantum Dots. *The Journal of Physical Chemistry A*, 123(34), 7401–7407. DOI: 10.1021/acs.jpca.9b05249 PMID: 31369271

Yang, X., Xi, J., Sun, Y., Zhang, Y., Zhou, G., & Wong, W. Y. (2019). A dopant-free twisted organic small-molecule hole transport material for inverted planar perovskite solar cells with enhanced efficiency and operational stability. *Nano Energy*, 64, 103946. DOI: 10.1016/j.nanoen.2019.103946

Yue, J., Sun, S., Liang, J., Zhong, W., Lan, L., Ying, L., Huang, F., Yang, W., & Cao, Y. (2016). Effects of pyridyl group orientations on the optoelectronic properties of regio-isomeric diketopyrrolopyrrole based π-conjugated polymers. *Journal of Materials Chemistry. C, Materials for Optical and Electronic Devices*, 4(13), 2470–2479. DOI: 10.1039/C6TC00051G

Zhang, H., Hu, Y., Chen, X., Yu, L., Huang, Y., Wang, Z., Wang, S., Lou, Y., Ma, X., Sun, Y., Li, J., Ji, D., Li, L., & Hu, W. (2022). A Centrosymmetric Organic Semiconductor with Donor–Acceptor Interaction for Highly Photostable Organic Transistors. *Advanced Functional Materials*, 32(20), 2111705. DOI: 10.1002/adfm.202111705

Zhang, Q., Wang, Y., Kan, B., Wan, X., Liu, F., Ni, W., Feng, H., Russell, T. P., & Chen, Y. (2015). A solution-processed high performance organic solar cell using a small molecule with the thieno [3,2-*b*]thiophene central unit. *Chemical Communications (Cambridge)*, 51(83), 15268–15271. DOI: 10.1039/C5CC06009E PMID: 26329677

Zhu, S., An, Z., Chen, X., Chen, P., & Liu, Q. (2015). Cyclic thiourea functionalized dyes with binary π-linkers: Influence of different π-conjugation segments on the performance of dye-sensitized solar cells. *Dyes and Pigments*, 116, 146–154. DOI: 10.1016/j.dyepig.2015.01.022

Chapter 7
Synthesis, Properties, and Bioactivity of Thiazolopyrimidines:
Preparation and Biological Activities of Thiazolopyrimidines

Navnath Tulshiram Hatvate
https://orcid.org/0000-0002-3388-3247
Institute of Chemical Technology, Jalna, India

Mukta Vivek Lele
https://orcid.org/0009-0008-7088-5170
Institute of Chemical Technology, Jalna, India

ABSTRACT

Thiazolopyrimidines are the fusion of thiazoles and pyrimidines heterocyclic rings, which has attracted much attention from medicinal and organic chemists because of its wide range of applications. Thiazolopyrimidines have applications in the pharmaceutical and agricultural fields, including analgesic, anti-inflammatory, herbicidal, antifungal, antimicrobial, antitumor, and antiparasitic properties. This chapter reviews the chemistry, different methods of synthesis and bioactivities of Thiazolopyrimidines. In addition, the patent and current clinical trials were also updated.

DOI: 10.4018/979-8-3693-7520-4.ch007

1.INTRODUCTION OF THIAZOLOPYRIMIDINES

Heterocycles are organic compounds containing two or more hetero atoms in their structure (Hatvate et al., 2018; Hatvate & Ghodse, 2020). They are popular due to the fact that they are easy to synthesize, resemble a wide range of similarities with the natural chemical structures of the body, and can also serve as a core component of drugs (Gadkari et al., 2021a, 2021b; Gadkari, Hatvate, et al., 2020; Gadkari, Jadhav, et al., 2020). Pyrimidines are one such heterocycle, which mainly resembles their structure with the DNA and RNA base pairs. Thus, pyrimidine is one of the major scaffolds required for genetic material and, thus, the further growth of individuals (Nammalwar & Bunce, 2024). 2,4 Dihydropyrimidines have gained recent attention due to their increased therapeutic value (Gadkari et al., 2021b). Thiazole /i.e., sulfur-containing heterocycle is known for its therapeutic applications, such as anti-inflammatory activity, anti-viral, and anti-tumor (Zarenezhad et al., 2021)(Parit et al., 2024). Thus, a heterocycle containing both thiazole and pyrimidine would give a synergistic activity, which could be further explored for all the above activities mentioned earlier. Several methods are available to synthesize the compounds, but green and nanoparticle-containing catalysts are the recent green chemistry techniques being developed for synthesizing these agents. Further, the synthesized thiazolopyrimidine (THPM) heterocycle can also be accessed to test other biological activities. Thus, THPMs are explored as potential therapeutic agents (NAGARAJAIAH et al., 2012)(Vasilkova et al., 2020).

1.1 Chemistry of Thiazolopyrimidines

As discussed earlier, THPM is a heterocyclic combination of thiazoles and pyrimidines. The structure of these molecules is as follows. This molecule can be substituted at various positions to obtain substituted THPM. The substitutions/ positions of different atoms in the nucleus are significant for the activity of THPM. As seen in Figure 2, the SAR of THPM reveals that the alkylated benzene ring is an active site for substitution and could give the THPM ring biological activity. Leaving groups can also be attached to the molecule, suggesting further scope for substituting it(Kendre et al., 2020). Further, the THPM molecule is reported to exist in a tautomeric form to attain stability (GOOD, 2015).

Figure 1. Thiazolopyrimidine and substituted THPM.

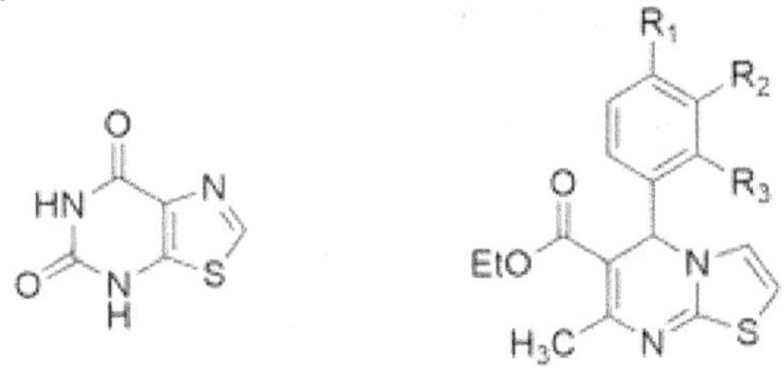

Figure 2. SAR of THPM

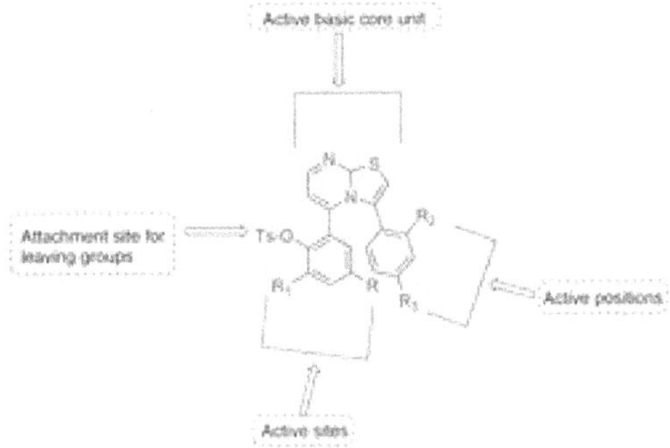

THPM exists as a tautomeric form in which electrons' resonance occurs within the ring.

Figure 3. Tautomeric forms of THPM.

2. SYNTHESIS OF THIAZOLOPYRIMIDINES

For the determination of the biological activity of THPM derivatives, most researchers synthesize the molecule using simple reactions or various other methods. Most of the researchers have also tried to synthesize the molecule using green chemistry techniques such as microwave irradiation. Some of the strategies used to synthesize this compound are discussed below.

2.1. From 3,4 dihydropyrimidine2-thione

In this synthesis, the reactants, 3,4-Dihydropyrimidines and ethyl-2-chloroacetate, were reacted in the presence of ethanol as a solvent for around 6 hours. A salt was formed after the evaporation of the solvent. The residue was dissolved in aqueous media and neutralized by the addition of base. Different techniques further purified the formed product. The appearance of the reported compound is yellow-colored product **1** (NAGARAJAIAH et al., 2012).

Figure 4. Reaction Scheme 1: Synthesis of THPM by 3,4 dihydropyrimidine2-thione

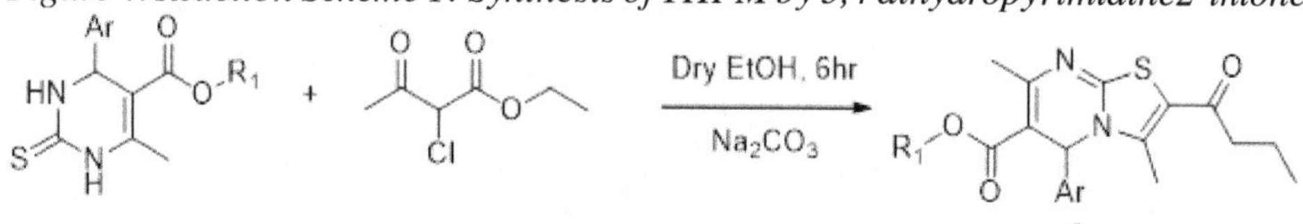

2.2 One-Pot Synthesis of THPM

Many organic chemists use one-pot synthesis to synthesize compounds, and THPM was also reported to be synthesized by the same. In this reaction, cyclohexanone and NBS were taken and allowed to react in the presence of PTSA in an organic solvent at ambient temperature. The formed intermediate was subjected to cyclization with THPM to yield the desired product **2**. (Sekhar et al., 2020).

Figure 5. Reaction Scheme 2: One-pot synthesis of substituted THPM

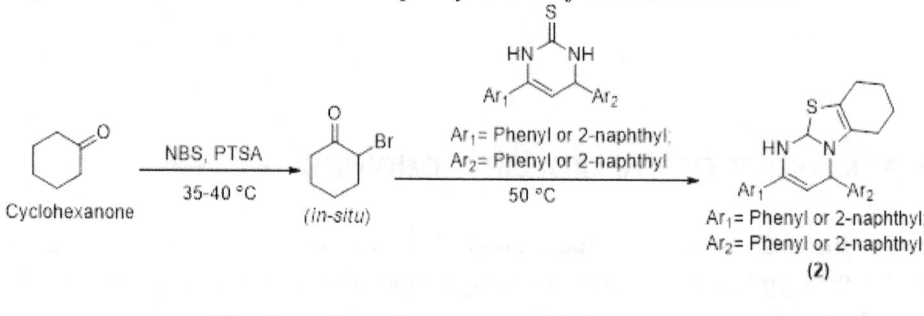

2.3 Synthesis of THPM with Ethyl Acetoacetate

A green chemistry approach for the synthesis of THPM is reported. This is a three-component condensation of ethyl acetoacetate with 1,3- thiazol-2-amine and aldehyde. Isopropanol is used as a solvent. The reaction is conducted under ultrasonic conditions at 20 °C for 2-3 hours, giving the desired heterocycle. Different aldehydes give different derivatives of **3** (Vasilkova et al., 2020).

Figure 6. Reaction Scheme 3: Synthesis of THPM by using ethyl acetoacetate

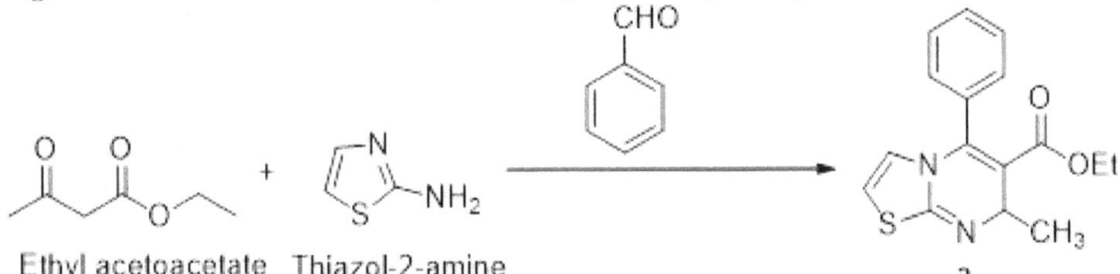

3

2.4 From dihydropyrimidine-2(1*H*)-thione

Aromatic aldehydes/ketones are reacted with thiourea in the presence of alcoholic KOH to give dihydropyrimidine-2(1*H*)-thione (**4**). This compound is a starting material for the generation of different compounds. On further treatment with chloroacetic acid/bromoacetic acid, different derivatives of THPM were formed (Gupta & Chaudhary, 2013).

Figure 7. Reaction Scheme 4: From dihydropyrimidine-2(1H)-thione

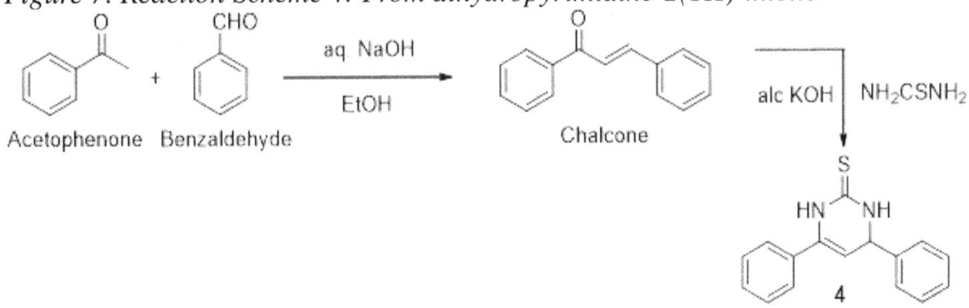

4

3. BIOLOGICAL ACTIVITY OF THPM

3.1 Anti-Convulsant Activity

A drug's anticonvulsant activity (ACV) is its ability to prevent seizures/convulsions by regulating the electrical activity in the nervous system. They are used as medications to treat seizures and epilepsy. The most commonly used ACV agents are phenobarbital and lamotrigine. THPM derivatives are found to be used as ACV agents. In a study conducted by various universities of the Department of Pharmaceutical Chemistry, Ukraine, negative results were.to ACV activity was obtained. The ACV study was carried out on male rats, and the reference ACV agents were phenobarbital and lamotrigine. Seizures were induced using the pentylenetetrazole-induced seizure model and were administered subcutaneously. The compound didn't have significant activity because of the condensed nature of THPM derivatives (Severina et al., 2019).

3.2 Anti-Cancer Activity

Cancer or tumor refers to the uncontrollable growth of cell mass or tissue. The disease is usually not curable if detected at a later stage and, hence, is one of the leading causes of death in the world. Thus, a wide range of results is being obtained in this field. THPM, a multifunctional moiety, has been examined for anti-cancer activity by various researchers.

Anti-cancer drugs inhibit the excessive growth of cells. Most of them bind to the nucleic acid and cause specific changes in DNA, preventing the cell from further dividing. Hence, DNA binding is checked for anti-cancer agents. In research, the anti-cancer activity of THPM was evaluated. Purine analogs were used as starting blocks for the synthesis of THPM. Various compounds were synthesized. Two compounds, **THPM 1,** had a notable DNA binding affinity of 28.38 ± 1. The positive reference used was doxorubicin. **THPM 1** was found to have the highest anti-cancer activity with GI_{50}, TGI, and LC_{50} values of 1.07, 6.61, and 34.7 μM, respectively. Further, it was concluded that the drugs give anti-cancer activity by binding to the DNA groves through hydrogen bonding. Molecular docking studies proved this (Al-Rashood et al., 2020)

Another study on the anti-cancer activity of THPM derivatives on the liver, prostate, and human colorectal cancer was conducted. For that, the cell cultures of the respective carcinomas, i.e., HCT-116, PC-3, and HepG-2, respectively, were taken. The cells were incubated, and their cytotoxic activity was studied using MTT (3-[4,5-dimethyl-2-thiazolyl)-2,5diphenyl-*2H*-tetrazolium bromide) assay. Before these, some cells were treated with doxorubicin as a positive control at 37 °C for two days while THPM derivatives were added to other plates. After that, MTT was given in each well and incubated for some hours. After incubation, a purple-colored formazan dye was formed and analyzed using a UV-Vis spectrophotometer at 590 nm. The cell viability of the treated cells was compared with that of the control group to interpret the results. Out of the results obtained, THPM 2 and THPM 3 showed significant activity against HCT-116 cancer cells with IC_{50} values of 60.9 ± 1.8 and 58.2 ± 5.1, respectively. **THPM 4** was tested for anti-cancer activity against PC-3 cancer cells, with 75.07 and 75.93 as IC_{50} values. Further, no significant activity against the HePG-2 cancer cell line was found for the THPM derivatives synthesized. The structural co-relation of the compound suggests the substituted diaryl THPM nucleus, which does not have oxo-substitution at C-6, was more active than the corresponding 6-oxo analog. Additionally, in the THPM system with arylidine function, substitution at the methylene carbon produced more active derivatives than their counterparts with the free CH_2 group, mainly when the trimethoxyphenyl moiety was present (Yousif et al., 2017). A similar study was conducted on the anti-cancer activity of THPM derivatives on the above cell lines. MTT assay was used, while doxorubicin was the positive control. 5 THPM derivatives were reported to give the desired anti-tumor activity against HCT-116 carcinoma cells, with their IC_{50} ranging from 92.2 to 120.1 nM. These compounds had a glycosyl moiety attached via a thioglycosidic linkage to the THPM ring system. The inhibition of activity further increases in the presence of xylopyranosyl moiety. The precise IC_{50} values in nM are tabulated below. Compounds 5 and 6 had potent activity against the PC-3 cell line, while **THPMs 5 and 6** had a moderate effect. Further, some had moderate activity against HeGp-2 cancer tissue. The IC_{50} of the compounds is tabulated below (Flefel et al., 2017)

Table 1. Anti-cancer activity of THPM

Compound	IC_{50} value PC-3 (nM) ± SD
THPM 5	223.9 ± 6.4
THPM 6	281.8 ±3.6

Figure 8. THPM derivatives showing anti-cancer activity

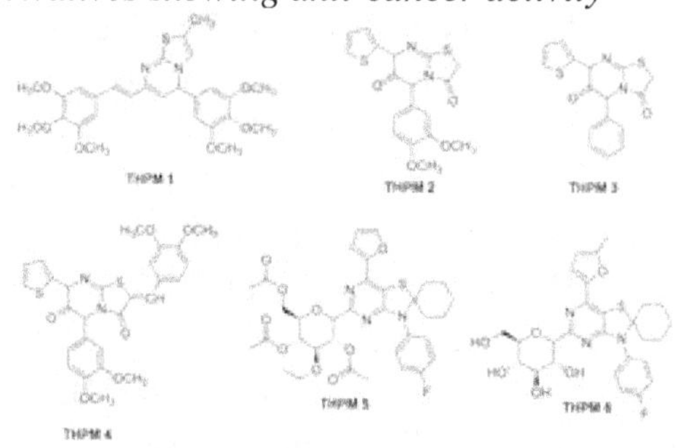

3.3 Anti-Inflammatory Activity

Inflammation is the primary indication of wounds and some of the major diseases. It is a phenomenon where the body's immune system is triggered and tends to produce Interleukins and cytokines. This is commonly treated using anti-inflammatory agents. Most of these agents are well-accepted by people. THPM is a newer class of derivatives that have been proven to have good anti-inflammatory activity (Saleh & Yousaf, 2018).

Animal models were used to study the anti-inflammatory activity of THPM derivatives. Rats were induced with carrageenan to produce inflammation. THPM derivatives and appropriate control groups were administered on the appearance of inflammation. The difference in paw height was measured, which served as a criterion for the anti-inflammatory effect of the compound. Several compounds gave the desired effect. Some of them are listed below. These compounds were reported to provide a percentage decrease in paw height by 67, 65, 68 and 66%, respectively. The action could be due to heterocycles (predominantly Nitrogen-containing rings such as piperazine and pyrimidine), oxadiazepine, and thiazole (Benzodifuranyl, 2020).

In a procedure similar to the one discussed above, anti-inflammatory activity was tested on Wistar rats. Rats were used to develop paw edema by injecting 0.1 mL of 1% (w/v) carrageenan in distilled water subcutaneously (s.c.) into the left hind paw's subplantar area. As a control, a group of six rats received an oral dose of DMSO in accordance with their weight. THPM was given in dosages of 20 mg/kg, p.o. Indomethacin was used as a reference. Rats' paw volumes were measured using a plethysmometer before and after receiving a 1% carrageenan injection at various intervals (1, 2, 3, and 4 hours). Each group's edema and inhibition rates were computed at the respective time intervals. The THPM derivative below had a significant anti-inflammatory activity compared to the reference and other derivatives used (Khalifa et al., 2015).

Figure 9. THPM derivatives showing anti-inflammatory activity

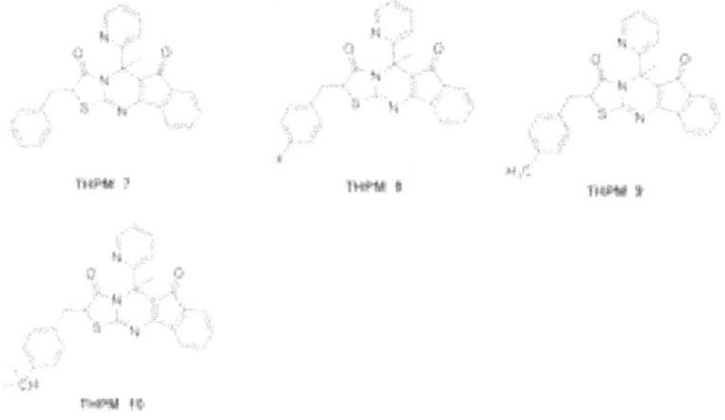

3.4 Anti-Bacterial Activity

Bacteria are a common cause of diseases in humans. Some of these bacteria cause harm to humans in the form of rashes, allergies, digestive system issues, and much more. Hence ant-bacterial are given to retard the growth of these microorganisms. These anti-bacterial drugs/chemicals are designed to control the growth of bacteria while being non-toxic to surrounding tissues. As a recent area of interest, THPM derivatives are also employed as anti-bacterial (Pereira et al., 2015).

In a study conducted using the micro broth dilution experiment, microbial strains *P. aeruginosa* were used to assess the anti-bacterial properties of THPM. As the most frequent causes of food poisoning in humans due to *S. aureus*, *Escherichia coli*, *P. aeruginosa*, *Pseudomonas,* and *S. pneumonia*, the anti-bacterial activity of these compounds was tested on them. The culture medium used was nutrient agar medium. The composition of nutrient agar was Beef Extract- 10 g, Peptone- 10 g, Sodium Chloride- 5 g, Distilled Water- 1000 ml. The standard drug used for the study was marketed as anti-bacterial ciprofloxacin. Two of the synthesized derivatives were found to have a required anti-bacterial activity comparable to the standard. Below is the table representing the preliminary screened antibacterial activity of the compounds against various bacterial strains. The values are expressed in mm as per their zone of inhibition. The authors predict that the activity shown is due to the presence of an indole-3-yl substituent at the fourth position of the 1,4 tetrahydropyrimidine ring (Jayanthi et al., 2022).

Table 2. THPM derivatives showing anti-bacterial activity against different microbial species

Compounds (Zone of inhibition in mm) Mean ± SEM	*E. coli*	*P. aeruginosa*	*S. aureus*	*Pseudomonas*	*B. subtilis*	*K. pneumoniae*
THPM 11	7.1	6.2	6.9	4.7	6.13	5.7
THPM 12	8.2	8.6	8.1	9.5	6.7	7.9
Ciprofloxacin (10µg/ml)	11.4	16.2	17.1	15.7	18.4	21.3

Figure 10. THPM derivatives showing anti-bacterial activity

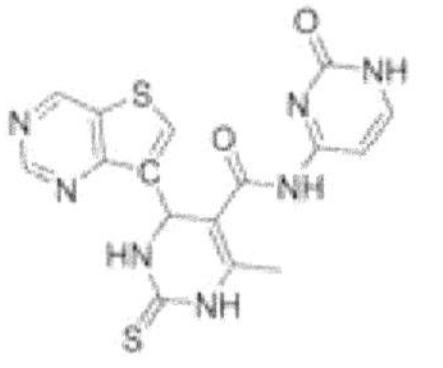
THPM 11

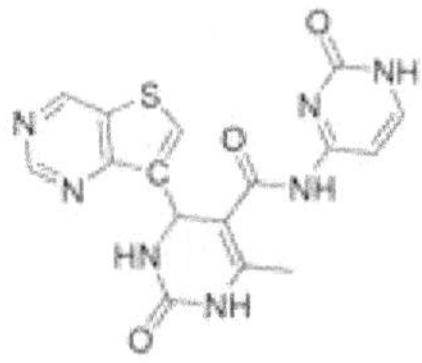
THPM 12

3.5 Anti-Microbial Activity

In a study conducted to check the Anti-microbial activity, the agar diffusion method was used at three different concentrations (0.5, 1.0, and 2.0 mg/mL) of THPM derivatives. They were tested against the two common bacterial strains, *Staphylococcus aureus* and *E. coli*. The nutrient medium consists of 10 g of peptone, 10 g of NaCl, 5 g of yeast extract, and 20 g of agar in 1000 mL of distilled water. Gentamycin, an anti-microbial, was considered a standard and filled in control vials to compare the activity of THPM derivatives. After a day of incubation at 37 °C, the zone of inhibition was measured. **THPM** 13 demonstrated a significant activity, while 14,15 and 16 gave satisfactory results. Similarly, the derivatives were screened against *Aspergillus niger* and *Penicillium chrysogenum* by the same method. The derivatives did not showed any significant activity compared to the standard. (NAGARAJAIAH et al., 2015)

Figure 11. THPM derivatives showing anti-microbial activity

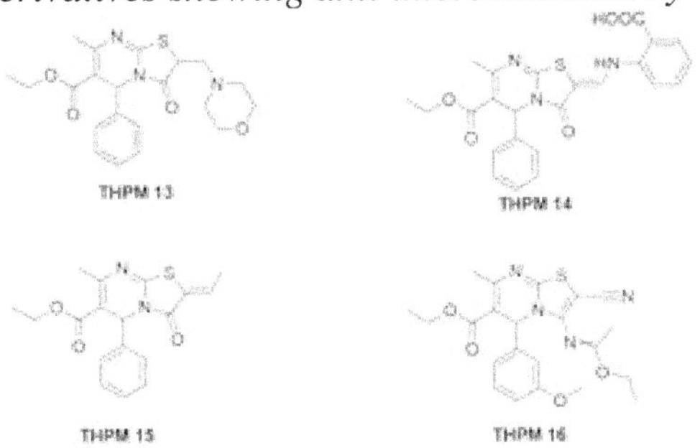

3.6 Anti-Fungal Activity

The anti-fungal activity of THPM compounds was studied in one of the research. This activity was studied on fungi species like *Aspergillus niger* and *Aspergillus ochraceus* by cup diffusion technique. Dithane M-45 was considered as a standard for the same. Out of the compounds tested, **THPM 17 and 18** had similar activity to the standard at a 0.25 mg/mL concentration. The typical structural features of the compounds that were considered to give activity include the presence of 2-thioxo-pyrimidine in the ring structure along with chloride or a dialkyl group in the nucleus. These substituted rings gave more activity than non-substituted compounds. The attachment of polycyclic compounds gave a potent activity. Further, it was discovered that the arylidene derivative of THPM and a halide had the potency of giving good anti-fungal activity (Ghorab et al., 1996).

Another novel approach, wherein a steroidal ring is attached to THPM moiety, was studied to explore its effectiveness as an anti-fungal agent. THPM was a pharmacophore in the D-ring, a steroid was taken as the fundamental skeleton, and various steroidal THPM derivatives were synthesized. Their activity was assessed against Candida species. CYP51 is necessary for both the production of cholesterol in mammals and the ergosterol production in fungi. As a result, the main target of therapeutic azole anti-fungal medications is fungus CYP51. A molecular docking analysis was performed to justify the observed biological activity and identify possible binding mechanisms. Additionally, an in-silico investigation was done to predict the pharmacokinetics. It was found that the synthesized compounds had promising anti-fungal activity and could be further explored in the field of anti-fungal agents (Iqbal et al., 2023)

Figure 12. THPM derivatives showing anti-fungal activity

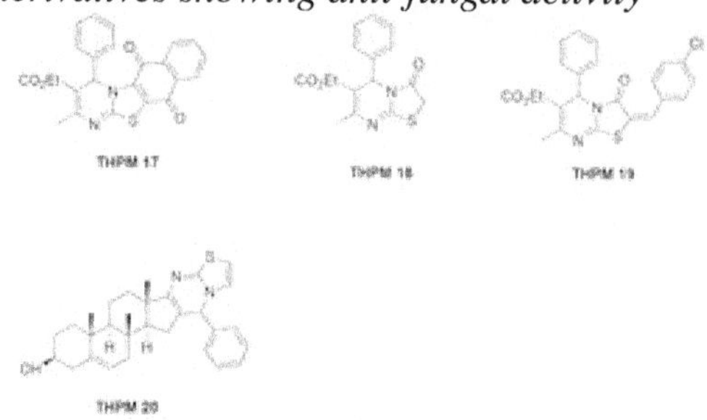

3.7 Antinociceptive Activity

When the body reacts to toxic stimuli, chemicals, drugs, injury, or adverse conditions, the action is called antinociceptive activity (Hill & Canals, 2022).

THPM derivatives were evaluated for their antinociceptive activity. The characteristic tail-flick technique was used in Wistar Albino mice, and the standard drug given was diclofenac sodium. The synthesized THPM derivatives were administered, and a control group was also kept. The reaction/response time on the animals was recorded at regular intervals of half hours up to 3 hours of treatment. The activity was calculated by a formula.

$$\text{Activity} = \frac{\textit{Reaction time after treatmet} - \textit{Reaction time before treatment}}{10 - \textit{Reaction time before treatment}} x\ 100$$

Where 10 is the cut-off time for tail flick

From all the synthesized derivatives, it was observed that six of the compounds had excellent activity. The characteristic feature of the compounds was that all of them were para-substituted. In the halogens, the fluoro derivatives had an edge in activity over other halogens (Selvam et al., 2012).

Table 3. Evaluation of antinociceptive activity of THPM derivatives

THPM (% antinociceptive activity)	30 min	1 h	2 h	3 h
THPM 21	22 ± 1.4	26 ± 1.7	30 ± 1.6	25 ± 1.1
THPM 22	24 ± 1.7	28 ± 1.7	32 ± 1.6	29 ± 1.5
THPM 23	28 ± 1.1	32 ± 1.3	35 ± 1.4	33 ± 1.4
THPM 24	26 ± 1.4	32 ± 1.7	34 ± 1.6	30 ± 1.0
THPM 25	25 ± 1.7	30 ± 1.7	33 ± 1.6	29 ± 1.4
THPM 26	19 ± 1.4	25 ± 1.7	28 ± 1.6	24 ± 1.0

Figure 13. THPM derivatives showing antinociceptive activity

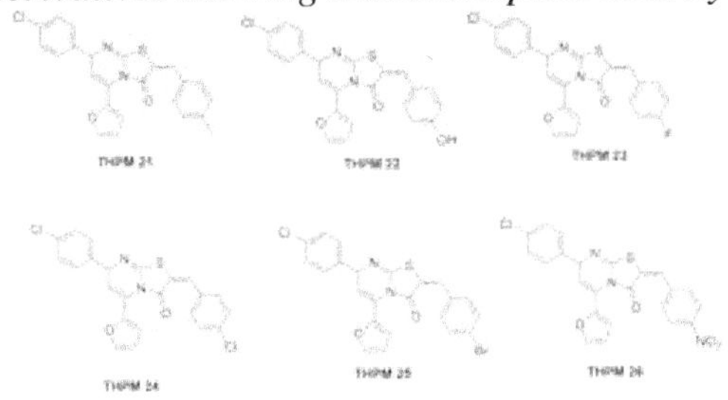

3.8 Anti-Parkinson's Activity

Parkinson's disease is a disease associated with the nervous system of the body. In this, the symptoms develop over a period of time, majorly being a tremor in hand. These tremors can worsen over a period of time and become uncontrollable, leading to stiffness of hands or difficulty in movement(Davie, 2008)

In a study conducted to evaluate the antiparkinson activity of THPM derivatives, a neuroleptic-induced catalepsy animal model is employed. This model relies on the animal response in which the animal is not able to gain its correct external posture. Mice used in a group of them were treated with haloperidol and should be able to travel below a rod within the stipulated time. A set of mice were also treated with levodopa (standard antiparkinson drug). Those induced with parkinson's but without any drug were not able to travel below the rod within the given time. THPM compounds were administered to a set of mice to evaluate the activity of THPM in comparison with haloperidol. Three of the compounds had a significant activity compared to the standard (Azam et al., 2010)

The SAR of these compounds denotes that the propyl group at **THPM** 29 had an edge over any aromatic/alkyl group. Further, if this propyl group was reduced to methyl or methylene blue, the activity recorded was less. However, in all three **THPMs,** the activity was compared to the standard, which proves that THPM derivatives have a significant scope as anti-parkinson agents.

Table 4. % catalepsy of anti-Parkinson's activity

Compound	**% catalepsy**
THPM 27	29.71%
THPM 28	27.98%
THPM 29	23.49%

Figure 14. THPM derivatives showing anti-Parkinson's activity

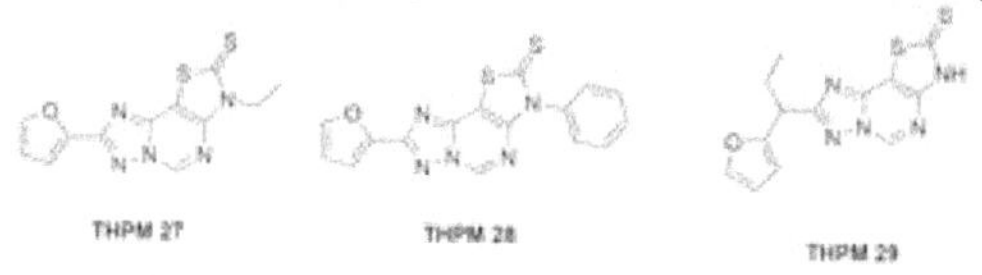

3.9 Anti-Viral Activity

The anti-viral activity of THPM derivatives was tested using a plaque reduction assay. A total of 17 compounds were tested, and two showed significant activity. They gave an inhibition of about 90% compared to the standard marketed drug acyclovir. Both the compounds gave an activity against HSV-1. This is mainly due to the presence of heterocyclic moieties in the compound. This moiety is known to alter the viral epitopes used to regulate cell DNA synthesis. The % inhibition of the **THPM** s was found to be 90%nd 95% at 2 and 5 μg/mL for **THPM** 30 at 2 μg/mL and 5 μg/mL, respectively (Mohamed et al., 2010).

A similar study was performed on THPM compounds to study their anti-viral activity, and the standard drug taken was acyclovir. Two compounds were reported to give notable anti-viral activity. Molecular docking studies were also performed to check the binding of the drugs to enzymes. HSV-1 thymidine kinase ligands were used as a target to check the binding affinity. **THPM** 31 had an inhibition of 87% at 15 μg/mL and 91% at 25 μg/mL, while **THPM** had an inhibition of 100% at both 50 μg/mL and 75 μg/mL (Mohamed et al., 2023).

Figure 15. THPM derivatives showing anti-viral activity

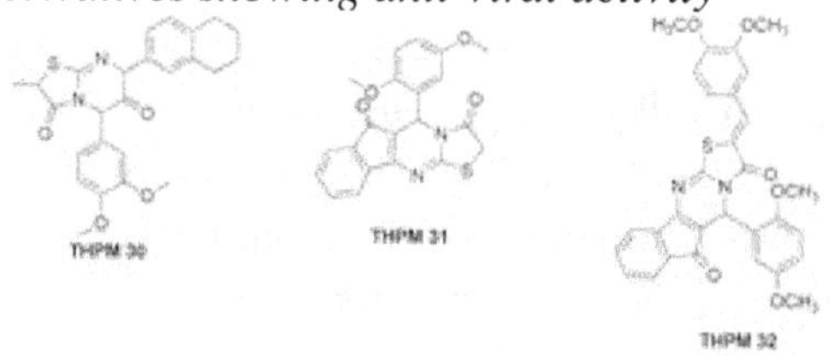

3.10 Anti-Psychotic Activity

Anti-psychotics are a class of drugs that are commonly administered to schizophrenic patients. The neurological disorder is commonly associated with positive, negative, and cognitive symptoms. The typical anti-psychotics are used to treat the positive and cognitive symptoms, while the negative symptoms are treated with atypical anti-psychotics. However, most of these agents are found to act on the dopaminergic receptor antagonist (Lewis, 2017).

In the study conducted, the anti-psychotic activity was tested on albino mice. As in schizophrenia, the subject loses its ability to think and act; the classical rota rod and a tail suspension test method were used. This test is performed to check the locomotor activity of the animal. The mice were divided into groups, one set being the control group, the other set being risperidone (standard drug), and the third set receiving THPM. The mice were kept in an actophotometer for 30 minutes to evaluate their locomotor activity. Next, they were subjected to a rota rod test and finally evaluated for a tail suspension test. Apart from this, QSAR studies were also performed on the prepared THPM. 5 compounds had comparable activity to the standard drug (Sawant, 2014).

Table 5. Evaluation of anti-psychotic activity

Compound	Locomotor activity (in sec)	Rota rod (performance time in sec)	Tail suspension (immobility time in sec)
THPM 33	118.24±0.84	48.49±1.25	71.33±1.85
THPM 34	122.26±1.04	44.33±1.85	67.53±2.17
THPM 35	120.37±0.72	46.62±1.73	69.45±1.15
THPM 36	121.21±1.53	45.53±1.55	72.66±2.90
THPM 37	122.21±1.50	43.74±1.73	70.66±2.02

Figure 16. HPM derivatives showing anti-psychotic activity

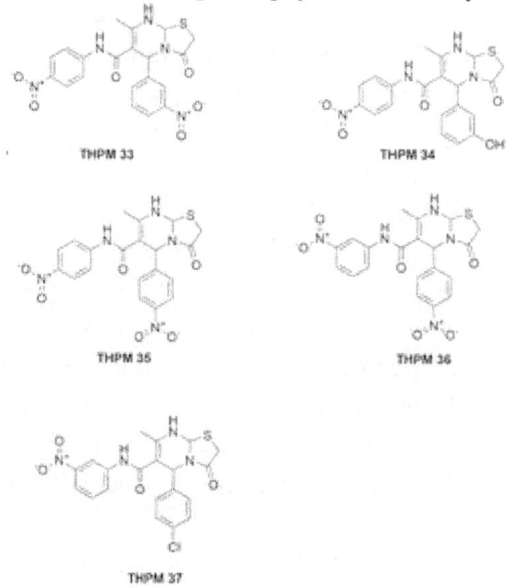

3.11 Corticotropin-Releasing Factor Receptor Antagonists

Depression is one of the most talked about mental diseases in the world: a hectic lifestyle and various other personal reasons cause it. Corticotropin-releasing factor is one of the major factors responsible for causing depression and anxiety. Hence, antagonists are prescribed for such patients (Paez-Pereda, Hausch, & Holsboer, 2011). THPM derivatives have also been found to have a good effect in the domain of medications. The THPM compounds were screened for their activity via docking studies. The docking studies gave satisfactory results. Their ability to pass the blood-brain barrier was also evaluated by screening their log P and topological polar surface area. The biological activity of these compounds was evaluated using radioiodinated CRFR1 and Tyr^{0}-sauvagine. The ability to inhibit binding to Tyr^{0}-sauvagine was checked. **THPM 38** gave more tha50% inhibition (Elgiushy et al., 2021).

Figure 17. THPM as Corticotropin-releasing factor receptor antagonists

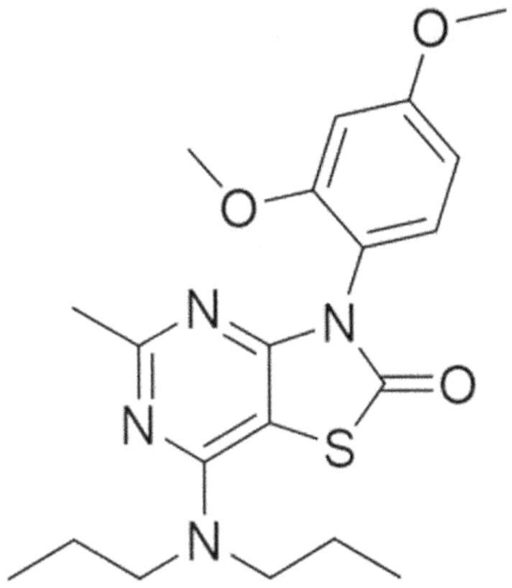

THPM 38

3.12 Analgesic Activity

Analgesic activities of the compounds have been discovered for ages. They are used as OTC medicines in the treatment of mild pain such as dental, muscle cramps, headaches, etc. They are also needed during operations, severe injuries, and medical emergencies. Hence, the need for analgesic agents is ever-increasing. THPM derivatives are reported to be useful as analgesic agents as well (Deshmukh, Morankar, & Kumbhare, 2014)

In a study to evaluate the analgesic activity of THPM derivatives, their central and peripheral analgesic activity were analyzed. Around 72 mice were selected and divided into groups and 1 group was administered with THPM derivative, one group tramadol as standard, and after 60 minutes, the animals were placed on a hot plate maintained at around 50 °C. The response time of the animal was measured as its tendency to lick its paw or hand in response to the thermal stimulus (Khalifa et al., 2015).The peripheral analgesic activity was tested using a writing test. The mice were treated with standard and THPM derivatives, using aspirin as a vehicle. After administration, mice were placed in a container, and several writhings were induced later by the IV-administered acetic acid. The protection was calculated using the formula(Khalifa et al., 2015).

% Protection= [(Control mean– Treated mean)/Control mean] × 100

The % protection time of central analgesic activity and reaction time after 90 mins was as follows

Table 6. Central analgesic activity of THPM

THPM derivative	% protection	Reaction time in seconds
THPM 39	20.5 ± 0.63	40.4
THPM 40	27.6 ± 0.98	89
THPM 41	20.1 ± 1.78	37
THPM 42	21.9 ± 1.00	50
THPM 43	19.0 ± 0.78	30
THPM 44	26.0 ± 2.20	78
THPM 45	28.8 ± 2.67	97
THPM 46	28.8 ± 2.58	97
THPM 47	24.6 ± 2.18	68
Tramadol	40.1 ± 2.28	174.6

The peripheral analgesic activity results were as follows

Table 7. Peripheral analgesic activity of THPM derivatives

THPM derivative	No of writhes/min	Protection in %
THPM 39	24.8 ± 1.02	69.31
THPM 40	39.6 ± 3.14	51
THPM 41	36.2 ± 2.58	55
THPM 42	44.8 ± 2.85	44
THPM 43	29.2 ± 2.50	63
THPM 44	26.2 ± 1.32	67
THPM 45	23.6 ± 2.16	70
THPM 46	38.8 ± 1.77	52
THPM 47	26.4 ± 2.46	67
Aspirin	17.4 ± 1.57	78

Figure 18. THPM derivatives showing Analgesic activity

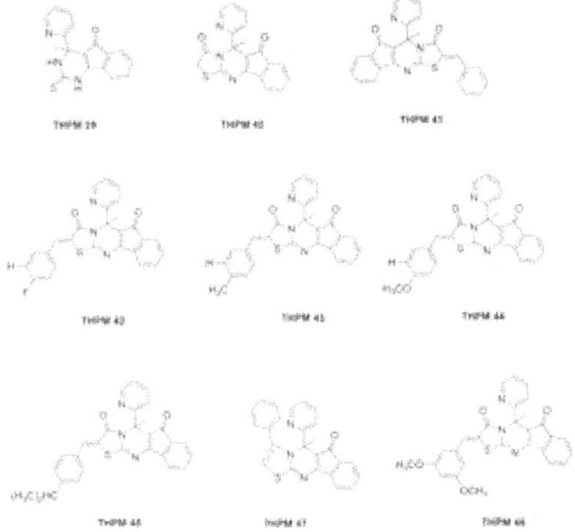

4. PATENT UPDATES ON THIAZOLOPYRIMIDINES

Table 8. Patent updates on THPM

Patent Id	Title	Assignee	Inventor/Author	Grant Date
EP2207781B1	Thiazolopyrimidine p13k inhibitor compounds and methods of use	F Hoffmann La Roche AG Genentech Inc	Georgette M. Castanedo, Janet L. Gunzner, Kimberly Malesky, Simon Mathieu, Alan G. Olivero, Daniel P. Sutherlin, Shumei Wang, Bing-Yan ZhuIrina, Chuckowree, Adrian Folkes, Sally Oxenford, Nan Chi Wan	2012-11-28
US8865658B2	Compositions containing, methods involving and uses of non-natural amino acids and polypeptides	Ambrx Inc	Zhenwei Miao, Junjie Liu, Thea Norman	2014-10-21
US7893060B2	Thiazolopyrimidines and their use as inhibitors of phosphatidylinositol-3 kinase	F Hoffmann La Roche AG	Timothy Colin Hancox, Neil Anthony Pegg, Mandy Christine Beswick, Toby Jonathan Blench, Elsa Amandine Dechaux, Janusz Jozef Kulagowski, Alan John Nadin, Stephen Price	2011-02-22
JP6644680B2	Insecticidal active mixtures containing carboxamide compounds		Szyklyak, Tatiana Gewehr, Markus	2020-02-12
US8158625B2	Bicyclic indole-pyrimidine PI3K inhibitor compounds selective for P110 delta, and methods of use	Genentech Inc, Hoffmann La Roche Inc	Georgette Castanedo, David Goldstein, Rama Kondru, Matthew Lucas, Wylie Palmer, Stephen Price, Brian Safina, Pascal Pierre Alexandre Savy, Eileen Mary Seward, Daniel P. Sutherlin, Zachary K. Sweeney	2012-04-17
JP6644679B2	Agricultural active mixtures containing carboxamide compounds		Gewer, Markus Szyklyak, Tatiana	2020-02-12
JP6918888B2	Use of pesticide-active carboxamide derivatives in soil and seed application, and treatment methods		R. Siklyak, Tatiana Gewehr, Markus Menon, Anil	2021-08-11

continued on following page

Table 8. Continued

Patent Id	Title	Assignee	Inventor/Author	Grant Date
TWI806073B	2-oxoimidazolidine-4-carboxamides as nav1.8 inhibitors		Isaac Yara Shapan, Ian M Bell, Christopher James Bongard, Christopher S Boogie, Jason M Cox, Michael J Kelly, Mark Eric Clayton, Hong Liu Jiang, Liu James J Bo Jins, Akshayi A. Samak, D. Vankist Wu Zhe	2023-06-21
JP7091596B2	Compounds that are active against nuclear receptors	Nuevolution AS	Grad, Sanne Schroeder-Nolager, Niels Gron-Ian Salvary, Gorjev, Alex Haarlene Teuber, Luigi Piero Staji,	2022-06-28
US9730944B2	Anti-viral activity of novel bicyclic heterocycles	Katholieke Universiteit Leuven	Steven De Jonghe, Ling-Jie, GaoPiet Herdewijn, Jean Herman, Miyeon Jang, Pieter Leyssen, Thierry Louat, Johan Neyts, Christophe Pannecouque, Bart Vanderhoydonck	2017-08-15
JP6047184B2	Triazines, pyrimidines, and pyridine analogs, and their use as therapeutics and diagnostic probes		Kumiljanovic, Vladimir Kumiljanovic, Natasha Zies, Bernd Weimann, Matthias	2016-12-21
KR102094007B1	Polymerizable compound, polymerizable composition, polymer, optically anisotropic body, and method for producing polymerizable compound		Gay Sakamoto Satoshi Kirikigumi Okuyama Kanako Taira	2020-03-26
US20070244133A1	Thienopyrimidines and Thiazolopyrimidines for Use in Medicine	AstraZeneca AB	Justin Bower, Alan Faull, Jon Winter	2007-10-18
JP7152369B2	Heterocyclic compounds and uses thereof	Takeda Pharmaceutical Co Ltd	Yuichi Kajita Tatsuya Ajigami Yuhei Miyanohana Tatsuju Koike Masahisa Ohno Norio Oyabu Masaki Ogino Kohei Takeuchi Yoshiteru Ito Norihito Toku Nagai Takahiro Sugimoto Toru Miya Saki Tsuneo Oda Yasutaka Hoashi Yasushi Hattori Keisuke Imamura	2022-10-12

continued on following page

Table 8. Continued

Patent Id	Title	Assignee	Inventor/Author	Grant Date
JP7135007B2	Heterocyclic compound	Takeda Pharmaceutical Co Ltd	Masahiro Ito Hideyuki Sugiyama Osamu Kubo Shiro Kikuchi Takeshi Yasui Yoshiko Kakegawa Zenichi Ikeda Toru Miya zaki Yasuhiro Arikawa Tomohiro Okawa Jinichi Yonemori Akinori Toida Takuto Kojima Yasuomi Asano Hironobu Sato Hironobu Mae zaki Shinobi Sasaki Hironori Kokubo Misaki Honma Minoru Sasaki Yasuhiro Imaeda	2022-09-12
EP3413851B1	Packaging	Clearside Biomedical Inc	Thomas Edward Godfrey, Rafael Victor Andino, Shelley Eckert Hancock, Keleigh Jo Strudthoff	2023-09-27
US11596545B2	Systems and methods for ocular drug delivery	Clearside Biomedical Inc	Rafael Victor Andino Thomas Edward GODFREY, Shelley Eckert HANCOCK, Samirkumar Patel, Keleigh Jo MAHN, Jesse YOO, Vladimir ZARNITSYN	2023-03-07
CN107750167B	Inhibitors of immune checkpoint modulators for the treatment of cancer and infections	Effector Therapeutics Inc	Kevin R. Webster, Vegas Goel	2021-05-25
US11203716B2	Compound, polymerizable compound, mixture, polymerizable liquid crystal composition, polymer, optical film, optically anisotropic product, polarizing plate, flat panel display device, organic electroluminescence display device, and anti-reflection film	Zeon Corp	Kei Sakamoto, Kumi Okuyama, Satoshi Kiriki	2021-12-21
EP4242202A1	Benzimidazole derivative and preparation method therefor and medical use thereof	Shenzhen Salubris Pharmaceuticals Co Ltd	Junjun Wu, Yinsuo L, UJianli WU, Ying Xiao, Wei Xing	2023-09-13

continued on following page

Table 8. Continued

Patent Id	Title	Assignee	Inventor/Author	Grant Date
US20230149292A1	Co-enzyme q10 formulations and methods of use	University of Miami	Sung Lan Hsia, Niven Rajin Narain, Jie Li, Kathryn J. Russell, Karune V. Woan, Indushekhar Persaud	2023-05-18
US11352673B2	Method for the diagnosis, prognosis and treatment of lung cancer metastasis	Roger Gomis, Evarist Planet	Institucio Catalana de Recerca, Estudis Avancats ICREA, Fundacio Privada Institut de Recerca Biomedica IRB	2019-04-25
AU2018200363B2	Dendrimer like amino amides possessing sodium channel blocker activity for the treatment of dry eye and other mucosal diseases	Parion Sciences Inc	Richard C. Boucher, Michael Ross Johnson, William Robert, Thelin	2019-10-17
US10899752B2	2,3-dihydro-4H-1,3-benzoxazin-4-one derivatives as modulators of cholinergic muscarinic M1 receptor	Takeda Pharmaceutical Co Ltd	Takahiro Sugimoto, Shinkichi Suzuki, Hiroki Sakamoto, Masami Yamada, Minoru Nakamura, Makoto Kamata, Kenichiro Shimokawa, Masataka Murakami, Jinichi Yonemori, Takuto Kojima	2021-01-26
US11892453B2	Method for the diagnosis, prognosis and treatment of prostate cancer metastasis	INBIOMOTION SL	Roger Gomis, Joël Jean-Mairet	2022-02-10
US11059780B2	Substituted pyrrolidine compound and use thereof	Takeda Pharmaceutical Co Ltd	Masaki Daini, Yuhei Miyanohana, Satoshi Mikami, Yuichi Kajita, Masaki Ogino, Tatsuki Koike, Kohei Takeuchi	2021-07-13

continued on following page

Table 8. Continued

Patent Id	Title	Assignee	Inventor/Author	Grant Date
AU2019201493B2	Method for the diagnosis, prognosis and treatment of breast cancer metastasis	Institucio Catalana de Recerca i Estudis Avancats ICREA, Fundacio Privada, Institut de Recerca Biomedica IRB	Anna Arnal Estape, Roger Gomis Cabre, Milica Pavlovic, Maria Tarragona Sunyer	2021-03-11
US10899752B2	2,3-dihydro-4H-1,3-benzoxazin-4-one derivatives as modulators of cholinergic muscarinic M1 receptor	Takeda Pharmaceutical Co Ltd	Takahiro Sugimoto, Shinkichi Suzuki, Hiroki Sakamoto, Masami Yamada, Minoru Nakamura, Makoto Kamata, Kenichiro Shimokawa, Masataka Murakami. Jinichi Yonemori, Takuto Kojima	2021-01-26

5. CONCLUSION AND FUTURE PERSPECTIVES

THPMs are a recent and an area of new interest. They have tremendous potential as various therapeutic agents, and their activities are comparable to or better than standard drugs. Various methods are reported to synthesize these agents, and most focus on the aspects of green principles as well. These agents contribute to anti-viral, anti-microbial, anti-psychotic, and certain immune conditions. There are various patents on the activity and synthetic procedures. Apart from this, there are some marketed drugs that contain this scaffold. Thus, THPMs find applications in multiple areas, and more of it will be explored further.

REFERENCES

Al-Rashood, S. T., Elshahawy, S. S., El-Qaias, A. M., El-Behedy, D. S., Hassanin, A. A., El-Sayed, S. M., El-Messery, S. M., Shaldam, M. A., & Hassan, G. S. (2020). New thiazolopyrimidine as anti-cancer agents: Synthesis, biological evaluation, DNA binding, molecular modeling and ADMET study. *Bioorganic & Medicinal Chemistry Letters*, 30(23), 127611. DOI: 10.1016/j.bmcl.2020.127611 PMID: 33068712

Azam, F., El-gnidi, B. A., Alkskas, I. A., Ahmed, M. A., Azam, F., El-gnidi, B. A., Alkskas, I. A., & Ahmed, M. A. (2010). *oxidative stress in mice neuroleptic-induced catalepsy and oxidative stress in mice. 6366*. DOI: 10.3109/14756361003671052

Benzodifuranyl, N. (2020). *Anti-Inflammatory and Analgesic Agents*. 1–25.

Elgiushy, H. R., Abou-Taleb, N. A., Holz, G. G., Chepurny, O. G., Pirmettis, I., Kakabakos, S., Karageorgos, V., Liapakis, G., Albohy, A., Abouzid, K. A. M., & Hammad, S. F. (2021). Synthesis, in vitro biological investigation, and molecular dynamics simulations of thiazolopyrimidine based compounds as corticotrophin releasing factor receptor-1 antagonists. *Bioorganic Chemistry*, 114, 105079. DOI: 10.1016/j.bioorg.2021.105079 PMID: 34174633

Flefel, E., El-Sayed, W., Mohamed, A., El-Sofany, W., & Awad, H. (2017). Synthesis and Anti-cancer Activity of New 1-Thia-4-azaspiro[4.5]decane, Their Derived Thiazolopyrimidine and 1,3,4-Thiadiazole Thioglycosides. *Molecules (Basel, Switzerland)*, 22(1), 170. DOI: 10.3390/molecules22010170 PMID: 28117699

Gadkari, Y. U., Hatvate, N. T., Takale, B. S., & Telvekar, V. N. (2020). Concentrated solar radiation as a renewable heat source for a preparative-scale and solvent-free Biginelli reaction. *New Journal of Chemistry*, 44(20), 8167–8170. DOI: 10.1039/D0NJ01351J

Gadkari, Y. U., Hatvate, N. T., & Telvekar, V. N. (2021a). Concentrated solar radiation-assisted one-pot/multicomponent synthesis of pyranopyrazole derivatives under neat condition. *Research on Chemical Intermediates*, 47(10), 4245–4255. DOI: 10.1007/s11164-021-04530-7

Gadkari, Y. U., Hatvate, N. T., & Telvekar, V. N. (2021b). Solar energy as a renewable energy source for preparative-scale as well as solvent and catalyst-free Hantzsch reaction. *Sustainable Chemistry and Pharmacy*, 21, 100444. DOI: 10.1016/j.scp.2021.100444

Gadkari, Y. U., Jadhav, N. L., Hatvate, N. T., & Telvekar, V. N. (2020). Concentrated Solar Radiation Aided Green Approach for Preparative Scale and Solvent-Free Synthesis of 3-Methyl-4-(hetero)arylmethylene Isoxazole-5(4H)-ones. *ChemistrySelect*, 5(39), 12320–12323. DOI: 10.1002/slct.202003348

Ghorab, M. M., Mohamed, Y. A., Mohamed, S. A., & Ammar, Y. A. (1996). SYNTHESIS AND RADIATION STABILITY OF NOVEL THIAZOLOPYRIMIDINES WITH EXPECTED ANTI-FUNGAL ACTIVITY. *Phosphorus, Sulfur, and Silicon and the Related Elements*, 108(1–4), 249–256. DOI: 10.1080/10426509608029657

GOOD. G. (2015). No Title No Title No Title. *Angewandte Chemie International Edition, 6(11), 951–952., 1*(April), 395–397.

Gupta, R., & Chaudhary, R. P. (2013). Synthesis, antimicrobial and DFT studies of novel fused thiazolopyrimidine derivatives. *Hc, 19*(3), 207–214. DOI: 10.1515/hc-2013-0024

Hatvate, N. T., & Ghodse, S. M. (2020). One-pot three-component synthesis of isoxazole using ZSM-5 as a heterogeneous catalyst. *Synthetic Communications*, 50(23), 3676–3683. DOI: 10.1080/00397911.2020.1815786

Hatvate, N. T., Ghodse, S. M., & Telvekar, V. N. (2018). Metal-free synthesis of 2-aminothiadiazoles via TBHP-Mediated oxidative C-S bond formation. *Synthetic Communications*, 48(3), 285–290. DOI: 10.1080/00397911.2017.1398330

Hill, R., & Canals, M. (2022). Experimental considerations for the assessment of in vivo and in vitro opioid pharmacology. *Pharmacology & Therapeutics*, 230, 107961. DOI: 10.1016/j.pharmthera.2021.107961 PMID: 34256067

Iqbal, A., Khan, A., Ahmedi, S., Manzoor, N., & Siddiqui, T. (2023). Synthesis, antifungal evaluation, and molecular docking studies of steroidal thiazolopyrimidines. *Steroids*, 193, 109186. DOI: 10.1016/j.steroids.2023.109186 PMID: 36736803

Jayanthi, M., Venkatesh, P., & Thirunavukkarasu, S. V. (2022). A novel and efficient synthesis of 1, 2, 3,4-tetrahydropyrimidine carboxamide derivatives by Biginelli reaction and their in vitro anti-bacterial activity. *Biomedicine (India)*, 42(2), 280–285. DOI: 10.51248/.v42i2.884

Kendre, B. V, Bhusare, S. R., & Hardas, P. S. (2020). *Antimicrobial and Anti-Inflammatory Activity Studies of Novel Thiazolopyrimidines. 10*(January).

Khalifa, N. M., Al, M. A., El, A., Amr, G. E., Baiuomy, A. R., & Abdel, R. F. (2015). *Synthesis and Biological Evaluation of Some Novel Fused Thiazolo [3, 2 a] Pyrimidines as Potential Analgesic and Anti Inflammatory Agents 1. 41*(2), 192–200. DOI: 10.1134/S1068162015020090

Lewis, V. A. (2017). Psychopharmacology. In *Pharmacology and Therapeutics for Dentistry* (pp. 133–155). Elsevier., DOI: 10.1016/B978-0-323-39307-2.00010-2

Mohamed, S. F., Abd-Elghaffar, H. S., Amr, A. E.-G. E., Elnaggar, D. H., Abou-Amra, E. S., Hosny, H. M., Mohamed, A. M., & Abd El-Shafy, D. N. (2023). New poly heterocyclic compounds based on pyrimidine-2-thiones: Synthesis, evaluation of putative anti-viral agents, DFT calculation, and molecular modeling. *Journal of Molecular Structure*, 1291, 136083. DOI: 10.1016/j.molstruc.2023.136083

Mohamed, S. F., Flefel, E. M., Amr, A. E.-G. E., & Abd El-Shafy, D. N. (2010). Anti-HSV-1 activity and mechanism of action of some new synthesized substituted pyrimidine, thiopyrimidine and thiazolopyrimidine derivatives. *European Journal of Medicinal Chemistry*, 45(4), 1494–1501. DOI: 10.1016/j.ejmech.2009.12.057 PMID: 20110135

Nagarajaiah, H., Khazi, I. M., & Begum, N. S.NAGARAJAIAH. (2012). Synthesis, characterization and biological evaluation of thiazolopyrimidine derivatives. *Journal of Chemical Sciences*, 124(4), 847–855. DOI: 10.1007/s12039-012-0271-z

Nagarajaiah, H., M Khazi, I. A., & Begum, N. S.NAGARAJAIAH. (2015). Synthesis of some new derivatives of thiazolopyrimidines and hydrolysis of its arylidene derivative. *Journal of Chemical Sciences*, 127(3), 467–479. DOI: 10.1007/s12039-015-0797-y

Nammalwar, B., & Bunce, R. A. (2024). Recent Advances in Pyrimidine-Based Drugs. *Pharmaceuticals (Basel, Switzerland)*, 17(1), 104. DOI: 10.3390/ph17010104 PMID: 38256937

Parit, S., Manchare, A., Ghodse, S., & Hatvate, N. (2024). Comparative review on homogeneous and heterogeneous catalyzed synthesis of 1,3-thiazole. *Synthetic Communications*, 54(23), 1–21. DOI: 10.1080/00397911.2024.2399187

Pereira, J. A., Pessoa, A. M., Cordeiro, M. N. D. S., Fernandes, R., Prudêncio, C., Noronha, J. P., & Vieira, M. (2015). Quinoxaline, its derivatives and applications: A State of the Art review. *European Journal of Medicinal Chemistry*, 97, 664–672. DOI: 10.1016/j.ejmech.2014.06.058 PMID: 25011559

Saleh, N., & Yousaf, Z. (2018). Tools and techniques for the optimized synthesis, reproducibility and scale up of desired nanoparticles from plant derived material and their role in pharmaceutical properties. In *Nanoscale Fabrication, Optimization, Scale-Up and Biological Aspects of Pharmaceutical Nanotechnology* (pp. 85–131). Elsevier. DOI: 10.1016/B978-0-12-813629-4.00003-6

Sawant, R. L. (2014). Synthesis, QSAR and docking studies of 5HT2A receptor antagonizing thiazolo[3,2-a]pyrimidines as anti-psychotic agents. *Marmara Pharmaceutical Journal*, 3(18), 109–109. DOI: 10.12991/mpj.2014187237

Sekhar, T., Thriveni, P., Venkateswarlu, A., Daveedu, T., Peddanna, K., & Sainath, S. B. (2020). One-pot synthesis of thiazolo[3,2-a]pyrimidine derivatives, their cytotoxic evaluation and molecular docking studies. *Spectrochimica Acta. Part A: Molecular and Biomolecular Spectroscopy*, 231, 118056. DOI: 10.1016/j.saa.2020.118056 PMID: 32006911

Selvam, T. P., Karthik, V., Kumar, P. V., & Ali, M. A. (2012). Design, synthesis, antinociceptive, and anti-inflammatory properties of thiazolopyrimidine derivatives. *Toxicological and Environmental Chemistry*, 94(7), 1247–1258. DOI: 10.1080/02772248.2012.703204

Severina, H. I., Skupa, O. O., Voloshchuk, N. I., Suleiman, M. M., & Georgiyants, V. A. (2019). Synthesis and anticonvulsant activity of 6-methyl-2-(2-oxo-2-arylethyl) thio) pyrimidin-4(3H)-one derivatives and products of their cyclization. *Farmatsiia (Sofia)*, 66(3), 141–146. DOI: 10.3897/pharmacia.66.e38137

Vasilkova, N. O., Nikulin, A. V., & Krivenko, A. P. (2020). Synthesis and Structure of Thiazolopyrimidine Derivatives. *Russian Journal of Organic Chemistry*, 56(6), 990–993. DOI: 10.1134/S1070428020060044

Yousif, M. N. M., El-Sayed, W. A., Abbas, H. A. S., Awad, H. M., & Yousif, N. M. (2017). Anti-cancer activity of new substituted pyrimidines, their thioglycosides and thiazolopyrimidine derivatives. *Journal of Applied Pharmaceutical Science*, 7(11), 21–32. DOI: 10.7324/JAPS.2017.71104

Zarenezhad, E., Farjam, M., & Iraji, A. (2021). Synthesis and biological activity of pyrimidines-containing hybrids: Focusing on pharmacological application. *Journal of Molecular Structure*, 1230, 129833. DOI: 10.1016/j.molstruc.2020.129833

Chapter 8
Thienopyrimidines Exploring the Chemistry and Bioactivity

Navnath Tulshiram Hatvate
https://orcid.org/0000-0002-3388-3247
Institute of Chemical Technology, Jalna, India

Trupti Sahebrao Shevkar
https://orcid.org/0009-0001-3992-6690
Institute of Chemical Technology, Jalna, India

Hemantkumar Akolkar
https://orcid.org/0000-0003-0882-1324
Abasaheb Marathe Arts and New Commerce Science College, India

ABSTRACT

Thienopyrimidine is a heterocyclic compound with a thiophene ring fused to a pyrimidine ring. Due to their diverse properties and applications, these compounds are found in pharmaceuticals, agrochemicals, and materials science. The potential biological activities of thienopyridine derivatives have been investigated and utilized in medicinal chemistry. It also has diverse pharmacological activity, including anti-Alzheimer, anti-cancer, anti-fungal, anti-microbial, anthelmintic, hypolipidaemic, and anti-diabetic activity. Many commercially available drugs, including pictilisib, relugolix, and olmutinib, contain thienopyrimidine rings. This chapter focuses on the chemistry, synthetic methods, and bioactivity of different thienopyrimidines. Moreover, this chapter includes a summary of the latest information on patents and clinical trials.

DOI: 10.4018/979-8-3693-7520-4.ch008

1. INTRODUCTION

The term "heteroatom" is derived from the Greek words "hetero," meaning distinct, and "kyklos," which means ring. Heteroatoms such as nitrogen, oxygen, Sulphur, and occasionally additional elements give heterocyclic compounds specific properties that distinguish them from carbocyclic compounds. Since years, medicinal chemists (Sonawane et al., 2024) and scientists have been interested with heterocyclic compounds, which have ring structures that include at least one heteroatom (Gour, 2023). This compounds are essential in chemistry and medicine, as they serve as the foundation for vital biomolecules, medicines, and agricultural chemicals (Hatvate et al., 2018; Makhova et al., 2020). Their quantities are continually expanding as a result of considerable synthetic research and utilisation (Qadir et al., 2022).

Heterocycles having the thienopyrimidine component are particularly interesting due to their unique pharmacological and biological properties. They are fundamentally similar and isoelectronic to purines, and several substituted thieno[2,3-*d*] pyrimidine **Figure 1a** derivatives have substantial and diverse biological functions (Abdel-Megid et al., 2016). Over the last two decades, there has been an abundance of interest in the synthesis of nitrogen and Sulphur-based heterocyclic molecules (Miaki et al., 2017). Because of the various biological functions, heterocyclic compounds with the thienopyrimidine moiety are becoming increasingly of the topic in scientific studies. Furthermore, thienopyrimidine-containing substances can be observed in a variety of natural products and pharmaceutical drugs (Tolba, Sayed, Sayed, et al., 2021).

Thienopyrimidines, nitrogen and sulfur-containing heterocycles, hold potential in a variety of medicinal applications, including anti-inflammatory activity (El-Shoukrofy et al., 2019), antibacterial (Patel et al., 2019), anticancer (Biradar et al., 2023), antialzheimer therapy (Triloknadh et al., 2018), antiviral therapy (Wang et al., 2022), antihyperlipidemic therapy (Shaukat Tamboli, Riyaj; Devidas Amrutkar, Rakesh; Sanchlal Jain, Kishor; Kumaradoss Kathiravan, 2013), antioxidant therapy (Elmongy et al., 2021), antimalarial activity (Barrows et al., 2020), antidiabetic activity (Islam et al., 2021), anti-tuberculosis activity (Amira T. A. Mohamed, 2019), anti-hypertensive activity (Ronald K. RussellJeffery B. Press, 1988) and antihistaminic action (Meenakshi Singh, 2012). Many investigations claimed that thienopyrimidines can alter the melanin formation in murine B16 cells (Tolba, Sayed, El-Dean, et al., 2021).

This chapter investigates recent advances in the synthesis, characteristics, and biological activities of thienopyrimidines, emphasising their potential as therapeutic agents in many disorders, as previously mentioned. It also investigates the structure-activity relationships (SAR) of these compounds for various biological activities, gives an extensive patent evaluation, and explores clinical trials performed for evaluating their medicinal potential.

1.1 Isomers of Thienopyrimidine

Thienopyrimidine is a bicyclic heterocyclic molecule composed of a five-membered thiophene ring joined to a six-membered pyrimidine ring having two nitrogen atoms. There are three isomeric forms of Thienopyrimidine: thieno[2,3-*d*]pyrimidine **Figure 1 a,** thieno[3,2-*d*]pyrimidine **Figure 1 b,** and thieno [3,4-*d*] pyrimidine **Figure 1 c** (Ibrahim et al., 1996). Some isomeric thienopyrimidines are crystalline, whereas the majority are coloured amorphous. Thieno[2,3-*d*]pyrimidines **Figure 1 a**, a bioisoster to purines, have a wide range of biological activities, likely due to interactions with physiological components. Thienopyrimidines, like thiophene-phenyl isosterism, are bio isosteres of quinazolines, which have numerous biological functions as detailed in scientific and patent literature (Ibrahim et al., 1996).

Figure 1. Isomers of thienopyrimidine

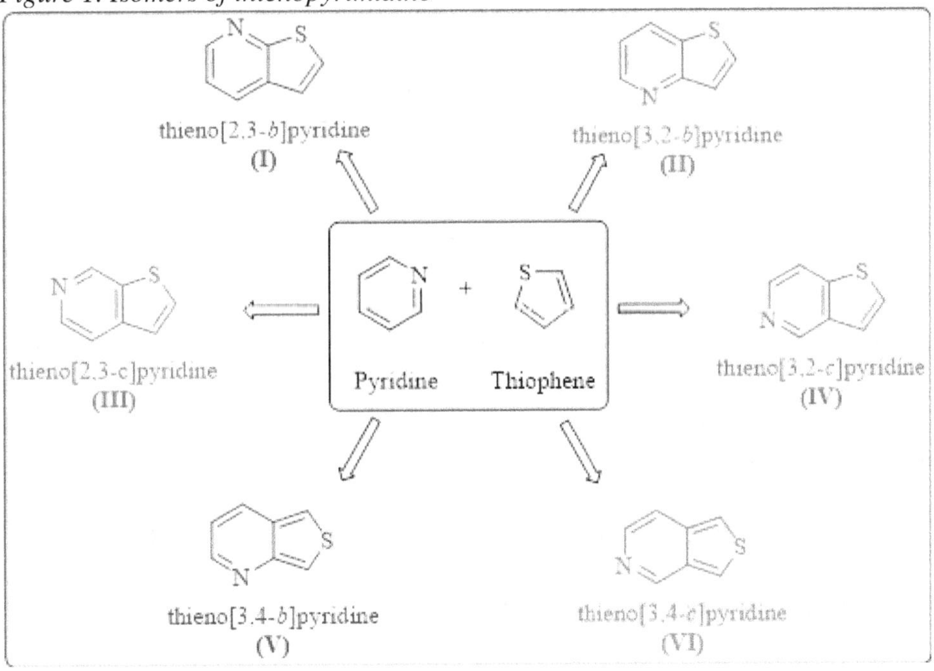

1.2 Chemistry of Thienopyrimidine

Baker *et al.,* reported the initial derivative of any of these ring systems in 1947: thieno[3,4-*d*] pyrimidines **Figure 1 c**. A year afterwards, McClelland and Stammers reported the synthesis of the first thieno[3,2-*d*]pyrimidine **Figure 1 b**. In 1953, Baker *et al.,* published the first thieno[2,3-*d*]pyrimidine **Figure 1a** synthesis (Varvounis et al., 1996). Fused pyrimidines have gained popularity due to their numerous biological applications. This is demonstrated by ongoing investigations on the chemistry of systems involving the pyrimidine ring and heterocycles, such as purines (Litvinov, 2006).

In thienopyrimidine ring, pyrimidine's weak electron density prevents electrophilic substitution but promotes nucleophilic substitution (Panneer Selvam et al., 2012). Thiophene, with delocalised electron pairs, behaves similarly to benzene and can undergo sulfonation, nitration, and halogenation (Prabhakar Kumar Varma, 2018). Nucleophilic substitution arises frequently in chloro thienopyrimidines, in which chlorine is substituted in a variety of nucleophiles, most commonly amino groups (Litvinov, 2006). In thieno[2,3-*d*]pyrimidine **Figure 1 a** and thieno[3,4-*d*]pyrimidine **Figure 1 c**, electrophilic substitution usually occurs at positions 6 or 7, whereas in thieno[3,2-*d*]pyrimidine **Figure 1 b**, greater pyrimidine annulation causes attack at position 7. Thienopyrimidine is commonly considered basic due to the nitrogen in the pyrimidine ring (Prabhakar et al., 2016).

2. SYNTHESIS OF THIENOPYRIMIDINES

Thienopyrimidine moiety synthesis comprises either annulating the pyrimidine nucleus on the parent thiophene ring or the thiophene nucleus on the parent pyrimidine ring. Additionally, they derived from acyclic molecules (Abdel-Megid et al., 2016). Several schemes for the synthesis, including both conventional and greener methods, are discussed in this section.

2.1 Conventional Methods for Synthesis of Thienopyrimidine

2.1.1 From Pyrimidine Derivatives

The most prevalent approaches for ring synthesis beginning with pyrimidine derivatives are given here.

From 4-chloropyrimidine:

Starting with 4-chloropyrimidines, as shown in **Scheme 1**, when CN or COOR are attached at position 4 **1**, the intramolecular cyclization of a properly 5-functionalized 4-mercapto-pyrimidine **2** is accomplished by Claisen-Schmidt, Thorpe-Ziegler, or Friedlander condensation. This process involves the relocation of the chlorine substituent on the pyrimidine ring **1** with mercaptoacetic acid esters or other mercapto derivatives featuring an electron withdrawing group (Z) (Wilding et al., 2017).

Figure 2. Scheme 1 Synthesis of thienopyrimidine from 4-chloropyrimidine

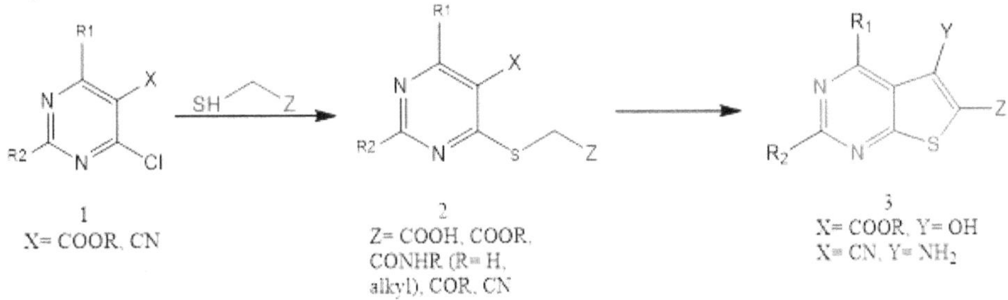

From 5-substituted pyrimidin-4(3)H-thiones:

Alkylation of 5-substituted (X=COOR, CN) pyrimidin-4(3H)-thiones **4** with chloroacetic acid derivatives (Z) can yield the mercaptopyrimidine intermediate **2**. Intramolecular cyclization can occur with or without isolating the intermediate. For Z = CN, use the Thorpe-Ziegler reaction, whereas for Z = COR, use the Claisen-Schmidt reaction or related condensation reactions **scheme 2** (Wilding et al., 2017).

Figure 3. Scheme 2 Synthesis of thienopyrimidine from 5-substitured pyrimidin-4(3) H-thiones

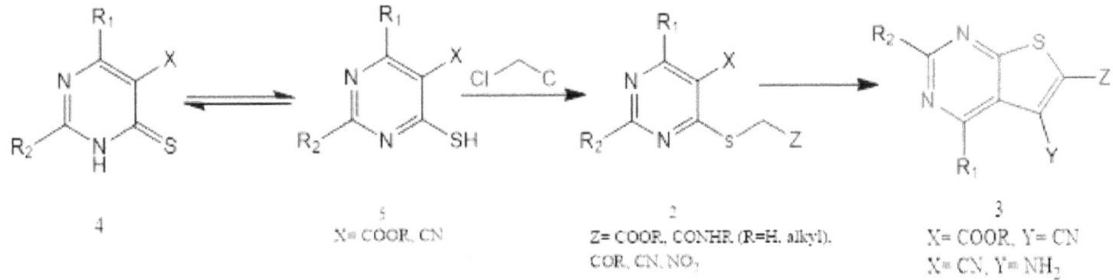

2.1.2 From 4-chlorothieno[2,3-*d*]pyrimidine

Figure 4. Scheme 3 synthesis of thienopyrimidine from 4-chlorothieno[2,3-d]pyrimidine

6 a

Palladium catalysts are used for the regioselective C-H arylation of thieno[2,3-*d*]pyrimidine. These reactions occur at the C6-position with aryl iodides **8** and at the C5-position with aryl boronic acids **9**, exhibiting significant regioselectivity **Scheme 4**. Mechanistic investigations imply that the palladium catalyst type influences regioselectivity, with cationic palladium favoring arylation at the C5-position (Yamada et al., 2020).

Figure 5. Scheme 4 Arylation of Thieno[2,3-d] pyrimidine

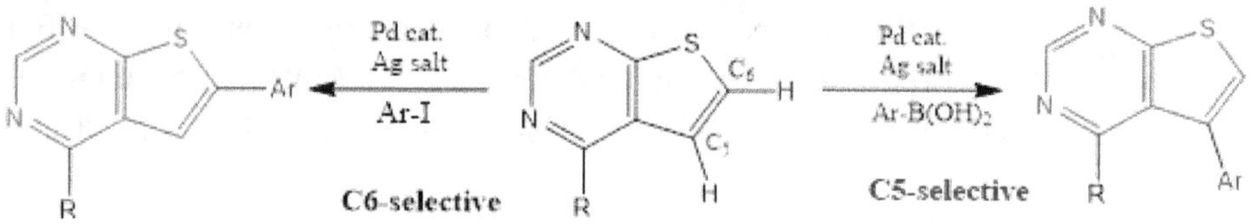

8 7 9

2.1.3 From 4-chloro-5-(trimethylsilylethynyl)pyrimidine

The use of trimethylsilylacetylene in the synthesis of thienopyrimidine was investigated. 4-chloro-5-iodo-pyrimidine and its alkyl homologs **10** interacted with TMSA to produce 5-trimethylsilylethynyl compounds **11**, which, unlike thieno[2,3-*d*] pyrimidines, transformed effectively under normal circumstances **Scheme 5.** This method can also be used to create 2-methylthiothieno[2,3-*d*]pyrimidines and 4-methoxy-2-methylthieno[2,3-*d*]pyrimidine (Kazushige Sogawa, 1970)

Figure 6. Scheme 5 Synthesis of thienopyrimidine from 4-chloro-5 (trimethylsilylethynyl)pyrimidine

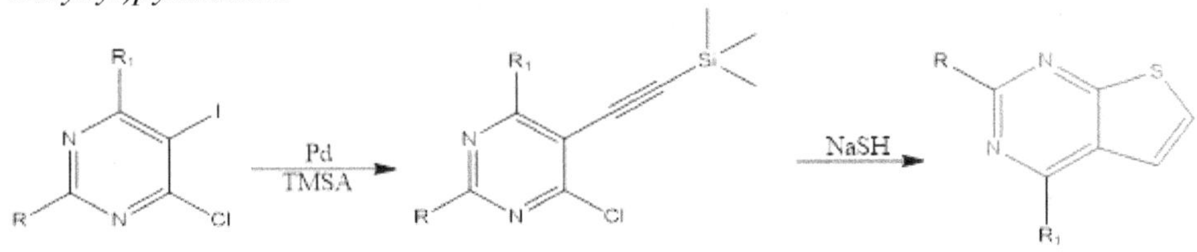

10
a: R= H, R'= H
b: R= H, R'= Me
c: R= Me, R'= H

11
d: R= Me, R'= Me
e: R= Iso-Pr, R': Me
f: R= SMe, R': Me

12
g: R= SMe, R'= H
h: R= Me, R'= OMe

2.1.4 From 4-Hydrazino-thieno[3,2-d]pyrimidine

Thieno[3,2-*d*]pyrimidine **b** is synthesized from the starting material 4-Hydrazino-thieno[3,2-*d*]pyrimidine **13**. The reaction occurs in the presence of oxygen and sodium ethanolate in ethanol, promoting the synthesis of the desired product **Scheme 6**. Sodium ethanolate acts as a base, causing the hydrazine group to deprotonate followed by the process to synthesized thieno[3,2-*d*]pyrimidine **b.** The reaction conditions, which include the solvent ethanol and the presence of oxygen, have significance to the effective synthesis of the target molecule (Quelques, 1971).

Figure 7. Scheme 6 Synthesis of thienopyrimidine from 4-Hydrazino-thieno[3,2-d] pyrimidine

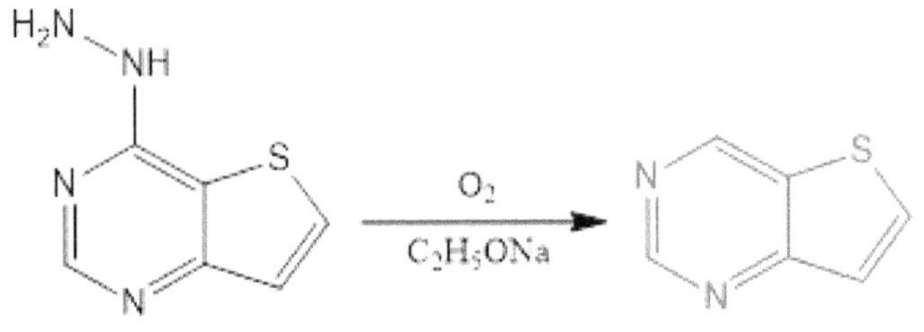

13

b

2.1.5 From Ketone

Thieno[3,2-*d*]pyrimidine **b** is synthesized from 3H-thieno[3,2-*d*]pyrimidine-4-one **14** via palladium on charcoal and trichlorophosphate. Trichlorophosphate chlorinates the ketone group, resulting in chlorothieno[3,2-*d*]pyrimidine **15**. Palladium on activated charcoal then catalyses the dichlorination and cyclisation, producing Thieno[3,2-*d*]pyrimidine **b** **Scheme 7**. Both reagents and catalysts are required for the transformation (*from ketone.pdf*, n.d.).

Figure 8. Scheme 7 Synthesis of thienopyrimidine from 3H-thieno[3,2-d]pyrimidine-4-one

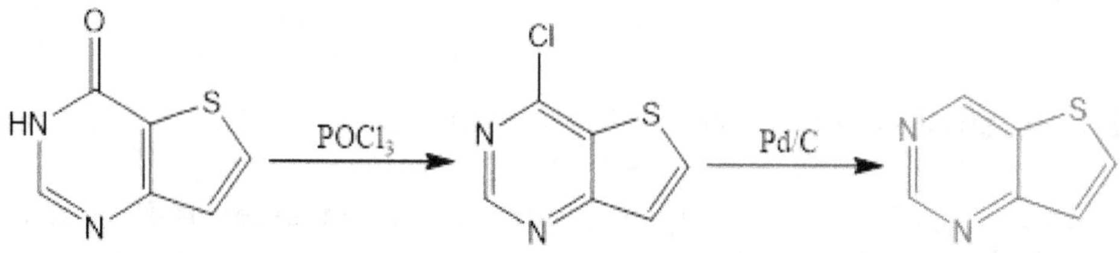

14 15 b

2.1.6 Synthesis from the Gewald Rection

The Gewald reaction, which involves the one-pot cyclocondensation of ketones or aldehydes with activated nitrile derivatives **17** and elemental Sulphur, is a flexible method for obtaining 2-aminothiophene derivatives. It is potentially simple to obtain 2-thioxo-6-aminothieno[3,2-d]pyrimidin-4-one **18** derivatives by using thiobarbituric acid **16** as the starting ketone. The aminothieno-fused derivatives and good yields were produced by treating piperidine with the appropriate alkyl cyanide **Scheme 8** (Contributor et al., n.d.).

Figure 9. Scheme 8 Synthesis of thienopyrimidine by Gewald reaction

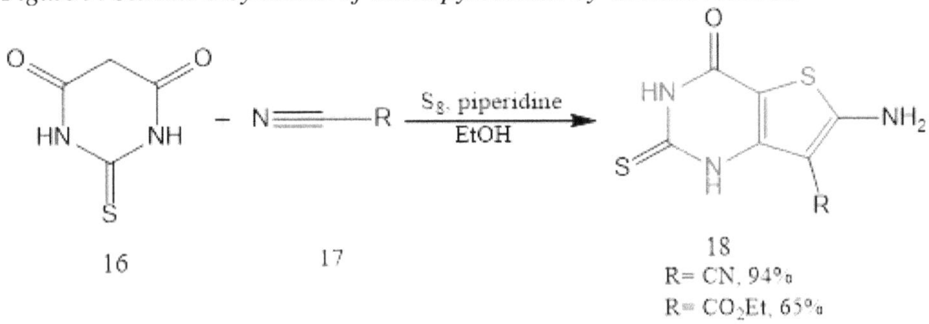

2.2 Greener Method for the synthesis of Thienopyrimidine

2.2.1 Microwave Assisted Synthesis

The microwave-assisted technique provides an environmentally friendly approach of synthesis thienopyrimidines and their derivatives considering it is rapid and produces a lower adverse impact on the environment throughout the synthesis process.

Thienopyrimidine is synthesised from the precursor 2-amino-5,5-dimethyl-7-oxo-4,5,6,7-tetrahydro-1-benzothiophene-3-carbonitrile **19** by an effective microwave irradiation method. Nitinkumar S. Shetty et al. accomplished this in a number of steps: the precursor was irradiated with formic acid for 15 minutes at 90°C to yield 6,6-dimethyl-6,7-dihydro-3H,5H-benzo[4,5]thieno[2,3-*d*]pyrimidine-4,8-dione **20** . Phosphorus oxychloride was added, and the mixture was irradiated at 95°C for 10 minutes to produce 4-chloro-6,6-dimethyl-5,6,7-benzothieno[2,3-*d*]pyrimidin-8-one **21**. Finally, a thienopyrimidine derivative **22** was produced **Scheme 9** by heating hydrazine hydrate at 50°C for 20 minutes using microwave radiation (Shetty, 2014).

Figure 10. Scheme 9 Synthesis of thienopyrimidine from 4-chloro-6,6-dimethyl-5,6,7-benzothieno[2,3-d]pyrimidin-8-one

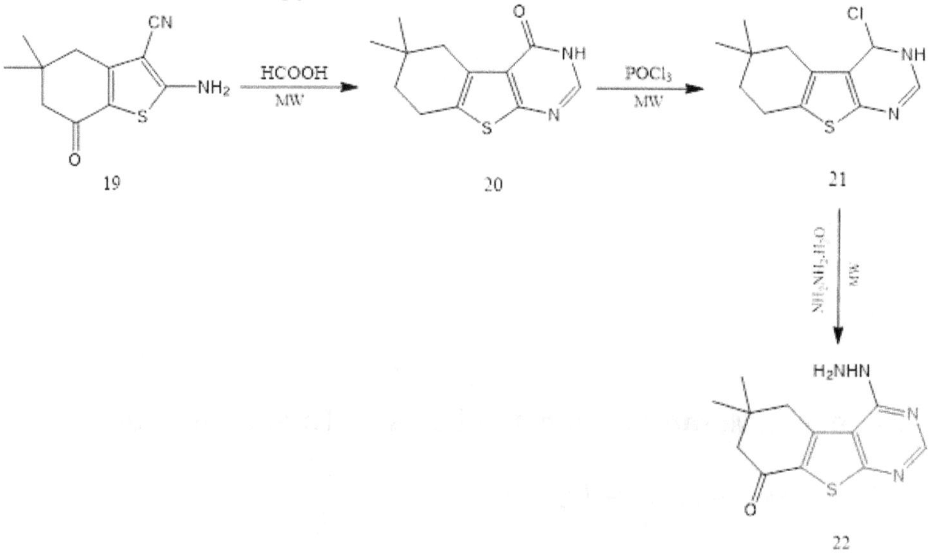

Many thienopyrimidine derivatives are synthesized using expensive, non-commercial, or poisonous ingredients, harsh environments, extended reaction periods, and sophisticated techniques. However, the employment of microwaves in solid-supported organic synthesis is a novel method. Kidwai, M. et al. found that microwave irradiation of solid support boosts yields and reduces reaction durations while removing the requirement for solvents during the reaction phase. However, solvents are still needed for product elution and agent adsorption. Using microwave technology, they synthesized a new 5-methyl-6-ethylcarboxylate-2-thioxo-thieno[3,2-*d*]pyrimidine-4(1H)-one **24** from 2-amino-3,5-diethyl carboxylate-4-methyl-thiophene **23** and monosubstituted thioureas on a K2CO3 solid support **Scheme 10** (Kidwai, M.* and Mishra, n.d.).

Figure 11. Scheme 10 Synthesis of thienopyrimidine from 2-amino-3,5-diethyl carboxylate-4-methylthiophene

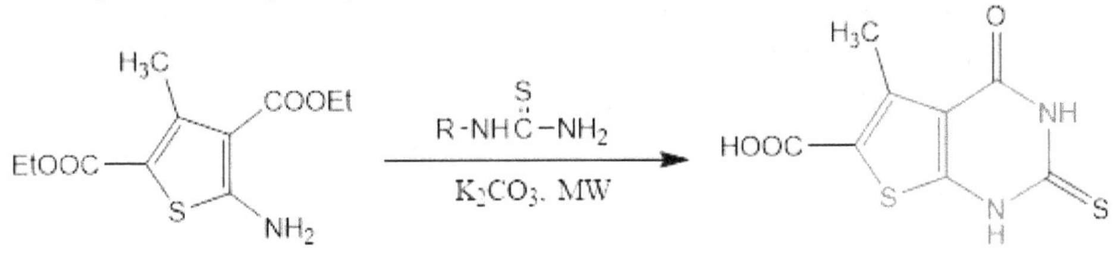

23

24

Thienopyrimidine derivatives **26** were synthesised by combining 2-amino-3-ethoxycarbonyl-4,5-disubstituted thiophenes and 2-amino-3-cyano-4,5-disubstituted thiophenes **Scheme 11**. Thienyl thiourea derivatives were then synthesised by condensing them with isothiocyanates under microwave irradiation for higher yields and quicker reaction times. Additional cyclisation, alkylation, carbamate synthesis, and nucleophilic aromatic substitution were carried out under microwave conditions. The structures were confirmed via IR, NMR, and MS spectra (Fatma E. M. El-baih, n.d.).

Figure 12. Scheme 11 Synthesis of thienopyrimidine by Microwave Irradiation

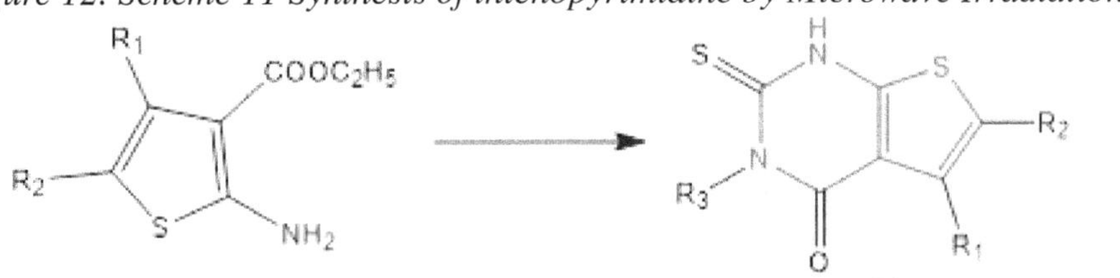

25

R_1 R_2
a $(CH_2)_4$
b $(CH_2)_2N(CH_3)CH_2$
c $(CH_2)_5$

26

R_1 R_2	R3
a$(CH_2)_4C_6H_5$	C_6H_5
b $(CH_2)_4$	p-Cl-C_6H_6
c $(CH_2)_4$	$(CH_2)_3CH_3$
d $(CH_2)_2N(CH_3)CH_2$	p-Cl-C_6H_6
e $(CH_2)_2N(CH_3)CH_2$	$(CH_2)_3CH_3$
f $(CH_2)_5$	C_6H_5
g $(CH_2)_5$	p-Cl-C_6H_6

3. CURRENT MARKETED DRUGS OF THIENOPYRIMIDINE

Sufugolix 27 (TAK-013), an orally active nonpeptide GnRH antagonist developed by Takeda Chemical Industries, has the potential to treat reproductive and hormone-related illnesses. It demonstrates insurmountable antagonism, probably because certain amino acid sequences in the human GnRH receptor form a "trap door" that prevents dissociation. Sufugolix was optimised as a thienopyrimidine derivative and modified to minimise cytochromemP450 inhibition while increasing GnRH antagonistic action (Szkudlinski, 2007). **Relugolix 28** (TAK-385), efficiently reduces luteinizing hormone levels in preclinical studies (Miwa et al., 2011). Relugolix is an orally active, non-peptide gonadotropin-releasing hormone (GnRH) receptor antagonist that has been registered in the US and Europe for the treatment of advanced prostate cancer (La et al., 2023).

Olmutinib (HM61713) is a third-generation EGFR-TKI that irreversibly binds to a cysteine in the kinase domain. It specifically targets EGFR L858R/T790M mutations and exon 19 deletions, with minimal effect on wild-type EGFR cells. Approved in South Korea in May 2016 for advanced NSCLC patients with EGFR T790M mutations, its global development was discontinued in September 2016 (B. Liao et al., 2017). **DDP 225 30**, Mitsubishi Tanabe Pharma, a thieno[2,3-*d*]pyrimidine derivative, aims to treat Alzheimer's disease by addressing attention deficits induced by cholinergic dysfunction (Kholoud I. Eissa, Mona M. Kamel, Lamia W. Mohamed, Ahmed S. Doghish, Radwan Alnajjar, Almed A. Al-Karmalawy, 2023). Epix Pharmaceutical introduced another thienopyrimidine derivative, **PRX-08066 31**, as a serotonin 5-HT subtype 2B receptor antagonist. It initially emerged as an alternative treatment for pulmonary hypertension and chronic obstructive pulmonary disease (COPD). Phase II clinical trials are now conducted. Additionally, PRX-08066 was discovered to inhibit fibroblasts, making it a prospective anticancer candidate (Kassab et al., 2024).

GSNS-314 32, a very powerful Aurora kinase inhibitor, has been tested in phase 1 dose escalation research to investigate safety and tolerability in patients with advanced solid malignancies. **SNS-314 33** is a highly effective and selective inhibitor of Aurora kinases A, B, and C. The chemical effectively inhibits HCT116 cell proliferation and soft agar colony formation after just 16 hours of exposure (Hansen et al., 2010). **GDC-0941 (Pictilisib) 34** serves as an orally bioavailable antagonist of class I PI3K that is 100-fold more potent against class I than class II, III, and IV family members. It is now in clinical trials for solid tumor indications, including breast cancer (Wallin et al., 2012). **GNE-490 35** and **GNE-493 36** were based on the structural alteration of GDC-0941. Two of these compounds inhibited the PI3K pathway with excellent efficacy and selectivity while maintaining favorable pharmacokinetic (PK) properties. They are now in phase II clinical trials for treating breast cancer (Ali et al., 2019).

CUDC-907 37, commonly known as Fimepinostat, is a synthetic, orally available small molecule that inhibits both PI3K and HDAC. Although the FDA has not approved CUDC-907 for any medical condition, it is being tested in clinical studies for a variety of tumors. Preclinical investigations indicate that it has therapeutic potential in hematopoietic malignancies and solid tumors, such as chronic lymphocytic leukemia, acute myeloid leukemia, thyroid cancer, and pancreatic adenocarcinoma (W. Liao et al., 2021). **Apitolisib 38** has also been demonstrated to reduce tumor growth in xenograft models, including those with activated PI3K or PTEN deletion, and to cause an exposure-related drop in Parkinson's disease biomarkers. In a FIH trial (NCT00854152), apitolisib demonstrated dose-proportional PK and tolerable tolerability at lower doses, resulting in 10 PRs in 120 patients with advanced solid tumors (Janku, 2017).

Figure 13. Current marketed drugs containing thienopyrimidine

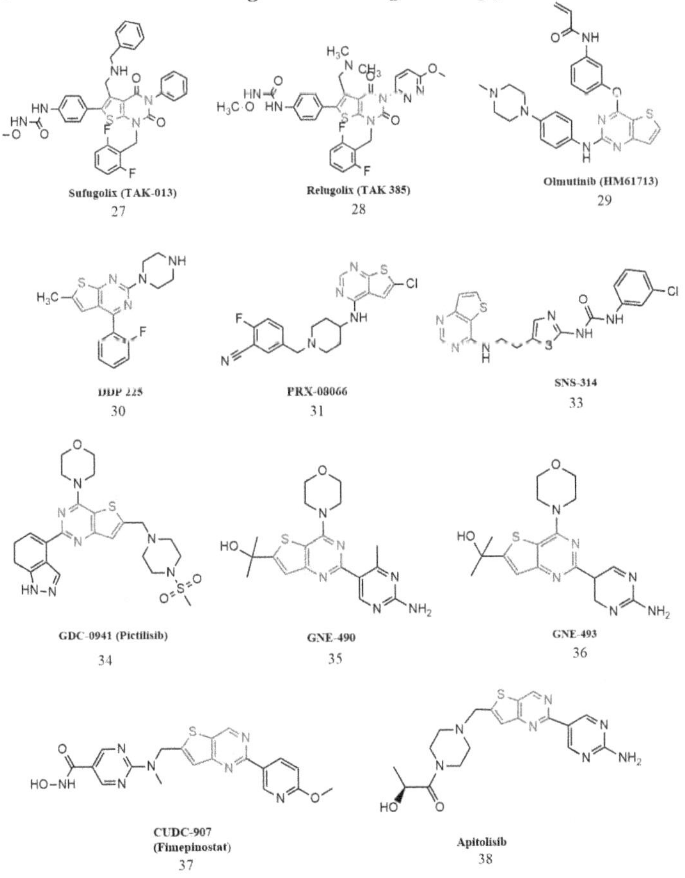

4. BIOLOGICAL ACTIVITIES OF THIENOPYRIMIDINE

Thienopyrimidine derivatives are frequently employed in organic chemistry due to their bioactive characteristics and broad biological activities. Numerous studies have shown that adding different functional groups within the thienopyrimidine ring improves its antibacterial, antifungal, anti-inflammatory, and other biological effects. Several thienopyrimidines, including fused thienopyrimidines, have displayed amazing anti-inflammatory and antimicrobial activity, as seen in **Figure 14**. The discussion about these activities begins here (Tolba et al., 2018).

Figure 14. Thienopyrimidine derivatives as anti-inflammatory agent and anti-microbial agent

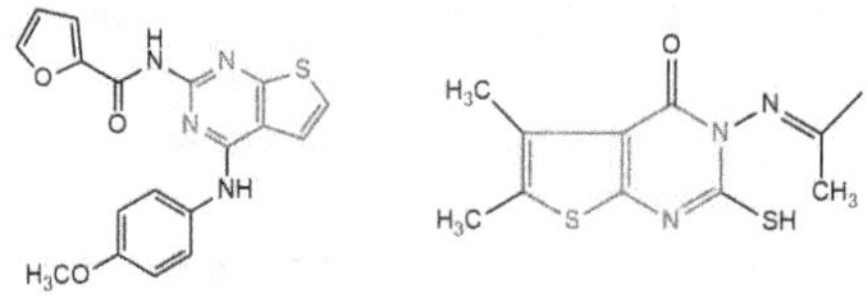

4.1 Anti-Inflammatory Activity

Inflammation is a necessary immune response, but abnormal reactions can result in diseases such as heart disease, cancer, and diabetes. Cyclooxygenases (COX-1 and COX-2) are important enzymes in inflammation, with COX-2 inhibitors providing effective but dangerous long-term treatment. Fused-thienopyrimidines have recently been identified as a promising anti-inflammatory scaffold (Thakur et al., 2024). Nonsteroidal anti-inflammatory medications (NSAIDs) are commonly used to alleviate arthritic pain and inflammation. However, they cause significant mortality and morbidity as a result of the complications associated with gastroduodenal ulcers, such as perforation and bleeding and many more like nephrotoxicity (Jackson et al., 2000).

Yuan Zhang et al., synthesized 26 prototypes to investigate their anti-inflammatory effects. Compounds **43, 44 and 45Figure 15** dramatically reduced nitric oxide and cytokine production in LPS-stimulated RAW264.7 cells without causing cytotoxicity. They also suppressed iNOS and COX-2 expression, NF-κB p65 nuclear translocation, and MAPK phosphorylation. Below compounds had anti-inflammatory properties comparable to Indomethacin in rat paw swelling trials. **43, 44 and 45** may inhibit NF-κB and MAPK pathways, suggesting their potential as anti-inflammatory medicines (Zhang et al., 2017).

Figure 15. Thienopyrimidine derivatives as anti-inflammatory agents

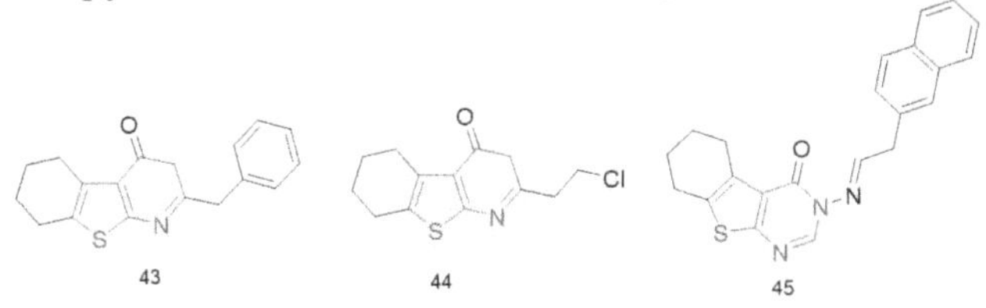

4.2 Anti-Diabetic Activity

Diabetes is a chronic disorder characterized by hyperglycemia due to insulin malfunction or poor glucose metabolism, resulting in low insulin levels in diverse organs. Type 1 diabetes (IDDM) is insulin-dependent, but Type 2 diabetes (NIDDM) is non-insulin-dependent and typically appears in adulthood (Manaithiya et al., 2021). DPP-IV inhibitors **46Figure 16** have emerged as innovative medicines that improve insulin production by increasing the half-life of GLP-1 and GIP. Wenhui Hu et al. produced compounds **Figure 16**, from that compounds **47, 48, 49, 50, 51 and 52** showing the most significant DPP-IV inhibition with an IC_{50} of 0.33 nM, promising *in vivo* efficacy, and a favorable pharmacokinetic profile, indicating clinical promise (Deng et al., 2011).

Figure 16. Thienopyrimidine derivatives as anti-diabetic agents

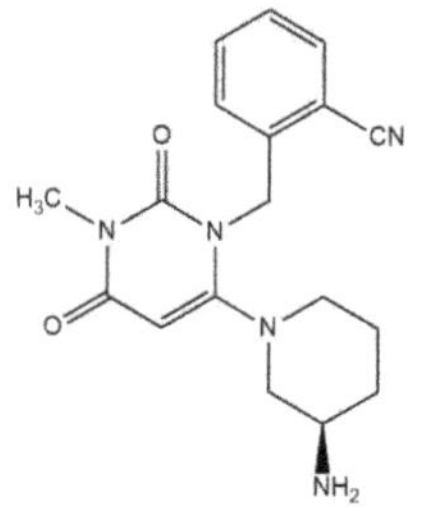

46

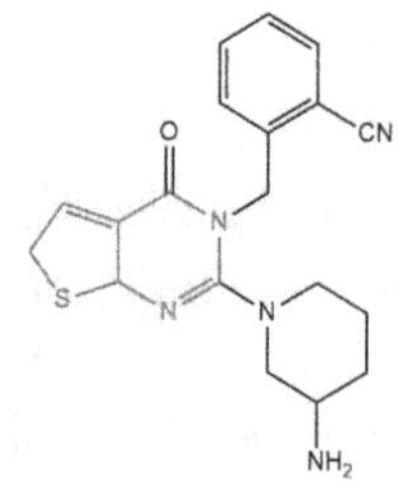

47

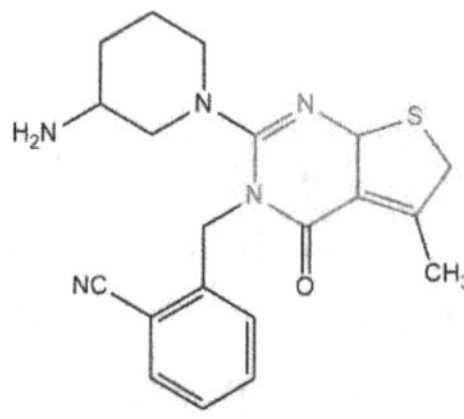

48

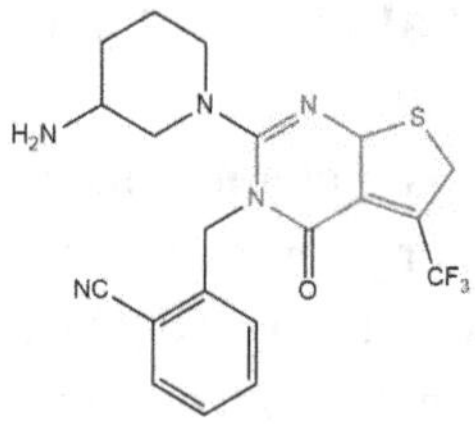

49

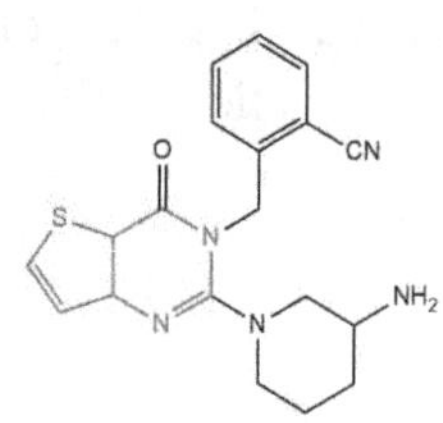

50

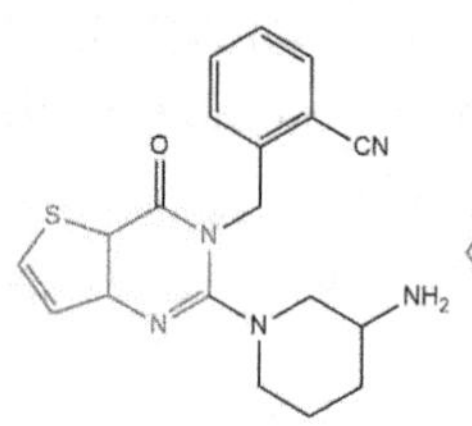

51

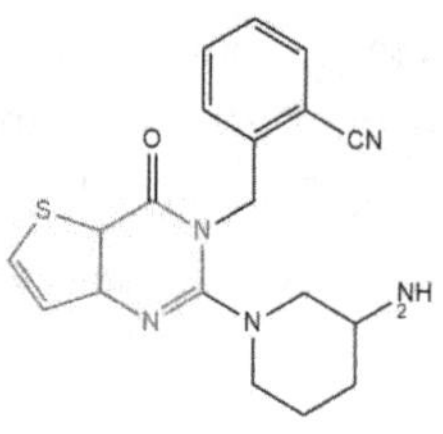

52

4.3 Anti-Cancer Activity

Thienopyrimidine-based heterocyclic compounds have a variety of pharmacological actions, including anticancer properties (M. A. Mohamed et al., 2021). Thienopyrimidines (TP), which consist of a thiophene ring fused with pyrimidine, are well-known bio isosteres of purines, an important component of the human metabolome. This scaffold's structural closeness to adenine makes it ideal for designing kinase inhibitors (Kinases et al., 2017).

Figure 17 A depicts several thienopyrimidine-containing pharmaceuticals with varying biological activity. Relugolix (TAK-385) **28**, a thienopyrimidine derivative, has passed phase III trials and is being evaluated as a GnRHR antagonist for the treatment of endometriosis and prostate cancer. Pictilisib (GDC-0941) **34**, a thieno[3,2-*d*]pyrimidine derivative that inhibits PI3K, is now in clinical trials for advanced solid tumours. Olmutinib **29**, an EGFR inhibitor, is a commercially available medicine utilised in the treatment of non-small cell lung cancer (Sayed, 2023). El-Dash et al., synthesized and characterized thienopyrimidine derivatives containing a thiosemicarbazide moiety, as well as tested there *in vitro* anticancer efficacy against prostate (PC-3) and colon (HCT-116) cancer cells. Compounds **53** and **54Figure 17** demonstrated more cytotoxicity than DOX, with compound **53** being 1.5 times more potent against PC-3 and 3 times more effective against HCT-116. Bromination at position 4 of the phenyl rings and the addition of an OCH3 group greatly increased activity. These findings indicated that thieno[2,3-*d*]pyrimidines with a thiosemicarbazide scaffold may be interesting candidates for further development as selective anticancer drugs (Salib et al., 2016).

Figure 17. Thienopyrimidine derivatives as anti-cancer agents

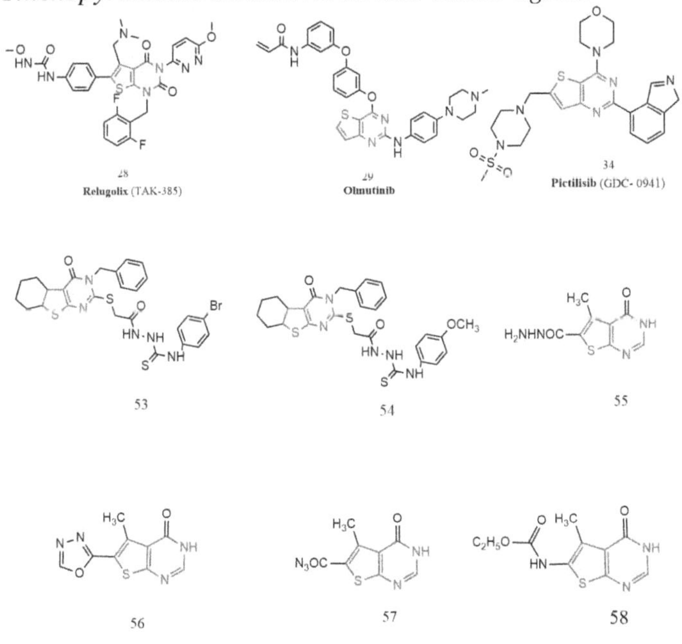

Novel thienopyrimidine derivatives were synthesized by reacting 5-methyl-4-oxo-3,4-dihydrothieno[2,3-*d*]pyrimidine-5-carbohydrazide with various carbonyl compounds, including anhydrides, acid chlorides, carbon disulfide, phenyl isothiocyanate, triethyl orthoformate, diethyl acetylene dicarboxylate, acetylacetone, pyrazole-4-carboxaldehyde, and isatin. The compounds demonstrated anti-tumor effectiveness against HePG2 and MCF-7 cell lines, compared to doxorubicin, as measured by the MTT assay. The study found that compounds **55, 56, 57** and **58** were the most active, while others showed little or no activity **Figure 18** . The effective dose (IC50) of the chemical resulted in 50% mortality in total cell count (A. T. A. Mohamed et al., 2020).

Thienopyrimidinones with thiosemicarbazide and 1,3,4-thiadiazole moieties were tested for their cytotoxicity against cancer cell lines HT-29, MDA-MB-231, HeLa, and HepG2, as well as the human diploid cell line Lep-3. Compound **59** had the maximum cytotoxicity against MDA-MB-231 (IC50 = 2.31×10^{-4} μM), while compound **62** was more active against HT-29 (IC50 = 0.001 μM). Compound **60** effectively inhibited HepG2 cells (IC50 = 0.99 μM) with minimal cytotoxicity against Lep-3 as mention in **Figure 18** (Mavrova et al., 2016). Mahasen M. Fouad et al., created novel thiophene and thienopyrimidine compounds, testing 23 compounds on five cancer cell lines: HepG-2, Hep-2, MCF-7, PC-3, and HeLa. Compounds **63, 64,** and **65** exhibited more anticancer activity than Doxorubicin **Figure 18**. These drugs also demonstrated strong DNA binding affinity and considerable inhibitory action against DNA polymerase, thymidylate synthase, and tyrosine kinase, implying that these mechanisms contribute to their anticancer properties (Fouad et al., 2018).

Novel thienopyrimidine compounds synthesized using a one-pot solvent-free approach shown strong anticancer activity. Among these, 2-(benzylamino)-5,6-dimethylthieno[2,3-*d*]pyrimidin-4(3H)-one **66Figure 18** shown high cytotoxicity, particularly against the MDA-MB-435 melanoma cell line. Structure-activity relationship investigations revealed that alterations to the pyrimidine and thiophene cores altered bioactivity, highlighting the possibility of optimizing these structures for improved anticancer efficacy (Shyyka et al., 2018).

Figure 18. Thienopyrimidine derivatives as anti-cancer agents

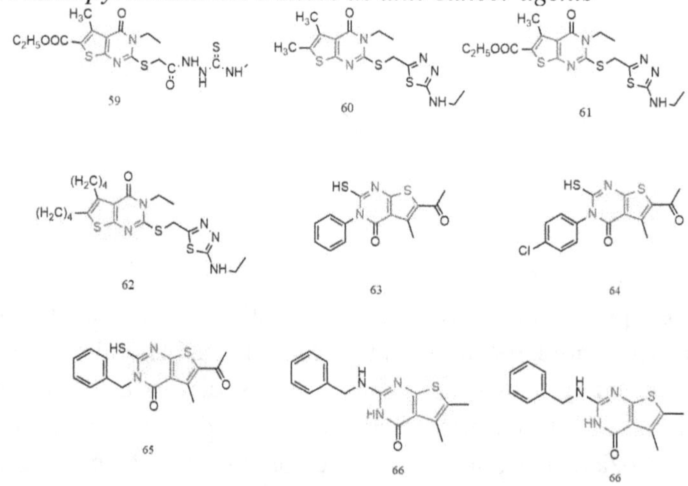

4.4 Anti-Oxidant Activity

Antioxidants, notably those containing thiophene or thienopyrimidine rings, protect against this damage and aid in the prevention of diseases such as atherosclerosis, stroke, diabetes, Alzheimer's, and cancer (Elmongy et al., 2021). C. Venkata Rao et al., synthesised *N*-substituted phenyl-5-methyl-6-(5-(4-substituted phenyl)-1,3,4-oxadiazol-2-yl)thieno[2,3-*d*]pyrimidin-4-amine derivatives **68** (a to I) and substituted phenylamino-5-methylthieno[2,3-*d*]pyrimidine-6-carboxylic acid derivatives **67** (A, B, C, and D) as illustrated in **Figure 19**. *In vitro* antioxidant screening revealed that compounds 2k, 2j, 2d, and 2e displayed high radical scavenging activity (Kotaiah et al., 2012)

Effective antioxidant activity was observed in thienopyrimidine derivatives, with notable SC_{50} values for compounds **69, 70,** and **71Figure 19** For comparison, ascorbic acid had an SC_{50} of 8.41 μM (Aly et al., 2010).

Figure 19. Thienopyrimidine derivatives as anti-oxidant agents

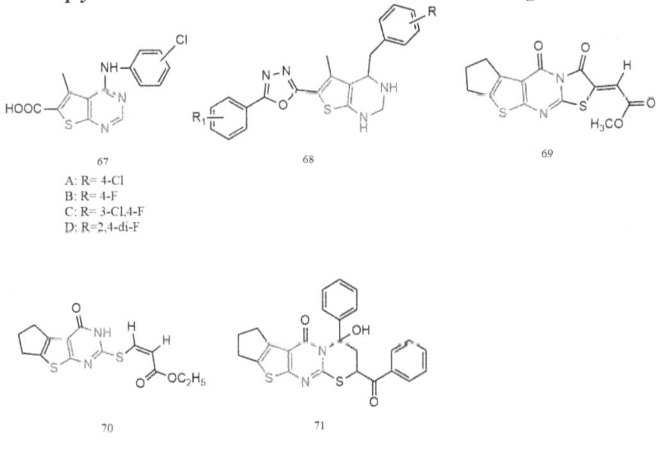

4.5 Diacylglycerol Transferase Inhibitors

Obesity, characterised by high triglyceride (TAG) levels, increases the risk of diabetes and cardiovascular disease. Diacylglycerol (DAG) is transformed to TAG *via* diacylglycerol acyltransferase (DGAT). DGAT-1 inhibitors are intended to treat metabolic problems. A new class of thieno[3,2-*d*]pyrimidine compounds was developed to inhibit DGAT-1. The cis-isomer **72** inhibited DGAT-1 in SF9 cells with high potency and selectivity, had a favorable pharmacokinetic profile, and accumulated primarily in the small intestine and liver as showed in **Figure 20**. In mouse and canine models, oral treatment of **72** lowered plasma TAG levels considerably during an oral lipid tolerance test (Dong Jin Hong, Seung Hyun Jung, Jisook Kim, Danbee Jung, Young Gil Ahn, 2019).

Figure 20. Thienopyrimidine derivative as Diacylglycerol transferase inhibitors

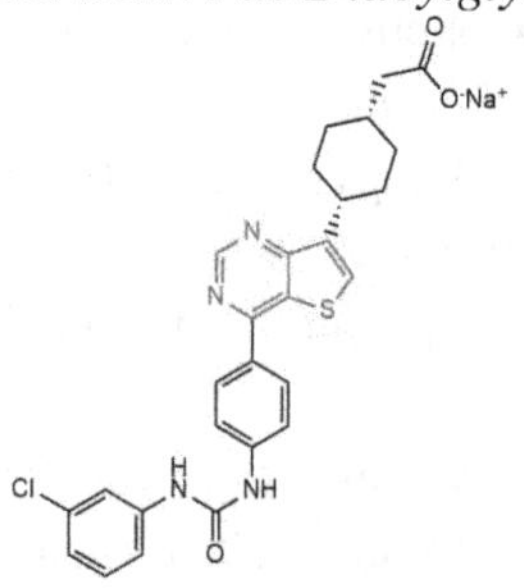

72

4.6 Anti-Alzheimer Activity

Alexey S. Chiriapkin et al. discovered 2-substituted 5,6,7,8-tetrahydrobenzo[4,5] thieno[2,3-*d*]pyrimidine-4(3H)-ones that exhibit significant anticholinesterase and anti-amyloid activity. Compounds **74, 75**, and **76** demonstrated the most potent anticholinesterase activity, whereas compounds **73, 77, 78, 79, 80, 81, and 82** were less effective **Figure 21**, comparable to Donepezil. Compounds **73, 75, 80,** and **81** demonstrated substantial amyloid inhibition, similar to GV-971 (Alexey S. chiriapkin, Ivan P. Kodonidi, dmitry I. pozdnyakov, 2022).

Figure 21. Thienopyrimidine derivatives as anti-alzheimer agents

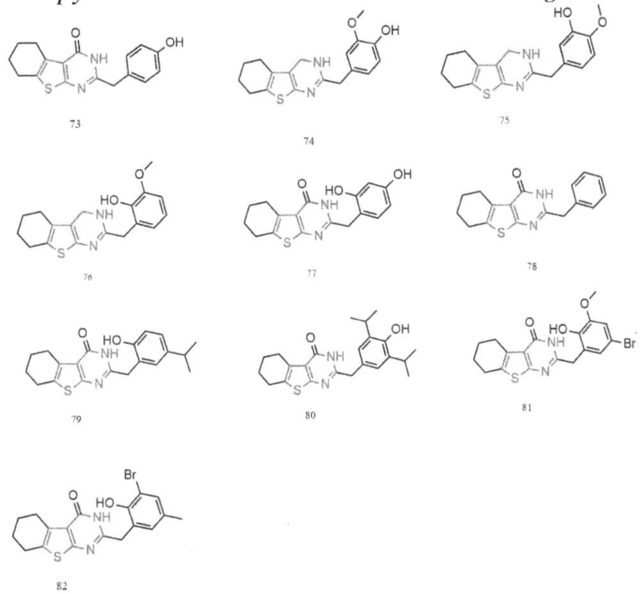

4.7 Anti-Viral Activity

Aymn E. Rashad et al., investigated thienopyrimidine compounds as antiviral agents against the avian H5N1 influenza. Compounds containing dihydronaphtho, naphtho[2,1-*b*]thiophene, and thieno[2,3-*d*]pyrimidine rings generated from 2-amino-4,5-dihydronaphtho[2,1-*b*]thiophene-1-carbonitrile were investigated. Compounds **83** and **84** demonstrated the most potent antiviral activity **Figure 22**. These compounds demonstrated the best efficacy, lowering viral presence by 92% at 5 µg/mL and 98% at 40 µg/mL. Dihydro derivatives outperformed aromatic compounds (Rashad et al., 2010).

Figure 22. Thienopyrimidine derivatives as anti-viral agents

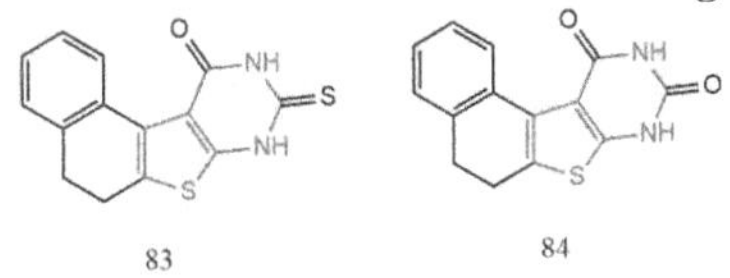

4.8 Anti-Bacterial Activity

Abdel-Rahman B.A. El-Gazzar et al., synthesized thienopyrimidine derivatives from thieno[2,3-*d*]pyrimidine-2,4-dithione and investigated their antibacterial properties. A number of substances showed antibacterial action, among them **85, 86, 87, 88, 89,** and **90** containing thienopyrimidine rings as mentioned in **Figure 23**. Some substances exhibited inhibition at concentrations as low as 128 µg/mL. De-acetylated S-glycosides shown better action against Pseudomonas putida, with MIC values of 64 µg/mL compared to ampicillin's 256 µg/mL (Hafez et al., 2010). Sidharth Chopra et al., developed thienopyrimidine derivatives as possible antibacterial and antimycobacterial medicines that target TrmD, which is essential for pathogens such as *Staphylococcus aureus*, *Pseudomonas aeruginosa*, and *Mycobacterium TB*. From a library of 200 chemicals, 22 were synthesised and evaluated. Compounds **91, 92, 93, 94, and 95Figure 23** effectively suppressed *M. tuberculosis* at MICs of 16-32 µg/mL. Additional study, including docking and ADME/T tests, revealed their promising antibacterial properties (Malasala et al., 2021).

Figure 23. Thienopyrimidine derivatives as anti-bacterial agents

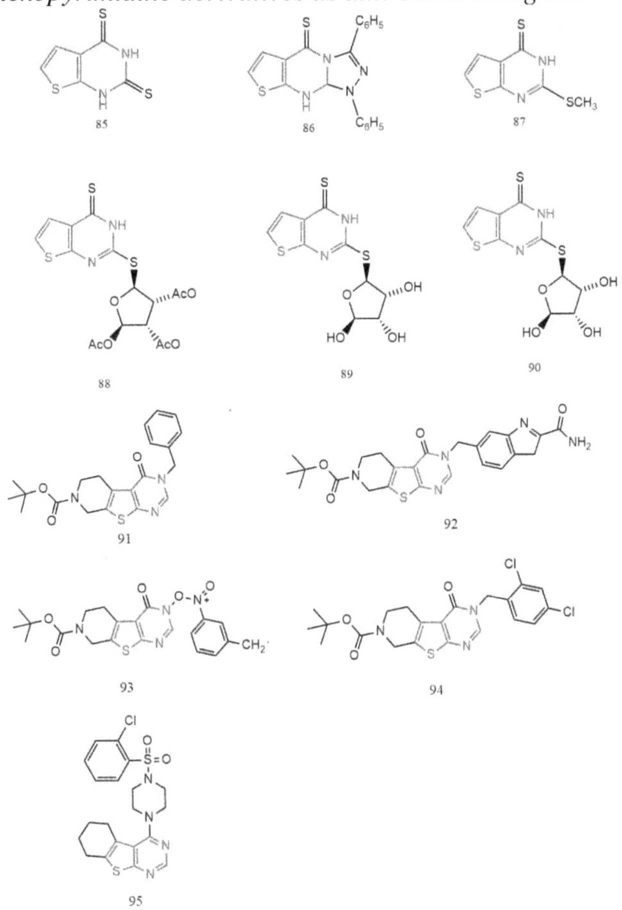

Rajesh H. Bahekar et al., produced two types of thieno[2,3-*d*]pyrimidine derivatives indicated in **Figure 24** N-[5-(2-furanyl)-2-methyl-4-oxo-4H-thieno[2,3-*d*] pyrimidin-3-yl]carboxamides **96**, **97**, and **98** and 3-substituted 5-(2-furanyl)2-methyl-3H-thieno[2,3-*d*]pyrimidin-4-ones **99**, **100**, and **101**. These compounds demonstrated strong antibacterial action against *E. coli*, *S. typhi*, *S. aureus*, and *B. subtilis*, as well as antimycobacterial activity against *M. tuberculosis* and *M. avium* (Chambhare et al., 2003).

Figure 24. Thienopyrimidine derivatives as anti-bacterial agents

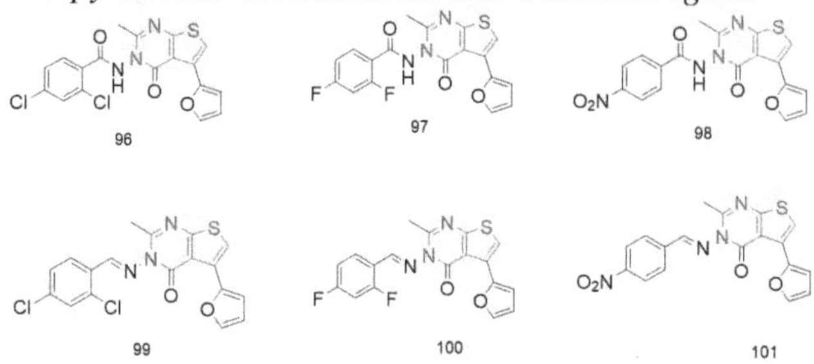

4.9 Anti-Hypertensive Activity

Thienopyrimidine derivatives, such as thieno[3,4-*d*]-, thieno[3,2-*d*]-, and thieno[2,3-*d*]pyrimidine-2,4-diones with (phenyl piperazinyl)alkyl substitution at *N*-3, showed significant antihypertensive effects in spontaneously hypertensive rats. Compounds **102, 103**, and **104,** which included a [(2-methoxyphenyl)piperazinyl] ethyl moiety at *N*-3 and hydrogen at *N*-1 **Figure 25,** had ED-50SBP values of 0.21, 0.19, and 1.0 mg/kg, respectively, lowering systolic blood pressure by 50 mmHg (Russell et al., 1988).

Figure 25. Thienopyrimidine derivatives as anti-hypertensive agents

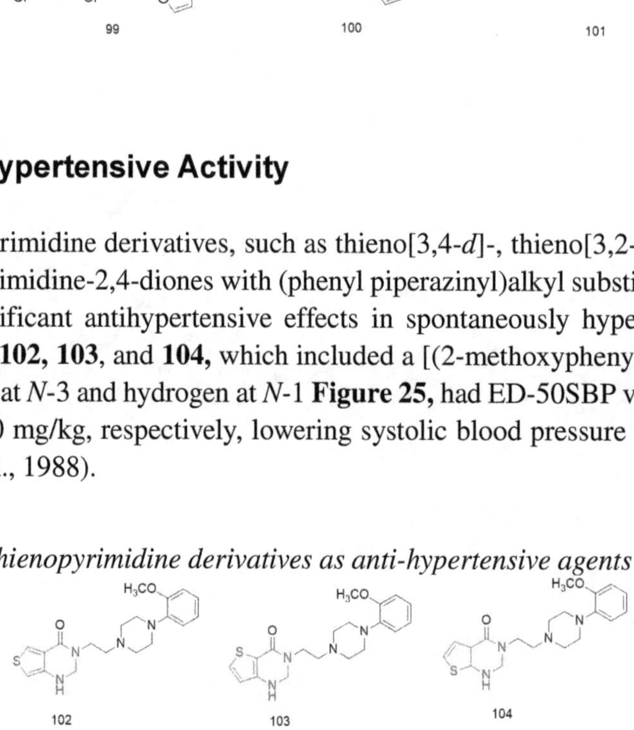

4.10 Anti-Hyperlipidaemic Activity

Docking experiments with the targets NPC1L1, ACL, CRP, LDM, SqS, and FXR demonstrated that 2-chloromethyl-5,6,7,8-tetrahydrobenzo(b)thieno[2,3-*d*] pyrimidin-4(3H)-one (LM-1554) **105Figure 26** binds more effectively with NPC1L1. Thienopyrimidines, particularly LM-1554, are effective antihyperlipidemic medicines that have shown promising outcomes in animal models when contrasted with gemfibrozil and ezetimibe, as well as large safety margins in toxicity studies (Khedkar et al., 2014).

Rangappa Srinath et al., Marikunte Venkataranganna et al., and Janardhan Saravanan et al., investigated thieno[2,3-d]pyrimidin-4(3H)-ones in a rat model of Triton 1339-induced hyperlipidemia. Compounds **106Figure 26** showed strong antihyperlipidemic activity, reducing total cholesterol, triglycerides, LDL cholesterol, and the atherogenic index while boosting HDL cholesterol (Srinath et al., 2011). Novel thienopyrimidine derivatives of azetidinone were synthesized to improve lipid-lowering properties comparable to ezetimibe. In Triton WR 1339-induced hyperlipidaemic rats, certain derivatives matched gemfibrozil's efficacy, with compound **107** significantly decreasing serum cholesterol and triglycerides and compound **108** significantly boosting HDL levels **Figure 26.** These findings indicate that structural alterations could enhance their antihyperlipidemic potency (Arya et al., 2013).

Figure 26. Thienopyrimidine derivative as anti-hyperlipidaemic agent

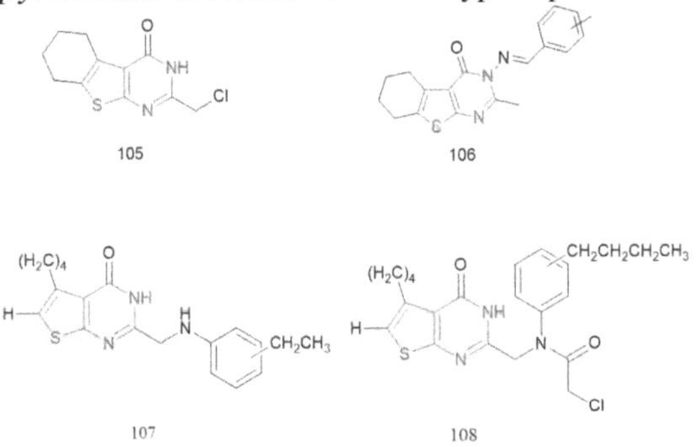

4.11 Anti-Tuberculosis Activity

Tuberculosis (TB) impacts around 2 billion people globally and is transmitted by droplets from infected individuals. *Mycobacterium tuberculosis*, the primary TB organism, causes significant mortality and grows quickly, with the World Health Organization reporting an outbreak every second. This bacteria develops primarily among macrophages (Chinsembu, 2016). Thienopyrimidine TP053 **109** is an effective antitubercular prodrug that targets both replicating and non-replicating *Mycobacterium tuberculosis* cells (**Figure 27**) (Chiarelli et al., 2020).

Figure 27. Thienopyrimidine derivatives as anti-tuberculosis agents

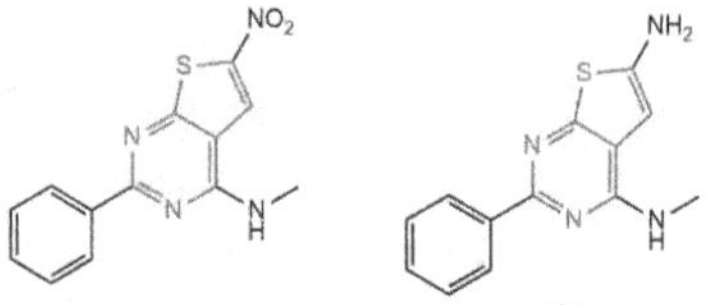

109
TP053

110
reduced TP053

4.12 Anti-Malarial Activity

In 2017, the WHO documented 219 million malaria infections and 435,000 fatalities, the majority of which were in children. Despite a 37% reduction in frequency since 2000, continued prevention and treatment are critical. Plasmodium falciparum is predominant in Africa, whereas Plasmodium vivax is more frequent elsewhere (Edwin G.Tse, 2019). Novel small molecule inhibitors of falcipain-2 (FP-2) from *Plasmodium falciparum* were developed, with IC_{50}s ranging from 1.46 to 11.38 μM. Compound **111Figure 28** showed approximately twice the potency of the initial compound. These results provided a strong basis for creating effective FP-2 inhibitors and advancing antimalarial drug development (Edwin G.Tse, 2019)

Thienopyrimidinones, such as gamhepathiopine **112**, exhibit potent antimalarial action at multiple phases of infection. M1 is efficacious at sub micromolar concentrations against both erythrocytic (EC50 = 0.045 μM) and hepatic (EC50 = 0.45 μM) forms of *Plasmodium falciparum*, including artemisinin-resistant parasites. Despite their rapid metabolization in mice, these data suggest that Thienopyrimidinones are a suitable chemotype for creating innovative multistage antimalarial medicines (Henriette bosson-vanga, Pierre Roques, Lucie Paloques, Francoise Benoit-Vical, Pierre Verhaeghe, 2021). Nicolas Masurier et al., Vincent Lisowski et al., and associates created 25 novel 4-amino-substituted analogues of Gamhepathiopine **112Figure 28**, a dual-action antimalarial. Among these, N2-(tert-butyl)-N4-(3-(dimethylamino) propyl).-6-(p-tolyl)thieno[3,2-*d*]pyrimidine-2,4-diamine **113** demonstrated improved physicochemical qualities and stability while retaining significant ant plasmodial action. These findings show that 4-aminothieno[3,2-*d*]pyrimidine derivatives have potential for future optimization in antimalarial medication development (Lagardère et al., 2023).

Figure 28. Thienopyrimidine derivatives as anti-malarial agents

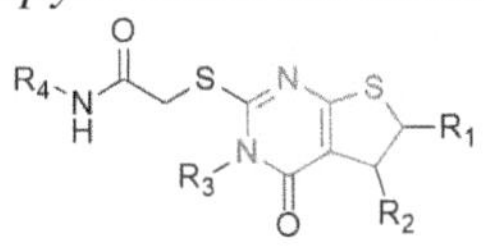

111

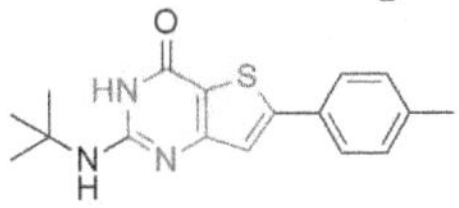

112
Gamhepathiopine

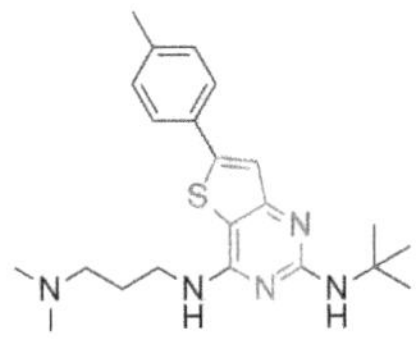

113

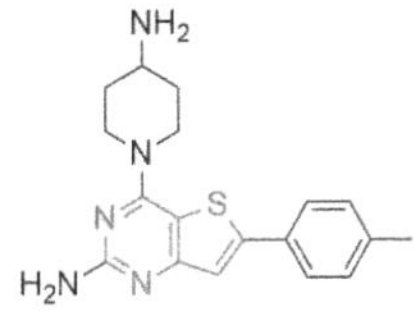

114

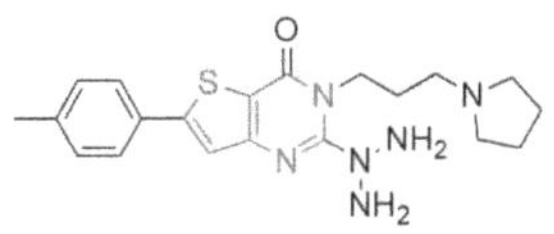

115

4.13 Anti-Trypanosomatid Activity

Trypanosomatid parasites are responsible for neglected diseases such as human African trypanosomiasis (HAT), Chagas disease, and leishmaniasis, which threaten millions of people worldwide. With complex life cycles and limited therapies, there is a critical need for innovative medications, particularly in low-resource rural health systems (Field et al., 2017).

Leishmaniases, caused by *Leishmania species*, are neglected tropical diseases with difficult treatment options due to toxicity, resistance, and expense. *N-Myristoyltransferase (NMT)* is a promising target, and thienopyrimidine inhibitors are effective against *Leishmania parasites*. Notably, PF-00349412 and IMP-083 **116** revealed significant *NMT* inhibition, while IMP-105 **117** showed selectivity for *Leish Kinetoplast DNA (kDNA) mania NMT* over human *NMTs***Figure 29**. Additional optimisation is required to improve potency and pharmacokinetic features for future *in vivo* applications (Andrew S. Bell, Zhiyoung Yu, jennie A. hutton, Megan H. Wright, James A. Brannigan, daniel Paape, shieley M. Roberts, Charlotte L. sutherell, Markus Ritzefeld, Anathony J. Wilkinson, Deborah F. Smith, Robin J. Leatherbarrow, 2020).

Optimizing chemotypes for other diseases can accelerate the discovery of new antiparasitic drugs. Research on the anilino quinazoline lapatinib and alkynyl thieno[3,2-*d*]pyrimidine GW837016X (NEU-391) **118Figure 29** revealed significant antiparasitic effects against *Trypanosoma brucei*, with NEU-391 increasing mouse lifespan by 50% and inhibiting parasite mitosis and cytokinesis. This indicated that NEU-391 is a promising candidate for further development (Woodring et al., 2018).

Figure 29. Thienopyrimidine derivatives as anti-trypanosomatid agents

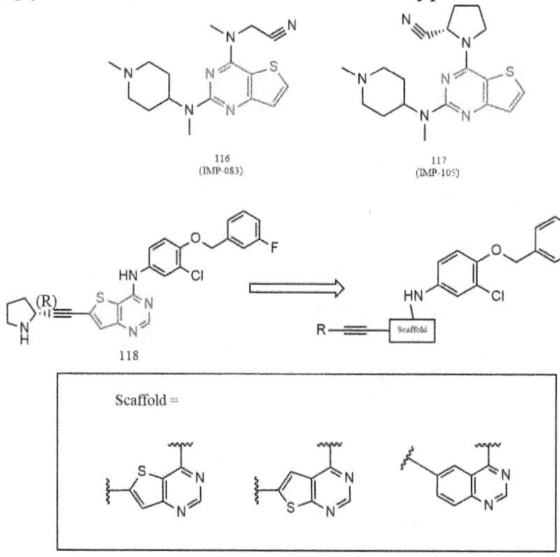

4.14 Anthelmintic Activity

Anthelmintics address helminth infections in a variety of organs, specifically complex parasites with muscular, neurological, digestive, and reproductive systems. The three primary forms of helminths are cestodes (tapeworms), nematodes (roundworms), and trematodes (flukes). These drugs are helpful in eliminating parasites from the host's body, including the gut, bile ducts, and tissues (Bhavsar et al., 2020).

New benzimidazole derivatives of thieno[2,3-*d*]pyrimidin-4-ones were tested for antiparasitic activity and demonstrated significant efficacy against *Trichinella spiralis larvae*. Compounds **119, 120**, and **121** which include a the *Kinetoplast DNA (kDNA)* trahydrobenzothienopyrimidin-4-one cycle **Figure 30**, had much greater anthelmintic activity than albendazole. These chemicals also showed 100% efficiency against *Lamblia muris* in mice, eliminating cysts from faeces. These findings indicated that adding a benzimidazole moiety improves interaction with biological targets, resulting in powerful anthelmintic and antiprotozoal activity (Mavrova et al., 2010).

Figure 30. Thienopyrimidine derivatives as anthelmintic agents

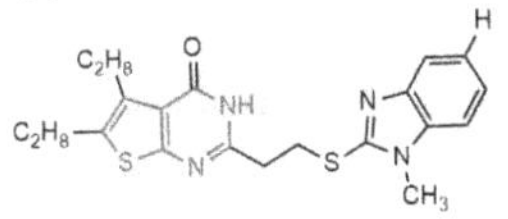

119

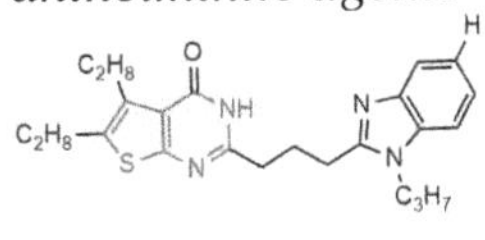

120

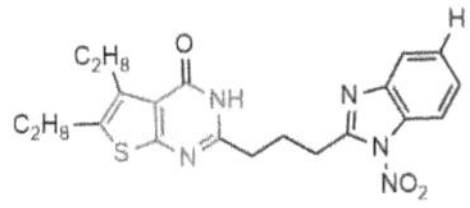

121

4.15 Anti-Convulsant Activity

A variety of condensed cyclopentanopyrido[3,2:4,5]thieno[3,2-*d*]pyrimidine derivatives **122, 123, 124, 125, 126** and **127** were synthesized, and they demonstrated considerable neurotropic activity as mentioned in **Figure 31**. Several compounds exhibited anticonvulsant properties, performing ethosuximide and producing sedative effects. Particularly, derivatives containing chlorine or acetamide substituents had the highest anticonvulsant effect, validating the thienopyrimidine derivatives' potential as anticonvulsant medicines (Dabaeva et al., 2020).

Figure 31. Thienopyrimidine derivatives as anti-convulsant agents

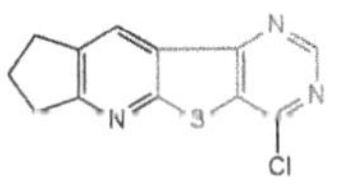

122

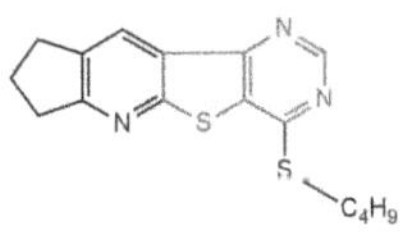

123

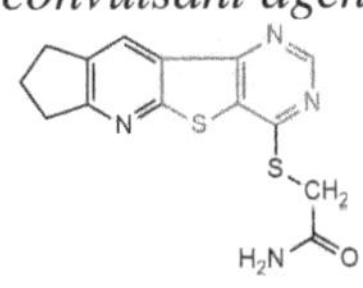

124

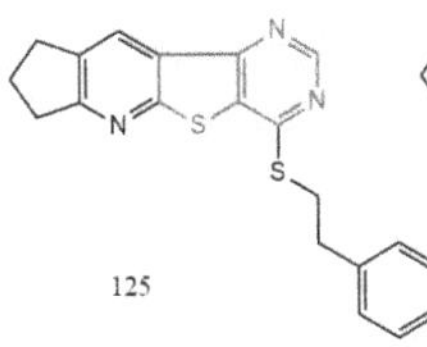

125

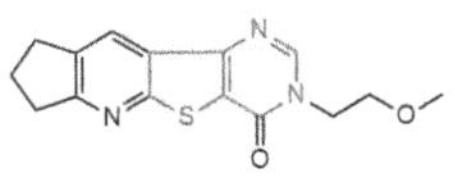

126

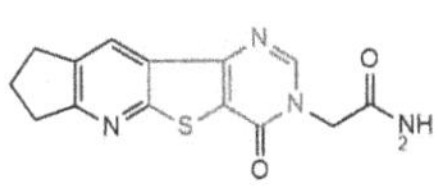

127

4.16 Anti-Psychotic Activity

Antipsychotic medicine, often known as neuroleptics, is used to treat schizophrenia and other psychoses. They are also utilized by specialists from many specialties to treat neurological problems, expanding their applicability beyond psychiatry (BLACK et al., 1985). Thienopyrimidine derivatives shown strong antipsychotic efficacy via many neurotropic pathways. Compound **128, 129 and 130** showed substantial anticonvulsant, anxiolytic, and depressive activity **Figure 32**, outperforming ethosuximide in PTZ-induced seizure models. Docking investigations demonstrated robust interactions with GABAA receptors and other important targets, indicated that these compounds are viable candidates for the development of effective antipsychotic drugs (Shushanik Sh. Dashyan, Eugene v. babaev, ervand g. paronikyan, Armen G. Ayvazyan, 2022).

Figure 32. Thienopyrimidine derivatives as anti-psychotic agents

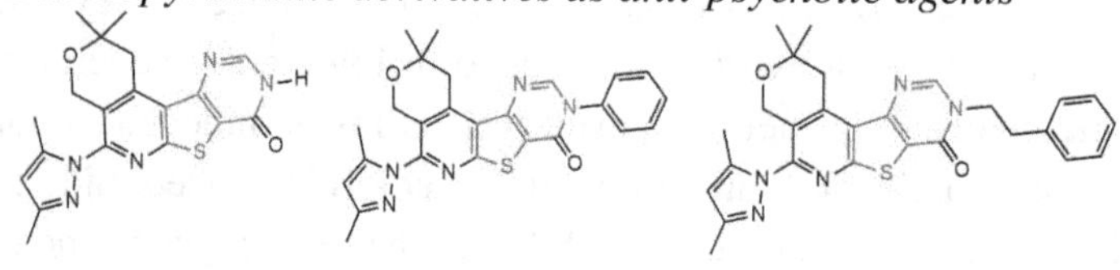

128 129 130

4.17 Anti-Arrhythmic Activity

New thiophene derivatives, including thienopyrimidines, were synthesized and studied utilizing IR, NMR, MS, and elemental analysis for the Anti-arrhythmic action. Compounds **131** and **132** demonstrated excellent antiarrhythmic effects **Figure 33**, outperforming conventional medicines such as procainamide and lidocaine. The findings supported the use of thienopyrimidines as effective antiarrhythmic medicines (Amr et al., 2010).

Figure 33. Thienopyrimidine derivatives as anti-arrhythmic agents

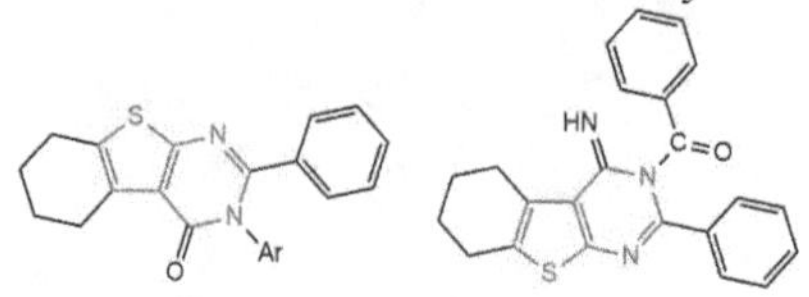

131 132

4.18 Anti-Histaminic Activity

Meenakshi Singh et al., and colleagues conducted a QSAR research on 38 thienopyrimidine derivatives **Figure 34** to identify new histamine H1-receptor antagonists. Using Genetic Function Approximation (GFA), they discovered essential descriptors—dipole, AlogP98, Jurs, and LUMO—that are required for H1-antihistaminic action. The study found that lipophilicity, together with thermodynamic and electrical characteristics, has a major impact on biological activity, offering robust prediction models for developing more potent H1-histamine inhibitors (Meenakshi Singh, Sushil K Singh, 2011).

Figure 34. Thienopyrimidine derivatives as anti-histaminic agents

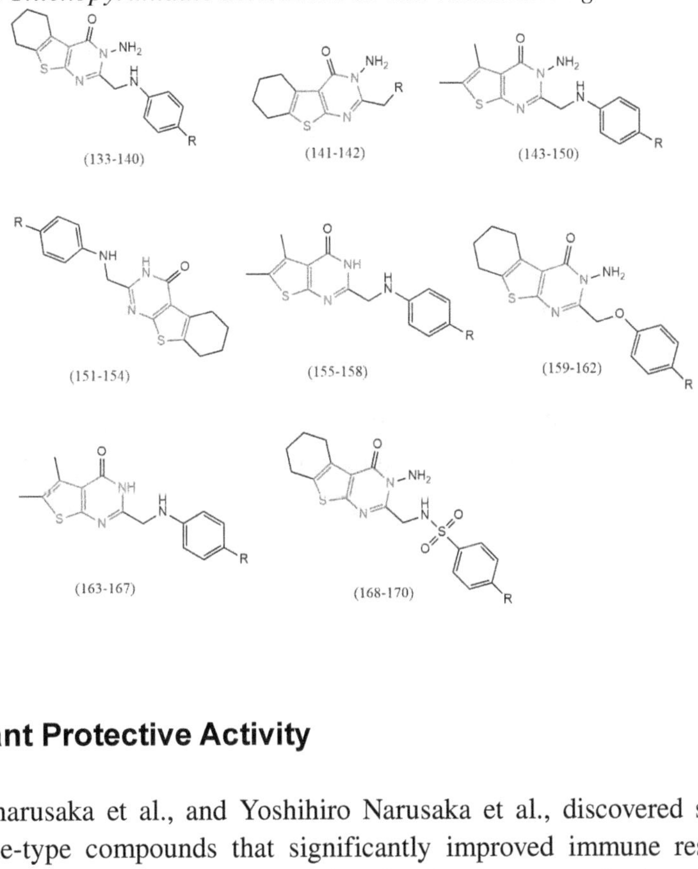

4.19 Plant Protective Activity

Mari narusaka et al., and Yoshihiro Narusaka et al., discovered six thienopyrimidine-type compounds that significantly improved immune responses in Arabidopsis thaliana. These compounds offered resistance to the fungal pathogen *Colletotrichum higginsianum* and the bacterial pathogen *Pseudomonas syringae pv. maculicola*. The thienopyrimidine core proved critical in increasing plant resistance, suggested these compounds potential candidates for novel plant activators and agrochemicals **Figure 35** (mari narusaka and Yoshihiro narusaka, 2017).

Figure 35. Thienopyrimidine derivatives as plant protective agents

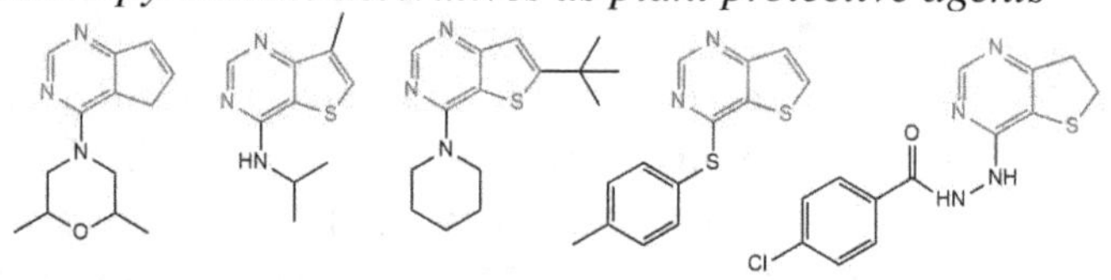

171 N2781 | 172 N2835 | 173 N2947 | 174 N2969 | 175 N2972

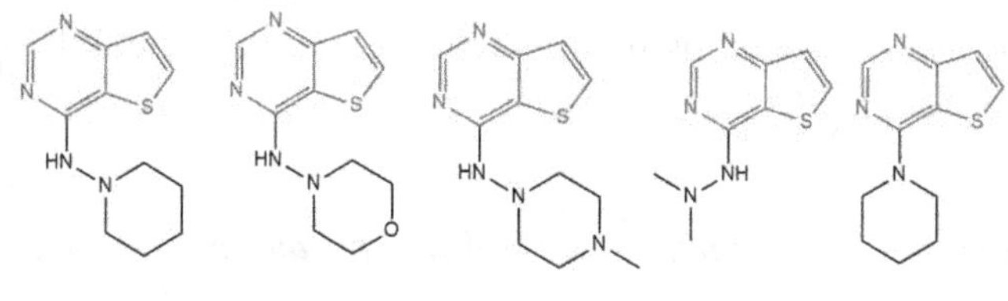

176 N2914C | 177 N2914A1 | 178 N2914A2 | 179 N2914A3 | 180 N2914A4

5. SAR OF BIOLOGICAL ACTIVITIES

The SAR of thienopyrimidines for various biological activities is provided below in **Figures 36-39**. This understanding allows for the rational design of novel derivatives with more effective and targeted biological characteristics increasing their potential as valuable scaffolds in the creation of novel therapeutic medicines. The research and development of the thienopyrimidine structure greatly advances drug discovery.

Figure 36. SAR of Thienopyrimidine ring showing the anti-plasmodial activity (Prisca Lagardere, Cyril Fersing, Nicolas Masurier, 2022)

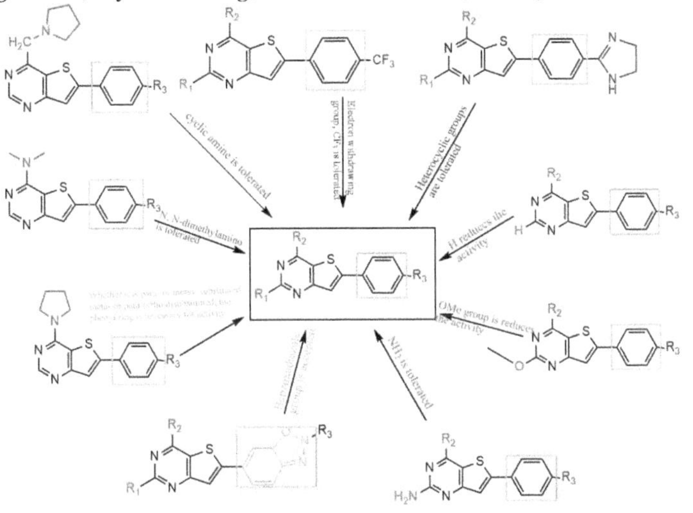

Figure 37. SAR of Thienopyrimidine

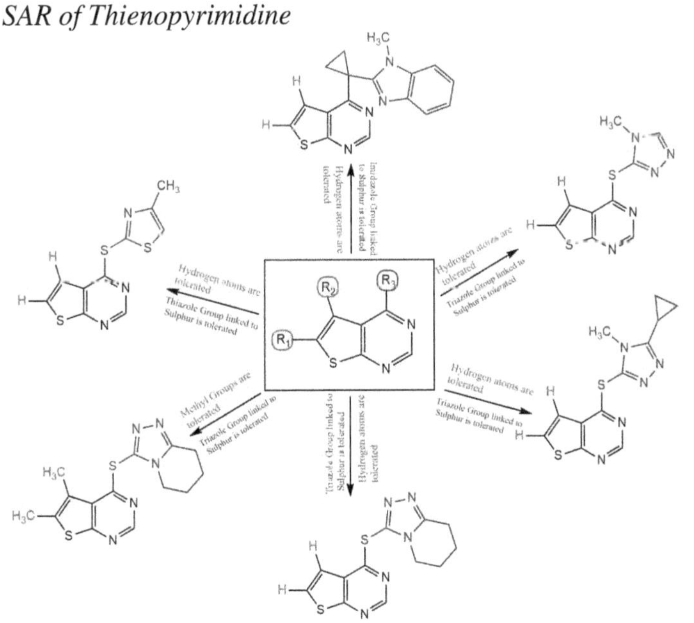

Figure 38. SAR of Thienopyrimidine ring as selective inhibitors of L.donovani and L.major NMT (Prisca Lagardere, Cyril Fersing, Nicolas Masurier, 2022).

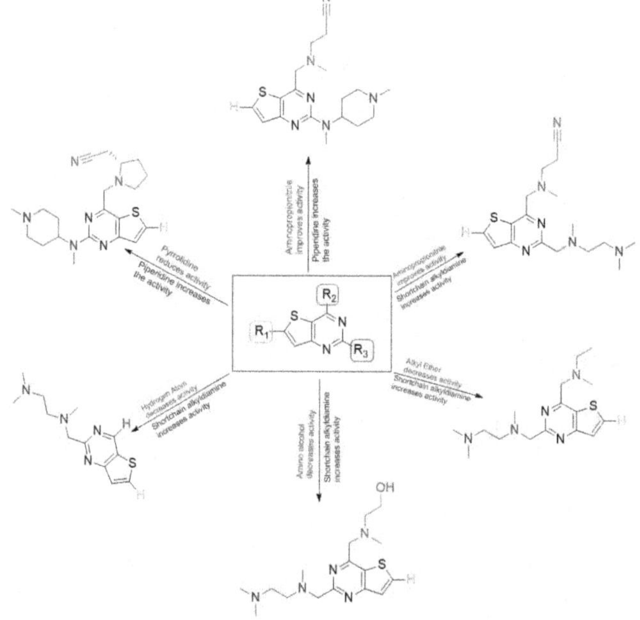

Figure 39 SAR of Thienopyrimidine indicating Anti-Cancer Activity (Eslam M.H. Ali, Mohammed S. Abdel-Maksoud, 2019).

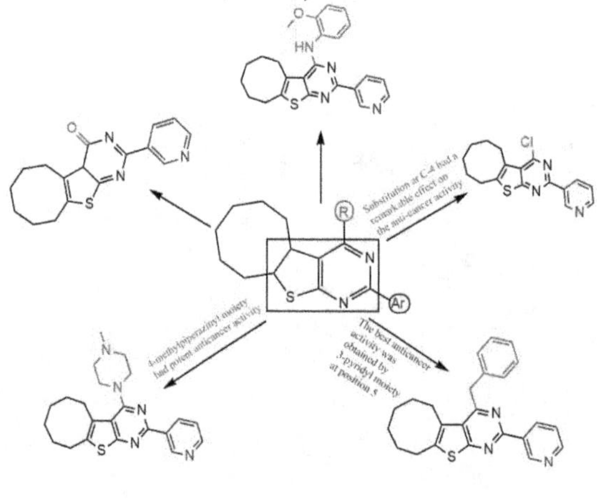

6. PATENT UPDATES ON THIENOPYRIMIDINE

Numerous patents have been filed concerning Thienopyrimidine, with the latest updates summarized in Table 1.

Table 1. Some of the patents on Thienopyrimidine

Patent Id	Title	Assignee	Inventor/Author	Grant date
CN-117126176-A	Thienopyrimidine compound and dopamine D thereof 2 Use of receptor allosteric antagonists	Changzhou University	Qian Mingcheng, Jin Jie, Sun Zhengyang, Li Yan, Jing Chenchen, Chen Xin, Hou Yanan, Zhao Shuai	28/11/2023
CN-118063485-A	Novel substituted thienopyrimidine compound and preparation method and application	Novel substituted thienopyrimidine compound and preparation method and application thereof	Chen Ye, Zhang Yadong, Liu Ju, Wang Zhi, Fu Haodong, Wang Yue	24/05/2024
CN-117126174-A	Cocrystals of thienopyrimidine compounds and uses thereof	Applicant Guangdong East Sunshine Pharmaceutical Co., Ltd.	Shan Yuefeng, Chen Liang, Ruan Sai, Huang Weiming	28/11/2024
CN-115746017-A	A thienopyrimidine compound, its preparation method and application	*In vivo* Chemical Technology (Guangzhou) Co., Ltd.	Zhang Jiongguang	07/03/2023
KR-20230072373-A	Composition for inhibiting LRRK2 containing a thienopyrimidine derivative as an active ingredient	Chung-Ang University Industry-Academic Cooperation Foundation	Min Kyung-hoon	24/05/2023
CN-116940575-A	Thienopyrimidinedione compounds and their applications	Shenzhen Kangzhe Pharmaceutical Co., Ltd.	Chen Xinhai Hu Boyu Zhou Kai Chen Zhaoguo Liu Bin Hu Guoping Li Jian Chen Shuhui	01/09/2022
TW-202233632-A	Thienopyrimidine derivative	Mainland China Chia Da Tianqing Pharmaceutical Group Co., Ltd.	Zhang Yinsheng, Liu Baomin, Huang Yu, Chen Zheng helped Zhu Yan cover Kuo	01/09/2022
WO-2021257828-A1	Substituted thienopyrimidines that interact with the ras superfamily for the treatment of cancers, inflammatory diseases, rasopathies, and fibrotic disease	SHY THERAPEUTICS, LLC	HADARI, Yaron, R; SCHMERTZLER, Michael; WILLIAMS, Theresa, M.; CARTA, Luca; HUTCHESON, Rebecca;	23/12/2021
EP-4073073-A1	Thienopyrimidine derivatives as lpa receptor 2 inhibitors	Chiesi Farmaceutici SpA	Gabriele Amari Elisabetta Armani Mafalda PAGANO Luca Raveglia	06/12/2023
CN-114516883-A	Thienopyrimidine compound, and pharmaceutical composition and application thereof	Suzhou Youli Biomedical Technology Co ltd	Chen Xuxing Chen Yanhong Zhao Zhao	20/05/2022

continued on following page

Table 1. Continued

Patent Id	Title	Assignee	Inventor/Author	Grant date
AU-2020327926-B2	Thienopyrimidine derivatives as ACC inhibitors and uses thereof	Gilead Sciences Inc	Jamie G. BATES Ana Z. GONZALEZ BUENROSTRO Hongyan Guo Xiaochun Han Brian J. Kirby Yurong Lai Michael L. Mitchell Gregg M. SCHWARZWALDER James G. Taylor Ting Wang	13/04/2023
CN-114206880-A	Benzothiophene, thienopyridine, and thienopyrimidine derivatives for modulation of STING	Ctxt Private Ltd	B. J. Morrow, J. G. Hubert, M. L. Dennis, A. N. Kuzup, P. A. Stuppel	18/03/2022
US-2020255447-A1	Thienopyrimidines and uses thereof	Memorial Sloan Kettering Cancer Center	Viviane Tabar Kosuke Funato	13/08/2020
CN-111057065-A	Preparation method and application of thienopyrimidine compound	Shenyang Pharmaceutical University	Zhao Yanfang, Hou Yunlei, Qin Mingze, Liu Yajing, Han Yufei, Tian Ye, Wang Ruxin	23/04/2023
EP-3891156-B1	New crystalline forms of a thienopyrimidine as mcl-1 inhibitor	Laboratoires Servier SAS Vernalis R&D Ltd	Emilie DE BAETS Julien Auvray Michael Lynch Nicolas LEBLANC	15/11/2023
CN-111187277-A	Thienopyrimidine derivative, preparation method and medical application thereof	Zhang Guobao, Hu Weimin, He Feng, Tao Weikang	Jiangsu Hengrui Medicine Co Ltd Shanghai Hengrui Pharmaceutical Co Ltd	12/04/2022
US-2022267346-A	Thienopyrimidine derivatives having stereo configurations and use thereof in medicine	Sunshine Lake Pharma Co Ltd	Zheng Gu Wanjun TANG Xinye YANG Xuli WANG Weihui YUAN Yunzeng CUI Wen Yang Jianyu LIU Yingjun Zhang	Pending
CN-111848678-A	Phosphorus-containing thienopyrimidine derivatives	Chia Tai Tianqing Pharmaceutical Group Co Ltd	Zhang Yinsheng, Liu Baomin, Huang Yu, Chen Zhengbang, Gai Kuo, Zhao Kaidi, You Xinyi, Tian Xiaomeng, Wang Xiaojin	Pending
BR-112020021153-A2	meglumine salts of thienopyrimidines	Cardurion Pharmaceuticals, Llc	Anuj K. Kumar Raymond E. Forslund	23/03/2021
CN-109879887-B	Thieno [3,2-*d*] pyrimidine derivative containing indole structure and application thereof	Liaoning University	Liu Ju, Chen Ye, Shi Jiantao, Gong Yilin, Ding Shi, Wang Yang, Li Jie	20/07/2021
CN-111171044-A	Thienopyrimidine compound and medical application thereof	Shenyang Research Institute of Chemical Industry Co Ltd	Chai Baoshan, Bu Yanxin, Wang Wanqiu, Wang Yunhua, Jiao Jiayuan, Guang Haihong, Zhao Xiancheng, Bi Yi	24/06/2022
CN-109180702-A	A kind of Thienopyrimidine ketone compound and application thereof	Sichuan University	Yang Shengyong Li Linli	30/11/2021

continued on following page

Table 1. Continued

Patent Id	Title	Assignee	Inventor/Author	Grant date
CN-111108109-A	Substituted dihydrothienopyrimidines and their use as phosphodiesterase inhibitors	Combination therapy Co.	J. Larson	Pending
CN-108586486-B	Preparation method of aryl-substituted thienopyrimidine compoud	Sichuan University	Chen Yuanwei	02/03/2021
JP-2018115214-A	Compositions comprising thienopyrimidine and thienopyridine compounds and methods of use thereof	University of Michigan	Grembecka Jolanta Grembecka Jolanta Cierpicki Tomasz Cierpicki Tomasz Borkin Dmitry Borkin Dmitry Hess Jay Hess Jason Duxin Sun Xiaoqin Li	25/03/2023
CN-108440559-A	Diaryl thienopyrimidines HIV-1 reverse transcriptase inhibitor and its preparation method and application	Shandong University	Liu Xinyong Kang Dongwei Zhanpeng	16/06/2020
WO-2018133858-A1	Thienopyrimidine derivative and use thereof in medicine	Sunshine Lake Pharma Co., Ltd.	Wanjun TANG Xinye YANG Zheng Gu Chenlu LI Zongyuan ZHANG Zhifu WAN Xiaojun Wang Yingjun Zhang	26/07/2018
CN-108424417-B	Thienopyrimidine derivative, preparation method and application in preparation of anti-tumor drugs	Henan Genuine Biotech Co Ltd	Chang Junbiao Du Jinfa	20/09/2019
EP-3567043-A1	Thienopyrimidine derivative, preparation method therefor, and application thereof in manufacturing of antitumor drugs	Henan Genuine Biotech Co Ltd	Junbiao Chang Jinfa Du	06/12/2023
TW-I780077-B	Thienopyrimidine compound, its preparation method, pharmaceutical composition and its application	Mainland China Nanjing Hongyun Biotechnology Co., Ltd.	Deng Xianming, Zhang Baoding, Liu Shuang, Dong Chao, Sun Xihuan, Huang Xiaoxing, Deng Zhou, Li Yunzhan, Lu Yue, Li Li, Hu Zhiyu	11/10/2022
EP-3498716-B1	Sulfonyl bearing thienopyrimidine compound, preparation method therefor, pharmaceutical composition, and application as anti-cancer agent	Xiamen University	Xianming Deng Baoding ZHANG Shuang Liu Chao Dong Xihuan SUN Xiaoxing HUANG Zhou DENG Yunzhan LI Yue Lu Li Li Zhiyu Hu	17/04/2024
CN-107253964-A	The preparation and application of the thienopyrimidines of the structure containing heteroaryl amide	Jiangxi Science and Technology Normal University	Zhu Wufu Zheng Pengwu Wang Lin Xiao Lanzhou Tang Qidong Liu Xiaobo Wang Caolin Zhao Bingbing	Pending

continued on following page

Table 1. Continued

Patent Id	Title	Assignee	Inventor/Author	Grant date
CN-107163061-A	The preparation and application of the thienopyrimidines of the structure containing pyrazoline	Jiangxi Science and Technology Normal University	Xu Shan, Zhu Wufu, Zheng Pengwu, Wang Qinqin, Xiao Zhen, Wang Caolin, Zhang Binliang, Jia Shuangzhijia	Pending
CN-106866699-B	A kind of diaryl thienopyrimidines HIV-1 reverse transcriptase inhibitor and its preparation method and application	Shandong University	Liu Xinyong Kang Dongwei Zhan Peng Wu Gaochan Huo Zhipeng	08/03/2023

7. ONGOING CLINICAL TRIALS

The clinical trial updates for thienopyrimidine analogues are shown in Table 2.

Table 2. Clinical trial study of thienopyrimidine

Sr. No.	Study Title	Study Type	Study Phase	Study Design	Condition	Status	ClinicalTrials.gov Identifier
1	Relugolix and Enzalutamide in Combination with Radiation Therapy for the Treatment of Very High-Risk Prostate Cancer, OPTIMAL Trial	Interventional	Phase 2	NA	Prostate Adenocarcinoma	Not yet recruiting	NCT06499870
2	Description of Relugolix Use in Patients with Prostate Cancer Within the VHA	Observational	-	NA	Prostate Cancer	Active, not recruiting	NCT06462014
3	Olmutinib 600 mg QD in Patients with T790M-positive NSCLC After Treatment With an EGFR-TKI	Interventional	Phase 1	NA	Non-Small Cell Lung Cancer	Terminated	NCT04510415
4	Olmutinib Trial in T790M (+) NSCLC Patients Detected by Liquid Biopsy Using BALF Extracellular Vesicular DNA	Interventional	Phase 2	NA	Non-Small Cell Lung Cancer	Completed	NCT03228277

continued on following page

Table 2. Continued

Sr. No.	Study Title	Study Type	Study Phase	Study Design	Condition	Status	ClinicalTrials.gov Identifier
5	Phase II Trial of HM61713 for the Treatment of 2nd Line T790M Mutation Positive Adenocarcinoma of the Lung	Interventional	Phase 2	NA	Non-Small Cell Lung Cancer	Terminated	NCT02485652
6	Phase II Trial to Evaluate the Efficacy and Safety of HM61713 as the 1st-line NSCLC Anticancer Therapy	Interventional	Phase 2	NA	Non-Small Cell Lung Cancer	Completed	NCT02444819
7	Safety and Tolerability Study of SNS-314 for Advanced Solid Tumors	Interventional	Phase 1	NA	Advanced Solid Tumors	Completed	NCT00519662
8	12 Weeks Treatment with DDP225 or Placebo in Patients with Chronic Functional Vomiting	Interventional	Phase 2	Randomized	Vomiting	Terminated	NCT00252993
9	8 Weeks Treatment with DDP225 in Patients With Diarrhea Predominant IBS	Interventional	Phase 2	Randomized	Irritable Bowel Syndrome	Completed	NCT00230581
10	An Open-Label Study to Evaluate the Safety and Efficacy of PRX-08066 in Patients with Pulmonary Hypertension and Chronic Obstructive Pulmonary Disease	Interventional	Phase 2	NA	COPD Pulmonary Hypertension	Terminated	NCT00677872
11	Safety and Effectiveness of PRX-08066 in Patients with Pulmonary Hypertension and Chronic Obstructive Pulmonary Disease	Interventional	Phase 2	Randomized	Pulmonary Hypertension Chronic Obstructive Pulmonary Disease	Completed	NCT00345774

continued on following page

Table 2. Continued

Sr. No.	Study Title	Study Type	Study Phase	Study Design	Condition	Status	ClinicalTrials.gov Identifier
12	A Study of the Effect of Formulation and Tablet Hardness on the Bioavailability of Pictilisib in Healthy Volunteers	Interventional	Phase 1	Randomized	Healthy Volunteer	Completed	NCT02092831
13	Study of GDC-0941 or GDC-0980 With Fulvestrant Versus Fulvestrant in Advanced or Metastatic Breast Cancer in Participants Resistant to Aromatase Inhibitor Therapy	Interventional	Phase 2	Randomized	Breast Cancer	Completed	NCT01437566
14	A Study of the Safety and Pharmacology of GDC-0941 in Combination with Erlotinib in Patients with Advanced Solid Tumors	Interventional	Phase 1	Non-Randomized	Non-Squamous Non-Small Cell Lung Cancer, Solid Cancers	Completed	NCT00975182
15	A Study of GDC-0941 in Participants with Locally Advanced or Metastatic Solid Tumors for Which Standard Therapy Either Does Not Exist or Has Proven Ineffective or Intolerable	Interventional	Phase 1	Non-Randomized	Solid Cancers	Completed	NCT00876109
16	Fimepinostat in Treating Brain Tumors in Children and Young Adults	Interventional	Early Phase 1	NA	Diffuse Intrinsic Pontine Glioma Recurrent Anaplastic Astrocytoma Recurrent Glioblastoma Recurrent Malignant Glioma Recurrent Medulloblastoma	Active, not recruiting	NCT03893487

continued on following page

Table 2. Continued

Sr. No.	Study Title	Study Type	Study Phase	Study Design	Condition	Status	ClinicalTrials.gov Identifier
17	CUDC-907 Treatment in People with Metastatic and Locally Advanced Thyroid Cancer	Interventional	Phase 2	NA	Thyroid Neoplasms Poorly Differentiated and Undifferentiated Thyroid Cancer Differentiated Thyroid Cancer	Terminated	NCT03002623
18	Trial of CUDC-907 in Children and Young Adults with Relapsed or Refractory Solid Tumors, CNS Tumors, or Lymphoma	Interventional	Phase 1	NA	Lymphoma Neuroblastoma Brain Tumor Solid Tumor	Active, not recruiting	NCT02909777
19	Study to Evaluate the Efficacy and Safety of CUDC-907 in Patients with RR DLBCL, Including Patients with MYC Alterations	Interventional	Phase 2	Randomized	Relapsed and/ or Refractory Diffuse Large B-cell Lymphoma Including With Myc Alterations	Completed	NCT02674750
20	CUDC-907 Treatment in People with Metastatic and Locally Advanced Thyroid Cancer	Interventional	Phase 2	NA	Thyroid Neoplasms Poorly Differentiated and Undifferentiated Thyroid Cancer Differentiated Thyroid Cancer	Terminated	NCT03002623
21	Open Label, Multi-center Study to Assess the Safety, Tolerability and Pharmacokinetics of CUDC-907 in Subjects With Advanced/ Relapsed Solid Tumors	Interventional	Phase 1	NA	Triple-Negative Breast Cancer High-grade Serous Ovarian Cancer Solid Tumors NUT Midline Carcinoma	Completed	NCT02307240

REFERENCES

Abdel-Megid, M., Elmahdy, K. M., Elkazak, A. M., Seada, M. H., & Mohamed, O. F. (2016). Chemistry of Thienopyrimidines and Their Biological Applications. *Journal of Pharmaceutical and Applied Chemistry*, 2(3), 103–127. DOI: 10.18576/jpac/020301

Alexey S. chiriapkin, Ivan P. Kodonidi, dmitry I. pozdnyakov, A. A. G. (2022). Synthesis, in vitro and docking studies of 2-substituted 5,6,7,8-tetrahydrobenzo[4,5]thieno[2,3-d]pyrimidine-4(3H)-one derivatives as agents for the treatment of Alzheimers disease. *Chimica Techno Acta, 9(2).*

Ali, E. M. H., Abdel-maksoud, M. S., & Oh, C. (2019). Bioorganic & Medicinal Chemistry Thieno [2, 3- d] pyrimidine as a promising scaffold in medicinal chemistry : Recent advances. *Bioorganic & Medicinal Chemistry*, 27(7), 1159–1194. DOI: 10.1016/j.bmc.2019.02.044 PMID: 30826188

Aly, A. A., Brown, A. B., Ramadan, M., Gamal-Eldeen, A. M., Abdel-Aziz, M., Abuo-Rahma, G. E. D. A. A., & Radwan, M. F. (2010). Thieno[2,3-d]pyrimidines in the synthesis of antitumor and antioxidant agents. *Archiv der Pharmazie*, 343(5), 301–309. DOI: 10.1002/ardp.200900245 PMID: 20232374

Amira T. A. Mohamed. (2019). *Synthesis and antitumor activity evaluation of some thienopyrimidine derivatives*. DOI: 10.1080/00397911.2019.1697822

Amr, A. E. G. E., Sherif, M. H., Assy, M. G., Al-Omar, M. A., & Ragab, I. (2010). Antiarrhythmic, serotonin antagonist and antianxiety activities of novel substituted thiophene derivatives synthesized from 2-amino-4,5,6,7-tetrahydro-N- phenylbenzo[b]thiophene-3-carboxamide. *European Journal of Medicinal Chemistry*, 45(12), 5935–5942. DOI: 10.1016/j.ejmech.2010.09.059 PMID: 20950897

Andrew, S. Bell, Zhiyoung Yu, jennie A. hutton, Megan H. Wright, James A. Brannigan, daniel Paape, shieley M. Roberts, Charlotte L. sutherell, Markus Ritzefeld, Anathony J. Wilkinson, Deborah F. Smith, Robin J. Leatherbarrow, and E. W. T. (2020). Novel Thienopyrimidine Inhibitors of Leishmania N-Myrostoyltransferase with On-Target Activity in Intracellular Amastigotes. *Journal of Medicinal Chemistry*.

Arya, N., Munde, M. K., Dwivedi, J., Jagdale, A. Y., & Jain, K. S. (2013). Design, synthesis, and antihyperlipidemic evaluation of novel 2-[1-(Substitutedphenyl)-4-oxo-azetidin-2-yl]-5,6-disubstitutedthieno[2,3-d] pyrimidin-4(3H)-ones. *Archiv der Pharmazie*, 346(8), 588–595. DOI: 10.1002/ardp.201300001 PMID: 23881664

Barrows, R. D., Hammill, J. T., Tran, M. C., Falade, M. O., Rice, A. L., Davis, C. W., Emge, T. J., Rablen, P. R., Kiplin Guy, R., & Knapp, S. (2020). Evaluation of 1,1-cyclopropylidene as a thioether isostere in the 4-thio-thienopyrimidine (TTP) series of antimalarials. *Bioorganic & Medicinal Chemistry*, 28(22), 115758. DOI: 10.1016/j.bmc.2020.115758 PMID: 33007559

Bhavsar, Z. A., Acharya, P. T., Jethava, D. J., & Patel, H. D. (2020). Recent advances in development of anthelmintic agents: Synthesis and biological screening. *Synthetic Communications*, 50(7), 917–946. DOI: 10.1080/00397911.2019.1695276

Biradar, M. S., Nargund, S. L., & Thapa, S. (2023). Synthesis and Cytotoxicity Assay of Aniline Substituted Thienopyrimidines for Anti-Colorectal Cancer Activity. *Results in Chemistry*, 5(February), 100926. DOI: 10.1016/j.rechem.2023.100926

Black, J. L., Richelson, E., & Richardson, J. W.BLACK. (1985). Antipsychotic Agents: A Clinical Update. *Mayo Clinic Proceedings*, 60(11), 777–789. DOI: 10.1016/S0025-6196(12)60421-2 PMID: 2865413

Chambhare, R. V., Khadse, B. G., Bobde, A. S., & Bahekar, R. H. (2003). Synthesis and preliminary evaluation of some N-[5-(2-furanyl)-2-methyl-4-oxo-4H-thieno[2,3-d]pyrimidin-3-yl]-carboxamide and 3-substituted-5-(2-furanyl)-2-methyl-3H-thieno[2,3-d]pyrimidin-4-ones as antimicrobial agents. *European Journal of Medicinal Chemistry*, 38(1), 89–100. DOI: 10.1016/S0223-5234(02)01442-3 PMID: 12593919

Chiarelli, L. R., Salina, E. G., Mori, G., Azhikina, T., Riabova, O., Lepioshkin, A., Grigorov, A., Forbak, M., Madacki, J., Orena, B. S., Manfredi, M., Gosetti, F., Buzzi, A., Degiacomi, G., Sammartino, J. C., Marengo, E., Korduláková, J., Riccardi, G., Mikušová, K., & Pasca, M. R. (2020). New Insights into the Mechanism of Action of the Thienopyrimidine Antitubercular Prodrug TP053. *ACS Infectious Diseases*, 6(2), 313–323. DOI: 10.1021/acsinfecdis.9b00388 PMID: 31729215

Chinsembu, K. C. (2016). Tuberculosis and nature's pharmacy of putative anti-tuberculosis agents. *Acta Tropica*, 153, 46–56. DOI: 10.1016/j.actatropica.2015.10.004 PMID: 26464047

Contributor, O., & Masurier, N. (n.d.). *Thienopyrimidine. 4*(Scheme 2).

Dabaeva, V. V., Bagdasaryan, M. R., Paronikyan, E. G., Dashyan, S. S., Paronikyan, R. G., Nazaryan, I. M., & Akopyan, A. G. (2020). Synthesis and Neurotropic Activity of Novel Condensed Cyclopentanopyrido[3′,2′:4,5]-Thieno[3,2-D]Pyrimidine Derivatives. *Pharmaceutical Chemistry Journal*, 54(7), 707–713. DOI: 10.1007/s11094-020-02259-y

De La, J., Curtis, C., Lawrence, D., Paul, G., Laurence, S., Ronald, B., Lu, S., Gatoulis, S. C., Brown, B., Migoya, E., & Shore, N. (2023). ORIGINAL RESEARCH ARTICLE A Phase I Clinical Trial Evaluating the Safety and Dosing of Relugolix with Novel Hormonal Therapy for the Treatment of Advanced Prostate Cancer. *Targeted Oncology*, 18(3), 383–390. DOI: 10.1007/s11523-023-00967-5 PMID: 37060432

Deng, J., Peng, L., Zhang, G., Lan, X., Li, C., Chen, F., Zhou, Y., Lin, Z., Chen, L., Dai, R., Xu, H., Yang, L., Zhang, X., & Hu, W. (2011). European Journal of Medicinal Chemistry The highly potent and selective dipeptidyl peptidase IV inhibitors bearing a thienopyrimidine scaffold effectively treat type 2 diabetes. *European Journal of Medicinal Chemistry*, 46(1), 71–76. DOI: 10.1016/j.ejmech.2010.10.016 PMID: 21106276

Dong, Jin Hong, Seung Hyun Jung, Jisook Kim, Danbee Jung, Young Gil Ahn, K. H. S. and K. H. M. (2019). Synthesis and biological evaluation of novel thienopyrimidine derivatives as diacylglycerol acyltransferase 1 (DGAT-1) inhibitors. *Journal of Enzyme Inhibition and Medicinal Chemistry*, •••, 35.

Edwin, G.Tse, M. K. and matthew H. T. (2019). The past, present and future of anti-malarial medicines. *Malaria Journal*. PMID: 30902052

El-Shoukrofy, M. S., Abd El Razik, H. A., AboulWafa, O. M., Bayad, A. E., & El-Ashmawy, I. M. (2019). Pyrazoles containing thiophene, thienopyrimidine and thienotriazolopyrimidine as COX-2 selective inhibitors: Design, synthesis, in vivo anti-inflammatory activity, docking and in silico chemo-informatic studies. *Bioorganic Chemistry*, 85(February), 541–557. DOI: 10.1016/j.bioorg.2019.02.036 PMID: 30807897

Elmongy, E., kedr, mohammed, abotaleb, nageh, & Abbas, S. (2021). Design and synthesis of new thienopyrimidine derivatives along with their antioxidant activity. *Egyptian Journal of Chemistry, 0*(0), 0–0. DOI: 10.21608/ejchem.2021.23721.2412

Eslam, M.H. Ali, Mohammed S. Abdel-Maksoud, C.-H. oh. (2019). Thieno[2,3-d] pyrimidine as a promising scaffold in medicinal chemistry: Recent advances. *Bioorganic & Medicinal Chemistry*. PMID: 30826188

Fatma E. M. El-baih, H. A. S. A.-B. and H. M. A.-H. (n.d.). *Synthesis of Some Thienopyrimidine Derivatives*.

Field, M. C., Horn, D., Fairlamb, A. H., Ferguson, M. A. J., Gray, D. W., Read, K. D., De Rycker, M., Torrie, L. S., Wyatt, P. G., Wyllie, S., & Gilbert, I. H. (2017). Anti-trypanosomatid drug discovery: An ongoing challenge and a continuing need. *Nature Reviews. Microbiology*, 15(4), 217–231. DOI: 10.1038/nrmicro.2016.193 PMID: 28239154

Fouad, M. M., El-Bendary, E. R., suddek, G. M., Shehata, I. A., & El-Kerdawy, M. M. (2018). Synthesis and in vitro antitumor evaluation of some new thiophenes and thieno[2,3-d]pyrimidine derivatives. *Bioorganic Chemistry*, 81(2), 587–598. DOI: 10.1016/j.bioorg.2018.09.022 PMID: 30248510

Gour, K. (2023). Organic Chemistry Heterocyclic compounds. *The Diverse World of Ringed Molecules.*, 6, 108–110. DOI: 10.37532/jmoc.2023.6(4).108-110

Hafez, H. N., Hussein, H. A. R., & El-Gazzar, A. R. B. A. (2010). Synthesis of substituted thieno[2,3-d]pyrimidine-2,4-dithiones and their S-glycoside analogues as potential antiviral and antibacterial agents. *European Journal of Medicinal Chemistry*, 45(9), 4026–4034. DOI: 10.1016/j.ejmech.2010.05.060 PMID: 20691339

Hansen, S. K., Harris, Æ. S. O., Hashash, Æ. A., Hoch, Æ. U., Hogan, Æ. J. N., Howlett, A. R., Jacobs, Æ. J. W., Lam, Æ. J. W., Ritchie, Æ. S. C., Romanowski, Æ. M. J., Silverman, J. A., Stockett, Æ. D. E., Teague, Æ. J. N., Zimmerman, Æ. K. M., & Taverna, Æ. Pietro. (2010). *SNS-314, a pan-Aurora kinase inhibitor, shows potent anti-tumor activity and dosing flexibility in vivo*. 707–717. DOI: 10.1007/s00280-009-1076-8

Hatvate, N. T., Ghodse, S. M., & Telvekar, V. N. (2018). Metal-free synthesis of 2-aminothiadiazoles via TBHP-Mediated oxidative C-S bond formation. *Synthetic Communications*, 48(3), 285–290. DOI: 10.1080/00397911.2017.1398330

Bosson-Vanga, H., Primas, N., Franetich, J. F., Lavazec, C., Gomez, L., Ashraf, K., & Mazier, D. (2021). A new thienopyrimidinone chemotype shows multistage activity against Plasmodium falciparum, including artemisinin-resistant parasites. *Microbiology Spectrum*, 9(2), e00274–e21.

Ibrahim, Y. A., Elwahy, A. H. M., & Kadry, A. M. (1996). Thienopyrimidines: Synthesis, reactions, and biological activity. *Advances in Heterocyclic Chemistry*, 65(65), 235–281. DOI: 10.1016/S0065-2725(08)60297-4

Islam, F., & Quadery, T. M. (2021). Therapeutic Potential, Synthesis, Patent Evaluation and SAR Studies of Thieno[3,2-d]pyrimidine Derivatives: Recent Updates. *Current Drug Targets*, 22(17), 1944–1963. DOI: 10.2174/1389450122666210526094047 PMID: 34042033

Jackson, L. M., & Hawkey, C. J. (2000). *COX-2 Selective Nonsteroidal Anti-Inflammatory Drugs Do They Really Offer Any Advantages ? 59*(6), 1207–1216.

Janku, F. (2017). Phosphoinositide 3-kinase (PI3K) pathway inhibitors in solid tumors : From laboratory to patients. *Cancer Treatment Reviews*, 59, 93–101. Advance online publication. DOI: 10.1016/j.ctrv.2017.07.005 PMID: 28779636

Jubeen, F., Iqbal, S. Z., Shafiq, N., Khan, M., Parveen, S., Iqbal, M., & Nazir, A. (2018). Eco-friendly synthesis of pyrimidines and its derivatives: A review on broad spectrum bioactive moiety with huge therapeutic profile. *Synthetic Communications*, 48(6), 601–625. DOI: 10.1080/00397911.2017.1408840

Kassab, A. E., Gedawy, E. M., & Sayed, A. S. (2024). Fused thiophene as a privileged scaffold: A review on anti-Alzheimer's disease potentials via targeting cholinesterases, monoamine oxidases, glycogen synthase kinase-3, and Aβ aggregation. *International Journal of Biological Macromolecules*, 265, 131018. DOI: 10.1016/j.ijbiomac.2024.131018 PMID: 38518928

Kazushige Sogawa, and M. D. P. (1970). NII-Electronic Library Service. *Chemical Pharmaceutical Bulletin, 43*, 2091. Retrieved from https://www.mendeley.com/research/geology-volcanic-history-eruptive-style-yakedake-volcano-group-central-japan/

Khedkar, V. M., Arya, N., Coutinho, E. C., Shishoo, C. J., & Jain, K. S. (2014). Docking study of novel antihyperlipidemic thieno[2,3-d]pyrimidine; LM-1554, with some molecular targets related to hyperlipidemia - an investigation into its mechanism of action. *SpringerPlus*, 3(1), 628. Advance online publication. DOI: 10.1186/2193-1801-3-628 PMID: 25392798

Eissa, K. I., Kamel, M. M., Mohamed, L. W., Doghish, A. S., Alnajjar, R., Al-Karmalawy, A. A., & Kassab, A. E. (2023). Design, synthesis, and biological evaluation of thienopyrimidine derivatives as multifunctional agents against Alzheimer's disease. *Drug Development Research*, 84(5), 937–961.

Kidwai, M.* and Mishra, A. D. (n.d.). *A Novel Synthesis of Thieno-pyrimidines Using Inorganic Solid Support. 24.*

Kinases, T., Potential, T., & Activities, A. (2017). *Review Article*. 1–24. DOI: 10.1002/ardp.201700242

Kotaiah, Y., Harikrishna, N., Nagaraju, K., & Venkata Rao, C. (2012). Synthesis and antioxidant activity of 1,3,4-oxadiazole tagged thieno[2,3-d]pyrimidine derivatives. *European Journal of Medicinal Chemistry*, 58, 340–345. DOI: 10.1016/j.ejmech.2012.10.007 PMID: 23149297

Prisca Lagardere, Cyril Fersing, Nicolas Masurier, and V. L. (2022). *Thienopyrimidine: A Promising Sacffold to Access Anti-Infective Agents. 15.*

Lagardère, P., Mustière, R., Amanzougaghene, N., Hutter, S., Casanova, M., Franetich, J. F., Tajeri, S., Malzert-Fréon, A., Corvaisier, S., Azas, N., Vanelle, P., Verhaeghe, P., Primas, N., Mazier, D., Masurier, N., & Lisowski, V. (2023). New antiplasmodial 4-amino-thieno[3,2-d]pyrimidines with improved intestinal permeability and microsomal stability. *European Journal of Medicinal Chemistry*, 249(January), 115115. Advance online publication. DOI: 10.1016/j.ejmech.2023.115115 PMID: 36680984

Liao, B., Lin, C., Lee, J., & Yang, J. C. (2017). Lung Cancer Optimal management of EGFR -mutant non-small cell lung cancer with disease progression on fi rst-line tyrosine kinase inhibitor therapy. *Lung Cancer (Amsterdam, Netherlands)*, 110(7), 7–13. DOI: 10.1016/j.lungcan.2017.05.009 PMID: 28676222

Liao, W., Yang, W., Xu, J., Yan, Z., Pan, M., Xu, X., Zhou, S., Zhu, Y., Lan, J., Zeng, M., Han, X., Li, S., Li, Y., Liang, K., Gao, Y., & Peng, Q. (2021). Therapeutic Potential of CUDC-907 (Fimepinostat) for Hepatocarcinoma Treatment Revealed by Tumor Spheroids-Based Drug Screening. *Frontiers in Pharmacology*, 12, 658197. Advance online publication. DOI: 10.3389/fphar.2021.658197 PMID: 34776939

Litvinov, V. P. (2006). The Chemistry of Thienopyrimidines. *Advances in Heterocyclic Chemistry*, 92(06), 83–143. DOI: 10.1016/S0065-2725(06)92003-0

Makhova, N. N., Belen'kii, L. I., Gazieva, G. A., Dalinger, I. L., Konstantinova, L. S., Kuznetsov, V. V., Kravchenko, A. N., Krayushkin, M. M., Rakitin, O. A., Starosotnikov, A. M., Fershtat, L. L., Shevelev, S. A., Shirinian, V. Z., & Yarovenko, V. N. (2020). Progress in the chemistry of nitrogen-, oxygen- and sulfur-containing heterocyclic systems. *Russian Chemical Reviews*, 89(1), 55–124. DOI: 10.1070/RCR4914

Malasala, S., Polomoni, A., Ahmad, M. N., Shukla, M., Kaul, G., Dasgupta, A., Chopra, S., & Nanduri, S. (2021). Structure based design, synthesis and evaluation of new thienopyrimidine derivatives as anti-bacterial agents. *Journal of Molecular Structure*, 1234, 130168. DOI: 10.1016/j.molstruc.2021.130168

Manaithiya, A., Alam, O., Sharma, V., Javed, M., & Mittal, S. (2021). Bioorganic Chemistry GPR119 agonists : Novel therapeutic agents for type 2 diabetes mellitus. In Bioorganic Chemistry (Vol. 113, Issue May). Elsevier Inc. DOI: 10.1016/j.bioorg.2021.104998

Narusaka, M., & Narusaka, Y. (2017). Thienopyrimidine-type compounds protect Arabidopsis plants against the hemibiotrophic fungal pathogen Colletotrichum higginsianum and bacterial pathogen Pseudomonas syringae pv. maculicola. *Plant Signaling & Behavior*, 12(3), e1293222.

Mavrova, A. T., Dimov, S., Yancheva, D., Rangelov, M., Wesselinova, D., & Tsenov, J. A. (2016). Synthesis, anticancer activity and photostability of novel 3-ethyl-2-mercapto-thieno[2,3-d]pyrimidin-4(3H)-ones. *European Journal of Medicinal Chemistry*, 123, 69–79. DOI: 10.1016/j.ejmech.2016.07.022 PMID: 27474924

Mavrova, A. T., Vuchev, D., Anichina, K., & Vassilev, N. (2010). Synthesis, antitrichinnellosis and antiprotozoal activity of some novel thieno[2,3-d]pyrimidin-4(3H)-ones containing benzimidazole ring. *European Journal of Medicinal Chemistry*, 45(12), 5856–5861. DOI: 10.1016/j.ejmech.2010.09.050 PMID: 20950896

Miaki, T., Kawamura, T., Sugahara, K., Alcoverro, M., Casas, J., Pardàs, M., Ahmadi, A., Abedin, M. A., Mattausch, H. J., Koide, T., Abadi, M., De Alfaro, L., Chack, S., Sharma, P., Mahesan, K. V., Bhargavi, S., Jayadevappa, D., Appia, V., Yezzi, A., . . . He, C. (2017). Main paper.pdf. In IEEE Transactions on Image Processing (Vol. 24, Issue 1, pp. 1–9). Retrieved from http://dx.doi.org/DOI: 10.1016/j.eswa.2013.05.056%0Ahttps://hal.inria.fr/hal-00912934/document#page=36%0Awww.ijirae.com%0Ahttp://www.springerlink.com/content/9662h2hwq81041g2

Miwa, K., Hitaka, T., Imada, T., Sasaki, S., Yoshimatsu, M., Kusaka, M., Tanaka, A., Nakata, D., Furuya, S., & Endo, S. (2011). pyrimidin-6-yl] phenyl } -3-methoxyurea (TAK-385) as a Potent, Orally Active. *Non-Peptide Antagonist of the Human Gonadotropin-Releasing Hormone Receptor.*, 1, 4998–5012. PMID: 21657270

Mohamed, A. T. A., Abou-Elregal, M. K., Youssef, A. S. A., Hemdan, M. M., Samir, S. S., & Abou-Elmagd, W. S. I. (2020). Synthesis and antitumor activity evaluation of some thienopyrimidine derivatives. *Synthetic Communications*, 50(3), 399–411. DOI: 10.1080/00397911.2019.1697822

Mohamed, M. A., Khattab, R. R., Wasfy, A. A. F., Abo-riya, M. A., El-Kalyoubi, S., & Hassan, N. A. (2021). Synthesis and Anticancer Activity Evaluation of Some Thienopyrimidine Derivatives. *Russian Journal of Bioorganic Chemistry*, 47(6), 1312–1323. DOI: 10.1134/S1068162021060194

Panneer Selvam, T., Richa James, C., Vijaysarathy Dniandev, P., & Karyn Valzita, S. (2012). A mini review of pyrimidine and fused pyrimidine marketed drugs. *Research in Pharmacy, 2*(4), 1–9. Retrieved from www.researchinpharmacy.com

Patel, P. S., Akbari, V. K., Modi, S. D., Belim, M. A., Tailor, R. B., Patel, H. D., & Dewani, B. (2019). Synthesis and Biological Evaluation of Schiff Base Involving Thieno [2, 3-D] Pyrimidine Moiety As Antimicrobial Agents. *Research Journal of Life Sciences, Bioinformatics, Pharmaceutical and Chemical Sciences*, 5(31), 31–41. DOI: 10.26479/2019.0505.03

Prabhakar, V., Babu Kondra, S., & Maddula, S. R. (2016). Synthesis, Structural Elucidation of Novel Thieno [2,3-d] Pyrimidine Core Unit Containing 1,2,4-Triazoles and Thiophenes as Potent Antimicrobial Activity. *Organic Chemistry: Current Research*, 05(03). Advance online publication. DOI: 10.4172/2161-0401.1000169

Qadir, T., Amin, A., Sharma, P. K., Jeelani, I., & Abe, H. (2022). A Review on Medicinally Important Heterocyclic Compounds. In The Open Medicinal Chemistry Journal (Vol. 16, Issue 1). DOI: 10.2174/18741045-v16-e2202280

Quelques, E. T. D. E. (1971). *Thi~nopyrimidine~ii. 27.*

Rashad, A. E., Shamroukh, A. H., Abdel-Megeid, R. E., Mostafa, A., El-Shesheny, R., Kandeil, A., Ali, M. A., & Banert, K. (2010). Synthesis and screening of some novel fused thiophene and thienopyrimidine derivatives for anti-avian influenza virus (H5N1) activity. *European Journal of Medicinal Chemistry*, 45(11), 5251–5257. DOI: 10.1016/j.ejmech.2010.08.044 PMID: 20828882

Ronald, K. RussellJeffery B. Press. (1988). Thiophene systems. 9. Thienopyrimidinedione derivatives as potential antihypertensive agents. *American Chemical Society, 31*(9).

Russell, R. K., Press, J. B., Rampulla, R. A., McNally, J. J., Falotico, R., Reiser, J. A., Bright, D. A., & Tobia, A. (1988). Thiophene Systems. 9. Thienopyrimidinedione Derivatives as Potential Antihypertensive Agents. *Journal of Medicinal Chemistry*, 31(9), 1786–1793. DOI: 10.1021/jm00117a019 PMID: 2842504

Salib, S. B., Khalil, O. M., Kamel, M. M., & El-Dash, Y. (2016). Synthesis and Antitumor Activity of Novel Thienopyrimidine Derivatives Containing Thiosemicarbazide Moiety. *OAlib*, 03(07), 1–7. DOI: 10.4236/oalib.1102876

Sayed, M. T. M. (2023). *CHEMISTRY Recent updates on thienopyrimidine derivatives as anticancer agents*. 659–681. DOI: 10.1007/s00044-023-03040-y

Shetty, N. S. (2014). A Facile Microwave-Assisted Synthesis of Novel Thienopyrimidines and Triazolothienopyrimidines and their Antimicrobial Activities. *International Journal of Advances in Chemical Engineering and Biological Sciences*, 1(1). Advance online publication. DOI: 10.15242/IJACEBS.C0114037

Shushanik, Sh. Dashyan, Eugene v. babaev, ervand g. paronikyan, Armen G. Ayvazyan, R. G. P. and L. S. H. (2022). Evaluation of Neurotropic Activity and Molecular Docking study of New Derivatives of pyrano[4”,3”:4′,5′]pyrido[3′,2′:4,5] thieno[3,2-d]pyrimidines on the basis of pyrano[3,4-c]pyridines. *Molecules (Basel, Switzerland)*. PMID: 35684318

Shyyka, O., Pokhodylo, N., Finiuk, N., Matiychuk, V., Stoika, R., & Obushak, M. (2018). Anticancer activity evaluation of new thieno[2,3-d]pyrimidin-4(3h)-ones and thieno[3,2-d]pyrimidin-4(3h)-one derivatives. *Scientia Pharmaceutica*, 86(3), 28. Advance online publication. DOI: 10.3390/scipharm86030028 PMID: 30012942

Meenakshi Singh, Sushil K Singh, M. T. C. (2011). A NOVEL QSAR MODEL FOR EVALUATION AND PREDICTING THE INHIBITION ACTIVITY OF H1-RECEPTOR ANTAGONISTS: A SERIES OF THIENOPYRIMIDNES. *Journal of Drug Delivery and Therapeutics*.

Singh, M., Singh, S. K., & Chhabria, M. T. (2012). A novel QSAR model for evaluating and predicting the inhibition activity of H1-receptor antagonists: A series of thienopyrimidine derivatives. *Journal of Drug Delivery and Therapeutics*, 2(1).

Sonawane, K., Said, R., & Hatvate, N. (2024). Synthesis, Properties, and Biological Applications of 1,3,4-Thiadiazoles. In S-Heterocycles (pp. 163–223). Royal Society of Chemistry. DOI: 10.1039/9781837674015-00163

Srinath, R., Venkataranganna, M., & Saravanan, J. (2011). Anti-hyperlipidemic Activity of Thieno - [2, 3-d] - Pyrimidin-4-(3H)-ones. *Macedonian Journal of Medical Sciences*, 4(4), 358–366. DOI: 10.3889/MJMS.1857-5773.2011.0189

Szkudlinski, M. W. (2007).. . *Challenges and Opportunities of Trapping Ligands*, 231–234. Advance online publication. DOI: 10.1124/mol.107.038208.Please

Shaukat Tamboli, Riyaj; Devidas Amrutkar, Rakesh; Sanchlal Jain, Kishor; Kumaradoss Kathiravan, M. (2013). *Synthesis and in vivo Antihyperlipidemic Potential of Novel Substituted Thieno[3,2-d]pyrimidines. 10.*

Thakur, S., Arora, S., Katiyar, M. K., Joshi, G., & Kumar, R. (2024). Fused Thienopyrimidines as Versatile Pharmacophores for the Development of Cyclooxygenase-2 Inhibitors. *ChemistrySelect*, 9(22), e202400514. Advance online publication. DOI: 10.1002/slct.202400514

Tolba, M. S., Kamal, A. M., Ahmed, E. M., & Hassanien, R. (2018). *Synthesis, reactions, and biological study of some new thienopyrimidine derivatives as antimicrobial and anti- inflammatory agents. July*, 1–10. DOI: 10.1002/jccs.201800292

Tolba, M. S., Sayed, A. M., Sayed, M., & Ahmed, M. (2021). Design, synthesis, biological evaluation, and molecular docking of some new Thieno[2,3-d] pyrimidine derivatives. *Journal of Molecular Structure*, 1246, 131179. DOI: 10.1016/j.molstruc.2021.131179

Tolba, M. S., Sayed, M., El-Dean, A. M. K., Hassanien, R., Abdel-Raheem, S. A. A., & Ahmed, M. (2021). Design, synthesis and antimicrobial screening of some new thienopyrimidines. *Organic Communications*, 14(4), 334–345. DOI: 10.25135/acg.oc.114.2109.2214

Triloknadh, S., Venkata Rao, C., Nagaraju, K., Hari Krishna, N., Venkata Ramaiah, C., Rajendra, W., Trinath, D., & Suneetha, Y. (2018). Design, synthesis, neuroprotective, antibacterial activities and docking studies of novel thieno[2,3-d]pyrimidine-alkyne Mannich base and oxadiazole hybrids. *Bioorganic & Medicinal Chemistry Letters*, 28(9), 1663–1669. DOI: 10.1016/j.bmcl.2018.03.030 PMID: 29602681

Varma, P. K. (2018). Therapeutic importance of synthetic thiophene. *Chemistry Central Journal*. PMID: 30564984

Varvounis, G., & Giannopoulos, T. (1996). Synthesis, chemistry, and biological properties of thienopyrimidines. *Advances in Heterocyclic Chemistry*, 66(66), 193–283. DOI: 10.1016/S0065-2725(08)60307-4

Wallin, J. J., Guan, J., & Prior, W. W. (2012). *GDC-0941, a Novel Class I Selective PI3K Inhibitor, Enhances the Efficacy of Docetaxel in Human Breast Cancer Models by Increasing Cell Death In Vitro and In Vivo GDC-0941, a Novel Class I Selective PI3K Inhibitor*. Enhances Increasing Cell Death In Vi., DOI: 10.1158/1078-0432.CCR-11-2088

Wang, Y., Luo, Y., Hu, D., & Song, B. (2022). Design, Synthesis, Anti-Tomato Spotted Wilt Virus Activity, and Mechanism of Action of Thienopyrimidine-Containing Dithioacetal Derivatives. *Journal of Agricultural and Food Chemistry*, 70(20), 6015–6025. DOI: 10.1021/acs.jafc.2c00773 PMID: 35576166

Wilding, B., & Klempier, N. (2017). Newest Developments in the Preparation of Thieno[2,3-d]pyrimidines. *Organic Preparations and Procedures International*, 49(3), 183–215. DOI: 10.1080/00304948.2017.1320513

Woodring, J. L., Behera, R., Sharma, A., Wiedeman, J., Patel, G., Singh, B., Guyett, P., Amata, E., Erath, J., Roncal, N., Penn, E., Leed, S. E., Rodriguez, A., Sciotti, R. J., Mensa-Wilmot, K., & Pollastri, M. P. (2018). Series of Alkynyl-Substituted Thienopyrimidines as Inhibitors of Protozoan Parasite Proliferation. *ACS Medicinal Chemistry Letters*, 9(10), 996–1001. DOI: 10.1021/acsmedchemlett.8b00245 PMID: 30344906

Yamada, S., Flesch, K. N., Murakami, K., & Itami, K. (2020). Rapid Access to Kinase Inhibitor Pharmacophores by Regioselective C-H Arylation of Thieno[2,3-d]pyrimidine. *Organic Letters*, 22(4), 1547–1551. DOI: 10.1021/acs.orglett.0c00143 PMID: 32009412

Zhang, Y., Luo, L., Han, C., Lv, H., Chen, D., Shen, G., Wu, K., Pan, S., & Ye, F. (2017). *Derivatives as Anti-Inflammatory Agents*. DOI: 10.3390/molecules22111960

Chapter 9
Thienopyridine Analogues:
A Versatile Scaffold in Medicinal Chemistry from Decades

Bonny Y. Patel
RK University, India

Khushal M. Kapadiya
https://orcid.org/0000-0001-9025-1177
RK University, India

ABSTRACT

Thienopyridine derivatives are important sulfur-containing compounds with a unique structure. There are six different Thienopyridine analogues that have shown various biological effects. These compounds are widely used in medicinal chemistry and some have been approved by the FDA for use as drugs. The most well-known applications of thienopyridine-based drugs are in treating cardiovascular and central nervous system (CNS) diseases. These compounds also have anti-inflammatory, anti-infective, antiviral, antiproliferative, antimicrobial, and antitumor properties. Thienopyridine derivatives also have positive effects on important characteristics of the compounds they are part of, such as solubility, lipophilicity, and hydrogen bonding capacity. This chapter will provide a brief overview of the various bioactivities of Thienopyridine derivatives.

DOI: 10.4018/979-8-3693-7520-4.ch009

1. INTRODUCTION

The pyridine scaffold is a critical core in the chemical structure of many approved drugs in the pharmaceutical market. Fused analogues of pyridine derivatives are often significantly more biologically intriguing than the corresponding monocyclic compounds. This is due to their varying pharmacophoric groups in different positions and their ability to interact with a wider range of receptors.

Pyridine frequently undergoes fusion with a variety of carbocycles and heterocycles, including benzene, naphthalene, pyrrole, furan, thiophene, pyrimidine, pyrazole, and imidazole. The chemical space of fused pyridine derivatives is exceptionally diverse, encompassing a wide array of compounds such as pyrrolopyridines, furopyridines, thienopyridines, oxazolopyridines, isoxazolopyridines, imidazopyridines, oxadiazolopyridines, thiazolopyridines, pyrazolopyridines, isothiazolopyridines, triazolopyridines, thiadiazolopyridines, tetrazolopyridines, and dithiolopyridines. These derivatives exhibit varying bioactivities and are of great importance in the field of medicinal chemistry. Specifically, thienopyridines, which are derivatives of pyridine fused to thiophene systems, have garnered attention due to their extensive applications and the diverse range of structural combinations they offer (Nkomba *et al.,* 2022).

Six isomeric thienopyridine structures are characterized by different annulation modes. The formation of thienopyridine can result in four distinct isomers: thieno[2,3-*b*]pyridine **(I)**, thieno[3,2-*b*]pyridine **(II)**, thieno[2,3-*c*]pyridine **(III)**, and thieno[3,2-*c*]pyridine **(IV)** when the fusion involves the 2- or 5-position of thiophene. Additionally, fusion at the 3- and 4-positions of thiophene leads to the creation of two isomeric thienopyridines, namely thieno[3,4-*b*]pyridine **(V)** and thieno[3,4-*c*]pyridine **(VI)**. (Fig 1). Extensive research has been conducted on thienopyridines **I-IV** and their derivatives, but there is limited available research on thieno[3,4-*b*]pyridines **(V)** and thieno[3,4-*c*]pyridines **(VI)**. These less explored isomers, **V** and **VI**, were first synthesized in 1970 and have been found to be significantly less stable compared to the other isomers (Shaw *et al.,* 2024).

Thienopyridine derivatives have high pharmacological potential, making them a privileged scaffold. The structural and isoelectronic characteristics of thienopyridine-containing compounds are similar to those of the vitamin types (B3 and B6), and they have become an attractive structural feature in producing many pharmaceutical drugs. The thienopyridine skeleton is crucial for ligands, materials, and medicinal components due to its pi-electron-enriched thiophene unit and pi-electron-deficient pyridine ring (Shaw *et al.,* 2024).

Figure 1. Various annulation mode of thienopyridine

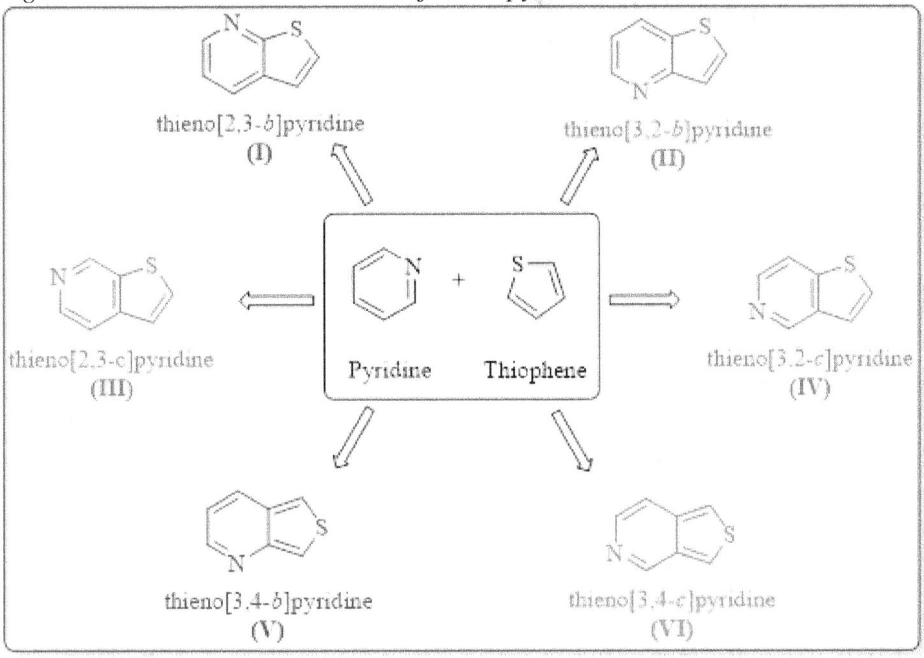

Thienopyridine compounds are a commonly used type of fused pyridine derivatives with significant pharmacological potential. These compounds have been utilized as precursors in the synthesis of agrochemicals and pharmaceuticals for many decades (Litvinov *et al.,* 2005). Notably, Ticlopidine, Clopidogrel, and Prasugrel are clinically available thienopyridine derivatives that act as non-competitive adenosine diphosphate receptor antagonists to effectively inhibit platelet aggregation (Figure 2) (Kam *et al.,* 2003; Makawana *et al.,* 2011). While Ticlopidine is seldom utilized presently due to its side effects, it can still be employed in preventing strokes and the closure of a new coronary stent when combined with aspirin.

Figure 2. Structures of thienopyridine scaffolds in therapeutic use

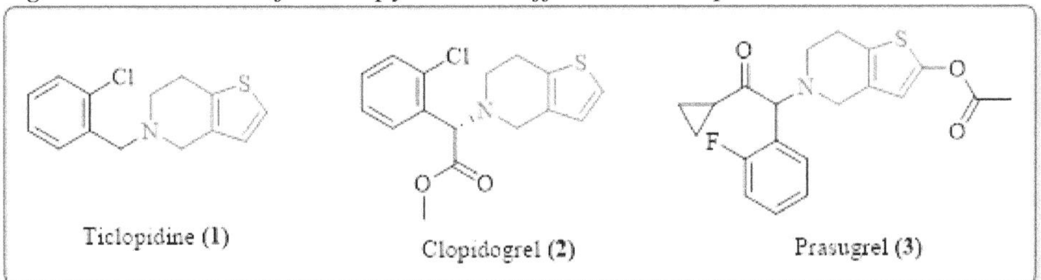

2. GENERAL SYNTHETIC METHODS OF VARIOUS THIENOPYRIDINES ISOMERS

A. THIENO[2,3-*b*]PYRIDINES

Thieno[2,3-*b*]pyridines were first mentioned in 1913. Nowadays, the chemistry of the [2,3-*b*]-isomer is best known, primarily due to the practical importance of many derivatives of thieno[2,3-*b*]pyridine (Steinkopf *et al.,* 1913). Thieno[2,3-*b*]pyridine is derived from fusing the C-2 and C-3 positions of the thiophene ring and pyridine ring, and it unequivocally demonstrates a broad spectrum of biological activities. These include anti-inflammatory, antibacterial, antidepressant, antiviral, antimicrobial, and antitumor effects (Mekky *et al.,* 2020; Madhusudana *et al.,* 2012; Amorim *et al.,* 2017). Moreover, Thieno[2,3-*b*]pyridine, a diketopyrrolopyrrole polymer, exhibits the properties of an n-type polymeric semiconductor, and is utilized in the development of organic field-effect transistors and solar cells (Chen *et al.,* 2018).

1. Synthetic Approaches Used for thieno[2,3-*b*]pyridines

1.1 Creating Pyridine Ring to Form thieno[2,3-*b*]pyridine System

The construction of the pyridine ring primarily relies on derivatives of 2-aminothiophene. Thieno[2,3-*b*]pyridines were synthesized by reacting 3-acetyl-2-aminthiophenes (1) with Phosphorus oxychloride ($POCl_3$) in Dimethylformamide (DMF) using the Vilsmeier-Haack type reaction. Abdelwahab et al. found that when the cyclization of 3-acetyl-2-(*N*-acetylamino)thiophenes (2) is carried out at 100 ºC, it results in the formation of 4-chloro-thienopyridines (3) (Scheme-1) (Abdelwahab *et al.,* 2016). On the other hand, conducting the reaction at a slightly lower temperature of around 65 ºC yields a mixture of both products: 4-chloro-3-formylthienopyridines (4, major, 80-85%) and 4-chloro-thienopyridines (3, minor, 0-14%) (Abdelwahab *et al.,* 2017).

Figure 3. Scheme-1 3-Acetyl-2-aminothiophenes undergo Vilsmeier-Haack type reaction to form thienopyridines

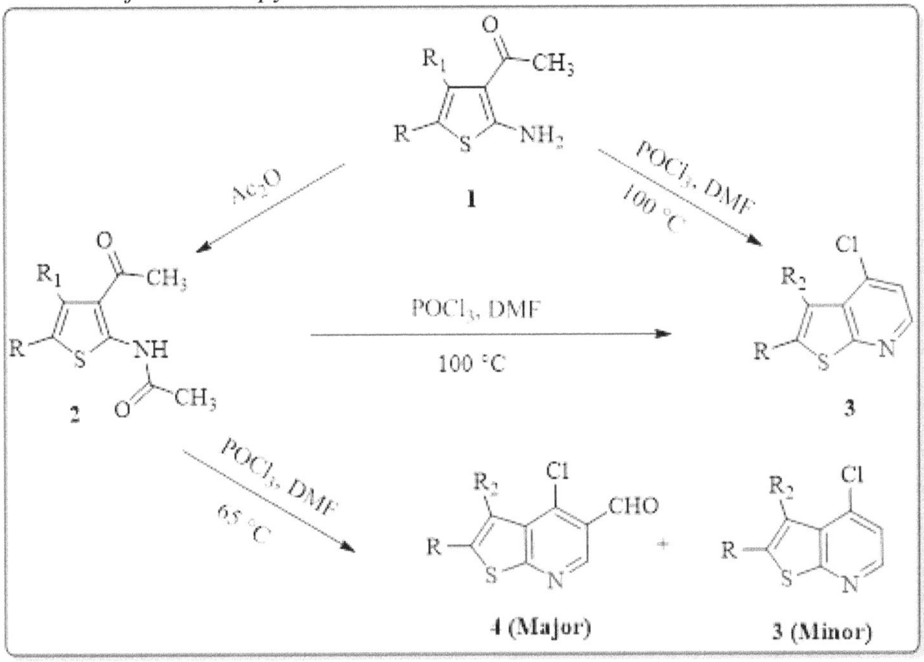

When it comes to synthesizing thieno[2,3-*b*]pyridine from 2-aminothiophenes, chemists commonly rely on two approaches: the Friedlander reaction (Method A) and the Gould-Jacobs reaction (Method B). These methods involve the use of acid-mediated or catalyzed cyclization, employing reagents such as TsOH, $AlCl_3$, $ZnCl_2$, $POCl_3$, $Yb(OTf)_3$, and $BF_3 \cdot Et_2O$ (as shown in Scheme-2) (Dotsenko *et al.*, 2020). In the Friedlander method, a [4 + 2] annulation process is employed, allowing the combination of 2-amino-3-cyano/carbonyl thiophenes (5) with a ketone (6). Conversely, the Gould-Jacobs approach utilizes a [3 + 3] annulation reaction, combining 2-amino thiophenes (8) with α,β-unsaturated nitriles or carbonyls. These methods serve as vital tools in the synthesis of thieno[2,3-b]pyridine, offering distinct pathways for achieving this important compound.

Figure 4. Scheme-2 Friedlander reaction and Gould-Jacobs reaction to form thienopyridines

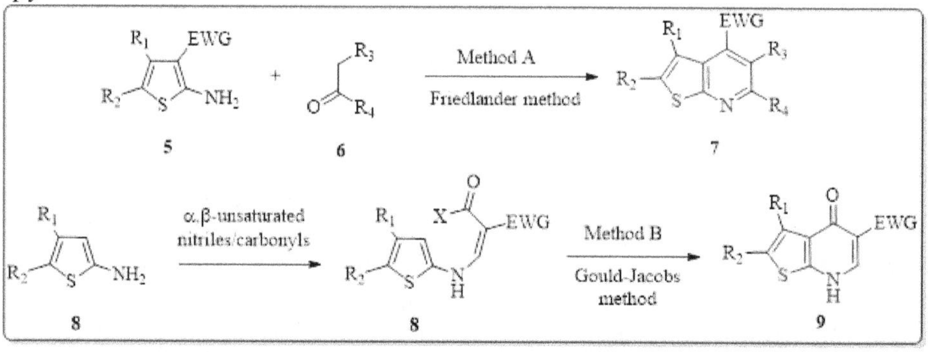

The compound tetrahydrobenzo[4,5]thieno[2,3-*b*]pyridine (11) was easily synthesized from tetrahydrobenzo[*b*]thiophene (10), methyl acetoacetate/ethyl acetoacetate, and Tin(IV) chloride (Scheme-3) (Lio *et al.,* 2005).

Figure 5. Scheme-3 Tetrahydrobenzo[4,5]thieno[2,3-b]pyridine from 2-aminothiophene derivative

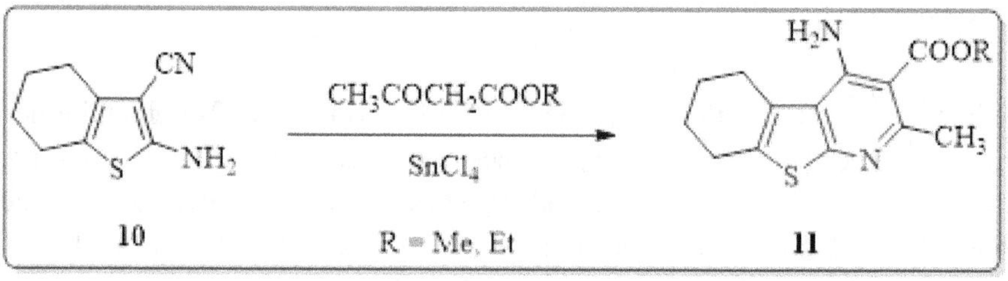

The pyridine fragment was created through intramolecular Bischler-Napieralski cyclization of 2-(acetamido)benzothiophene (12) in the past decade. The highest yield of the desired isoquinoline derivative (13) was obtained using P_2O_5-$POCl_3$ as a condensing agent (Scheme-4) (Deprets *et al.,* 2000).

Figure 6. Scheme-4 Thieno[2,3-b]pyridine using intramolecular Bischler–Napieralski cyclization

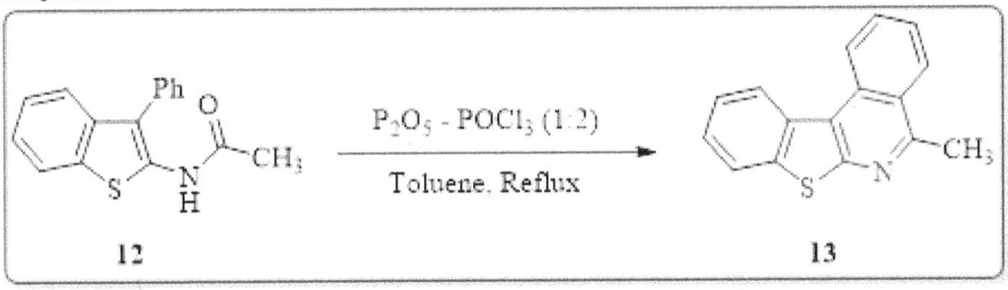

Thieno[2,3-*b*]pyridine derivatives were produced by causing 2-aminothiophenes with a substituent containing the electrophilic γ-carbon atom at position 3 to undergo intramolecular cyclization. Amines (14) were created through a multicomponent cascade heterocyclization process involving phenyl isothiocyanate, CH-acids, and alkylating agents (El-Taweel *et al.,* 2001; Monarch *et al.,* 1992). The cyclization of compounds (14) results in the formation of 6-oxo-(imino)-6,7-dihydrothieno[2,3-*b*] pyridines (15) as the final products (Scheme-5) (Mohareb *et al.,* 1992).

Figure 7. Scheme-5 Thieno[2,3-b]pyridine derivatives using intramolecular cyclization using electrophilic γ-carbon atom

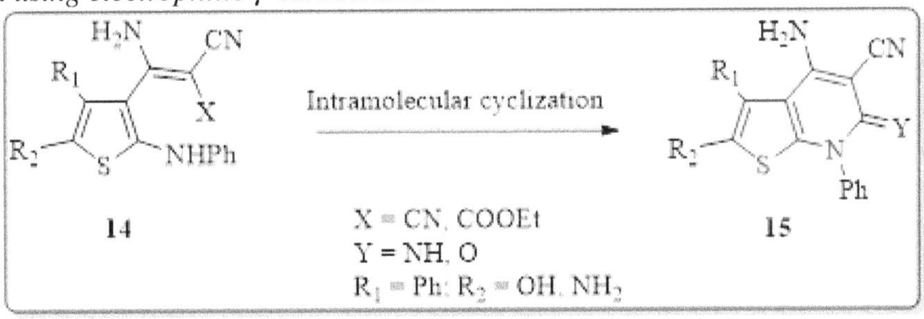

1.2 Creating Thiophene Ring to Form thieno[2,3-*b*]pyridine System

The synthesis of a thienopyridine framework involves fusing a thiophene ring onto a pyridine structure using cyclization methods. One common method is the intramolecular Thorpe-Ziegler cyclization of precursor 2-(alkylthio)-pyridine-3-carbonyls or carbonitriles (17), which are synthesized from readily available 3-cyanopyridine-2(1*H*)-thiones (16) and α-halo carbonyls or α-haloacetonitriles (Mohamed *et al.,* 2007). The stability is influenced by the electron-withdrawing effect of the substituent Y, and the concentration of the carbanion intermediate determines the cyclization rate. It has been empirically established that the activity of the electron-withdrawing substituents Y increases in the following series (Litvinov *et al.*, 1992).

NO_2 > Ph-C=O > CN > COOR > $CONH_2$ > H

The reaction conditions varied widely depending on the basic medium used, while acid catalysts were rarely used. Sometimes, ring closure occurs spontaneously when attempting to alkylate thiones (16) in the presence of equimolar amounts of bases; in these processes, intermediates (17) cannot be isolated. Generally, thienopyridines can be easily synthesized at room temperature, although heating is required in certain reactions. This method has a wide range of applications and has been used to prepare various derivatives of thienopyridines and their related systems, such as quinoline, isoquinoline, and pyridine. Furthermore, a one-pot procedure combines nucleophilic substitution followed by thiophene ring closure achieved through a base-mediated reaction of 2-chloropyridine-3-carbonitriles (18) with thioglycolates (Scheme-6).

Figure 8. Scheme-6 Thieno[2,3-b]pyridine derivatives using Thorpe-Ziegler cyclization

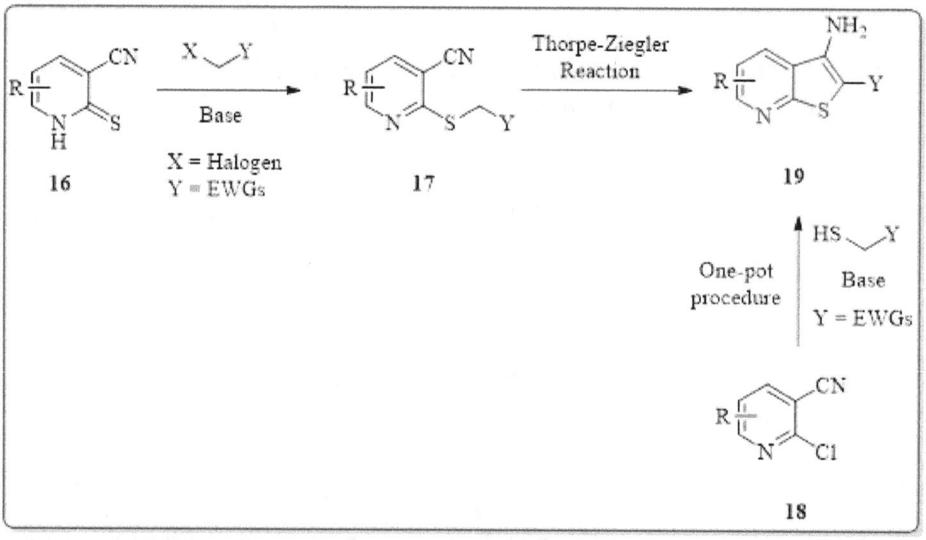

The Thorpe reaction was effectively used in synthesizing partially hydrogenated thienopyridines from tetrahydro- or dihydropyridine-2-thiolates, the corresponding pyridinethiones, or their S-alkyl derivatives. Reactions with tetrahydro derivatives (20) result in the retention of the initial saturated structure, producing lactams nucleus (21) as the final products (Scheme-7) (Krivokolysko *et al.*, 1999; Dotsenko *et al.*, 2003).

Figure 9. Scheme-7 Thieno[2,3-b]pyridine derivatives using Thorpe reaction modification

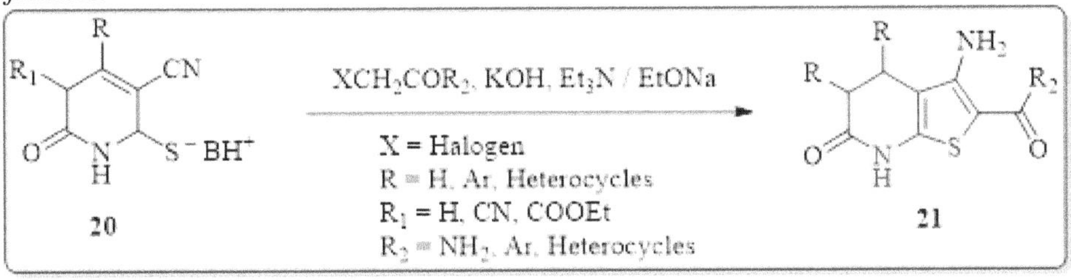

One of the examples of Nucleofuges, 1,2-bipyridinium chloride (22), reacts with ethyl thioglycolate, eliminating pyridine to form sulfide intermediate (23), which isomerizes to thienopyridine (24) in the Thorpe reaction conditions (Scheme-8) (Yakovlev *et al.*, 1998).

Figure 10. Scheme-8 Thieno[2,3-b]pyridine derivatives using nucleofuges example

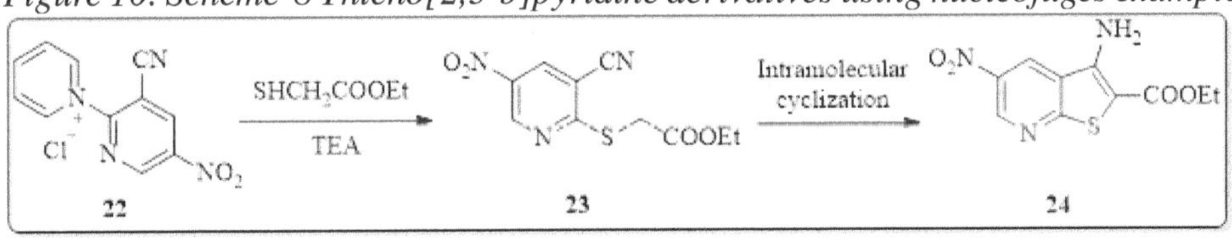

1.3 Creating thieno[2,3-*b*]pyridine Systems by the Construction of Both Heterocyclic Rings

The Zhang research group successfully executed a one-pot sequential synthesis of substituted thieno[2,3-*b*]pyridine (27) from enaminone (25) and 2-cyanothioacetamide (26) under a basic medium (K_2CO_3) in DMF, followed by a reaction with bromoacetonitrile. This reaction unequivocally proceeds by generating a pyridine-2(1*H*)-thione intermediate from enaminone and 2-cyanothioacetamide, followed by a [4 + 2] cyclization with bromoacetonitrile, ultimately resulting in a tetrasubstituted final thieno[2,3-*b*]pyridine derivative (Scheme-9) (Alinaghizadeh *et al.*, 2016).

Figure 11. Scheme-9 Thieno[2,3-b]pyridine derivatives using enaminone and 2-cyanothioacetamide

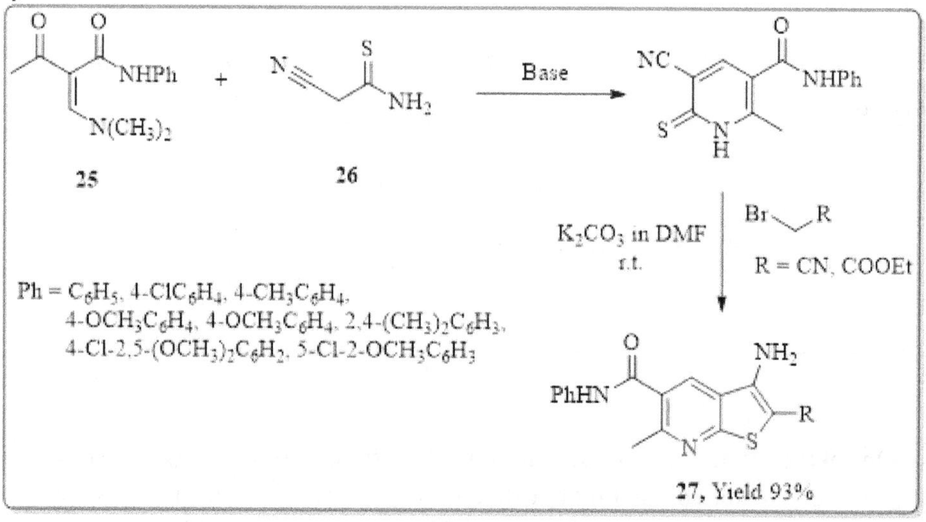

Karpov *et al.* demonstrated a cascade reaction which shown in Scheme-10 of thioglycolates (29) with readily available potassium 2-acyl-1,1,3,3-tetracyanopropenides (28) in a pyridine medium to yield highly functionalized thieno[2,3-*b*] pyridine products (30) (Grigor'ev *et al.*, 2017).

Figure 12. Scheme-10 Thieno[2,3-b]pyridine derivatives using cascade cyclization of 2-acyl-1,1,3,3-tetracyanopropenides

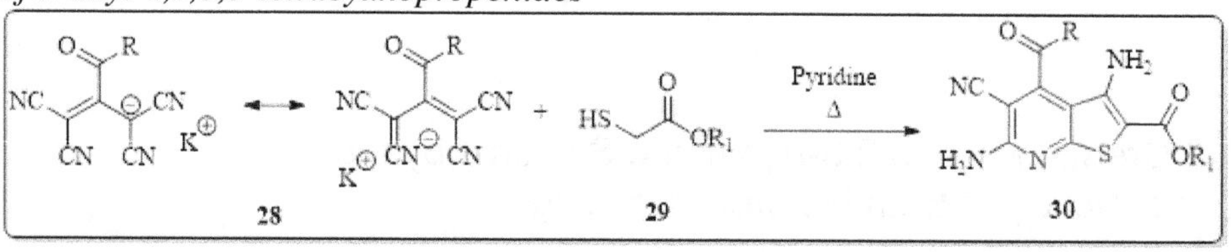

B. THIENO[3,2-*b*]PYRIDINES

Thieno[3,2-*b*]pyridines are formed by fusing the C-2 and C-3 positions of a pyridine ring with the C-3 and C-2 positions of a thiophene ring in a clockwise direction. However, there is a lack of literature on the preparation and properties of thieno[3,2-*b*]pyridine derivatives (II) compared to the [2,3-*b*] isomer. Compounds containing the thieno[3,2-*b*]pyridine structure are renowned for their exceptional antibacterial activity, akin to their [2,3-*b*] isomer (Bodige *et al.*, 2019). Furthermore, these compounds exhibit diverse pharmacological properties, such as serving as ligands of immunoregulators, antagonists of α-1-adrenoreceptor, γ-aminobutyric acid receptors, leukotriene, endothelin receptor, blockers of calcium channels, inhibitors of topoisomerase, and binding to neuron growth factor to p75NGF receptors. Besides these, thieno[3,2-*b*]pyridines demonstrate potent activity with low IC_{50} values against VEGFR/c-Met and serve as highly efficient inhibitors of Src kinase activity (Zhang *et al.*, 2016; Boschelli *et al.*, 2004).

2. Synthetic Approaches Used for thieno[3,2-*b*]pyridines

2.1 Creating Pyridine Ring to Form thieno[3,2-*b*]pyridine System

The pyridine ring was formed using the Gould-Jacobs reaction, following the methods reported in 1991. This process involves the intramolecular cyclization of enamino diester (compound 31) to produce the thieno[3,2-*b*]pyridine derivative (compound 32) (Scheme-11). This derivative has demonstrated antihypertensive and antibacterial properties (Malicorne *et al.*, 1991). Moreover, Batchu *et al.* developed a method that uses iodine to help with an intramolecular electrophilic aromatic cyclization of thiophen-2-yl-acrylates derivative (33), resulting in the production of thieno[3,2-*b*]pyridines (34). This process was carried out by carrying out the reaction in chloroform at room temperature using stoichiometric amounts of iodine and K_2CO_3 (Scheme-11) (Batchu *et al.*, 2012).

Figure 13. Scheme-11 Thieno[3,2-b]pyridine derivatives using Gould–Jacobs reaction and electrophilic aromatic cyclization of thiophen-2-yl-acrylates

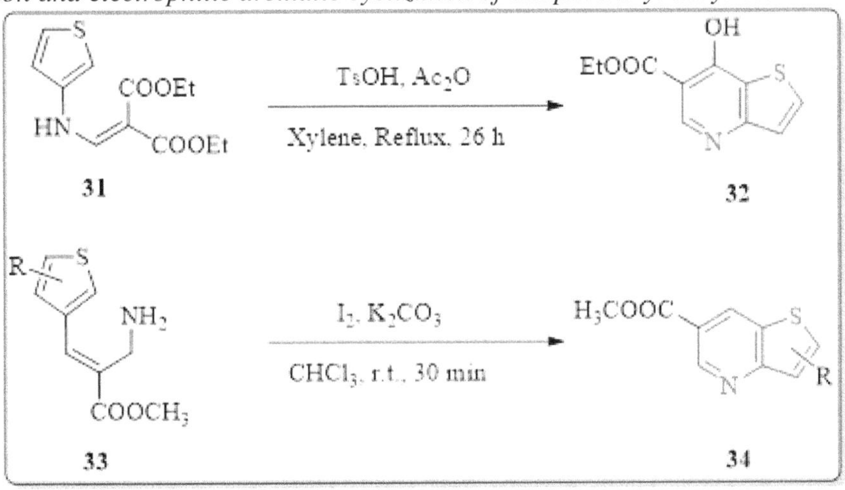

Thieno[3,2-*b*]pyridine derivatives are commonly synthesized from readily available 3-amino thiophenes. The Masarat group's work robustly describes the preparation of thieno[3,2-*b*]pyridines (36) by heating 3-aminothiophen-2-yl substrates (35) in a solution of *N,N*-dimethylformamide and dimethyl acetal (DMF-DMA) in xylene (Scheme 12). Notably, the use of the Thorpe–Guareschi reaction with sodium ethoxide (NaOEt) in ethanol resulted in the production of thieno[3,2-*b*]pyridinones (37) instead of thieno[3,2-*b*]pyridine (Masaret *et al.*, 2021; Zubarev *et al.*, 2013).

Figure 14. Scheme-12 Thieno[3,2-b]pyridine derivatives obtained from 3-amino thiophenes

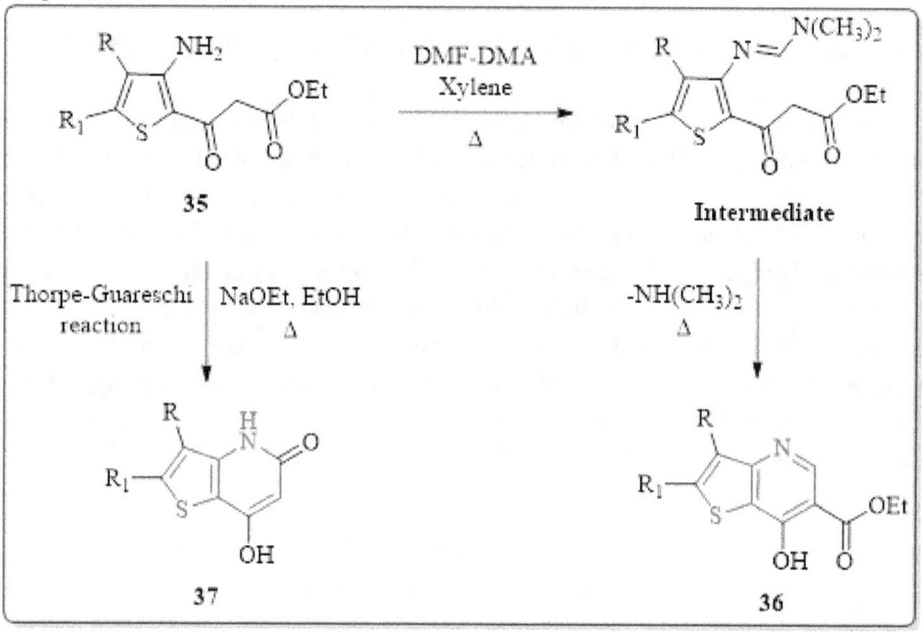

Additionally, Mohareb *et. al.* developed the cyclocondensation reaction of substituted thiophene (38) which results in diol derivatives of thieno[3,2-*b*]pyridine (39). This is a good example of intramolecular cyclization of 3-aminothiophenes containing a γ-functionalized substituent at position 2 (Scheme 13) (Mohareb *et al.*, 1991).

Figure 15. Scheme-13 Diol derivative of thieno[3,2-b]pyridine synthesized from 3-amino thiophenes containing a γ-functionalized substituent

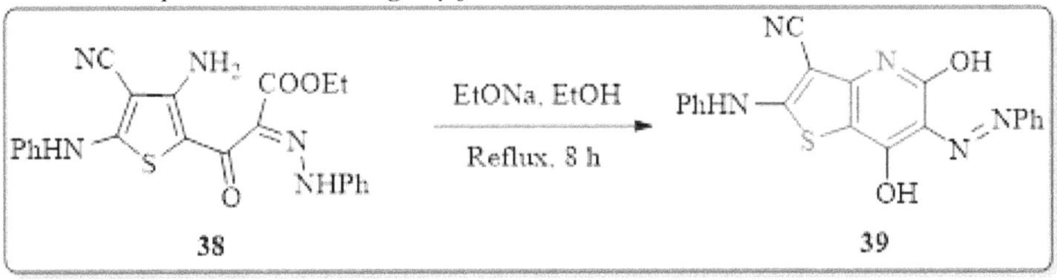

2.2 Creating Thiophene Ring to Form thieno[3,2-*b*]pyridine System

The method involves the reaction of 3-halo-2-cyanomethylpyridine or ester (40, Y = CN or COOEt) with heterocumulenes such as phenyl isothiocyanate or carbon disulphide (CS_2). This reaction results in the formation of the dianions of ketene S,N-acetal or ketene S,S-acetal, leading to the production of fused thieno[3,2-*b*] pyridine structures 41 and 42, respectively. The low yields of the final products can be attributed to the low reactivity of the halogen atom at position 3 of the pyridine ring (Scheme-14) (Bremner *et al.,* 1998).

Figure 16. Scheme-14 Cyclization method for thieno[3,2-b]pyridine using hetero-cumulenes

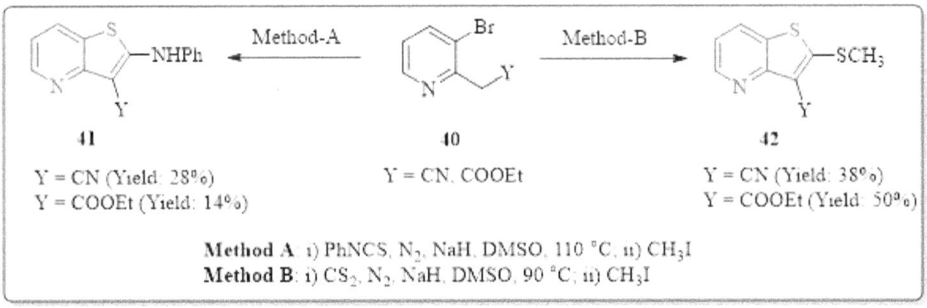

A common method for producing thieno[3,2-*b*]pyridine (44) products involves combining 3-halo-2-alkynylpyridines (A) with sodium sulfide (Na_2S) (Peixoto *et al.,* 2012). To create the final compound 44, an intermediate is formed from 3-chloro-2-bromopyridines (43) through the selective coupling of alkynes using the Sonogashira coupling method. The intermediate 3-halo-2-alkynylpyridines is then heated overnight at 130 °C in DMF with Na_2S to yield the expected product 44 in a good yield (Scheme-15).

Figure 17. Scheme-15 Thieno[3,2-b]pyridine obtained from the cyclization of 3-chloro-2-alkynylpyridines using Sonogashira coupling

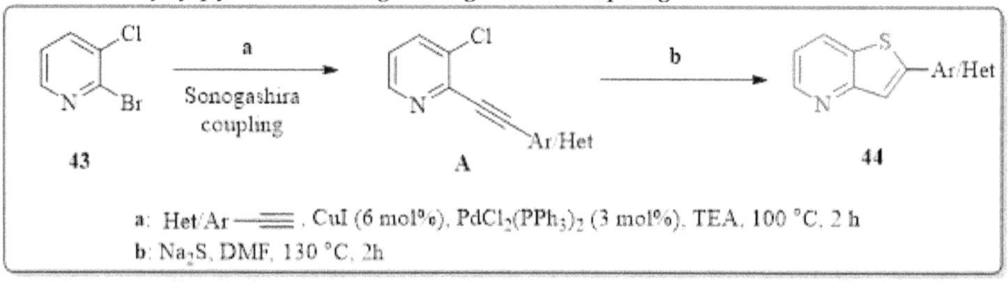

Fort and his coworker used a diverse approach to discover a method for triggering cyclization of 2-alkynyl-3-methylthiopyridines (45) with an electrophile. This resulted in the formation of thieno[3,2-*b*]pyridine derivatives (46) (Scheme 16). Importantly, they found that using iodine as an electrophile led to higher product yields compared to using bromine or N-bromosuccinimide. (NBS) (Comoy *et al.,* 2006).

Figure 18. Scheme-16 Thieno[3,2-b]pyridine generated from triggering cyclization of 2-alkynyl-3-methylthiopyridines with an electrophile

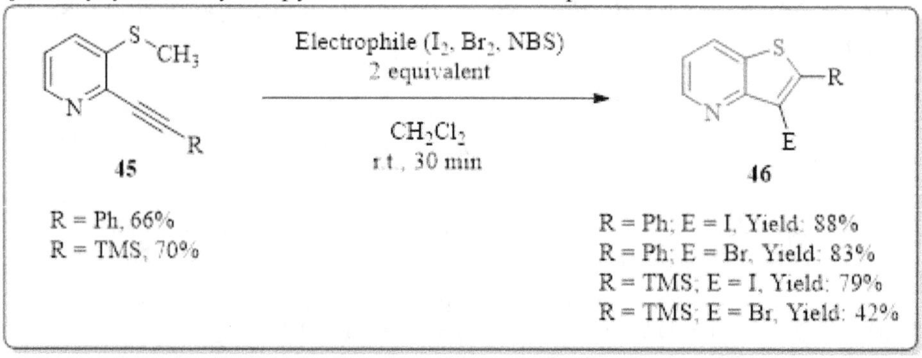

2.3 Creating thieno[3,2-b]pyridine Systems by the Construction of Both Heterocyclic Rings

The synthesis of thieno[3,2-*b*]pyridines involves a cascade heterocyclization using easily available reagents: sodium 2,2-dicyano-1-(alkylamino)ethene-1-thiolate (compound 47) and 2-(2-bromo-1-phenylethylidene)malononitrile (compound 48). This process shows promise, as it leads to the consecutive formation of the thiophene ring and then the pyridine ring in the presence of a basic catalyst (Scheme-17). Ultimately, this results in the production of polyfunctionalized thieno[3,2-*b*]pyridine (compound 49) (Ivanov *et al.,* 1996).

Figure 19. Scheme-17 Thieno[3,2-b]pyridines involves a cascade heterocyclization

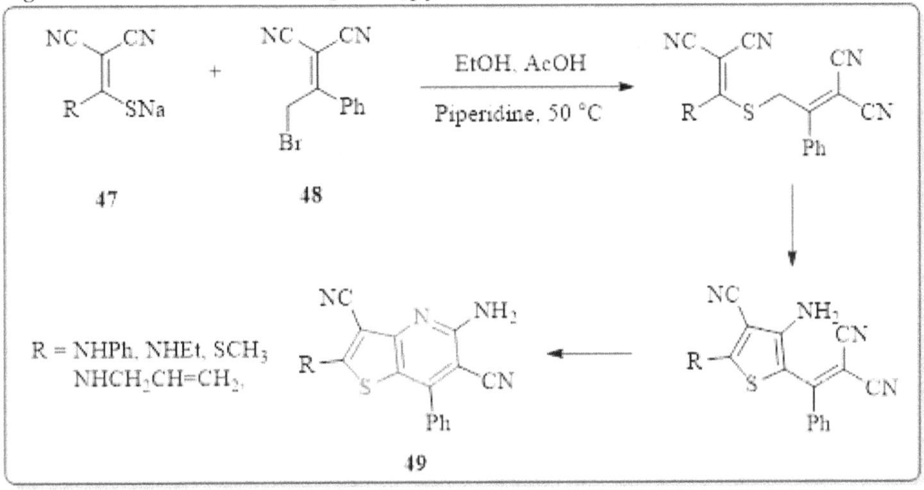

C. THIENO[2,3-*c*]PYRIDINES

Studies on the synthesis and transformations of thieno[2,3-*c*]pyridines have been of great interest to chemists. They are used as building blocks for making complex structures and have diverse biological activities. These activities include cardioprotective, antiproliferative, anticonvulsant, neuroprotective, fungicidal, vasodilatory, antiallergic, antitumor, antiviral, anxiolytic, anti-inflammatory, hypnotic, analgesic, and neurotropic actions (Sabnis *et al.,* 1999; Yurovskaya *et al.,* 1999). Additionally, some thieno[2,3-*c*]pyridine derivatives act as inhibitors of topoisomerase II, kappa-receptor agonists, 5-lipoxygenase inhibitors, modulators of neurotransmitter functions with antidepressant activity, tachykinin antagonists, drugs that enhance erythropoiesis, cGMP-specific phosphodiesterase 5 inhibitors, antagonists of cell adhesion molecule expression, dilators of the nephrovascular tract, platelet activation factor antagonists, allosteric modulators of adenosine receptors, inhibitors of interleukin 2 production, and appetite suppressants (RZhKhim, 2002; Litvinov *et al.,* 2000). Thieno[2,3-*c*]pyridines synthesis methods can be broadly categorized based on constructing the pyridine or thiophene rings.

3. Synthetic Approaches Used for thieno[2,3-*c*]pyridines

3.1 Creating Pyridine Ring to Form thieno[2,3-*c*]pyridine System

The ring closure reaction of thienylacrylamide, or acetal of *N*-(thiophen-3-ylmethyl)ethanamine, is commonly used for preparing thieno[2,3-*c*]pyridines and has been employed for many years (Engstrom *et al.,* 2009). A mild microwave condition along with a promising Pictet-Spengler type cyclization allows for the cyclic-condensation reaction of 2-thienyl-ethylamine (50) with an aldehyde to form the thieno[2,3-*c*]pyridine structure (51). Another Pictet-Spengler reaction involves the use of thiotryptophane esters and amides (52), which undergo cyclocondensation with formaldehyde to produce bio-active benzothienopyridines derivatives (53) in their tosylates or hydrochlorides (Scheme-18) (M´esangeau *et al.,* 2005; Kawakubo *et al.,* 1993).

Figure 20. Scheme-18 Synthesis of thieno[2,3-c]pyridines by Pictet-Spengler reaction

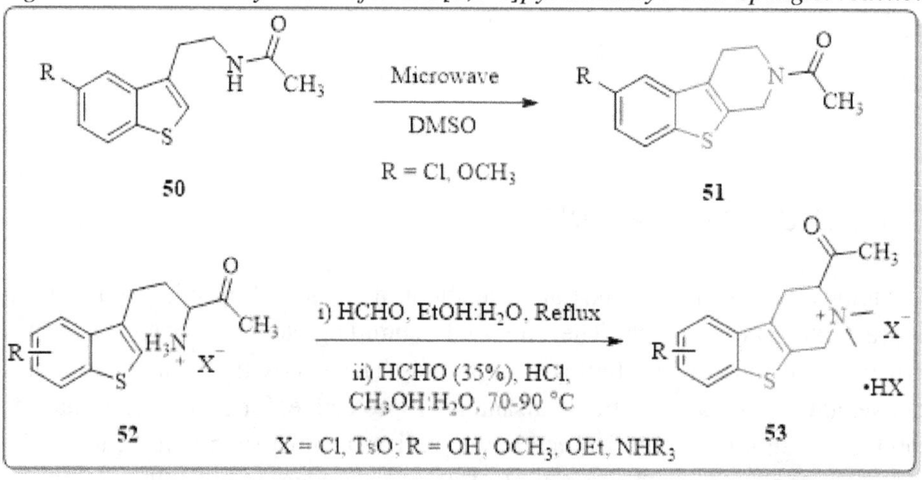

Dingqiao Yang and his research group developed a one-pot, microwave-assisted Sonogashira coupling reaction. They used 3-bromothiophene-2-carbaldehyde (54), terminal alkynes, and ammonium acetate to produce various thieno[2,3-*c*]pyridines (55) (Scheme 19). During the reaction, the coupling between bromothiophene carbaldehyde (54) and alkyne generates an (alkynyl)thiophene carbaldehyde intermediate which further cyclizes to give thienopyridines. This occurs *via* imination of the aldehyde in the presence of ammonium acetate (Yang *et al.,* 2012). Furthermore, in the presence of a ruthenium (II) catalyst, the oxidative coupling of (hetero)arylmethanamine derivatives (56) with internal alkyne (57) resulted in the production of thieno[2,3-*c*]pyridines 58, featuring a wide variety of substituents. The research group noted that when methanol solvent was present, an additional product 59 was formed due to the nucleophilic substitution of a methoxide ion to the pyridine ring. Additionally, product 60 was generated in trace amounts in certain cases. (Scheme 19) (Villuendas *et al.,* 2013).

Figure 21. Scheme-19 Metal-catalyzed synthesis of thieno[2,3-c]pyridines using coupling reaction concept

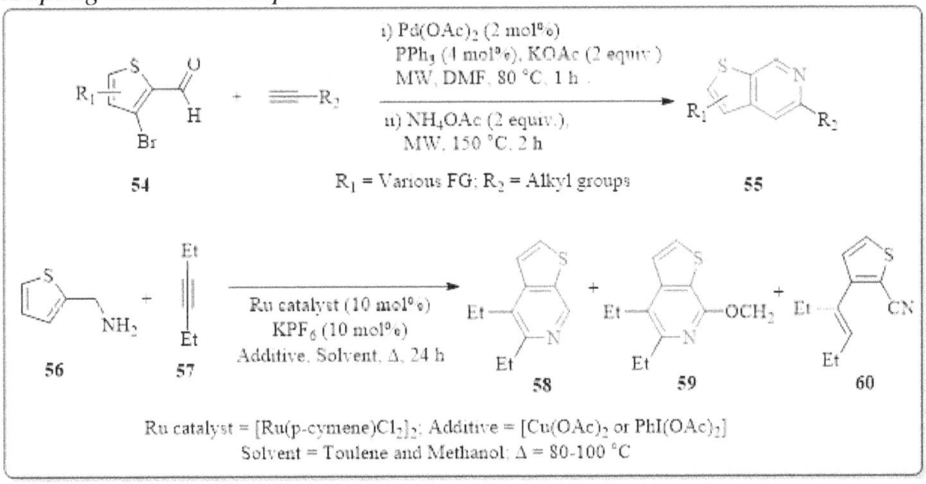

3.2 Creating Thiophene Ring to Form thieno[2,3-*c*]pyridine System

A limited number of methods exist for constructing thieno[2,3-*c*]pyridines based on the thiophene fragment. The Multicomponent Gewald reaction is the most renowned method, involving N-substituted piperidin-4-ones (61), elemental sulfur, and methylene-active nitriles in the presence of secondary amines (Castanedo *et al.,* 2001). This method consistently yields tetrahydrothienopyridines (62) at rates ranging from 60 to 80%. Common organic solvents are predominantly used, along with basic catalysts such as TEA, DEA, piperidine, and morpholine under moderate to occasionally high heating (40-60 °C or 70-80 °C) (Scheme-20). Moreover, the products prepared are vital precursors for synthesizing a variety of poly-fused systems.

Figure 22. Scheme-20 Multicomponent Gewald reaction for the synthesis of thieno[2,3-c]pyridines

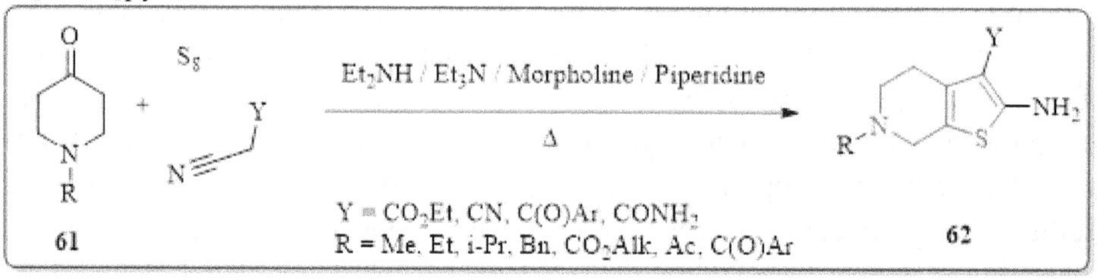

Two separate research groups, Bakar and Wright, have developed a cyclocondensation method using isonicotinic acid and its derivatives (63 & 65). This process is catalyzed by bases like NaH or CH_3ONa and provides a convenient way to close the thiophene ring to produce the corresponding 3-hydroxy thieno[2,3-*c*]pyridine derivatives (64 & 66) in good yields (Scheme-21) (Baker *et al.,* 1988; Wright *et al.,* 1993).

Figure 23. Scheme-21 Synthesis of thieno[2,3-c]pyridines using isonicotinic acid and its derivatives

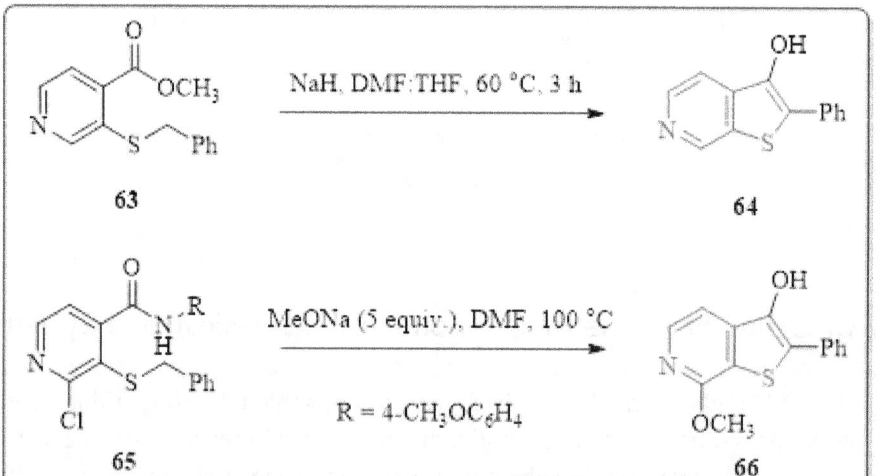

3.3 Creating thieno[2,3-*c*]pyridine Systems by the Construction of Both Heterocyclic Rings

The simultaneous construction of pyridine and thiophene rings to synthesize thieno[2,3-*c*]pyridine derivatives offers significant benefits in terms of efficiency and waste reduction, making it highly desirable in synthetic organic chemistry. The β-Sulfinylamide (67) underwent Pummerer rearrangement in the presence of acetic anhydride and tosylic acid, yielding a fused tricyclic intermediate A. This intermediate further oxidized to the corresponding fused lactam final product (68) in the presence of $NaIO_4$ and methanol, albeit with a relatively lower yield (Scheme-22) (Padwa *et al.,* 1999).

Figure 24. Scheme-22 Thieno[2,3-c]pyridines through Pummerer cyclization using imidosulfoxides

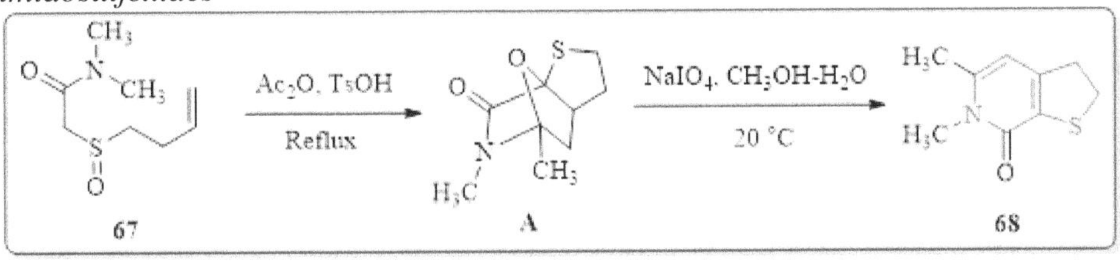

D. THIENO[3,2-*c*]PYRIDINES

The chemistry of this class of compounds has attracted interest in the field of organic chemistry due to the high biological activities of many thieno[3,2-*c*]pyridine derivatives, which are synthesized using well-known procedures. Ticlopidine, Clopidogrel, and Prasugrel are clinically available thieno[3,2-*c*]pyridine molecules that act as non-competitive adenosine diphosphate receptor antagonists. They effectively inhibit platelet aggregation and function as antithrombotic drugs in the body (see Figure 2) (Farid *et al.,* 2010). Thieno[3,2-*c*]pyridine derivatives have demonstrated various activities including antibacterial, antimicrobial, analgesic, antitumor, anti-inflammatory, hyperglycemic, vasodilatory, hypotensive, and anticonvulsant properties. Additionally, they exhibit unique features such as being 5-HT3-receptor antagonists, drugs for treating epilepsy, inhibitors of fibrinogen binding, HIV-1 protease inhibitors, selective dopamine agonists, serotonine antagonists, metalloprotease inhibitors, farnesyl transferase inhibitors, elastase inhibitors, and nephrovascular dilators. Some thieno[3,2-*c*]pyridines have shown potential in treating cardiac disorders as cardio and cerebroprotective drugs for neurologic diseases (Karande *et al.*, 2013; Li *et al.*, 2021; Litvinov *et al.*, 2007). 4-Functionalized thieno[3,2-*c*] pyridines serve as key ligands, alongside acetylacetonate (acac), in phosphorescent iridium(III) complexes for advancing LEDs technologies, in addition to their pharmaceutical application (Lee *et al.*, 2019; Fan *et al.*, 2013).

4. Synthetic Approaches Used for thieno[3,2-*c*]pyridines

4.1 Creating Pyridine Ring to Form Thieno[3,2-*c*]pyridine System

In 1993 Wikel and research group was developed modified the Pomeranz-Fritsch reaction for the synthesis of thieno[3,2-*c*]pyridine. This compound can be prepared in four steps starting from thiophene-3-carbaldehyde (69). The final product was then transformed into the corresponding tosylate (70) for storage (Scheme-23) (Wikel *et al.*, 1993).

Figure 25. Scheme-23 Thieno[3,2-c]pyridines through modified Pomeranz–Fritsch reaction

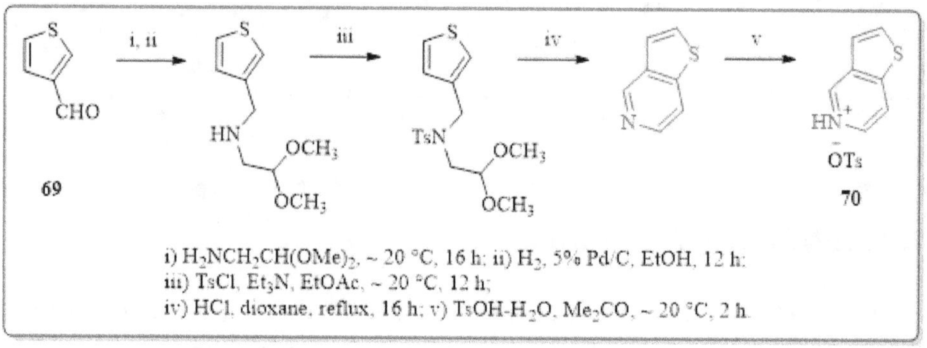

The Stille cross-coupling reaction of 2-tributylstannyl-3-thiophenecarbaldehyde (71) with various thienylcarbamates (A-C) produces the thieno[3,2-*c*]pyridines compounds (72-74) in varying yields (Scheme-24) (Gronowitz *et al.*, 1988).

Figure 26. Scheme-24 Thieno[3,2-c]pyridines through the Stille cross-coupling reaction

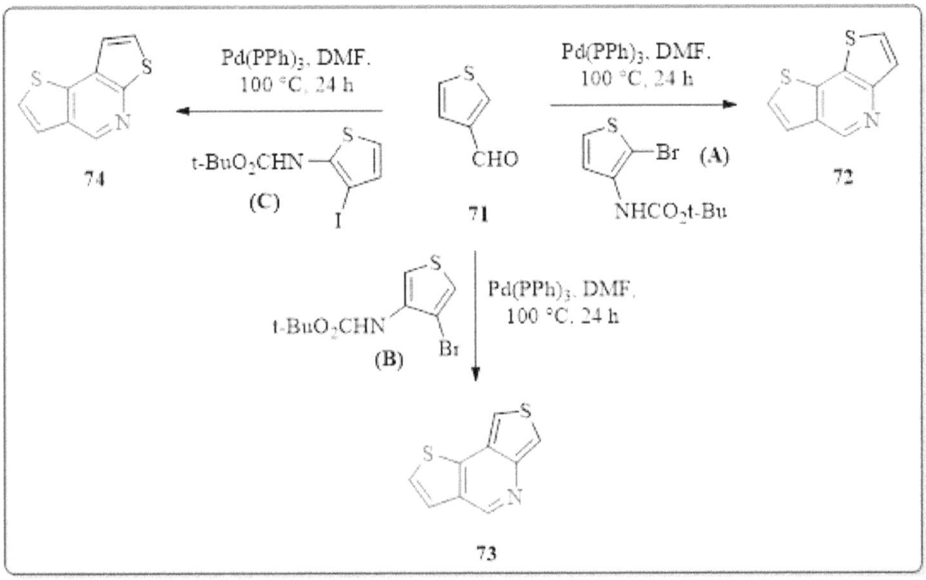

The Pictet-Spengler reaction is widely accepted for synthesizing thienopyridines. This method offers a convenient and straightforward way to produce highly effective drugs like Ticlopidine and Clopidogrel. As per the Scheme-25, the precursor of Ticlopidine, 4,5,6,7-tetrahydrothieno[3,2-*c*]pyridine (76), or the drug itself (77), can be created by reacting 2-(2-thienyl)ethylamine derivative (75) (or its acid chloride) with formaldehyde in the presence of HCl at high temperature (Cen *et al.*, 1997).

Figure 27. Scheme-25 Thieno[3,2-c]pyridines through the Pictet-Spengler reaction

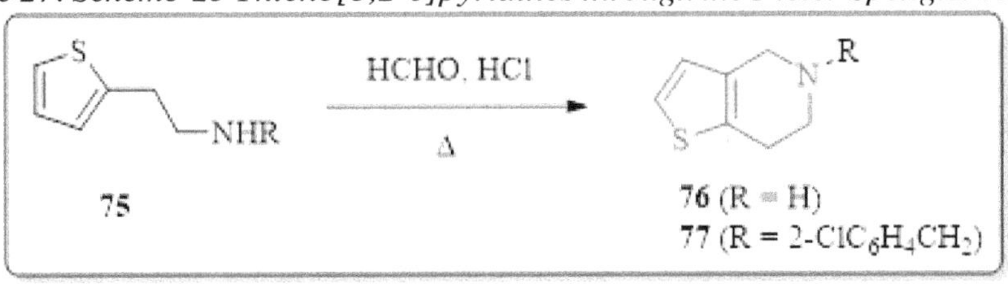

4.2 Creating Thiophene Ring to Form thieno[3,2-*c*]pyridine System

The methods for transforming pyridine to thieno[3,2-*c*]pyridine are not well developed. Most procedures involve replacing the substituent at position 4 of the pyridine ring with a sulfur-containing fragment, followed by cyclization of the resulting product. When 4-chloropyridines (78) react with methyl thioglycolate, they produce 2-methyl ester derivatives of thieno[3,2-*c*]pyridines 79 and 80 in one step by replacing the chlorine atom using Thorpe or Thorpe-Dieckmann cyclization (Scheme-26) (Dunn *et al.,* 1987).

Figure 28. Scheme-26 Thieno[3,2-c]pyridines by Thorpe or Thorpe-Dieckmann cyclization

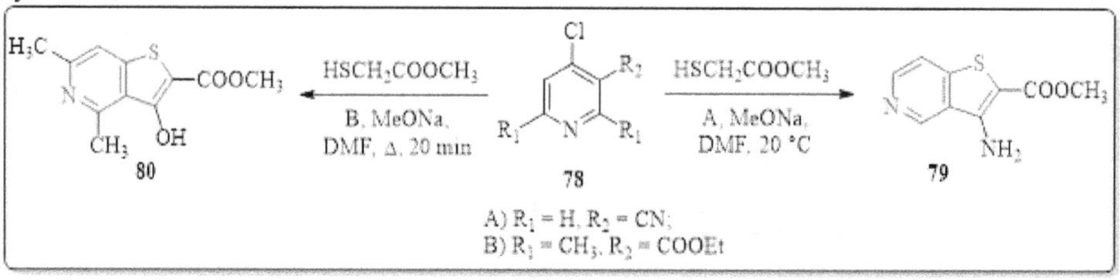

The reaction of tetrachloropyridine (81) with sodium hydrosulfide (NaSH) only produced the 4-mercapto derivative (82). This derivative was then converted into thieno[3,2-*c*]pyridine-2-carboxylate (83) with a 77% yield using 5-alkylation and Dieckmann cyclization. This is another example of the synthetic approach to thieno[3,2-*c*]pyridines of this type (Scheme-27) (Sipyagin *et al.,* 1996).

Figure 29. Scheme-27 Thieno[3,2-c]pyridines from tetrachloropyridine

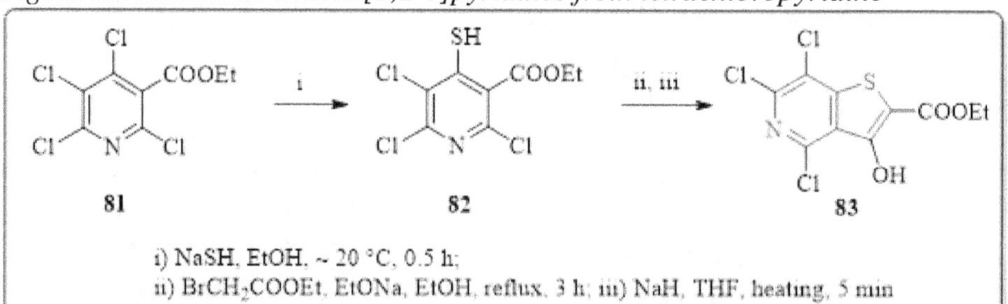

4.3 Creating thieno[3,2-*c*]pyridine Systems by the Construction of Both Heterocyclic Rings

In 2016, Sahu *et. al.* reported a one-pot as well as step-by-step synthesis of aminothieno[3,2-*c*]pyridin-4-one-2-carboxylates (86) using thienopyran intermediate (85). The cyclic condensation of methyl mercaptoacetate with the 2*H*-pyran-2-one-3-carbonitrile substrate (84) produces the thienopyran intermediate. Follow-up treatment with hydrazine under various conditions yields the final product with affordable yields (Scheme 28). In a two-step process, compound 85 undergoes ring opening using sodium methoxide base at room temperature to produce tetrasubstituted thiophene (compound 87). Subsequently, compound 87 undergoes a cyclocondensation reaction with hydrazine hydrate to yield the final product (compound 86) (Sahu *et al.*, 2016).

Figure 30. Scheme-28 Highly functionalized thieno[3,2-c]pyridines by one-pot and step-by-step methods

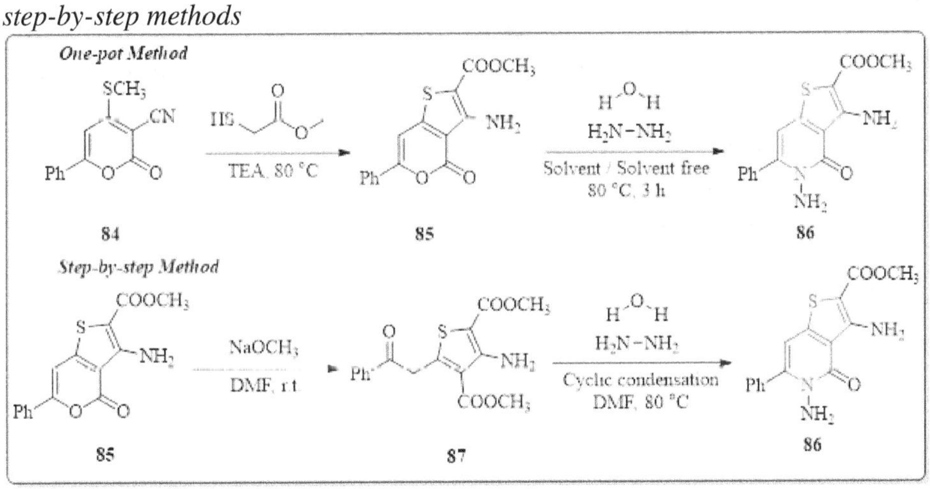

5. Application of Various Thienopyridines Isomers

A vital sulfur-containing link distinguishes the class of chemicals known as thienopyridines, which have a broad range of therapeutic uses in vascular disorders. Except for their dipole moments caused by π-electrons, isomeric thienopyridines have essentially equivalent physicochemical characteristics. For thieno[2,3-*b*]pyridine (0.39 D), the π-electron contribution to the dipole moments is minor, while for thieno[3,2-*b*]pyridine (1.65 D), it is greatest. On the other hand, the thienopyridine series' total dipole moments exhibit the opposite pattern. These characteristics have undoubtedly affected the potential of thienopyridines as medicines. The therapeutic qualities of the several thienopyridines covered in this article, however, are thought to be attributable to their substitutions because of the slight variations in these characteristics.

5.1 Antibacterial and Antifungal Activity

In 2020, Zaki and colleagues developed thieno[2,3-*b*]pyridines that had exceptional antibacterial action against *Bacillus cereus*, *Staphylococcus aureus*, *Pseudomonas aeruginosa*, and *Escherichia coli* (Scheme-29) (Zaki *et al.*, 2020). These compounds limit the proliferation of bacterial cells by blocking DNA gyrase topoisomerase, an enzyme that prevents DNA separation in bacterial cells during replication. With an inhibition zone value of 9 to 18 mm and a minimum inhibitory concentration (MIC) value ranging from 7 to 14 µg/mL, all of the compounds 88(65, 89(66, 90(68, 91(69), and 92(70) exhibit moderate action against these bacteria. In comparison to the commercial drug ofloxacin, which is assessed as 3 µg/mL and 22–29 mm against these bacteria, their MIC values and inhibitory zone area are comparable. Furthermore, these compounds were tested for their antifungal qualities against four distinct fungi: *Aspergillus flavus*, *Candida albicans*, *Geotrichum candidium*, and *Trichophyton rubrum*. Chloroacetyl derivative (90) had the best efficacy against *C. albicans* out of all the substances examined; it also showed modest action against *G. candidium* and *A. flavus*. *T. rubrum*, on the other hand, showed resistance to compound 90.

Figure 31. Scheme-29 Thieno[2,3-b]pyridine derivatives with anti-bacterial and antifungal potency

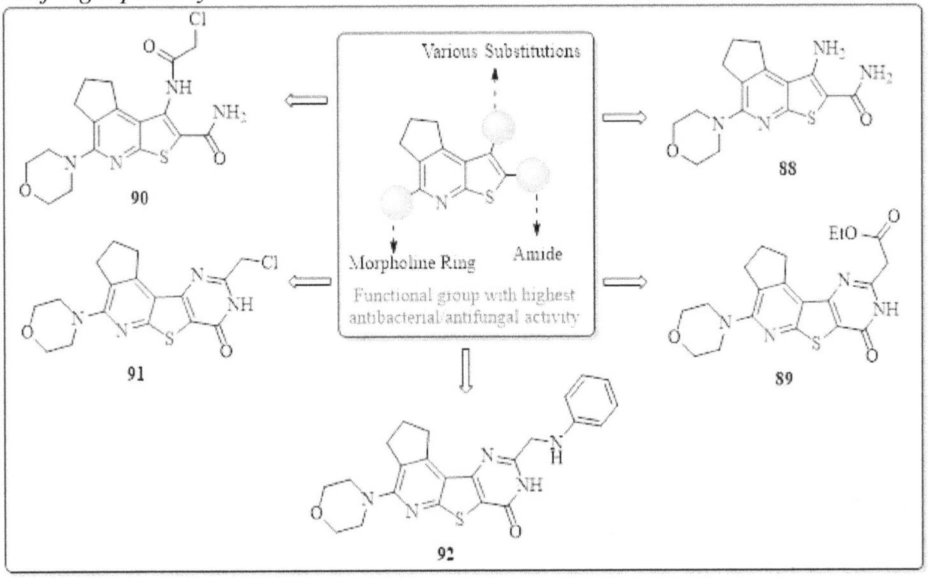

Using amoxicillin trihydrate as the reference medication, El-Deen *et al.* synthesized pyridothienopyrimidine derivatives with success and tested them in vitro against a panel of six Gram-negative and positive bacterial strains: *B. subtilis*, *B. cereus*, *S. aureus*, *E. coli*, *P. aeruginosa*, and *S. typhimurium*. These pyridothienopyrimidine compounds have strong antibacterial activity, especially against Gram-negative bacteria, as seen by their MIC values, which vary from 7.81 to 62.5 µg/mL. Compounds 93 and 94 showed greater efficacy against Gram-positive strains of bacteria, especially *S. aureus* (MIC = 15.63 µg/mL), than against gram-negative versions. Among the investigated bacterial strains, the 4'-(4-methylpiperazin-1-yl) derivative 94 exhibited noteworthy potency with MIC values ranging from 7.81 to 15.63 µg/mL, matching or exceeding those of amoxicillin (Scheme-30) (El-Deen *et al.*, 2020).

Figure 32. Scheme-30 Synthesis of pyridothienopyrimidine derivatives with antibacterial efficacy

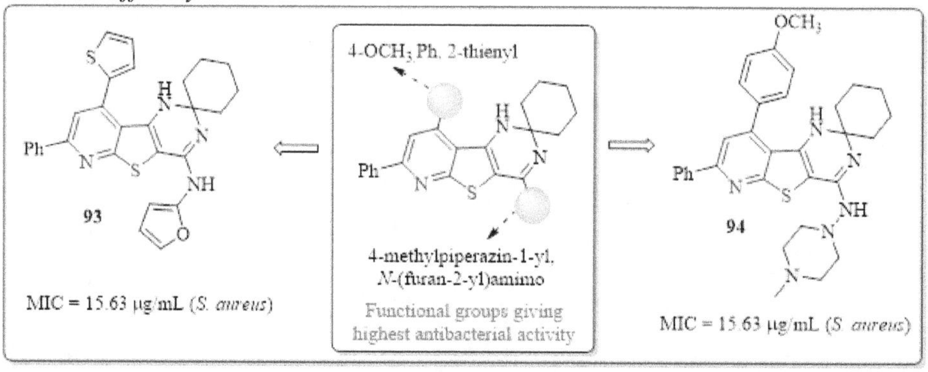

A ZnO-TiO_2 nanoparticle that is employed as a catalyst in the process of creating new tetrahydrothienopyridine series bearing 1,3,4-oxadiazole scaffolds (96). This is achieved by utilizing aldehyde derivatives and cyclic condensation of acetohydrazide of dihydrothieno[2,3-*c*]pyridine (95) under microwave irradiation (Scheme 31) (Sangshetti *et al.,* 2013). Since the two core moieties in this study are known to have potential antibacterial potency, linking the two heterocyclic systems was expected to boost the antimicrobial activity. The antibacterial qualities of these novel compounds were evaluated in four distinct bacterial strain tests: *B. subtilis, P. aeruginosa, S. aureus*, and *E. coli*. *A. Niger* and *C. albicans* were the two fungi that were employed to test the antifungal properties. The MIC values were determined using the conventional agar method, and they were compared to the MIC of ampicillin, ciprofloxacin, and Miconazole for their antibacterial and antifungal effects, respectively.

Electron withdrawing groups, such as fluoro and chloro, showed distinct effects on both fungal and bacterial growth. Significant action against all investigated bacterial strains is produced by 2,6-dichloro (96a) and 2,4-dichloro (96b) substituted derivatives as compared to 4-Cl substituted analogs. When compared to ciprofloxacin (MIC value 25, 50, and 50), the *para*-fluorophenyl derivative (96c) shows the most powerful activity against *E. coli, S. aureus*, and *B. subtilis* (MIC values 25, 25, and 50 μg/mL, respectively). However, the most promising fungal inhibitory action against *C. albicans* is likewise displayed by these compounds, with a MIC value of 20 μg/mL.

Figure 33. Scheme-31 Synthesis of antimicrobial oxadiazole fused thieno[2,3-c] pyridines

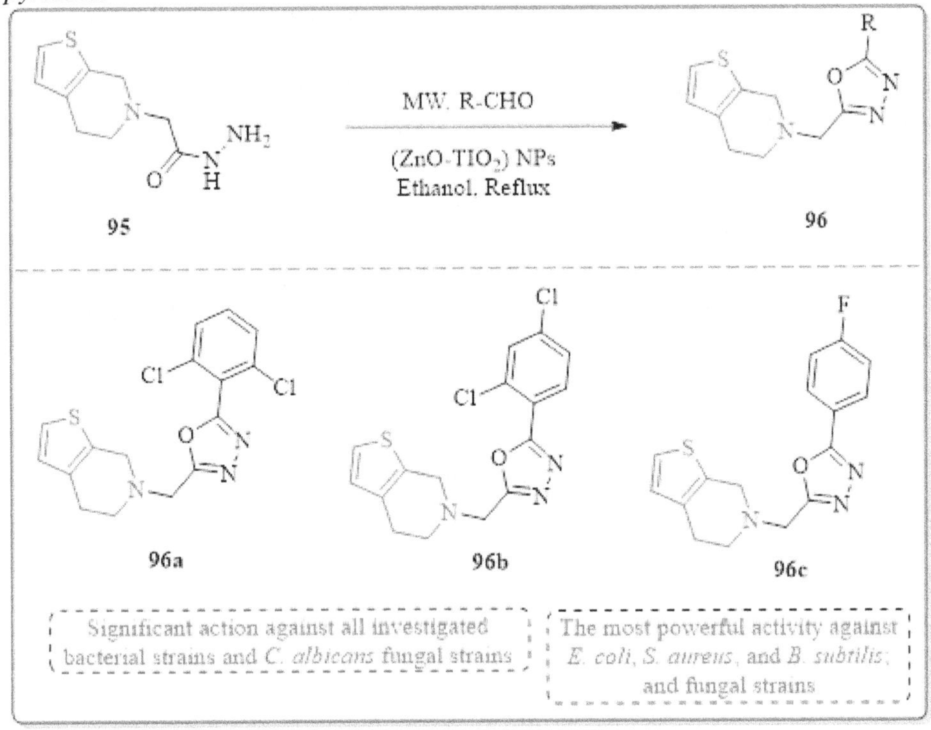

5.2 As an Antiplatelet Agent

Drugs based on thienopyridines are also widely used as antiplatelet medications. A few medications that are frequently used to reduce a patient's risk of stroke and heart attack are Ticlopidine (1), Clopidogrel (2), and Prasugrel (3) (Scheme 32). Essentially, platelets are members of ADP GPCRs and have two purine-based receptors, P2Y1 and P2Y12. Aggregation and a change in form are the results of an interaction between the Gq-PLC-IP3-Ca^{2+} and the ADP-activated P2Y1 receptor. Conversely, in response to ADP activation, the P2Y12 receptor binds to Gi and blocks adenylyl cyclase, which in turn reduces cyclic AMP levels and cyclic AMP-dependent inhibition of platelet actuation.

Small compounds with varying N-atom substitutions, 4,5,6,7-tetrahydrothieno[3,2-*c*]pyridine bases make up all three drugs. Ticlopidine lowers the chance of blood clot formation by inhibiting platelet activation (Shreen *et al.*, 2021). Due of its perceived decreased risk of hematological side effects such neutropenia and thrombocytopenic purpura, Clopidogrel has primarily supplanted Ticlopidine in clinical use. This antiplatelet drug works by permanently blocking platelets' P2Y12 subtype of ADP-receptor, which lowers platelet aggregation and the chance of thrombotic events (Shah *et al.,* 2013; Pereira *et al.,* 2019). Furthermore, compared to Clopidogrel, Prasugrel suppresses platelet aggregation more quickly, consistently, and dangerously. *In-vivo* and *in-vitro* research shows that Prasugrel reduces platelet aggregation ten times better than Clopidogrel. After being absorbed by the gastro-intestinal (GI) tract, esterases found in the gut wall, liver, and plasma hydrolyze Prasugrel into a thiolactone (R-95913). This thiolactone is oxidized by cytochrome P450 enzymes, specifically CYP3A and CYP2B6 isoenzymes, creating the active thiol metabolite R-138727 which further converted into the main active metabolite (R-106583). This metabolite then irreversibly binds to the platelet surface P2Y12 ADP receptors, which are connected to G-proteins (Testa *et al.,* 2010).

Figure 34. Theme-32 Thieno[3,2-c]pyridine based anti-platelet drugs and its metabolism process

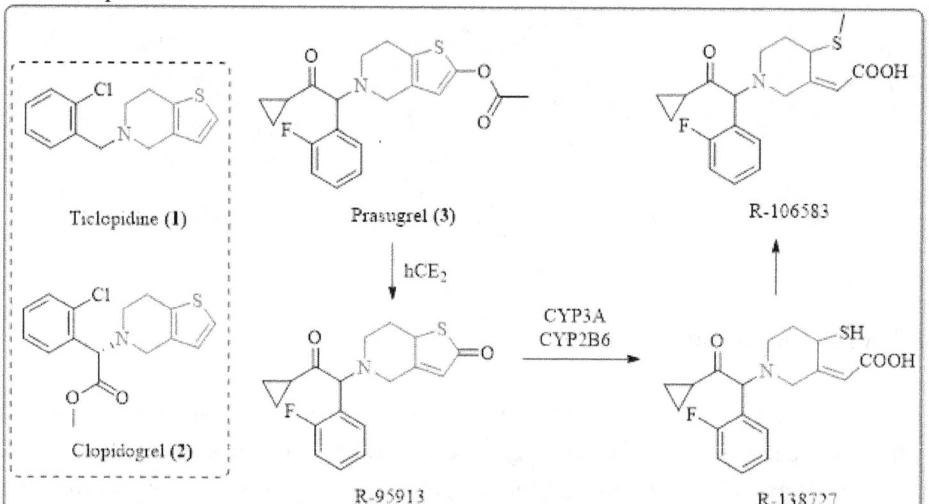

5.3 Anticancer Activity

The capacity of thienopyridines to either kill or stop the growth of cancer cells is known as their cytotoxic action. There are many different types of cancer on the market, and different hetero-fused medications, such as thienopydines, have been employed to treat them. Among these, non-small cell lung cancer (NSCLC) accounts for the bulk of lung cancer cases (>80%), and the majority of patients' first diagnosis was made when their disease was already advanced. The gene EGFR, or epidermal growth factor receptor, codes for a particular protein required for cell division and upkeep (Li *et al.,* 2021). Research may find that cancer treatments are more successful if changes in the EGFR gene in tumor tissue are examined. It has been observed that AZD9291 (97, Osimertinib) exhibits the ability of its amino-pyrimidine nucleus and carbonyl side chain to interact with EGFR. It is important to remember that a thiophene ring of thienopyridines present at the drug molecule in a unique modification may potentially form new hydrogen bonds and promote increased drug-EGFR interaction. The 4,5,6,7-tetrahydrothieno[3,2-*c*]pyridine structure (98) was used as a ligand for efficient interaction with $EGFR^{T790M}$ (Scheme 33) (Tu *et al.,* 2017) in accordance with the theory. Comparing the analogous study to the conventional drug, which has 97.3% inhibitory activity at 0.1 μM and 99.5 at 1.0 μM doses, it demonstrates 97.9% inhibitory activity at 0.1 μM and 99.2 at 1.0 μM concentrations, respectively.

Figure 35. Scheme-33 Tetrahydrothieno[3,2-c]pyridine for binding $EGFR^{T790M}$ effectively as a ligand

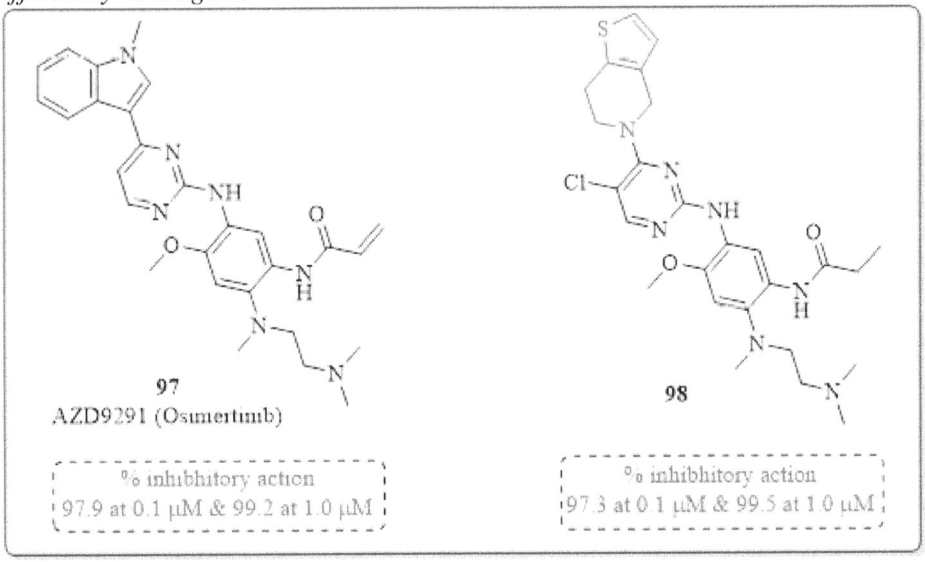

Recepteur d'origine nantais (RON) splice mutations are interesting targets for anticancer therapy, particularly in primary ovarian cancer, as they may indicate oncogenic driver mutations. Eleven thieno[3,2-*b*]pyridine based compounds were identified by Ryu et al. as strong inhibitors of these oncoproteins. Along with ADME/DMPK experiments, this group evaluated its inhibitory actions against several RON splice variants (RON-155, -160, -165) (Ryu *et al.,* 2023). Compound 99 exhibits a respectable level of action against RON species because of its 2-pyridinyl substitution at the thienopyridine ring's C_2. Scheme 34 notes that the activity of the molecule is further enhanced by the introduction of [(2-methoxyethyl)amino]ethyl (100)(IC_{50} = 0.48 μM), (4-methylpiperazin-1-yl)ethyl (101)(IC_{50} = 0.26 μM), (3-methoxyazetidin-1-yl)ethyl (102)(IC_{50} = 0.16 μM), and morpholinoethyl (103)(IC_{50} = 0.39 μM) groups on the pyridine ring. *In-vitro* tests have demonstrated the higher effectiveness of compounds 100, 101, and 102; nevertheless, compound 103 has more encouraging pharmacokinetics in rats, improved ADME profiles, and comparable cytotoxicity. Moreover, compound 103 also demonstrated favorable exposure (12.60 h/μg/mL) and quick absorption (1 h) in an oral pharmacokinetic (PK) trial, which specifically led to increased bioavailability. Using BALB/c-nude female mice as xenograft models, the drug (103) was evaluated *in-vivo* for its ability to inhibit tumor development (TGI) by evaluating its dose response in KM12C, SW620, and T84 colorectal cancer cell-derived models. In the KM12C model, compound 103 was administered orally (OA) at doses of 1, 3, 10, and 30 mg/kg. The result was a dose-dependent decrease in the volume of the tumor, with corresponding tumor growth inhibition (TGI) values of 29.4%, 35.0%, 74.0%, and 85.2%, respectively.

In-vivo studies on mouse xenograft models, including the PDX model, revealed that compound 103 had remarkable anticancer effectiveness against RON splice variants compared to drug candidate 104 (BMS-777607). Clinical trial failures have been mostly attributed to off-target inhibitor effects caused by structural similarities in kinase drug pockets. Compound 103, however, showed impressive kinase selectivity, blocking over 80% of just 13 kinases, including RON (Scheme-34).

Figure 36. Scheme-34 Anticancer activity of thieno[3,2-b]pyridines against RON splice variants

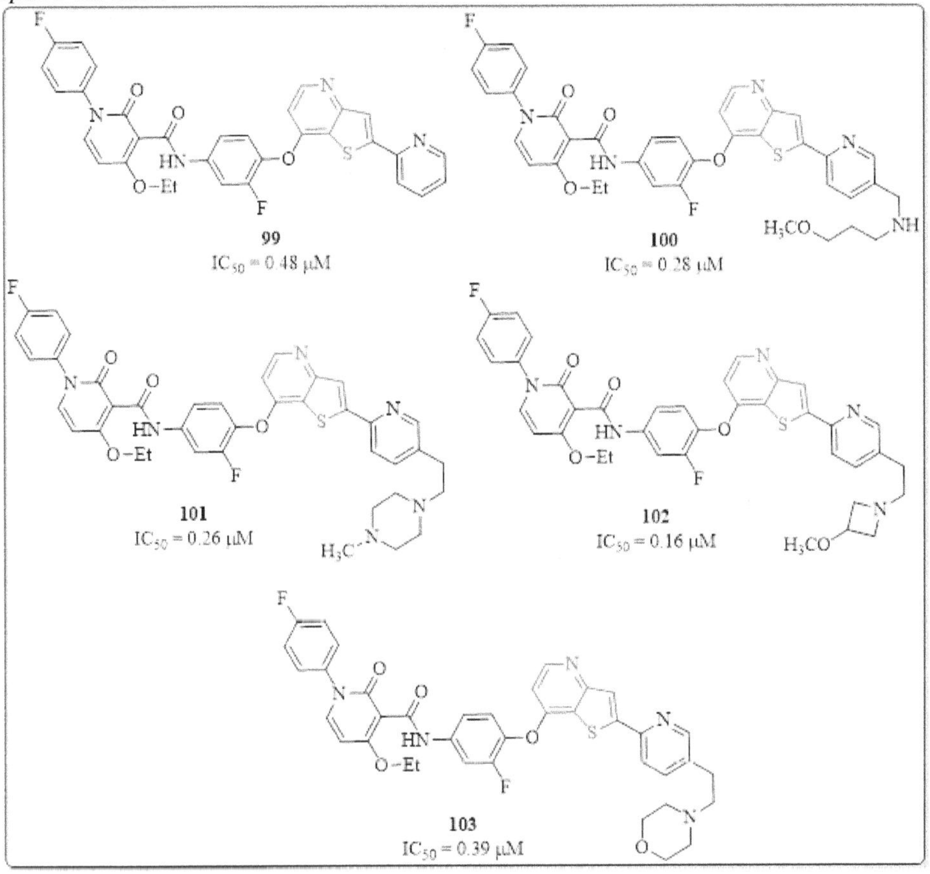

5.4 Antiviral Activity

The Mayaro virus (MAYV), which belongs to the Alphavirus genus in the Togaviridae family, is spread by arthropods. This fetal virus, known as MAYV, causes acute sickness and chronic joint pain, which frequently results in physical disability and, in rare circumstances, even death (Paschoalino *et al.*, 2023). For Mayaro fever, there are still no licensed antiviral therapies, even though many MAYV infections are chronic. Nevertheless, recent studies have identified a number of thieno[2,3-*b*] pyridine compounds with promising antiviral efficacy against MAYV. *p*-Methyl functionalized compound 104 showed the least toxicity (CC50 = 2500 μM) when compared to ribavirin (CC50 = 523 μM). It also had a good active concentration

(IC50 = 20.0 μM), a high selectivity index (SI) of value 125, and comparable activity to *m*-CH_3, *m*-NO_2, and m-F functionalized compounds 105, 106, and 107. However, these compounds are more cytotoxic (CC50 values of 450, 760, and 400 μM, respectively) (Scheme-35) (Amorim *et al.,* 2017).

Figure 37. Scheme-35 Antiviral activity of thieno[2,3-b]pyridines-5-carbonitriles against Mayaro virus

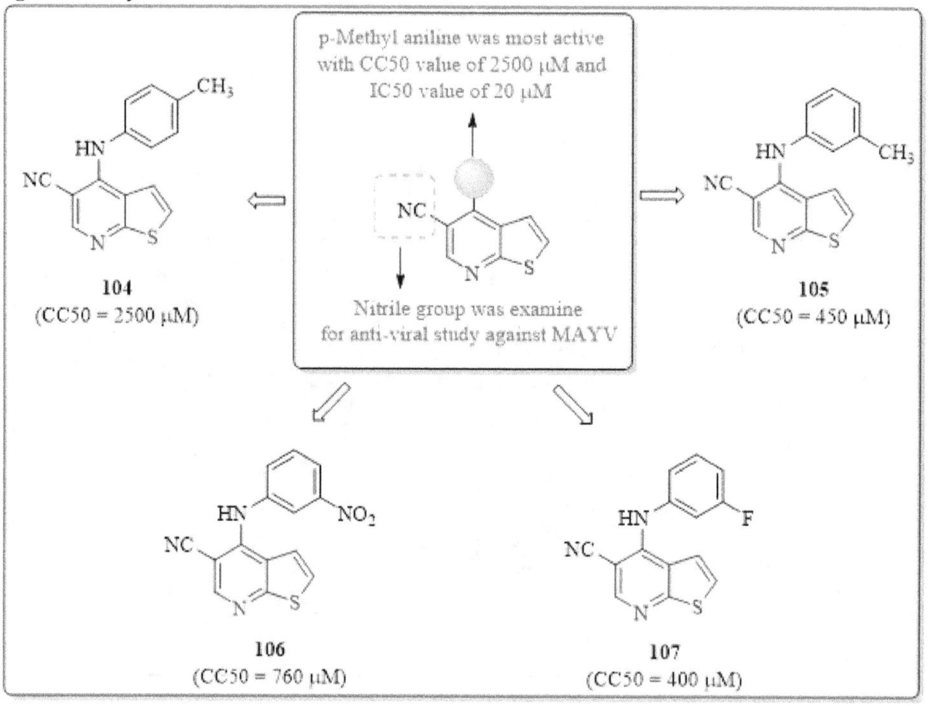

Using a similar methodology, the C_5 ester-substituted thienopyridines that were previously substituted with nitrile-substituted thienopyridines (104–107) were analyzed for antiviral characteristics and their ability to inhibit the Herpes Simplex Virus 1 (HSV-1). By amination of ethyl 4-chlorothieno[2,3-*b*]pyridine-5-carboxylate with various aromatic amines, a series of five compounds based on thieno[2,3-*b*] pyridine-5-carboxylate analogs were produced. Surprisingly, compound 108 showed a capacity to suppress HSV-1 infection by roughly 86% as opposed to 82% for acyclovir. Three of the derivatives, numbered 108–110, demonstrated effective antiviral activity, and none of the derivatives, at any concentration, showed cytotoxic effects on host cells, suggesting that these molecules may target a particular stage of the virus replication cycle (Scheme-36) (Bernardino *et al.,* 2004).

Figure 38. Scheme-36 Antiviral activity of thieno[2,3-b]pyridines-5-carboxylates against HSV-1 virus

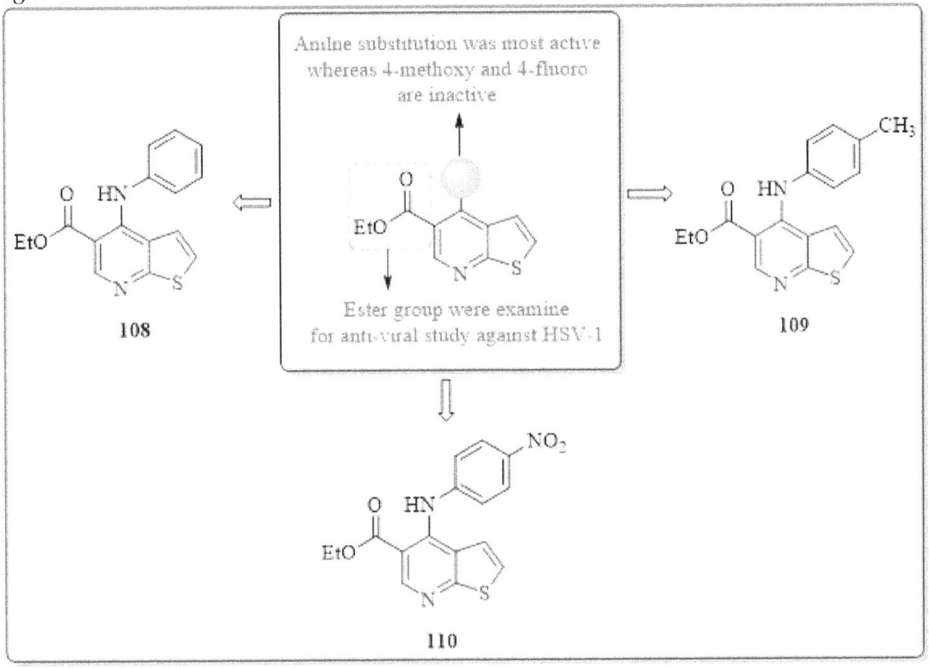

5.5 Miscellaneous

5.5.1 As a COX-2 enzyme inhibitor

The enzyme cyclooxygenase (COX) controls the synthesis of prostanoids from arachidonic acid, such as thromboxane, and prostaglandins, such as prostacyclin. Human COX enzyme comes in two main forms: COX-1 and COX-2. Pain and inflammation are reduced by medications that block COX-2. Safer COX-2 inhibitors must be developed, nevertheless, as clinical trials have shown that some COX-2 inhibitors raise the risk of heart attacks and strokes. Of the two, COX-2 is normally dormant in the majority of cells during normal circumstances but becomes activated during an inflammatory response. Excess prostaglandins from COX-2 overexpression raise the risk of colorectal cancer (Liu *et al.,* 2007; Greenhough *et al.,* 2009). In a study, many hydrazones using components of thienopyridine and evaluated their potential as inhibitors of COX-2. In both experimental and docking tests against the COX-2 enzyme, bromo (111) and methyl (112) substituted hydrazones demonstrated encouraging COX-2 inhibitory efficacy, particularly when compared to the well-known COX-2 inhibitor Celecoxib. Compound 112 demonstrated a strong network of contacts, including a hydrogen bond with ARG 44. In contrast, compound 113 revealed hydrogen bonds with CYS 47, CYS 41, and ALA 151, in that order (Scheme-37).

Figure 39. Scheme-37 COX-2 inhibition activity of thieno[2,3-b]pyridines

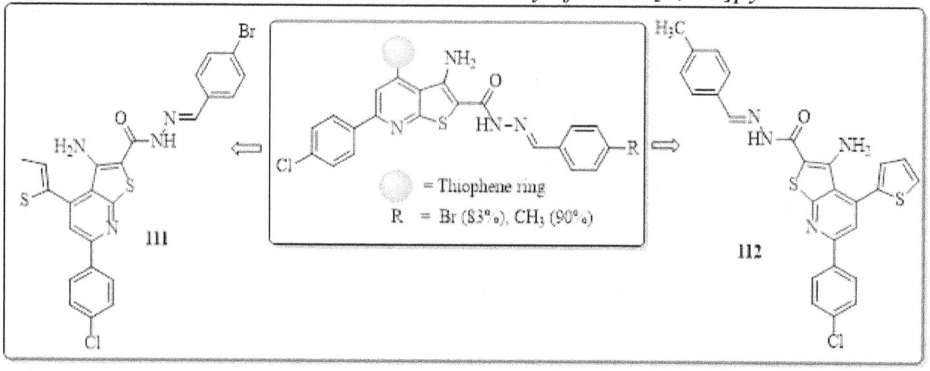

5.5.2 As a ligand for adenosine receptors (ARs)

The nucleoside adenosine acts as the natural agonist of adenosine receptors (ARs), a significant subset of G protein-coupled receptors (GPCRs). There are four distinct kinds of adenylyl cyclase (ARs): A1, A2A, A2B, and A3. Of these, A1 and A3 are engaged in adenylyl cyclase activity inhibition, while A2A and A2B are involved in stimulation. These receptors are widely dispersed in the human body and are involved in a variety of physiological processes. They can be found in the brain, eyes, lungs, kidneys, heart, liver, joints, and blood cell tissues (Abbracchio *et al.,* 1999). As a result, a variety of illnesses, such as asthma, anxiety, cancer, heart related diseases, psychosis, sleep disorders, and many other pathological conditions thought to be connected to variations in the body's adenosine levels, may be treated by targeting ARs (Klotz *et al.,* 2000).

Furthermore, anticonvulsant medications are used to treat epilepsy, a chronic neurological illness caused by increased neuronal activity in the brain that is typically characterized by recurrent, spontaneous convulsions. Similar to adenosine, this naturally occurring anticonvulsant inhibits excitatory brain activity by binding to A1 adenosine receptors (Rombo *et al.,* 2018). However, major peripheral side effects that primarily affect the cardiovascular system and cause secondary problems like bradycardia and hypotension, as well as CNS side effects like sedation, pose challenges to the application of adenosine-based A1 receptor in the treatment of epilepsy (Ribeiro *et al.,* 2002).

Recent studies employing A1 adenosine receptor agonists to treat neurological diseases like epilepsy based on thieno[2,3-*b*]pyridine have been published. The Swiss ADME program was utilized to examine the pharmacokinetic profiles, physiochemical characteristics, and drug-likeness of produced compounds 114a-d in order to determine their potential therapeutic value. Out of all the substances, 114c showed a reasonably acceptable inhibition constant when used as an agonist on the A1 adenosine receptor (rA1Ki = 61.9 nM). The molecule did not meet the ideal values for the saturation parameter, despite meeting the ideal values for lipophilicity, size, solubility, and flexibility. Further structural optimization is required despite its potent and selective A1 receptor agonist characteristics; this will probably involve reducing lipophilicity, size, and polarity (Scheme-38) (Nkomba *et al.,* 2022).

Figure 40. Scheme-38 Adenosine A1 receptor of thieno[2,3-b]pyridines derivatives

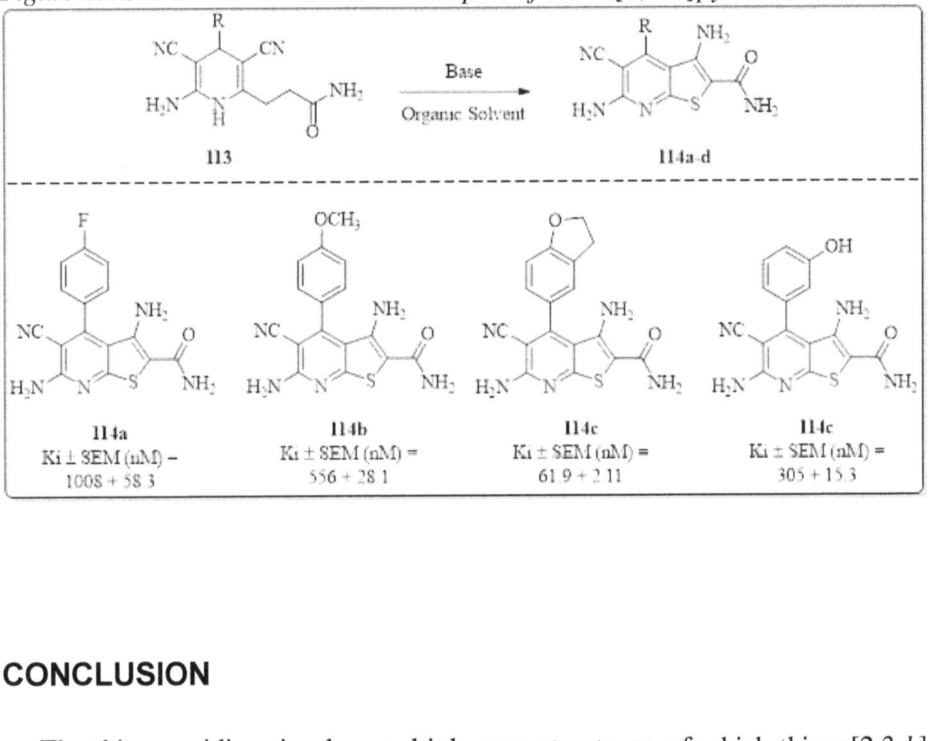

CONCLUSION

The thienopyridine ring has multiple core structures, of which thieno[2,3-*b*], thieno[2,3-*c*], thieno[3,2-*b*], and thieno[3,2-*c*] have been intensively studied due to their ease of synthesis and biological efficacy. When joined or bonded to the thienopyridine ring, nitrogen heterocycles with five or six members - such as pyridine, pyrimidine, piperazine, and morpholine as well as diazole, oxadiazole, and triazole have greatly enhanced the biological activity of these compounds. The N-functionalized amide group at the C_2 position is the most frequently encountered bioactive functional group within the thienopyridine ring in bioactive compounds. While the C_4 and C_6 locations have been functionalized with aryl and arylamine groups to increase bioactivity, the C_3 position has often been investigated with amine groups. In conclusion, the new synthetic methods shown here expand the structural repertory of these molecules and open the door to tailored designs with better qualities. Fused thienopyridines, a class of compounds that exhibit a broad variety of pharmacological activity, including antiviral, antifungal, antibacterial, and antiplatelet characteristics, hold great promise in the field of medicinal chemistry. For increased efficacy and resistance, commercial antiplatelet medications based on thienopyridine, such as ticalopidine, clopidogrel, and prasugrel, are modifiable in a variety of ways. Additionally, fused thienopyridines have lately gained importance in the drug development process as possible anticancer agents with therapeutic relevance as well as epilepsy treatments. Due to the combination of targeted therapeutic uses and sophisticated chemistry, fused thienopyridines are predicted to become increasingly significant as the research advances. This will open up new avenues for resolving difficult issues in a range of medical research contexts.

REFERENCES

Abbracchio, M. P., & Cattabeni, F. (1999). Brain adenosine receptors as targets for therapeutic intervention in neurodegenerative diseases. *Annals of the New York Academy of Sciences*, 890(1), 79–92. DOI: 10.1111/j.1749-6632.1999.tb07983.x PMID: 10668415

Abdelwahab, A. B., Hanna, A. G., & Kirsch, G. (2016). Synthesis of novel 3-Acetyl-2-aminothiophenes and Investigation of their behaviour in the reaction with Vilsmeier–Haack reagent. *Synthesis*, 2881–2888.

Abdelwahab, A. B., Hanna, A. G., & Kirsch, G. (2017). Targeted synthesis of 4-Chloro-3-formylthieno [2,3-*b*] pyridine and/or 4-chlorothieno [2,3-*b*] pyridine by reaction between N-protected 3-Acetyl-2-aminothiophenes and Vilsmeier–Haack reagent. *Synthesis*, 49(13), 2971–2979. DOI: 10.1055/s-0036-1588992

Alinaghizadeh, F., Zahedifar, M., Seifi, M., & Sheibani, H. (2016). Cascade synthesis of thieno[2,3-*b*]pyridines by using intramolecular cyclization reactions of 3-cyano-2-(organylmethylthio) pyridines. *Journal of the Brazilian Chemical Society*, 27, 663–669. DOI: 10.5935/0103-5053.20150309

Amorim, R., de Meneses, M. D., Borges, J. C., da Silva Pinheiro, L. C., Caldas, L. A., Cirne-Santos, C. C., de Mello, M. V., de Souza, A. M., Castro, H. C., de Palmer Paixao, I. C., Campos, R. D., Bergmann, I. E., Malirat, V., Bernardino, A. M. R., Rebello, M. A., & Ferreira, D. F. (2017). Thieno[2,3-*b*]pyridine derivatives: A new class of antiviral drugs against Mayaro virus. *Archives of Virology*, 162(6), 1577–1587. DOI: 10.1007/s00705-017-3261-0 PMID: 28213871

Amorim, R., de Meneses, M. D., Borges, J. C., da Silva Pinheiro, L. C., Caldas, L. A., Cirne-Santos, C. C., de Mello, M. V., de Souza, A. M., Castro, H. C., de Palmer Paixão, I. C. N., Campos, R. M., Bergmann, I. E., Malirat, V., Bernardino, A. M. R., Rebello, M. A., & Ferreira, D. F. (2017). Thieno[2, 3-b]pyridine derivatives: A new class of antiviral drugs against Mayaro virus. *Archives of Virology*, 162(6), 1577–1587. DOI: 10.1007/s00705-017-3261-0 PMID: 28213871

Baker, R. K., Rupprecht, K. M., Pessolano, A. A., & Durette, P. L. (1988) US4767766 A, United States.

Batchu, H., Bhattacharyya, S., & Batra, S. (2012). Iodine-mediated intramolecular electrophilic aromatic cyclization in allylamines: A general route to synthesis of quinolines, pyrazolo[4,3-*b*] pyridines, and thieno[3,2-*b*]pyridines. *Organic Letters*, 14(24), 6330–6333. DOI: 10.1021/ol303109m PMID: 23230893

Bernardino, A. M., Pinheiro, L. C., Ferreira, V. F., & Azevedo, A. R. (2004). Synthesis and antiviral activity of new 4-(phenylamino)thieno[2,3-*b*]pyridine derivatives. *Heterocyclic Communications*, 10(6), 407–410. DOI: 10.1515/HC.2004.10.6.407

Bodige, V., Ravula, P., Gulipalli, K. C., Endoori, S., Cherukumalli, P., Jn, N. S., & Seelam, V. (2019). Design, synthesis, antitubercular and antibacterial activities of pyrrolo[3,2-*b*]pyridine-3-carboxamide linked 2-methoxypyridine derivatives and in silico docking studies. *Synthetic Communications*, 49(17), 2219–2234. DOI: 10.1080/00397911.2019.1618874

Boschelli, D. H., Wu, B., Sosa, A. C. B., Durutlic, H., Ye, F., Raifeld, Y., Golas, J. M., & Boschelli, F. (2004). Identification of 7-phenylaminothieno-[3,2-*b*]pyridine-6-carbonitriles as a new class of Src kinase inhibitors. *Journal of Medicinal Chemistry*, 47(27), 6666–6668. DOI: 10.1021/jm049237m PMID: 15615514

Bremner, D. H., Dunn, A. D., Wilson, K. A., Sturrock, K. R., & Wishart, G. (1998). The synthesis of thienopyridines from ortho-halogenated pyridine derivatives. Part 3. *Synthesis*, 1998(8), 1095–1097. DOI: 10.1055/s-1998-2124

Castanedo, G. M., & Sutherlin, D. P. (2001). Synthesis of tetrasubstituted thiophenes on solid-support using the Gewald reaction. *Tetrahedron Letters*, 42(41), 7181–7184. DOI: 10.1016/S0040-4039(01)01470-8

Cen, J., & Zhongguo Yiyao Gongye Zazhi, (1997). *Chemical Abstract*, 127, 331417.

Chen, H. Y., Nikolka, M., Wadsworth, A., Yue, W., Onwubiko, A., Xiao, M., White, A. J., Baran, D., Sirringhaus, H., & McCulloch, I. (2018). A thieno[2,3-*b*] pyridine-flanked diketopyrrolopyrrole polymer as a n-type polymer semiconductor for allpolymer solar cells and organic field-effect transistors. *Macromolecules*, 51(1), 71–79. DOI: 10.1021/acs.macromol.7b00934

Comoy, C., Banaszak, E., & Fort, Y. (2006). New efficient access to thieno[3,2-*b*] pyridine derivatives via regioselective lithiation of 3-methylthiopyridine. *Tetrahedron*, 62(25), 6036–6041. DOI: 10.1016/j.tet.2006.04.008

Deprets, S., & Kirsch, G. (2000).*European Journal of Organic Chemistry*, 1353. Advance online publication. DOI: 10.1039/C6RA17315B

Dotsenko, V. V., Buryi, D. S., Lukina, D. Y., & Krivokolysko, S. G. (2020). Recent advances in the chemistry of thieno[2,3-*b*]pyridines 1. Methods of synthesis of thieno[2,3-*b*]pyridines. *Russian Chemical Bulletin*, 69(10), 1829–1858. DOI: 10.1007/s11172-020-2969-2

Dotsenko, V. V., Krivokolysko, S. G., & Litvinov, V. P. (2003). Synthesis of Hexa- and Octahydropyrido[3'2':4,5]thieno[3,2-*d*]pyrimidines. *Chemistry of Heterocyclic Compounds*, 39(1), 110–112. DOI: 10.1023/A:1023033027915

Dunn, A. D., & Norrie, R. (1987). Nucleophilic displacements in pyridine rings. *Journal of Heterocyclic Chemistry*, 24(1), 85–89. DOI: 10.1002/jhet.5570240118

El-Taweel, F. M. A., Elagamey, A. A., El-Kenawy, A., & Waly, M. A. (2001). Novel synthesis of thiazole, coumarin,pyridine,thiophene and thieno[2,3-*b*]pyridine derivatives. *Phosphorus, Sulfur, and Silicon and the Related Elements*, 176(1), 215–225. DOI: 10.1080/10426500108055120

Engstrom, K. M., Baize, A. L., Franczyk, T. S., Kallemeyn, J. M., Mulhern, M. M., Rickert, R. C., & Wagaw, S. (2009). Improved synthesis of 3-Substituted-4-amino[3,2-*c*]thienopyridines. *The Journal of Organic Chemistry*, 74(10), 3849–3855. DOI: 10.1021/jo9003772 PMID: 19371092

Fan, C., Miao, J., Jiang, B., Yang, C., Wu, H., Qin, J., & Cao, Y. (2013). Highly efficient, solution-processed orange–red phosphorescent OLEDs by using new iridium phosphor with thieno[3, 2-*c*]pyridine derivative as cyclometalating ligand. *Organic Electronics*, 14(12), 3392–3398. DOI: 10.1016/j.orgel.2013.10.013

Farid, N. A., Kurihara, A., & Wrighton, S. A. (2010). Metabolism and disposition of the thienopyridine antiplatelet drugs ticlopidine, clopidogrel, and prasugrel in humans. *Journal of Clinical Pharmacology*, 50(2), 126–142. DOI: 10.1177/0091270009343005 PMID: 19948947

Ger. Pat. 10031389; RZhKhim, 19O93P (2002).

Greenhough, A., Smartt, H. J., Moore, A. E., Roberts, H. R., Williams, A. C., Paraskeva, C., & Kaidi, A. (2009). The COX-2/PGE 2 pathway: Key roles in the hallmarks of cancer and adaptation to the tumour microenvironment. *Carcinogenesis*, 30(3), 377–386. DOI: 10.1093/carcin/bgp014 PMID: 19136477

Grigor'ev, A. A., Karpov, S. V., Kayukov, Y. S., Nasakin, O. E., Gracheva, I. A., & Tafeenko, V. A. (2017). A new route to highly substituted thieno[2,3-*b*]pyridines via cascade heterocyclization of 2-acyl-1,1,3,3-tetracyanopropenide salts. *Chemistry of Heterocyclic Compounds*, 53(2), 230–235. DOI: 10.1007/s10593-017-2044-6

Gronowitz, S., Hornfeldt, A. B., & Yang, Y. H. (1988).*Chemica Scripta*, 28, 281.

Ivanov, V. L., Artemov, V. A., Shestopalov, A. M., Nesterov, V. N., Struchkov, Yu. T., & Litvinov, V. P. (1996). 2-Bromo-1-Phenylethylidenemalononitrile In Synthesis Of Functionally Substituted Thieno[3,2-*b*]Pyridines And Thiazolo[4,5-*b*]Pyridines. *Khimiya Geterotsiklichesikh Soedinenii*, 413.

Karande, V., Darandale, S. N., & Shinde, D. B. (2013). Microwave assisted nano (ZnO–TiO2) catalyzed synthesis of some new 4,5,6,7-tetrahydro-6-((5-substituted-1, 3, 4- oxadiazol-2-yl) methyl) thieno[2,3-*c*]pyridine as antimicrobial agents. *Bioorganic & Medicinal Chemistry Letters*, 23(7), 2250–2253. DOI: 10.1016/j.bmcl.2013.01.041 PMID: 23434418

Kawakubo, H., Takagi, S., Yamaura, Y., Katoh, S., Ishimoto, Y., Nagatani, T., Mochizuki, D., Kamata, T., & Sasaki, Y. (1993). (R)-1,2,3,4-tetrahydro[1]benzothieno[2,3-c]pyridines: Novel optically active compounds with strong 5-HT1A receptor binding ability exhibiting anticonflict activity and lessening of memory impairment. *Journal of Medicinal Chemistry*, 36(23), 3526–3532. DOI: 10.1021/jm00075a006 PMID: 7902439

Klotz, K. N. (2000). Adenosine receptors and their ligands. *Naunyn-Schmiedeberg's Archives of Pharmacology*, 362(4-5), 382–391. DOI: 10.1007/s002100000315 PMID: 11111832

Krivokolysko, S. G., Dyachenko, V. D., Nesterov, V. N., Sharanin, Yu. A., & Struchkov, Yu. T. (1999).*Russian Journal of Organic Chemistry*, 35, 966.

Lee, S. B., & Ha, Y. (2019). Synthesis and characterization of phosphorescent iridium(III) complexes containing thienopyridine ligands. *Molecular Crystals and Liquid Crystals (Philadelphia, Pa.)*, 679(1), 65–70. DOI: 10.1080/15421406.2019.1597547

Li, J., An, B., Song, X., Zhang, Q., Chen, C., Wei, S., Fan, R., Li, X., & Zou, Y. (2021). Design, synthesis and biological evaluation of novel 2,4-diaryl pyrimidine derivatives as selective $EGFR^{L858R/T790M}$ inhibitors. *European Journal of Medicinal Chemistry*, 212, 113019. DOI: 10.1016/j.ejmech.2020.113019 PMID: 33429247

Li, Y., Chang, Y., Fu, J., Ding, R., Zhang, L., Liang, T., Liu, Y., Liu, Y., & Hu, J. (2021). Design, synthesis and biological evaluation of aminopyrimidine derivatives bearing a 4, 5, 6, 7-tetrahydrothieno[3,2-*c*]pyridine as potent EGFR inhibitors. *European Journal of Medicinal Chemistry*, 226, 113845. DOI: 10.1016/j.ejmech.2021.113845 PMID: 34534838

Lio, J.-Ch., He, H.-W., Tan, X.-S., & Ding, M.-W. (2005). *Russian Chemical Bulletin*, Kam, P.C. & Nethery, C.M. (2003). The thienopyridine derivatives (platelet adenosine diphosphate receptor antagonists), pharmacology and clinical developments. *Anaesthesia*, 58(1), 28–35. DOI: 10.1046/j.1365-2044.2003.02960.x PMID: 15601269

Litvinov, V. P., Dotsenko, V. V., & Krivokolysko, S. G. (2005). Thienopyridines: Synthesis, properties, and biological activity. *Russian Chemical Bulletin*, 54(4), 864–904. DOI: 10.1007/s11172-005-0333-1

Litvinov, V. P., Dotsenko, V. V., & Krivokolysko, S. G. (2007). The chemistry of thienopyridines. *Advances in Heterocyclic Chemistry*, 93, 117–178. DOI: 10.1016/S0065-2725(06)93003-7

Litvinov, V. P., Rodinovskaya, L. A., Sharanin, Yu. A., Shestopalov, A. M., & Senning, A. (1992). Advances in the Chemistry of 3-Cyanopyridin-2(1*H*)-Ones, -Thiones, And -Selenones. *Sulfur Reports*, 13(1), 1–142. DOI: 10.1080/01961779208048951

Litvinov, V. P., Roman, S. V., & Dyachenko, V. D. (2000). Naphthyridines. Structure, physicochemical properties and general methods of synthesis. *Russian Chemical Reviews*, 69(3), 218. DOI: 10.1070/RC2000v069n03ABEH000553

Liu, J., Seibold, S. A., Rieke, C. J., Song, I., Cukier, R. I., & Smith, W. L. (2007). Prostaglandin endoperoxide H synthases: Peroxidase hydroperoxide specificity and cyclooxygenase activation. *The Journal of Biological Chemistry*, 282(25), 18233–18244. DOI: 10.1074/jbc.M701235200 PMID: 17462992

M'esangeau, C., Yous, S., P'er'es, B., Lesieur, D., & Besson, T. (2005). Pictet–Spengler heterocyclizations via microwave-assisted degradation of DMSO. *Tetrahedron Letters*, 46(14), 2465–2468. DOI: 10.1016/j.tetlet.2005.02.043

Madhusudana, K., Shireesha, B., Naidu, V. G., Ramakrishna, S., Narsaiah, B., Rao, A. R., & Diwan, P. V. (2012). Anti-inflammatory potential of thienopyridines as possible alternative to NSAIDs. *European Journal of Pharmacology*, 678(1–3), 48–54. DOI: 10.1016/j.ejphar.2011.12.019 PMID: 22209879

Makawana, J. A., Patel, M. P., & Patel, R. G. (2011). Synthesis and in vitro antimicrobial evaluation of penta-substituted pyridine derivatives bearing the quinoline nucleus. *Medicinal Chemistry Research*, 21(5), 616–623. DOI: 10.1007/s00044-011-9568-6

Malicorne, G., Bompart, J., Giral, L., & Despaux, E. (1991). Synthèse et activité anti-bactérienne d'acides 4,7-dihydro-4-éthyl-7-oxothiéno[3,2-*b*]pyridine-6-carboxyliquesSynthesis and antibacterial activity of 4,7-dihydro-4-ethyl-7-oxothieno(3,2-b)pyridine-6-carboxylic acids. *European Journal of Medicinal Chemistry*, 26(1), 3–11. DOI: 10.1016/0223-5234(91)90207-4

Masaret, G. S. (2021). Convenient synthesis and anticancer evaluation of novel pyrazolyl-thiophene, thieno[3,2-*b*]pyridine, pyrazolo[3,4-*d*]thieno[3,2-*b*]pyridine and pyrano[2,3-*d*]thieno[3,2-*b*]pyridine derivative. *Journal of Heterocyclic Chemistry*, 58(6), 1344–1358. DOI: 10.1002/jhet.4262

Mekky, A. E., Sanad, S. M., Said, A. Y., & Elneairy, M. A. (2020). Synthesis, cytotoxicity, in-vitro antibacterial screening and in-silico study of novel thieno[2,3-*b*]pyridines as potential pim-1 inhibitors. *Synthetic Communications*, 50(15), 2376–2389. DOI: 10.1080/00397911.2020.1778033

Mohamed, O. S., Al-Taifi, E. A., El-Emary, T. I., & Bakhite, E. A. (2007). Studies on the synthesis of some new cyanopyridine-thione and thieno[2,3-*b*]pyridine derivatives. *Phosphorus, Sulfur, and Silicon and the Related Elements*, 182(5), 1061–1082. DOI: 10.1080/10426500601096369

Mohareb, R. M. (1992). Phenyl isothiocyanate in heterocyclic synthesis: Novel synthesis of thiazoles, thieno[2,3-*b*]pyridine, thiophene and thieno[3,2-*c*]pyridazine derivatives. *Monatshefte für Chemie*, 123(4), 341–347. DOI: 10.1007/BF00810946

Mohareb, R. M., Aziz, S. I., & El-Saka, I. (1991).*Sulfur Letters*, 13, 229.

Mohi El-Deen, E. M., Abd El-Meguid, E. A., Karam, E. A., Nossier, E. S., & Ahmed, M. F. (2020). Synthesis and biological evaluation of new pyridothienopyrimidine derivatives as antibacterial agents and *Escherichia coli* topoisomerase II inhibitors. *Antibiotics (Basel, Switzerland)*, 9(10), 695. DOI: 10.3390/antibiotics9100695 PMID: 33066400

Monarch, R. M. (1992).*Italian Chemical Society*, 122, 147.

Nkomba, G., Terre'Blanche, G., Janse van Rensburg, H. D., & Legoabe, L. J. (2022). Design, synthesis and evaluation of amino-3,5-dicyanopyridines and thieno[2,3-*b*] pyridines as ligands of adenosine A_1 receptors for the potential treatment of epilepsy. *Medicinal Chemistry Research*, 31(8), 1277–1297. DOI: 10.1007/s00044-022-02908-9 PMID: 35634433

Padwa, A., Heidelbaugh, T. M., & Kuethe, J. T. (1999). Synthesis of substituted 2-pyridones via the pummerer Cyclization-Deprotonation-Cycloaddition cascade of imidosulfoxides. *The Journal of Organic Chemistry*, 64(6), 2038–2049. DOI: 10.1021/jo982315r PMID: 11674298

Paschoalino, M., Marinho, M. D., Santos, I. A., Grosche, V. R., Martins, D. O., Rosa, R. B., & Jardim, A. C. (2023). An update on the development of antiviral against Mayaro virus: From molecules to potential viral targets. *Archives of Microbiology*, 205(4), 106. DOI: 10.1007/s00203-023-03441-y PMID: 36881172

Peixoto, D., Begouin, A., & Queiroz, M. J. (2012). Synthesis of 2-(hetero)arylthieno[2,3-*b*] or [3, 2-*b*]pyridines from 2,3-dihalopyridines,(hetero)arylalkynes, and Na_2S Further functionalizations. *Tetrahedron*, 68(35), 7082–7094. DOI: 10.1016/j.tet.2012.06.057

Pereira, N. L., Rihal, C. S., So, D. Y. F., Rosenberg, Y., Lennon, R. J., Mathew, V., Goodman, S. G., Weinshilboum, R. M., Wang, L., Baudhuin, L. M., Lerman, A., Hasan, A., Iturriaga, E., Fu, Y. P., Geller, N., Bailey, K., & Farkouh, M. E. (2019). Clopidogrel pharmacogenetics state-of-the-art review and the TAILOR-PCI study. *Circulation: Cardiovascular Interventions*, 12(4), 1–11. DOI: 10.1161/CIRCINTERVENTIONS.119.007811 PMID: 30998396

Ribeiro, J. A., Sebastiao, A. M., & De Mendonça, A. (2002). Adenosine receptors in the nervous system: Pathophysiological implications. *Progress in Neurobiology*, 68(6), 377–392. DOI: 10.1016/S0301-0082(02)00155-7 PMID: 12576292

Rombo, D. M., Ribeiro, J. A., & Sebastião, A. M. (2018). Role of adenosine receptors in epileptic seizures. In Borea, P., Varani, K., Gessi, S., Merighi, S., & Vincenzi, F. (Eds.), *The Adenosine Receptors* (pp. 309–350). Humana Press. DOI: 10.1007/978-3-319-90808-3_13

Ryu, H., Kim, H., Park, I., Lee, M., Park, Y. S., Jin, D. H., Hur, S. C., Park, J., & Lee, H. (2023). Discovery of novel, thienopyridine-based tyrosine kinase inhibitors targeting tumorigenic RON splice variants. *ACS Medicinal Chemistry Letters*, 14(9), 1198–1207. DOI: 10.1021/acsmedchemlett.3c00206 PMID: 37736180

Sabnis, R. W., Rangnekar, D. W., & Sonawane, N. D. (1999). 2-aminothiophenes by the gewald reaction. *Journal of Heterocyclic Chemistry*, 36(2), 333–345. DOI: 10.1002/jhet.5570360203

Sahu, S. N., Singh, S., Shaw, R., Ram, V. J., & Pratap, R. (2016). One-pot and stepwise synthesis of thieno[3,2-*c*]pyridin-4-ones. *RSC Advances*, 6(88), 85515–85520. DOI: 10.1039/C6RA17315B

Sangshetti, J. N., Dharmadhikari, P. P., Chouthe, R. S., Fatema, B., Lad, V., Karande, V., Darandale, S. N., & Shinde, D. B. (2013). Microwave assisted nano (ZnO–TiO2) catalyzed synthesis of some new 4, 5, 6, 7-tetrahydro-6-((5-substituted-1, 3, 4-oxadiazol-2-yl)methyl) thieno[2,3-*c*]pyridine as antimicrobial agents. *Bioorganic & Medicinal Chemistry Letters*, 23(7), 2250–2253. DOI: 10.1016/j.bmcl.2013.01.041 PMID: 23434418

Shan, J., & Sun, H. (2013). The discovery and development of prasugrel. *Expert Opinion on Drug Discovery*, 8(7), 897–905. DOI: 10.1517/17460441.2013.793668 PMID: 23614463

Shaw, R., Tewari, R., Yadav, M., Pandey, E., Tripathi, K., Rani, J., Althagafi, I., & Pratap, R. (2024). Recent advancements in the synthesis of fused thienopyridines and their therapeutic applications. *European Journal of Medicinal Chemistry Reports*, 12, 100185. DOI: 10.1016/j.ejmcr.2024.100185

Sheth, S., Brito, R., Mukherjea, D., Rybak, L. P., & Ramkumar, V. (2014). Adenosine receptors: Expression, function and regulation. *International Journal of Molecular Sciences*, 15(2), 2024–2052. DOI: 10.3390/ijms15022024 PMID: 24477263

Shreen, S., Uddin, M. B., Ali, M. S., & Sultana, Z. (2021). The role of antiplatelet agents in ischemic events. *Journal of Drug Delivery and Therapeutics*, 11(4-S), 187–194. DOI: 10.22270/jddt.v11i4-S.4926

Sipyagin, K. A. M. (1996).*Geterotsikl. Soedin.*, 3, 364.

Steinkopf, W., & Lutzkendorf, G. (1913).*Chem. Ztg.*, 36, 379.

Testa, L., Biondi Zoccai, G. G., Valgimigli, M., Latini, R. A., Pizzocri, S., Lanotte, S., Laudisa, M. L., Brambilla, N., Ward, M. R., Figtree, G. A., Bedogni, F., & Bhindi, R. (2010). Current concepts on antiplatelet therapy: Focus on the novel thienopyridine and nonthienopyridine agents. *Advances in Hematology*, 2010(1), 595934. DOI: 10.1155/2010/595934 PMID: 21151515

Tu, Y., Wang, C., Xu, S., Lan, Z., Li, W., Han, J., Zhou, Y., Zheng, P., & Zhu, W. (2017). Design, synthesis, and docking studies of quinazoline analogues bearing aryl semicarbazone scaffolds as potent EGFR inhibitors. *Bioorganic & Medicinal Chemistry*, 25(12), 3148–3157. DOI: 10.1016/j.bmc.2017.04.001 PMID: 28428040

Villuendas, P., & Urriolabeitia, E. P. (2013). Primary amines as directing groups in the Ru-catalyzed synthesis of isoquinolines, benzoisoquinolines, and thienopyridines. *The Journal of Organic Chemistry*, 78(11), 5254–5263. DOI: 10.1021/jo400344m PMID: 23650873

Wikel, J. H., Denney, M. L., & Vasileff, R. T. (1993). A convenient synthesis of the thieno[3,2- *c*]pyridine nucleus. *Journal of Heterocyclic Chemistry*, 30(1), 289–290. DOI: 10.1002/jhet.5570300150

Wright, S. W., & Corbett, R. L. (1993). A synthesis of 2-aryl-3-hydroxybenzothiophenes and analogs by the base promoted cyclization of N-phenyl-2-(benzylthio) benzamides. *Tetrahedron Letters*, 34(18), 2875–2878. DOI: 10.1016/S0040-4039(00)60469-0

Yakovlev, M. Yu., Kadushkin, A. V., Solov'eva, N. P., Anisimova, O. S., & Granik, V. G. (1998). ω-Substituted-2-(polyenamino)- or annelated nitropyridines from 1-(3-cyano-5-nitropyridyl-2)-pyridinium salts. *Tetrahedron*, 54(21), 5775–5780. DOI: 10.1016/S0040-4020(98)00265-8

Yang, D., Burugupalli, S., Daniel, D., & Chen, Y. (2012). Microwave-assisted one-pot synthesis of isoquinolines, furopyridines, and thienopyridines by palladium-catalyzed sequential Coupling–Imination–Annulation of 2-bromoarylaldehydes with terminal acetylenes and ammonium acetate. *The Journal of Organic Chemistry*, 77(9), 4466–4472. DOI: 10.1021/jo300494a PMID: 22494334

Yurovskaya, M. A., & Mit'kin, O. D. (1999). Functionalization of pyridines. Reactions with formation of a bond between carbon and group IV, V, or VI heteroatom. *Chemistry of Heterocyclic Compounds*, 4, 437.

Zaki, R. M., El-Dean, A. M., Radwan, S. M., & Ammar, M. A. (2020). Synthesis, reactions, and antimicrobial activity of novel heterocyclic compounds containing cyclopenta[*d*]thieno[2,3-*b*]pyridine moiety and related fused heterocycles. *Russian Journal of Bioorganic Chemistry*, 46(1), 85–96. DOI: 10.1134/S1068162020010148

Zhang, J., Jiang, V., Jiang, Y., Guo, M., Zhang, S., Li, J., He, J., Liu, J., Wang, J., & Ouyang, L. (2016). Recent advances in the development of dual VEGFR and C-met small molecule inhibitors as anticancer drugs. *European Journal of Medicinal Chemistry*, 108, 495–504. DOI: 10.1016/j.ejmech.2015.12.016 PMID: 26717201

Zubarev, A. A., Shestopalov, A. M., Larionova, N. A., Rodinovskaya, L. A., & Shestopalov, A. A. (2013). New regio-selective method of combinatorial synthesis of substituted thiophenes, thieno[3,2-*b*]pyridines and other heterocycles via combination of 'domino'-type reactions. *Tetrahedron*, 69(46), 9648–9655. DOI: 10.1016/j.tet.2013.09.025

Chapter 10
Synthesis, Properties, and Biological Activity of Benzothiadiazine

Navnath Tulshiram Hatvate
https://orcid.org/0000-0002-3388-3247
Institute of Chemical Technology, Jalna, India

Rutuja Rambhau Ukhade
https://orcid.org/0009-0007-4907-4691
Institute of Chemical Technology, Jalna, India

Hemantkumar Akolkar
https://orcid.org/0000-0003-0882-1324
Abasaheb Marathe Arts and New Commerce Science College, India

ABSTRACT

Heterocyclic compounds are ubiquitous in biologically active compounds, including many drugs. Benzothiadiazine is a class of sulphur and nitrogen-containing heterocyclic compounds comprised of a benzene ring fused with thiadiazine. Benzothiadiazine rings appear in many marketed drugs, such as bendroflumethiazide, chlorothiazide, cyclothiazide, hydrochlorothiazide, and diazoxide, which are popularly known for their diuretic activity. The ring has diverse pharmacological activity, including anticancer, antifungal, antimicrobial, anthelmintic, hypolipidaemic, and antidiabetic activity. This chapter focuses on the chemistry, synthetic methods, and bioactivity of different benzothiadiazine. Furthermore, the updates on patents and clinical trials have also been summarized in the present chapter.

DOI: 10.4018/979-8-3693-7520-4.ch010

1. INTRODUCTION

Heterocycles are crucial due to their biological, pharmacological, chemical and other essential activity (Bhowmik et al., 2024; Gadkari et al., 2021; Hatvate et al., 2020; Hatvate et al., 2018). The ring containing N, O, S and other heteroatoms has been proven effective therapeutically, agriculturally and in material sciences (Hatvate et al., 2024; Parit et al., 2024).

The benzothiadiazine skeleton is a prominent structure frequently found in numerous drugs and pharmaceuticals, including bendroflumethiazide, chlorothiazide, cyclothiazide, hydrochlorothiazide, and diazoxide. These compounds belong to the thiazide class of diuretics, which are widely employed in managing hypertension. Structurally, benzothiadiazine-1,1-dioxides are heterocyclic compounds that consist of a sulfur atom and two nitrogen atoms fused to a benzene ring. This distinctive arrangement imparts a broad spectrum of biological and pharmacological activities (Ittamalla et al., 2024; Khelili et al., 2012). Some of the benzothiadiazine ring-containing drugs are shown in **Figure 1.**

Figure 1. Some of the Benzothiadiazine ring-containing drugs

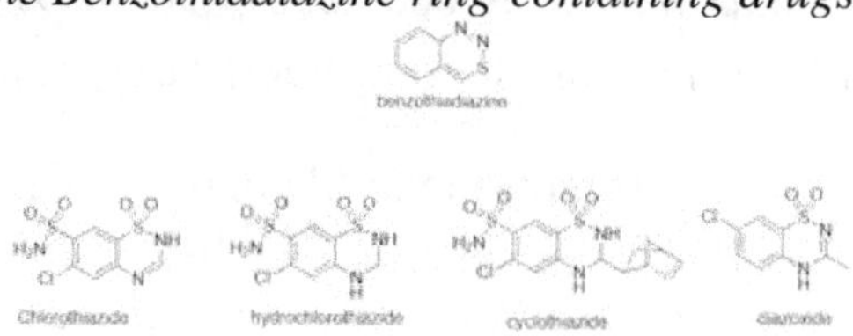

Benzothiadiazine derivatives are known for their diverse therapeutic effects, including antimicrobial, anti-cancer, antiviral, anti-diabetic, and antihypertensive properties. Furthermore, certain derivatives act as ATP-sensitive potassium channel openers, a mechanism that has significant implications in various physiological and pathological processes. Beyond these pharmacological roles, the cyclic sulfonamide or sultam moiety, part of the benzothiadiazine structure, is gaining increasing attention in medicinal chemistry due to its importance in drug design. This functionality enhances the interaction between the drug and biological targets, improving the efficacy and selectivity of these compounds (Francotte et al., 2010; Tedesco et al., 2006).

Overall, the benzothiadiazine scaffold has proven to be a versatile and valuable framework in the development of therapeutic agents, with continued research exploring its potential in treating a wide range of diseases. The following sections will delve deeper into the chemical properties, structure-activity relationships, and therapeutic applications of benzothiadiazine derivatives.

1.1 Isomers of Benzothiadiazine

The benzothiadiazine ring system, a notable heterocyclic structure, can exist in different isomeric forms depending on the substituent's positions and the arrangement of the atoms within the ring. These isomers primarily differ in the relative positions of the sulfur and nitrogen atoms within the thiadiazine ring and the placement of the dioxide group ($–SO_2$).

1,2-Benzothiadiazine (Benzothiadiazine-1,2-dioxide): In this isomer, the thiadiazine ring is fused to the benzene ring such that the sulfur and nitrogen atoms are adjacent to the carbon atoms of the benzene ring. The thiadiazine ring features the dioxide group ($–SO_2$), which plays a significant role in determining the chemical and biological properties of the molecule. An example of this isomeric form is chlorothiazide.

1,3-Benzothiadiazine (Benzothiadiazine-1,3-dioxide): In the 1,3-isomer, the fusion of the thiadiazine ring to the benzene ring occurs in a different orientation, where the positioning of the sulfur and nitrogen atoms, as well as the dioxide group ($–SO_2$), differs from the 1,2-isomer. This alternate configuration impacts the reactivity and biological activity of the molecule. A notable example of this isomer is hydrochlorothiazide.

The orientation of the dioxide group ($–SO_2$) within the thiadiazine ring significantly influences the electron distribution and reactivity of the molecule. This reactivity directly affects how the compound interacts with biological targets, potentially altering its efficacy in pharmaceutical applications. Isomeric forms may exhibit different solubility profiles in various solvents due to differences in the overall molecular structure. These solubility variations can influence the formulation and delivery of these compounds in pharmaceutical preparations, affecting bioavailability and pharmacokinetics. The stability of benzothiadiazine derivatives is also dependent on the isomeric form. Some isomers may be more prone to chemical degradation under certain environmental conditions, impacting their shelf life, stability during storage, and overall therapeutic efficacy. Isomerism can lead to significant changes in biological activity, as the spatial arrangement of atoms can affect the binding affinity of the molecule to its target receptors or enzymes. A change in isomeric form may alter the pharmacodynamics of the compound. In bioisosterism, even minor structural changes, such as the shift between 1,2- and 1,3-isomers, can lead to differences in biological response, including potency, selectivity, and side effects.

1.2 Chemistry of Benzothiadiazine

Benzothiadiazines are an essential class of heterocyclic compounds containing nitrogen and sulfur atoms within their ring structure. These compounds have been studied extensively due to their diverse biological activities, including their use as diuretics, anti-diabetic agents, and, in some cases, as potential anti-cancer compounds.

It is a heterocyclic ring with a benzene ring fused into a 1,2,4-thiadiazine ring, and the general formula for benzothiadiazines can be denoted as $C_7H_6N_2S$. The ring is aromatic, and the benzene ring provides aromatic stability to the molecule. The thiadiazine ring introduces electron-withdrawing effects due to the S and N atoms, affecting the electronic distribution within the molecule. Due to its unique ring structure, the ring shows good reactivity and can undergo various chemical reactions, including substitutions and condensations.

Benzothiadiazine ring is involved in a vast category of diuretics and antihypertensive agents for various complications; the figure shows some marketed products containing benzothiadiazine ring e.g. Chlorothiazide, hydrochlorothiazide, cyclothiazide, diazoxide (Gyűjtő et al., 2020).

The position of the dioxide group and other substituents on the benzene ring can influence the reactivity of the benzothiadiazine ring. For example, electron-withdrawing groups can make the compound more reactive toward nucleophiles or electrophiles.

The development of new compounds known as mono-, di-, and trialkyl-substituted 7-chloro-3,4-dihydro-*2H*-1,2,4-benzothiadiazine 1,1-dioxides, which are being studied for their potential as positive allosteric modulators of AMPA receptors. These receptors play a crucial role in fast excitatory signaling in the brain and are linked to cognitive functions such as learning and memory. The authors synthesized 16 new compounds and evaluated their effects on AMPA receptors using various assays. The results showed that certain compounds, particularly 7b, 7d, and 7e, exhibited strong activity as AMPA receptor potentiators, with effective concentrations ranging from 2.7 to 4.3 μM in fluorescence assays. Interestingly, the compound 7f, which contained a cyclopropyl group, was significantly less potent (EC 2× = 60 μM) compared to its 4-methyl analog, indicating an unexpected structure-activity relationship. The study also revealed that the presence of a methyl group at the 3-position of the benzothiadiazine scaffold influenced the binding mode and potency of the compounds, suggesting that modifications at this position could enhance their effectiveness as cognitive enhancers (Larsen et al., 2016).

2. SYNTHESIS OF BENZOTHIADIAZINE

The synthesis of benzothiadiazines typically involves the condensation of ortho-disubstituted aromatic amines (such as 2-aminobenzenesulfonamides) with carbon disulfide, followed by cyclization and oxidation. Various synthetic routes have been developed to introduce different substituents on the benzene ring or the thiadiazine ring to modify the biological activity of these compounds.

2.1 From Metal Catalyzed Reactions

Daoshan Yang and collaborators developed a method to synthesize nitrogen heterocycle-fused benzimidazole and 1,2,4-benzothiadiazine-1,1-dioxide derivatives. Their approach involved a one-pot synthesis technique, utilizing metal copper as a key catalyst and *L*-proline as an additional catalyst. The reaction mechanism follows the Ullmann-type coupling, a well-known process for forming carbon-nitrogen bonds.

To optimize the reaction conditions, the researchers systematically varied several parameters. They employed CuI as the catalyst and L-proline as the ligand. Sodium methoxide was used as the base, and the reaction was carried out using DMF as the solvent. The temperature was optimized at 130 °C. Under these conditions, the reaction proceeded efficiently, yielding the desired products with an impressive yield of 89%.

The proposed mechanism for the reaction begins with the interaction between L-proline and CuI, resulting in the formation of a five-membered ring complex. This intermediate undergoes oxidative addition with 2-bromo-N-phenylbenzenesulfonamide, leading to the formation of a copper-coordinated complex. The reaction is then treated with imidazole under basic conditions provided by cesium carbonate, leading to another intermediate complex's formation. This intermediate subsequently undergoes a double reductive elimination, releasing the metal complex and yielding the final benzothiadiazine-containing compound as shown in **Synthetic Sheme 1** (Yang et al., 2015).

Figure 2. Synthetic Scheme 1: Copper-catalyzed synthesis of benzothiadiazine ring

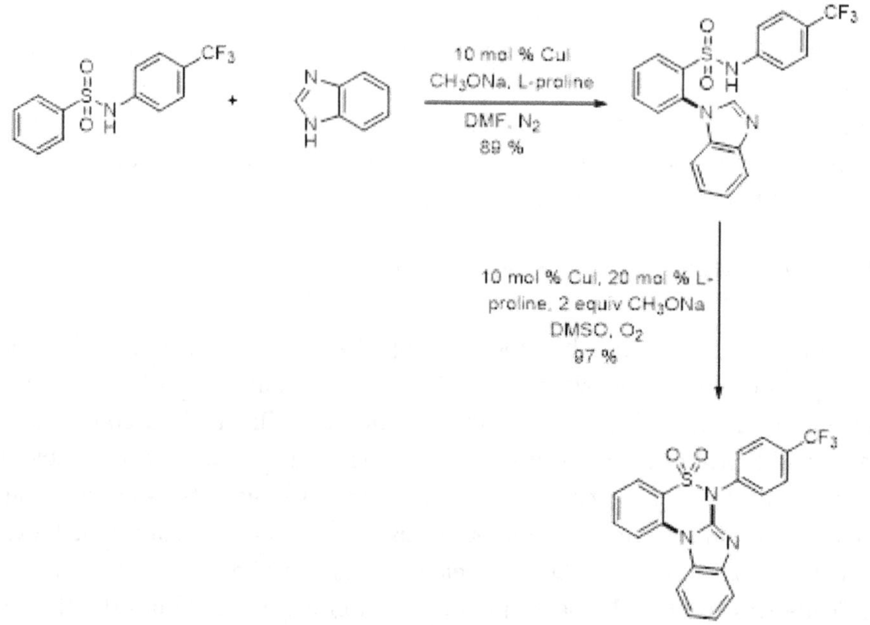

Peng Shi and colleagues synthesized a benzothiadiazine derivative through a rhodium-catalyzed reaction involving C-H amidation followed by cyclization to form a heterocyclic ring. The synthetic approach employed sulfoximines and dioxazolones as key substrates. The catalytic system comprised 3 mol% of $[Cp^*RhCl_2]_2$ (pentamethylcyclopentadienyl rhodium dichloride dimer) and 12 mol% of $AgSbF_6$ (silver hexafluoroantimonate), with 1,2-dichloroethane (DCE) serving as the solvent. Additionally, 1.0 equivalent of pivalic acid was used as an additive, significantly enhancing the reaction yield to 91%.

The proposed reaction mechanism begins with the interaction of the two catalysts, forming a cationic rhodium complex. This intermediate then reacts with the sulfoximine substrate, forming a five-membered ring with the elimination of HX. Subsequently, this rhodium complex interacts with dioxazolone, generating a rhodium nitrenoid intermediate. The nitrenoid undergoes a rearrangement to form a C-N bond, yielding the amidated product. The rhodium catalyst is regenerated in this process, facilitating a subsequent step where dehydrative ring cyclization occurs, ultimately producing the benzothiadiazine structure as given below in **Synthetic Scheme 2** (Shi et al., 2020).

Figure 3. Synthetic Scheme 2: Rhodium catalyzed synthesis of benzothiadiazine

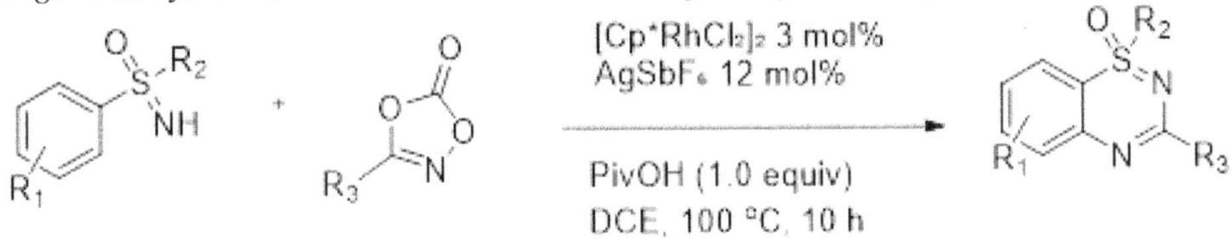

Ramsingh D. Padmaja et al. developed a method for synthesizing benzothiadiazine ring-containing compounds using a one-pot telescopic approach to achieve chemoselective synthesis. The study focused on synthesizing a series of compounds by substituting the benzo[e]pyrido/pyrazino/pyridazino[1,2,4]thiadiazine dioxide framework. The reaction scheme was carefully optimized by varying parameters such as solvent choice, temperature conditions, and reaction time. Optimal conditions were identified, yielding an impressive 90% product yield when acetonitrile was used as the solvent, with the reaction carried out at room temperature for 10 hours.

The synthetic process is based on a one-pot telescoped reaction, where 2-aminopyridine reacts with a substituted 2-chlorobenzene sulfonyl chloride under basic conditions using potassium carbonate (K_2CO_3) and acetonitrile as the solvent. The reaction is conducted at room temperature for 10 hours, forming an N-sulfonylated intermediate. This intermediate is then subjected to reflux conditions with the addition of 5 mol% of copper(I) iodide (CuI), forming benzothiadiazine-5,5-dioxides with excellent yields.

The proposed mechanism for this reaction involves a Ullmann-type C-N coupling catalyzed by CuI. The process begins with the nucleophilic attack of the pyridine nitrogen's lone pair on the sulfonyl group, eliminating HCl and forming the N-sulfonylated product. Subsequent in situ addition of CuI facilitates coordination of the copper metal, followed by a specific rearrangement, culminating in the removal of the copper atom and the formation of a six-membered benzothiadiazine ring as per **Synthetic Scheme 3** (Padmaja et al., 2019).

Figure 4. Synthetic Scheme 3: Ullmann-type coupling catalyzed by copper iodide

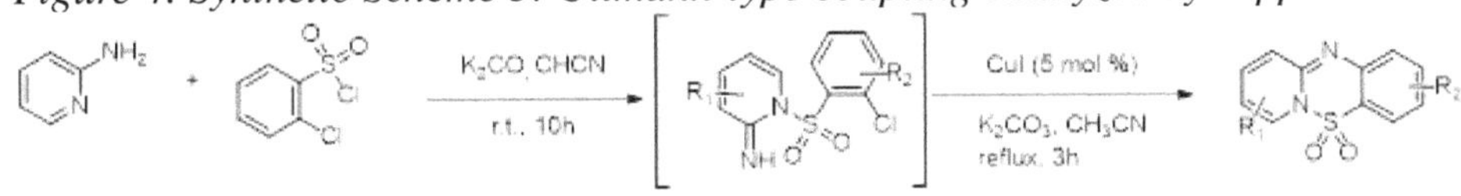

2.2 From Azido Sulphonamide

Rene Hommelsheim et al. synthesized 3-amino-1,2,4-benzothiadiazine 1,1-dioxide from amino sulfonamides, utilizing a palladium-catalyzed domino reaction involving 2-azido sulfonamides and isocyanides at room temperature for 12 hours. The reaction conditions were optimized by testing various parameters, with entry 15 in their study identified as the most effective. This entry employed $Pd(dba)_2$ as a pre-catalyst at a 1.5 mol% concentration, using DMF as the solvent, and reduced the reaction time to 2 hours while achieving an excellent yield of 95% for the 3-amino-substituted 1,2,4-benzothiadiazine 1,1-dioxides. In a modified approach by the same researchers, Cs_2CO_3 was added to the reaction mixture, conducted at 60 °C with a continuous supply of argon. Both processes required this inert atmosphere for optimal results. The proposed catalytic cycle for this domino reaction begins with the 2-azido sulfonamide forming a metal-ligand complex with the active palladium catalyst derived from the pre-catalyst. Coordination with isocyanide follows, leading to the elimination of nitrogen gas (N_2) and the formation of a palladium imido complex. This complex undergoes rearrangement, resulting in nitrene transfer and the formation of an intermediate product. The active palladium catalyst is then released as the intermediate decordinate to form carbodiimide, which subsequently undergoes intramolecular cyclization through nucleophilic attack, yielding the final product, either 3 or 5 in their respective manuscript as shown in **Synthetic Scheme 4** (Hommelsheim et al., 2024).

Figure 5. Synthetic Scheme 4: Synthesis of benzothiadiazine from azido sulfonamide catalyzed by palladium compound

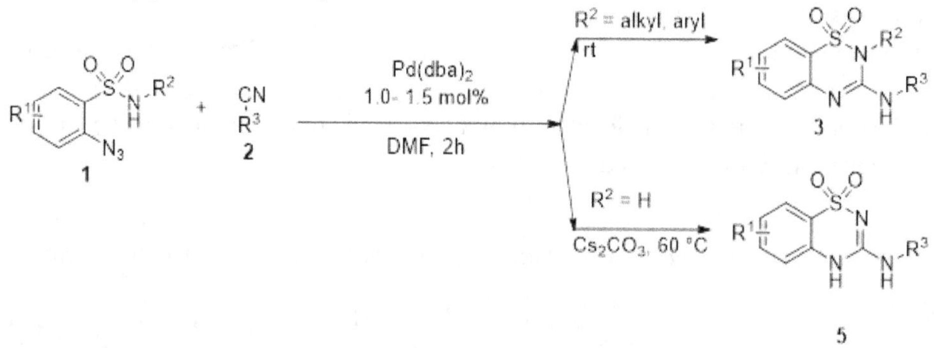

Christopher Blackburn et al. and colleagues synthesized 3-amino-1,2,4-benzothiadiazine 1,1-dioxides through an aza-Wittig/heterocumulene annulation of *o*-azidobenzenesulfonamides using resin-supported triphenylphosphine (TPP)

as shown in **Synthetic Scheme 5**, followed by a reaction with isocyanate, resulting in a high yield. The reaction parameters were optimized, and a series of derivatives were synthesized. The purity and yield of these derivatives varied depending on the reaction conditions. Among the series, compound 9 (product 5i) exhibited a particularly high yield of 85% (Blackburn et al., 2005).

Figure 6. Synthetic Scheme 5: Synthesis of benzothiadiazine from azido sulfonamide catalyzed by resin-supported triphenylphosphine

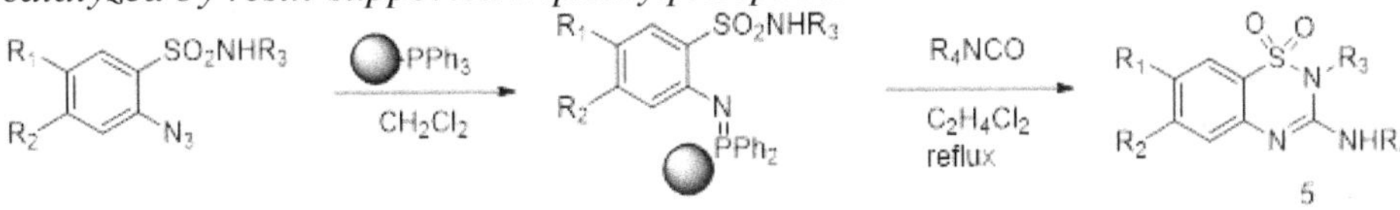

2.3 Through Hetero-Cyclization

Alexander A. Shalimov et al. and his team synthesized 1,2,4-benzothiadiazine 1,1-dioxides via the heterocyclization of N (chlorosulfonyl)imidoyl chlorides with anilines. The process began with the synthesis of imidoyl chloride, which was obtained from acetyl isocyanate and chlorosulfonic acid. Subsequent chlorination of this intermediate produced imidoyl chloride with a yield of 80%. The following hetero-cyclization reaction **Synthetic Scheme 6,** was catalyzed by either diisopropylethylamine or aluminium chloride. Through the variation of reaction parameters, the researchers successfully derived a long series of molecular derivatives of these compounds (Sinitsa et al., 2016).

Figure 7. Synthetic Scheme 6: Synthesis of benzothiadiazine through hetero-cyclization

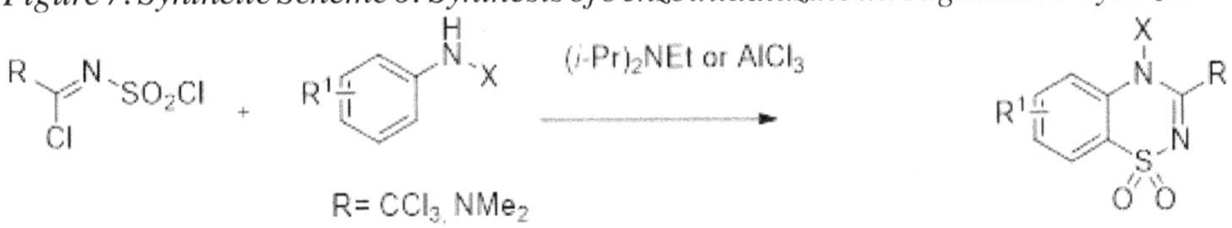

2.4 From Alkylation

Imre Gyűjtő et al. focuses on the synthesis of 4-methyl-*2H*-1,2,3-benzothiadiazine 1,1-dioxides and their transformations through alkylation and reduction processes. The authors describe how certain derivatives of these compounds exhibit pharmacological potential, particularly as positive allosteric modulators of the AMPA receptor, which is crucial for cognitive functions and learning processes. The synthesis involves several key steps, including ortho lithiation of acetophenone ketals, introduction of the chlorosulphonyl group, and subsequent ring closure with hydrazine monohydrate or acetohydrazide to form variously substituted benzothiadiazine dioxides. The paper also highlights the successful N(2)-alkylation and reduction of the C=N double bond, leading to the formation of *N*-(2)-alkyl-4-methyl-3,4-dihydro derivatives, which were further modified through reductive alkylation with aldehydes to achieve N(3)-alkylation.

Figure 8. Synthetic Scheme 7: Synthesis of benzothiadiazine by alkylation

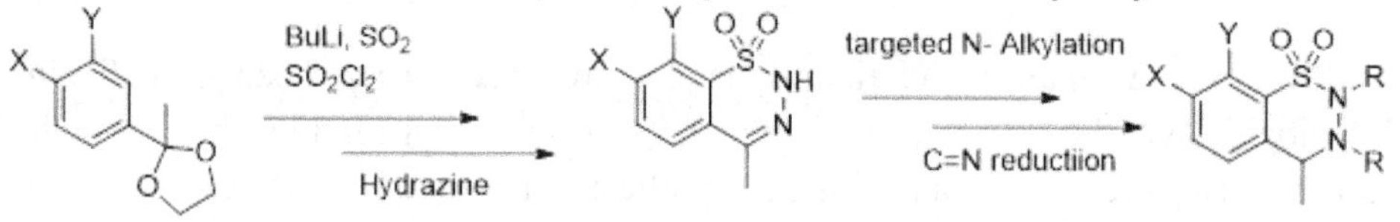

The synthesis method begins with the ortho lithiation of acetophenone ketals, followed by introducing a chlorosulphonyl group. This is achieved by treating the lithiated ketals with sulphur dioxide and sulphuryl chloride. The resulting intermediates undergo ring closure with hydrazine or acetohydrazide, yielding the desired benzothiadiazine dioxides. Further transformations include N(2)-alkylation using alkyl halides in the presence of t-BuOK, and reduction of the C=N bond using sodium borohydride in trifluoroacetic acid, resulting in various alkylated derivatives shown in **Synthetic Scheme 7** with significant biological activity (Gyűjtő et al., 2019).

2.5 From Direct Arylation

Wright J et al. describe the diazotization of 2-aminobenzophenones (compound 6) as treating the amino group with nitrous acid and converting it into a diazonium salt. This reactive intermediate is then subjected to a reaction with sulfur dioxide in the presence of copper(II) chloride, which facilitates the formation of *o*-benzoylbenzenesulfonyl chlorides (compound 7). These sulfonyl chlorides are key intermediates and undergo a cyclization reaction upon treatment with hydrazine. This process leads to forming 4-aryl-substituted target compounds (compound 5), completing the synthesis of the desired heterocyclic products, as shown in **Synthetic Scheme 8**This method provides an efficient route to synthesizing complex aryl-substituted derivatives (Wright, 1968).

Figure 9. Synthetic Scheme 8: Synthesis of benzothiadiazine through direct arylation

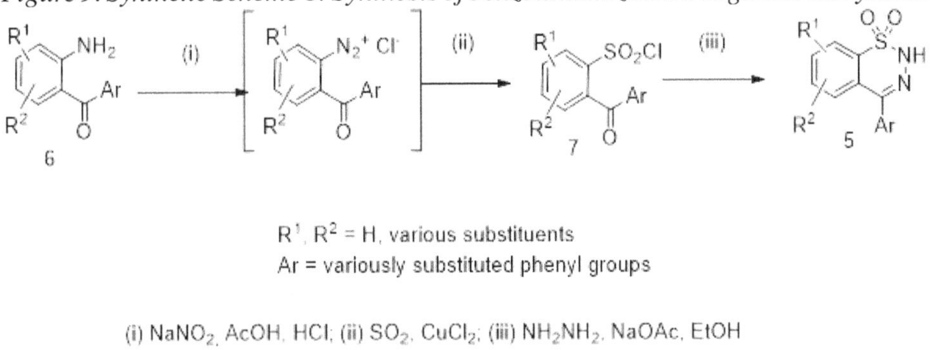

2.6 From Asymmetric Solid Phase

Peng Du et al. conducted the asymmetric synthesis of 3,4-dihydro-2*H*-1,2,4-benzothiadiazine-1,1-dioxides utilizing scandium(III)-inda-pybox as the catalyst. The reaction involved the intermolecular condensation and intramolecular asymmetric amine addition cascade sequences of 2-aminobenzenesulfonamide and aldehydes. Through extensive optimization of reaction parameters, they determined that compound 20 achieved the highest yield of 81%. This reaction was catalyzed by the Lewis acid $Sc(OTf)_3$ in DCM as the solvent under a temperature of 20 °C, given below in **Synthetic Scheme 9.**

The reaction produced a mixture of R and S enantiomers, focusing on obtaining the R enantiomer of compound 3a. The results demonstrated that compound 3a was obtained with a 92% enantiomeric excess, indicating a high degree of enantiomeric purity (Du et al., 2016).

Figure 10. Synthetic Scheme 9: Scandium/Pybox catalyzed asymmetric synthesis of benzothiadiazine

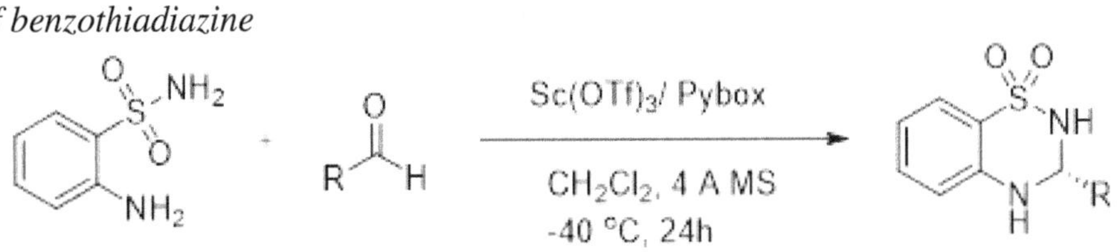

2.7 From Two-Pot Asymmetric Synthesis

The α-aminonitrile intermediates generated through multicomponent (MC) Strecker strategies have proven to be highly versatile, serving as valuable building blocks for synthesizing more complex and structurally interesting compounds. Notably, Fitch and his colleagues demonstrated the utility of these intermediates by developing an efficient two-pot synthesis method for the asymmetric preparation of benzothiadiazine derivatives for example given in **Synthetic Scheme 10**. This approach highlights the potential of α-aminonitrile intermediates in constructing advanced heterocyclic frameworks, which are critical for pharmaceutical chemistry and drug design applications (Fitch et al., 2005; Marqués-López et al., 2015).

Figure 11. Synthetic Scheme 10: Two-pot asymmetric synthesis of benzothiadiazine

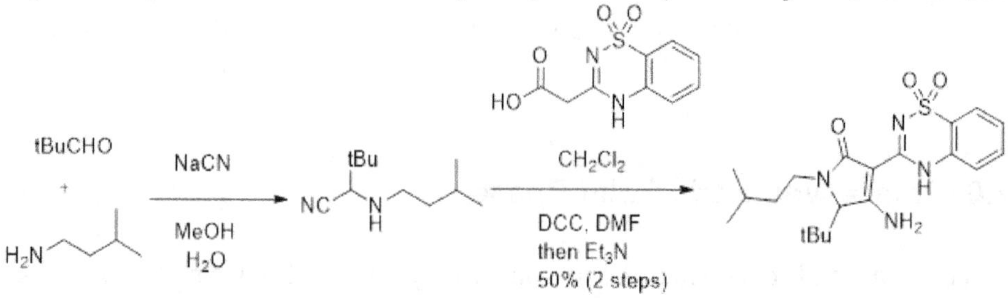

2.8 From Three-Component Synthesis

Jinho Kim et al. and colleagues successfully synthesized the benzothiadiazine ring using a three-component coupling approach. This method involved the intramolecular N-arylation of an intermediate, which was generated through the reaction of ammonium chloride, terminal alkynes, and sulfonyl azide in the DCM as solvent and TEA used as base. The entire process was catalyzed by copper, with copper playing a critical role in the three-component coupling and the subsequent cyclization step, facilitating the formation of the benzothiadiazine ring shown in **Synthetic Scheme 11.**The product obtained was purified by flash chromatography, and the obtained product was a white solid with 90% purity (Kim et al., 2008).

Figure 12. Synthetic Scheme 11: Three-component synthesis of benzothiadiazine

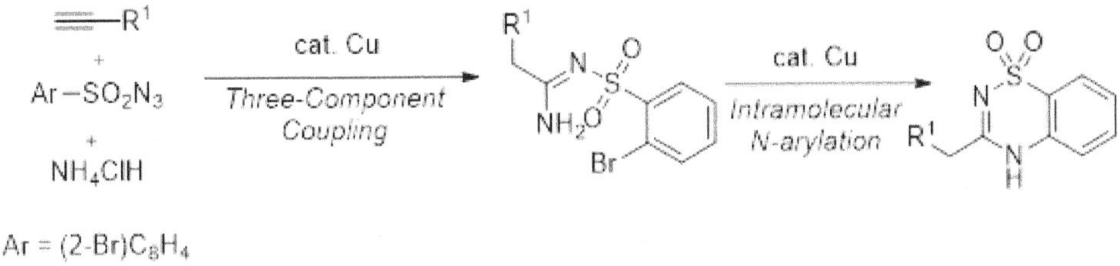

3. CURRENT MARKETED DRUGS OF BENZOTHIADIAZINE

Table 1. Currently marketed drugs of Benzothiadiazine

Drug name	Class	Indication	Action
Hydrochlorothiazide	Thiazide diuretic	Hypertension, edema	Reduces blood pressure by inhibiting sodium reabsorption in the distal tubules of the kidney, promoting water excretion.
Chlorothiazide	Thiazide diuretic	Hypertension, heart failure, edema	Similar to hydrochlorothiazide; inhibits sodium reabsorption in the distal tubules.
Bendroflumethiazide	Thiazide diuretic	Hypertension, edema	Acts by reducing blood volume, decreasing vascular resistance, and promoting diuresis.
Cyclothiazide	Thiazide-like diuretic	Hypertension, sometimes used experimentally for neuroprotection	Blocks sodium-chloride transport in the nephron.
Diazoxide	Potassium channel opener (non-diuretic)	Hypertensive emergencies, hyperinsulinism	Opens ATP-sensitive potassium channels, leading to relaxation of vascular smooth muscle and reduction of insulin secretion.
Methyclothiazide	Thiazide diuretic	Hypertension, edema	Inhibits sodium reabsorption and increases excretion of sodium and water.

The **Table 1** gives the information about currently marketed drugs containing Benzothiadiazine- ring, which has gibven along with its class, indication and action.

4. BIOLOGICAL ACTIVITIES OF BENZOTHIADIAZINE

4.1 Diuretic Activity

Benzothiadiazine derivatives have been very commonly used as diuretics in day-to-day life for a very long time. 1,2,4-benzothiadiazine-1,1-dioxide derivatives like chlorothiazide, bendroflumethiazole and hydrochlorothiazide are used in hypertension treatment (Hu et al., 2021). Marketed drugs for the scaffold are chlorothiazide, hydrochlorothiazide, cyclothiazide and diazoxide are shows diuretic and antihypertensive property (Gyűjtő et al., 2020).

Ole B. Tvaermose Nielsen et al. synthesized the benzothiadiazine diuretics, such as 3-benzylamino-4-methyl-5-sulfamylbenzoic acid, are characterized by their unique ring structure, which contributes to their pharmacological properties. Diuretics exhibit significant diuretic effects, often superior to other classes like benzisothiazoles. For instance, the 4-arylmethyl-5-sulfamylbenzoic acid derivatives showed excellent potency after both intravenous (iv) and oral administration. The diuretic activity is influenced by structural modifications. Compounds with a benzyl substituent at the 4-position tend to be more effective than those with phenylethyl groups. Specific side chains, such as n-butylamino, enhance oral absorption and overall efficacy. In studies comparing various derivatives, benzothiadiazine 79 was found to be devoid of diuretic activity, highlighting the importance of structural features in determining efficacy. The synthesis and screening of these compounds have been crucial in understanding their diuretic properties, leading to the identification of promising candidates for further investigation (Nielsen et al., 1973).

4.2 Anti-Cancer Activity

Sumit Chhabra et al. synthesized the derivatives of 1,2,4-benzothiadiazine-1,1-dioxides, showing anti-cancer activity as they screened it against 60 different human tumour cells, where they synthesized a total of 10 molecules to check their cytotoxic activity and from that BTZ-4 shows more potential to act as anti-cancer (Chhabra et al., 2023).

Thoukhir B Shaikh et al. evaluated the quinazoline and quinazolino-benzothiadiazole derivatives for their anti-cancer and anti-mitotic activity against a panel of five cancer and one regular cell line. They got the results as fifteen of the synthesized molecules showed a significant effect against them with a growth inhibitory effect of 0.15-5.0 μM, and the molecules successfully gained induced cell cycle arrest in the G2/M phase. It has been concluded that the tubulin polymerization % inhibition rate of the ATS-7 named compound is comparatively higher than others compared with the standard nocodazole (Shaik et al., 2020).

Badar I. Huwaimel et al. synthesized a series of halogenated benzothiadiazine derivatives and evaluated their anti-cancer activity by targeting mitochondrial respiratory complex II. The study focused on compounds with higher inhibition of complex II as potential anti-cancer agents. The percentage inhibition of complex II at a concentration of 100 μM was found to be 81±6.9% for compound 12b, 55±15.2% for compound 12c, 64±15.3% for compound 20c, 51±13.9% for compound 24b, and 55±8.2% for compound 24c these compounds are given in

Figure 13. These values were compared with known inhibitors, diazoxide and atpenin A5.

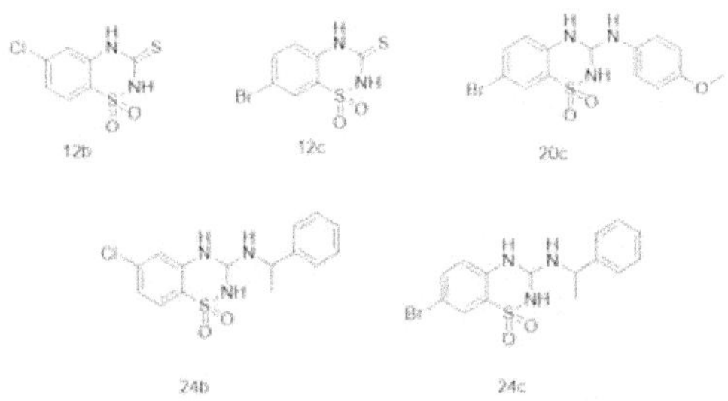

Figure 13 Benzothiadiazine ring-containing compounds showing anti-cancer activity.

Additionally, the compounds were screened against 22Rv1 prostate cancer cells and MDA-MB-468 triple-negative breast cancer cells. The activity of these derivatives was also compared with 5-fluorouracil, showing that the derivatives exhibited more potent anti-cancer activity than 5-fluorouracil (Huwaimel et al., 2021).

4.3 Antimicrobial Activity

Benzothiadiazine derivatives have demonstrated potential antibacterial activity against certain Gram-positive and Gram-negative bacteria. Research has substantiated that these derivatives possess antibacterial properties.

Ahmed Kamal et al. synthesized a series of 10-substituted 5,5-dioxo-5,10-dihydro[1,2,4]triazolo[1,5-*b*][1,2,4]benzothiadiazine aryl sulfonamide derivatives. The structures of these compounds were confirmed through spectral data analysis, elemental analysis, and various quantum chemical calculations. The series includes compounds 10a-j and 13a-f. Among these, compounds 10f, 10i, and 13e demonstrated activity against E. coli, while compounds 13a, 13d, and 13e were active against *B. subtilis* shown in **Figure 14** (Kamal et al., 2007).

Figure 14. Benzothiadiazine ring-containing compounds showing antimicrobial activity 1

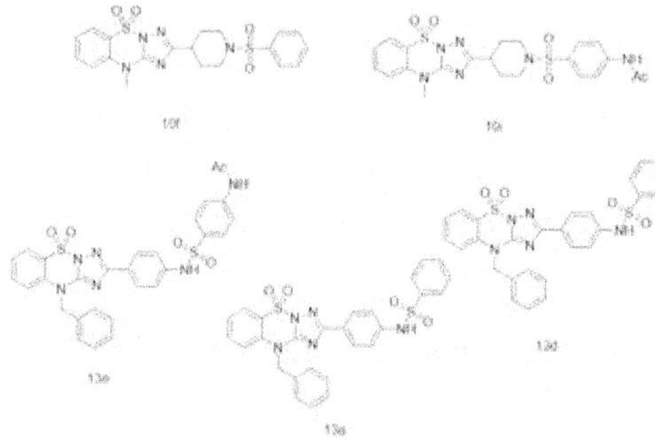

Nilay Bhatt et al. synthesized a novel series of *N*-alkyl-2-(3,5-dimethyl-1,1-dioxido-*2H*-1,2,6- thiadiazin-4-yl)benzamides by the coupling of the thiadiazo derivatives with different amine derivatives in the presence of carbonyldiimidazole (CDI). Bioassay results for the synthesized series 2a-j indicate that all compounds exhibit notable activity against bacterial strains. This family of molecules is particularly active against Gram-positive bacterial species such as *S. aureus* and *B. subtilis*. Among the series, compounds 2c, 2e, and 2g shown in **Figure 15,** show the most significant results, with zones of inhibition against *S. aureus* measuring 30 mm, 28 mm, and 21 mm, respectively, and against *B. subtilis* measuring 27 mm, 28 mm, and 23 mm, respectively. These results were compared with the antibiotic ciprofloxacin (Bhatt et al., 2013).

Figure 15. Benzothiadiazine ring-containing compounds showing antimicrobial activity 2.

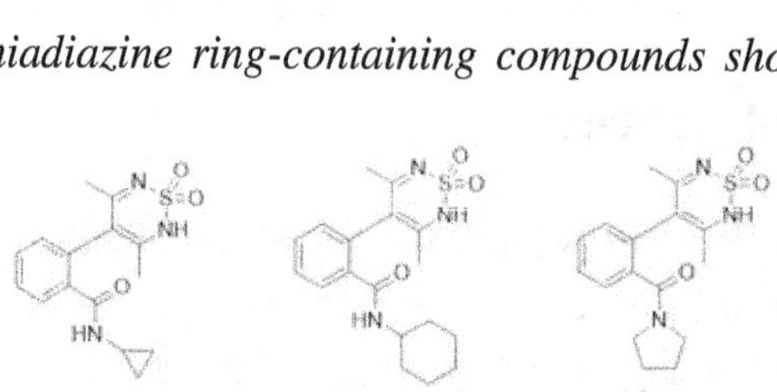

4.4 Anti-Tubercular Activity

Ahmed Kamal et al. synthesized the derivatives of benzothiadiazine-1,1-dioxide and screened against *mycobacterium tuberculosis* as well as bacteria and fungi. They synthesized a series 10a-f, 11a-c, and 12a-f, whose structural characteristics were confirmed through NMR and other spectral analyses. In vitro screening of these derivatives revealed that compound 10c exhibits a promising minimum inhibitory concentration (MIC) value of 2 μg/mL. Compounds 10b, 10e, and 12b given in **Figure 16** also display MIC values of 4 μg/mL, indicating effective bacterial inhibition. These results were compared with the antibiotic rifampicin. Compound 4 demonstrated the highest efficacy in the study, exhibiting a minimum inhibitory concentration (MIC) of 1 μg/mL. This was determined through in vivo studies against the *M. tuberculosis* strain (Kamal et al., 2010).

Figure 16. Benzothiadiazine ring-containing compounds showing anti-tubercular activity 1.

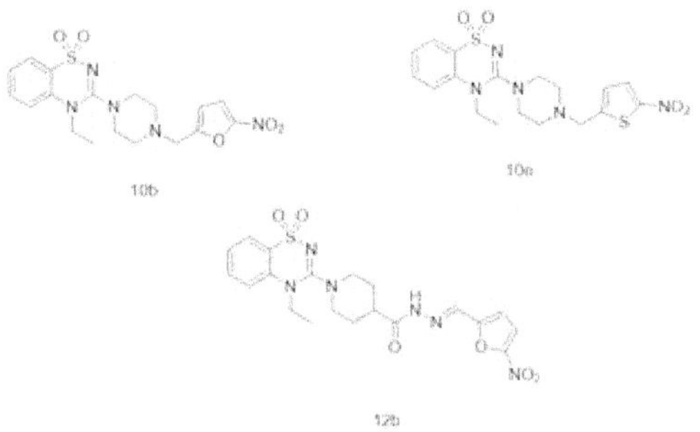

Narender Addu et al. synthesized 1,2,4-benzothiadiazine-1,1-dioxide derivatives using sulfonic acid-functionalized Wang resin. These compounds were screened against *Mycobacterium tuberculosis* H37Rv strains through in silico and in vitro studies. The synthesized series, 3a-r, showed that compounds 3c, 3d, and 3e from **Figure 17** exhibited significant interactions with the bacteria, as confirmed by the performed studies (Addu et al., 2023).

Figure 17. Benzothiadiazine ring-containing compounds showing anti-tubercular activity 2

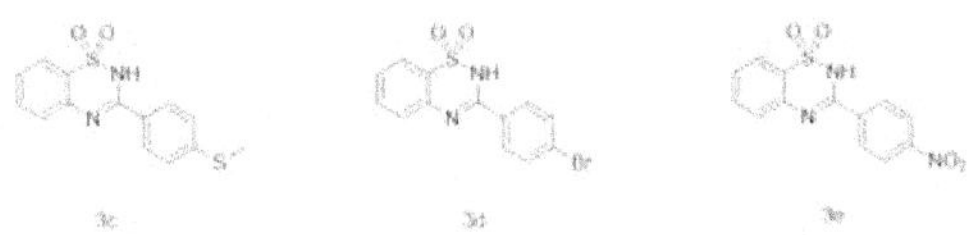

4.5 Antifungal Activity

L Fengyun et al. designed and synthesized novel 3,4-dihydro-*2H*-1,2,4-benzothiadiazine 1,1-dioxide-based strobilurins. A total of 27 compounds from series 7a-aa were synthesized and screened against various fungi, including Cercospora arachidicola, Alternaria solani, Physalospora piricola, Gibberella zeae, and five other fungal species. In vitro studies demonstrated that the synthesized compounds exhibited effective fungicidal activity compared to azoxystrobin. The initial assessment of fungicidal action at 50 µg/mL revealed that compounds 7b and 7k had superior activity compared to others and were selected for further bioassay in greenhouse conditions (*in vivo*), with comparisons to azoxystrobin and trifloxystrobin. The *in vivo* study, conducted at 300 µg/mL, indicated that compound 7b exhibited excellent fungicidal activity, whereas compound 7k showed moderate activity against some of the fungi, structures of these superiar compounds given in **Figure 18** (Li et al., 2020).

Figure 18. Benzothiadiazine ring-containing compounds showing antifungal activity 1

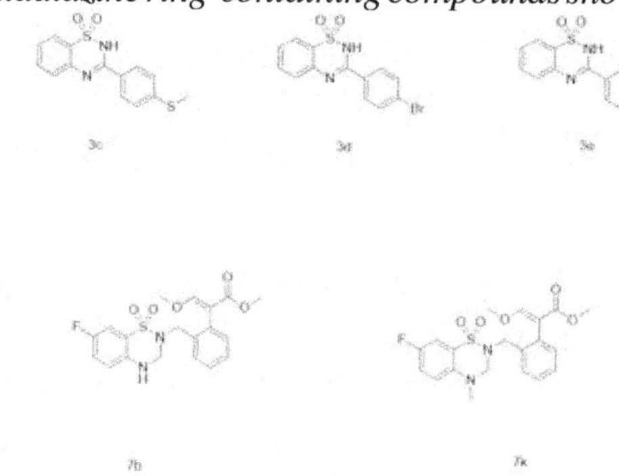

Ahmed Kamal and colleagues synthesized a series of compounds derived from benzothiadiazole-1,1-dioxide. These compounds were evaluated for their antifungal activity using a zone of inhibition assay against Candida albicans. The inhibition zones were measured at 100 µg/mL and 50 µg/mL concentrations, with amphotericin B (AM-B) serving as the reference standard. Among the synthesized compounds, 10a, 10d, 11c, 12b, 12c, 12d, 12e, and 12f of **Figure 19** demonstrated significant antifungal activity compared to AM-B. A subset of compounds was selected for further studies (Kamal et al., 2010).

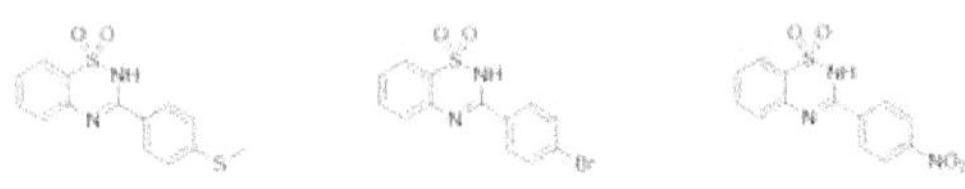

Figure 19. Benzothiadiazine ring-containing compounds showing antifungal activity 2

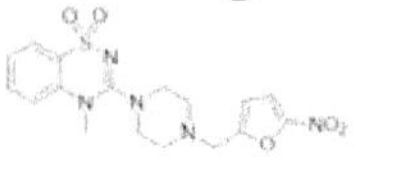

10a

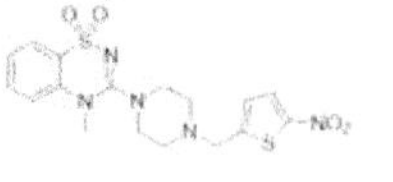

10d

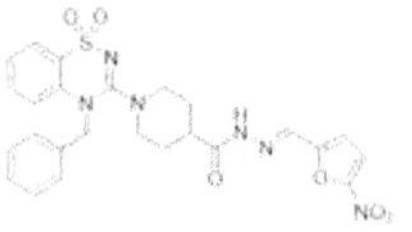

12c

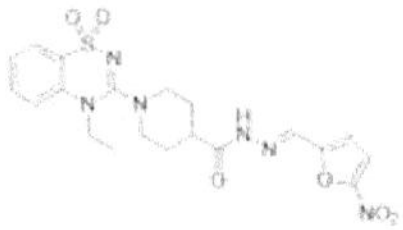

12b

4.6 Anti-Inflammatory Activity

L. Costantino et al. conducted a study in which they synthesized four derivatives of 2,6-bis-(1,1-dimethylethyl)phenol from molecules 3 and 4 containing a benzothiadiazine ring. These derivatives were then evaluated for their anti-inflammatory properties, specifically against carrageenan-induced paw oedema in rats. The anti-inflammatory activity of these compounds was compared to the standard drugs indomethacin and vitamin E, which are well-established anti-inflammatory agents. Among the synthesized compounds, molecule 4 exhibited significant anti-inflammatory effects. When administered at a 5 mg/kg dose, compound 4 led to a 40% increase in paw oedema after 8 hours, with a standard deviation of ±4.9%. This suggests that compound 4 demonstrates promising anti-inflammatory potential compared to the controls show in **Figure 20**, with results notably close to the reference anti-i(*Antiinflamatory.Pdf*, n.d.)udy (*Antiinflamatory.Pdf*, n.d.).

Figure 20. Benzothiadiazine ring-containing compounds showing anti-inflammatory activity

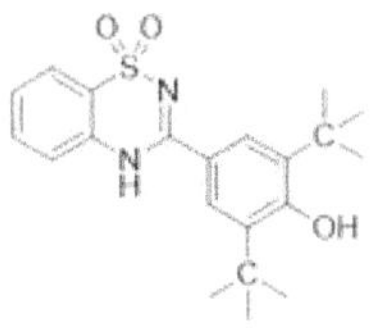

3

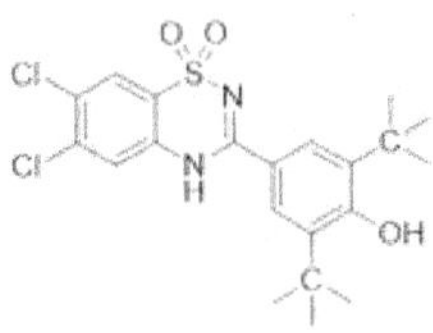

4

4.7 Antiviral Activity

Peng Zhan et al. performed the antiviral activity of benzo/heterothiadiazine dioxides, particularly thiadiazine derivatives, have demonstrated significant antiviral activity against various viruses, including HCMV, VZV, HCV, and HIV. The reverse transcriptase (RT) of HIV-1 is a primary target for antiviral drugs, with NNRTIs showing higher safety and selectivity compared to other inhibitors. Some derivatives, such as 1-[(2-Acetoxyethoxy)methyl]-4-amino-6-benzyl-(5H)-imidazo[4,5-c]-1,2,6-thiadiazine 2,2-dioxide, exhibit activity against both HIV-1 and HIV-2, indicating their potential as new leads for anti-HIV agents. Compounds like Compound 38 and Compound 39 from **Figure 21** have shown potent inhibition of HCV polymerase NS5B, with IC_{50} values below 5 nM and 1.7 nM, respectively, showcasing their effectiveness. While some compounds displayed minor activity, the research highlights that structural modifications of these derivatives could lead to more potent antiviral agents. The study emphasizes the necessity for further optimization to enhance their efficacy against drug-resistant strains (Zhan et al., 2008).

Figure 21. Benzothiadiazine ring-containing compounds showing antiviral activity

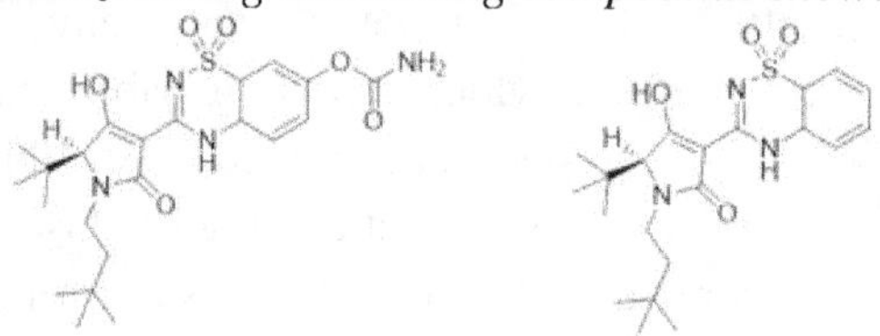

38 39

4.8 Antihypertensive Activity

Ashley E. Thomson et al. provide clinical observation on the antihypertensive activity of diazoxide, a benzothiadiazine derivative, which has been demonstrated through various studies. Upon intravenous administration, diazoxide effectively lowered blood pressure in hypertensive subjects, with reductions averaging 26/16 mm Hg while simultaneously increasing cardiac output, which ranged from 0.75 to 5.71 litres per minute. This indicates that diazoxide primarily reduces peripheral vascular resistance rather than negatively affecting cardiac output. Notably, the drug's antihypertensive effects were not dependent on sodium depletion or reduced intravascular volume, as evidenced by the maintenance of blood pressure reduction even in the presence of positive sodium and water balance. In clinical trials, oral doses of diazoxide (0.2 to 0.5 g per day) resulted in significant blood pressure reductions for periods ranging from eight to 50 weeks in 26 out of 30 hypertensive subjects. However, weight gain of 1 to 5 kg was observed in most patients, alongside some experiencing peripheral edema. The addition of chlorothiazide (0.5 g/day) after diazoxide treatment led to diuresis without further lowering blood pressure, highlighting the complex interplay between these medications. Overall, while diazoxide is effective in managing hypertension, its long-term use may be complicated by side effects such as salt retention and gastrointestinal disturbances, from the article compounds showing antihypertensive activity are given in **Figure 22** (THOMSON et al., 1962).

Figure 22. Benzothiadiazine ring-containing compounds showing antihypertensive activity

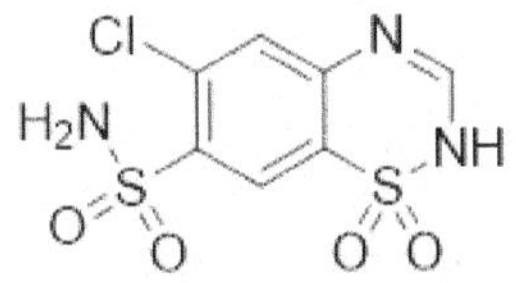

4.9 Antioxidant Activity

Kamel Harrouche et al. synthesized a series of sulfonyl thiourea moieties containing 1,2,4-benzothiadiazine-1,1-dioxide derivatives and conducted structural analyses. Antioxidant studies were performed and compared with ascorbic acid and quercetin, including DPPH assay, ABTS assay, H_2O_2 scavenging capacity, and reducing power assay. Derivative 12a exhibited significant antioxidant activity with EC_{50} values of 0.49 ± 0.03 mM in the DPPH assay and 0.39 ± 0.03 mM in the ABTS assay. In the reducing power assay, compounds 12i and 12d demonstrated the highest activity. In contrast, in the H_2O_2 scavenging capacity assay, compounds 12b and 12f showed superior antioxidant activity, nine times more potent than ascorbic acid. Compounds 12a, 12b, and 12i from **Figure 23** exhibited notable antioxidant properties (Harrouche et al., 2019).

Figure 23. Benzothiadiazine ring-containing compounds showing antioxidant activity

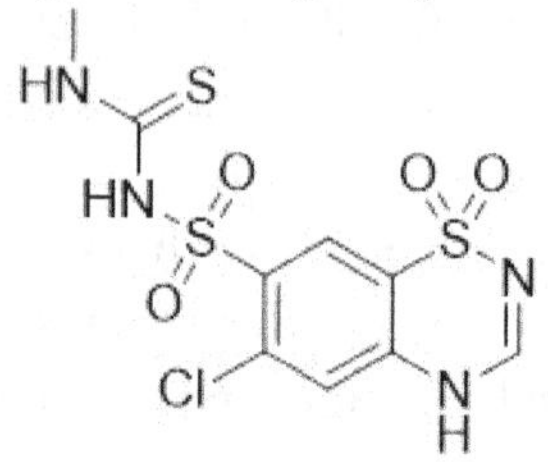

4.10 Anti-Diabetic Activity

Xin Chen and colleagues synthesized a series of acetic acid derivatives of 3,4-dihydro-*2H*-1,2,4-benzothiadiazine 1,1-dioxide, which function as aldose reductase inhibitors and exhibit anti-diabetic properties. The anti-diabetic efficacy of these derivatives was assessed through in vitro studies targeting the enzymes ALR1 and ALR2. Among the synthesized compounds, 9i, 9j, and 9m demonstrated promising inhibitory activity, with IC50 values against ALR2 of 0.066 μM (0.054-0.077 μM), 0.041 μM (0.034-0.048 μM), and 0.032 μM (0.024-0.040 μM), respectively. These values are considered competitive compared to fidarestat, a known aldose reductase inhibitor, which exhibits an IC50 value of 0.029 μM (0.020-0.037 μM) against ALR2. Furthermore, the *in vivo* efficacy of these compounds was evaluated in diabetic rats, where they demonstrated significant ALR2 inhibitory activity. Specifically, compounds 9i, 9j, and 9m were shown to reduce elevated sorbitol levels in the sciatic nerve of diabetic rats by 68%, 61%, and 62%, respectively. The corresponding nerve sorbitol concentrations for compounds 9i, 9j, and 9m were measured at 0.1378 ± 0.0154 μmol/L, 0.1455 ± 0.0114 μmol/L, and 0.1444 ± 0.0240 μmol/L. These results are notably close to the efficacy of fidarestat, which achieved a sorbitol concentration of 0.1039 ± 0.0105 μmol/L in the same experimental setup, compounds showing promising activity are given in **Figure 24** (Chen et al., 2010).

Figure 24. Benzothiadiazine ring-containing compounds showing anti-diabetic activity

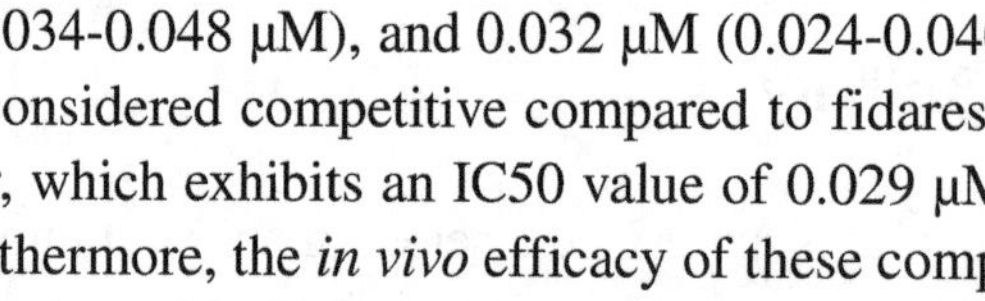

4.11 Antileishmanial Activity

Daisy K. Mangwegape et al. synthesized a series of benzothiadiazine-1,1-dioxide derivatives and evaluated their antileishmanial activity in vitro. These derivatives, varying based on their linkages to the active group, were tested against *Leishmania donovani* 15, *Leishmania donovani* 9515, and Leishmania major IR-173. Their efficacy was compared to amphotericin B, a standard antileishmanial agent. The derivatives exhibited varying degrees of inhibitory activity, with compounds 1, 2b, 2c, and 2d showing the highest activity against *L. donovani* 15, with IC50 ± SD values of 7.33 ± 0.92 μM, 9.82 ± 2.34 μM, 6.52 ± 0.35 μM, and 10.15 ± 2.57 μM, respectively. Against *L. donovani* 9515, compounds 2a, 2b, and 2c demonstrated better activity with IC50 ± SD values of 12.03 ± 2.81 μM, 12.48 ± 2.99 μM, and 15.67 ± 3.68 μM, respectively. For L. major IR-173, compounds 1e, 2b, and 2d were most effective, with IC50 ± SD values of 0.20 ± 0.00 μM, 0.40 ± 0.10 μM, and 0.41 ± 0.00 μM, respectively. Compounds showing antileishmanial activity are given in **Figure 25** Cytotoxicity testing on Vero cells revealed that none of the compounds exhibited significant cytotoxic effects when compared to the standard value of emetine, indicating a favourable safety profile (Mangwegape et al., 2021).

Figure 25. Benzothiadiazine ring-containing compounds showing antileishmanial activity

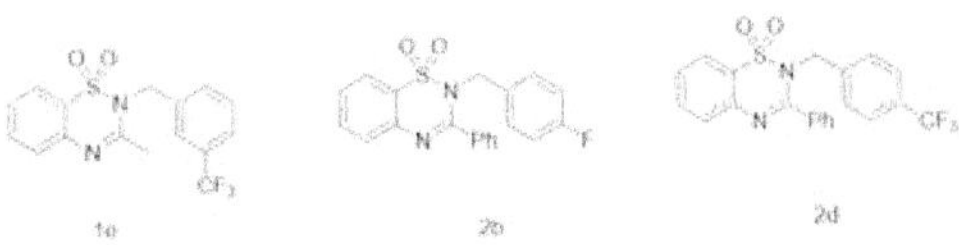

1e 2b 2d

Nadine Henning et al. synthesized a series of benzothiazidazine-1,1-dioxide ethylene glycol derivatives and evaluated their in vitro efficacy against Leishmania species. A total of 12 compounds were synthesized and screened for antipromastigote activity and cytotoxicity. Among these compounds, compound 9 demonstrated superior antileishmanial activity with an IC50 of 153.40±7.92 μM, compared to the standard antileishmanial agent amphotericin B. Compound 1 exhibited the highest cytotoxicity with an IC50 of 323.10±215.24 μM compared to the standard cytotoxic agent emetine (Henning et al., 2023).

5. PATENT UPDATES ON BENZOTHIADIAZINE

Table 2. Patent updates on Benzothiadiazine

Patent ID	Title	Assignee	Inventor/ Author	Grant Date
US20150111896A1	Hematopoietic protection against chemotherapeutic compounds using selective cyclin-dependent kinase 4/6 inhibitors	University of North Carolina at Chapel Hill	Norman Edward Sharpless, Jay C. Strum, John E. Bisi, Patrick J. Roberts, Matthew R. Ramsey	23/4/2015
US11505572B2	Therapeutics for the treatment of glaucoma	Mayo Foundation for Medical Education and Research University of Minnesota	Michael P. Fautsch, Peter Dosa, Michael A. Walters	22/11/2022
WO2008045558A2	Benzothiadiazine compounds and their use	Epix Delaware, Inc.	Merev Fichman, Rosa E. Melendez, Pradyumna Mohanty, Jian Lin, Dilara Mccauley	17/4/2008
CN107417641B	Benzothiadiazine derivative and preparation method and application thereof	Shandong University		31/3/2020
US2809194A	Thiadiazine type natriuretic agents			8/10/1957
US7396829B2	Nitric oxide enhancing diuretic compounds, compositions and methods of use	Nicox SA	David S. garvey, L. Gordan Letts, Richard A. EARL, Maiko Ezawa, Xinqin Fang, Ricky d. Gaston.	8/7/2008
US4559340A	Antihypertensive agents	Merck Sharp and Dohme Corp	Bernard R. Neustadt, David R. Andrews, Paul E. Mcnamara	17/12/1985

Some of the patents of benzothiadiazine ring-containing compounds have been given in **Table 2.**

6. CONCLUSION AND FUTURE PERSPECTIVE

In conclusion, benzothiadiazine compounds, as sulfur- and nitrogen-containing heterocycles, have proven to be versatile scaffolds in medicinal chemistry, playing a central role in various marketed drugs, particularly as diuretics and antihypertensive known for their presence in many marketed drugs, such as bendroflumethiazide, chlorothiazide, cyclothiazide, hydrochlorothiazide, and diazoxide, primarily recognized for their diuretic and antihypertensive properties. The chapter covers diverse synthetic methods, including metal-catalyzed reactions, hetero-cyclization, direct arylation, alkylation, and three-component synthesis, allowing for the creation of benzothiadiazine derivatives with tailored biological properties. These compounds exhibit a wide range of pharmacological activities, such as anti-cancer, antimicrobial, antifungal, antihypertensive, and anti-diabetic effects, among others. Recent advances in patents reflect ongoing interest in optimizing these compounds for enhanced therapeutic potential. Looking ahead, benzothiadiazines hold promise in targeted drug design, multifunctional drug development, and green chemistry applications, with future research likely to expand their role in treating various conditions, including cancer, diabetes, and neurodegenerative diseases.

REFERENCES

Addu, N., Miriyala, H., Kapavarapu, R., Kolli, S. K., & Pal, M. (2023). Wang-OSO3H catalyzed one-pot sonochemical synthesis of 1,2,4-benzothiadiazine-1,1-dioxide derivatives: Their in silico / in vitro assessments against MtbCM. *Journal of Molecular Structure*, 1283, 135313. DOI: 10.1016/j.molstruc.2023.135313

Bhatt, N., Vyas, K., Nimavat, K., Govender, T., Kruger, H. G., Maguire, G. E. M., & Bhatt, P. (2013). Synthesis of novel N-alkyl-2-(3,5-dimethyl-1,1-dioxido-2H-1,2,6-thiadiazin-4-yl)benzamides as potential antimicrobial agents. *Journal of Pharmacy Research*, 6(3), 355–360. DOI: 10.1016/j.jopr.2013.03.008

Bhowmik, D., Kamandar, P., Gupta, J., Mete, S., & Hatvate, N. (2024). An efficient, green and micellar catalyzed preparative-scale synthesis of 3,4-Disubstituted isoxazole-5(4H)-ones in the water. *Sustainable Chemistry for the Environment*, 5, 100070. DOI: 10.1016/j.scenv.2024.100070

Blackburn, C., Achab, A., Elder, A., Ghosh, S., Guo, J., Harriman, G., & Jones, M. (2005). Synthesis of 3-Amino-1,2,4-benzothiadi- azine 1,1-Dioxides via a Tandem Aza-Wittig/Heterocumulene Annulation. *The Journal of Organic Chemistry*, 70(24), 10206–10209. DOI: 10.1021/jo051843h PMID: 16292875

Chen, X., Zhu, C., Guo, F., Qiu, X., Yang, Y., Zhang, S., He, M., Parveen, S., Jing, C., Li, Y., & Ma, B. (2010). Acetic acid derivatives of 3,4-dihydro-2H-1,2,4-benzothiadiazine 1,1-dioxide as a novel class of potent aldose reductase inhibitors. *Journal of Medicinal Chemistry*, 53(23), 8330–8344. DOI: 10.1021/jm100962a PMID: 21062005

Chhabra, S., & Shah, K. (2023). Synthesis and anticancer activity of 1,2,4-Benzothiadiazine-1,1-dioxides. *Results in Chemistry*, 5, 100873. DOI: 10.1016/j.rechem.2023.100873

Du, P., Zhou, H., Sui, Y., Liu, Q., & Zou, K. (2016). Asymmetric synthesis of 3,4-dihydro-2H-1,2,4-benzothiadiazine-1,1- dioxides catalyzed by scandium(III)-inda-Pybox. *Tetrahedron*, 72(12), 1573–1578. DOI: 10.1016/j.tet.2016.02.009

Fitch, D. M., Evans, K. A., Chai, D., & Duffy, K. J. (2005). A Highly Efficient, Asymmetric Synthesis of Benzothiadiazine-Substituted Tetramic Acids: Potent Inhibitors of Hepatitis C Virus RNA-Dependent RNA Polymerase. *Organic Letters*, 7(24), 5521–5524. DOI: 10.1021/ol052371w PMID: 16288546

Francotte, P., Goffin, E., Fraikin, P., Lestage, P., Van Heugen, J.-C., Gillotin, F., Danober, L., Thomas, J.-Y., Chiap, P., Caignard, D.-H., Pirotte, B., & de Tullio, P. (2010). New Fluorinated 1,2,4-Benzothiadiazine 1,1-Dioxides: Discovery of an Orally Active Cognitive Enhancer Acting through Potentiation of the 2-Amino-3-(3-hydroxy-5-methylisoxazol-4-yl)propionic Acid Receptors. *Journal of Medicinal Chemistry*, 53(4), 1700–1711. DOI: 10.1021/jm901495t PMID: 20108934

Gadkari, Y. U., Hatvate, N. T., & Telvekar, V. N. (2021). Solar energy as a renewable energy source for preparative-scale as well as solvent and catalyst-free Hantzsch reaction. *Sustainable Chemistry and Pharmacy*, 21, 100444. DOI: 10.1016/j.scp.2021.100444

Gyűjtő, I., Porcs-Makkay, M., Lukács, G., Pusztai, G., Garádi, Z., Tóth, G., Nyulasi, B., Simig, G., & Volk, B. (2019). Synthesis of 4-methyl-2H-1,2,3-benzothiadiazine1,1-dioxides and their further transformation via alkylation and reduction steps. *Synthetic Communications*, 49(24), 3475–3485. DOI: 10.1080/00397911.2019.1673777

Gyűjtő, I., Simig, G., Porcs-Makkay, M., & Volk, B. (2020). Synthesis and Chemistry of 1,2,3-Benzothiadiazine 1,1-Dioxide Derivatives: A Comprehensive Overview. *Chemistry (Basel, Switzerland)*, 2(3), 674–690. DOI: 10.3390/chemistry2030043

Harrouche, K., Lahouel, A., Belghobsi, M., Pirotte, B., & Khelili, S. (2019). Synthesis, characterization, and investigation of the antioxidant activity of some 1,2,4-benzothiadiazine-1,1-dioxides bearing sulfonylthioureas moieties. *Canadian Journal of Chemistry*, 97(12), 824–832. DOI: 10.1139/cjc-2019-0194

Hatvate, N., & Gupta, J. (2024). An efficient, green and solvent-free three-component synthesis of 2,4,6-triarylpyridines utilizing ZSM-5 as a heterogeneous catalyst. *Synthetic Communications*, 54(20), 1760–1770. DOI: 10.1080/00397911.2024.2403779

Hatvate, N. T., & Ghodse, S. M. (2020). One-pot three-component synthesis of isoxazole using ZSM-5 as a heterogeneous catalyst. *Synthetic Communications*, 50(23), 3676–3683. DOI: 10.1080/00397911.2020.1815786

Hatvate, N. T., Ghodse, S. M., & Telvekar, V. N. (2018). Metal-free synthesis of 2-aminothiadiazoles via TBHP-Mediated oxidative C-S bond formation. *Synthetic Communications*, 48(3), 285–290. DOI: 10.1080/00397911.2017.1398330

Henning, N., Kannigadu, C., & Aucamp, J. (2023). *Probing benzothiadiazine-1, 1-dioxide ethylene glycol derivatives against Leishmania : synthesis and in vitro efficacy evaluation*. 1–14.

Hommelsheim, R., van Nahl, R., Hanek, L. M., Ward, J. S., Rissanen, K., & Bolm, C. (2024). Palladium-Catalyzed Domino Reaction for the Synthesis of 3-Amino 1,2,4-Benzothiadiazine 1,1-Dioxides. *Advanced Synthesis & Catalysis*, 366(4), 717–724. DOI: 10.1002/adsc.202300875

Hu, Z., Martí, J., & Lu, H. (2021). Structure of benzothiadiazine at zwitterionic phospholipid cell membranes. *The Journal of Chemical Physics*, 155(15), 154303. Advance online publication. DOI: 10.1063/5.0065163 PMID: 34686044

Huwaimel, B. I., Bhakta, M., Kulkarni, C. A., Milliken, A. S., Wang, F., Peng, A., Brookes, P. S., & Trippier, P. C. (2021). Discovery of Halogenated Benzothiadiazine Derivatives with Anticancer Activity**. *ChemMedChem*, 16(7), 1143–1162. DOI: 10.1002/cmdc.202000729 PMID: 33331124

Ittamalla, C., Chary, D. Y., Balasubramanian, S., & Reddy, B. V. S. (2024). Rh(III)-Catalyzed C–H Annulation of 3-Aryl-2 H -benzo[e][1,2,4]thiadiazine-1,1-dioxides with Vinylene Carbonates. *European Journal of Organic Chemistry*, 27(4), e202300983. Advance online publication. DOI: 10.1002/ejoc.202300983

Kamal, A., Khan, M. N. A., Reddy, K. S., Rohini, K., Sastry, G. N., Sateesh, B., & Sridhar, B. (2007). Synthesis, structure analysis, and antibacterial activity of some novel 10-substituted 2-(4-piperidyl/phenyl)-5,5-dioxo[1,2,4]triazolo[1,5-b][1,2,4]benzothiadiazine derivatives. *Bioorganic & Medicinal Chemistry Letters*, 17(19), 5400–5405. DOI: 10.1016/j.bmcl.2007.07.043 PMID: 17719222

Kamal, A., Shetti, R. V. C. R. N. C., Azeeza, S., Ahmed, S. K., Swapna, P., Reddy, A. M., Khan, I. A., Sharma, S., & Abdullah, S. T. (2010). Anti-tubercular agents. Part 5: Synthesis and biological evaluation of benzothiadiazine 1,1-dioxide based congeners. *European Journal of Medicinal Chemistry*, 45(10), 4545–4553. DOI: 10.1016/j.ejmech.2010.07.015 PMID: 20705368

Khelili, S., Kihal, N., Yekhlef, M., de Tullio, P., Lebrun, P., & Pirotte, B. (2012). Synthesis and pharmacological activity of N-(2,2-dimethyl-3,4-dihydro-2H-1-benzopyran-4-yl)-4H-1,2,4-benzothiadiazine-3-carboxamides 1,1-dioxides on rat uterus, rat aorta and rat pancreatic β-cells. *European Journal of Medicinal Chemistry*, 54, 873–878. DOI: 10.1016/j.ejmech.2012.05.011 PMID: 22647221

Kim, J., Lee, S. Y., Lee, J., Do, Y., & Chang, S. (2008). Synthetic Utility of Ammonium Salts in a Cu-Catalyzed Three-Component Reaction as a Facile Coupling Partner. *The Journal of Organic Chemistry*, 73(23), 9454–9457. DOI: 10.1021/jo802014g PMID: 18956843

Larsen, A. P., Francotte, P., Frydenvang, K., Tapken, D., Goffin, E., Fraikin, P., Caignard, D.-H., Lestage, P., Danober, L., Pirotte, B., & Kastrup, J. S. (2016). Synthesis and Pharmacology of Mono, Di, and Trialkyl-Substituted 7 Chloro-3,4-dihydro 2H 1,2,4-benzothiadiazine 1,1-Dioxides Combined with X-ray Structure Analysis to Understand the Unexpected Structure–Activity Relationship at AMPA Receptors. *ACS Chemical Neuroscience*, 7(3), 378–390. DOI: 10.1021/acschemneuro.5b00318 PMID: 26771108

Li, F., Gong, J., Liu, J., Li, Y., & Li, Z. (2020). Design and Synthesis of Novel 3,4-Dihydro-2H-1,2,4-benzothiadiazine 1,1-Dioxides-based Strobilurins as Potent Fungicide Candidates. *Chemical Research in Chinese Universities*, 36(6), 1190–1195. DOI: 10.1007/s40242-020-0160-9

Mangwegape, D. K., Zuma, N. H., Aucamp, J., & N'Da, D. D. (2021). Synthesis and in vitro antileishmanial efficacy of novel benzothiadiazine-1,1-dioxide derivatives. *Archiv der Pharmazie*, 354(5), 2000280. Advance online publication. DOI: 10.1002/ardp.202000280 PMID: 33491807

Marqués-López, E., & Herrera, R. P. (2015). Bucherer–Bergs and Strecker Multicomponent Reactions. In Multicomponent Reactions (pp. 331–357). Wiley. DOI: 10.1002/9781118863992.ch10

Nielsen, O. B. T., Nielsen, C. K., & Feit, P. W. (1973). Aminobenzoic Acid Diuretics. 5.1 3-Amino-4-arylmethyl-5-sulfamylbenzoic Acid Derivatives and Related Compounds. *Journal of Medicinal Chemistry*, 16(10), 1170–1177. DOI: 10.1021/jm00268a022 PMID: 4749473

Padmaja, R. D., Balamurali, M. M., & Chanda, K. (2019). One-Pot, Telescopic Approach for the Chemoselective Synthesis of Substituted Benzo[e]pyrido/pyrazino/pyridazino[1,2- b][1,2,4]thiadiazine dioxides and Their Significance in Biological Systems. *The Journal of Organic Chemistry*, 84(18), 11382–11390. DOI: 10.1021/acs.joc.9b00869 PMID: 31436096

Parit, S., Manchare, A., Ghodse, S., & Hatvate, N. (2024). Comparative review on homogeneous and heterogeneous catalyzed synthesis of 1,3-thiazole. *Synthetic Communications*, 54(23), 1–21. DOI: 10.1080/00397911.2024.2399187

Shaik, T. B., Malik, M. S., Routhu, S. R., Seddigi, Z. S., Althagafi, I. I., & Kamal, A. (2020). Evaluation of Anticancer and Anti-Mitotic Properties of Quinazoline and Quinazolino-Benzothiadiazine Derivatives. *Anti-cancer Agents in Medicinal Chemistry*, 20(5), 599–611. DOI: 10.2174/1871520620666191224122204 PMID: 31884931

Shi, P., Tu, Y., Wang, C., Kong, D., Ma, D., & Bolm, C. (2020). Synthesis of Benzothiadiazine-1-oxides by Rhodium-Catalyzed C–H Amidation/Cyclization. *Organic Letters*, 22(22), 8842–8845. DOI: 10.1021/acs.orglett.0c03212 PMID: 33170727

Sinitsa, A. D., & Onys, P. P. (2016). *Heterocyclization of N - (chlorosulfonyl) imidoyl chlorides with anilines, a new method of synthesis of 1, 2, 4-benzothiadiazine 1, 1-dioxides. 52*(4), 267–274. DOI: 10.1007/s10593-016-1873-z

Tedesco, R., Shaw, A. N., Bambal, R., Chai, D., Concha, N. O., Darcy, M. G., Dhanak, D., Fitch, D. M., Gates, A., Gerhardt, W. G., Halegoua, D. L., Han, C., Hofmann, G. A., Johnston, V. K., Kaura, A. C., Liu, N., Keenan, R. M., Lin-Goerke, J., Sarisky, R. T., & Duffy, K. J. (2006). 3-(1,1-dioxo-2H-(1,2,4)-benzothiadiazin-3-yl)-4-hydroxy-2(1H)-quinolinones, potent inhibitors of hepatitis C virus RNA-dependent RNA polymerase. *Journal of Medicinal Chemistry*, 49(3), 971–983. DOI: 10.1021/jm050855s PMID: 16451063

THOMSON, A. E., NICKERSON, M., GASKELL, P., & GRAHAME, G. R. (1962). Clinical observations on an antihypertensive chlorothiazide analogue devoid of diuretic activity. *Canadian Medical Association Journal*, 87, 1306–1310. PMID: 13981225

Wright, J. B. (1968). The preparation of 2 H -1,2,3-benzothiadiazine-1,1-dioxides, 11 H -11 a -dihydrobenzimidazo[1,2- b] [1,2] benzisothiazole-5,5-dioxides, 6 H -dibenzo[c,g] [1,2,5] thiadiazocine-5,5-dioxides and 5 H -Dibenzo[c,g] [1,2,6] thiadiazocine-6,6-dioxides. *Journal of Heterocyclic Chemistry*, 5(4), 453–459. DOI: 10.1002/jhet.5570050403

Yang, D., An, B., Wei, W., Tian, L., Huang, B., & Wang, H. (2015). Copper-Catalyzed Domino Synthesis of Nitrogen Heterocycle-Fused Benzoimidazole and 1,2,4-Benzothiadiazine 1,1-Dioxide Derivatives. *ACS Combinatorial Science*, 17(2), 113–119. DOI: 10.1021/co500125n PMID: 25549532

Zhan, P., Liu, X., & Clercq, E. (2008). Recent Advances in Antiviral Activity of Benzo/Heterothiadiazine Dioxide Derivatives. *Current Medicinal Chemistry*, 15(15), 1529–1540. DOI: 10.2174/092986708784638898 PMID: 18537628

Chapter 11
Benzothiazepine:
Recent Advances in Synthesis and Biological Application

Sarita G. Kundlikar
https://orcid.org/0009-0002-9627-7233
New Arts, Commerce, and Science College, India

Hemantkumar N. Akolkar
https://orcid.org/0000-0003-0882-1324
New Arts, Commerce, and Science College, India

Dilip R. Thube
New Arts, Commerce, and Science College, India

Nirmala R. Darekar
Radhabai Kale Mahila Mahavidyalaya, India

Vijay M. Khedkar
Vishwakarma University, India

ABSTRACT

Benzothiazepines are the heterocyclic compounds with a benzene ringfused with sulphur and nitrogen containing seven membered thiazepine ring. Benzothiazepines can be of 1,2-, 1,4- and 1,5- derivatives depending on the varied positions of sulphur and nitrogen in thiazepine ring. These benzothiazepine derivatives has druggable potential in therapeutic medicine for drug development and drug discovery.Benzothiazepine derivatives are endowed with various pharmacological activities like antipsychotic agents, calcium channel antagonists, cytochrome bc1 inhibitors, antidepressants, anticonvulsants etc. Benzothiazepine moiety is present in variety of drugs like Diltiazem (Calcium channel Blockers), Thiazesim (Antidepressant),

DOI: 10.4018/979-8-3693-7520-4.ch011

Clentiazem (Calcium Channel Blockers), GW - 577 (Bile Acid Protein Inhibitor) etc. Due to the biological importance of benzothiazepine molecule in heterocyclic compounds and medicinal chemistry, we have decided to present different synthetic routes for the synthesis of benzothiazepine derivatives and variety of pharmacological activities are reported in the literature.

1.1. INTRODUCTION

The study of the medicinal chemistry is majorly related to discovery, development and identification of mechanism of action of biologically active compounds at the molecular level, (Mjos & Orvig, 2014). Synthesis of novel compounds in organic synthesis plays a crucial role in this tedious process of drug discovery. Molecular framework of the structure has versatile binding extensions that enable to be projected as a novel, potent and selective ligands for a range of different biological targets, (Evans *et al.,* 1988).

Benzothiazepines are the heterocyclic compounds with a benzene ring fused with sulphur and nitrogen containing seven membered thiazepine ring. Benzothiazepines can be of 1,2-, 1,4- and 1,5- derivatives **(Figure 1)** depending on the varied positions of sulphur and nitrogen in thiazepine ring.

Figure 1. Benzothiazepines structures

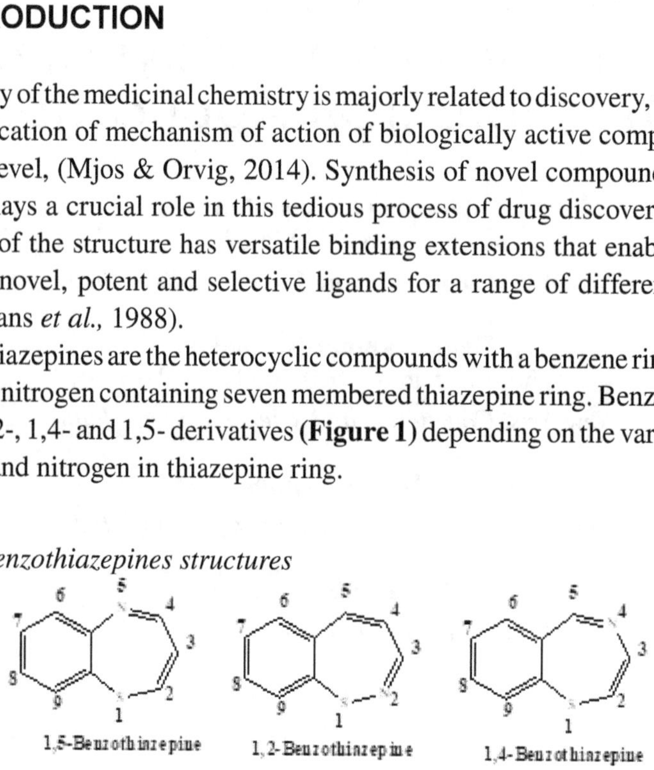

1,5-Benzothiazepine 1,2-Benzothiazepine 1,4-Benzothiazepine

1,5-Benzothiazepines are associated with renowned biological activities in the central nervous system and other therapeutic actions, (Morton *et al.,* 2000). Over years of thorough research in the area of this heterocyclic core, 1,5-Benzothiazepines have firmly stood as a "drug prejudice core" due to variety of biological activities like anti-HIV (Di Santo *et al.,* 2006), Ca+2 antagonist (Campiani *et al.,* 1996), antimicrobial (Saini & Joshi, 2008), CNS depressant (Falco *et al.,* 2006), antifungal (Wenhong & Huazue, 2013), anti-inflammation (Panda & Chowdhary, 2008), antiplatelet aggregation (Genton *et al.,* 1977), antipsychotic agents, (Bajaj *et al.,* 2003) (Kumar, 2013) (Rabjohn, 2005) (Rajendraprasad *et al.,* 2009).

These benzothiazepine derivatives has druggable potential in therapeutic medicine (Crossley, Rajani, & Yuan, 2023) for drug development and drug discovery, (Ogunnupebi *et al.,* 2023) (Phippen & McErlean, 2011). Benzothiazepine derivatives are endowed with various pharmacological activities like antipsychotic agents (Kumar, 2013), calcium channel antagonists (Hitoshi & Jasper, 2022), cytochrome bc_1 inhibitors (Dong *et al.,* 2011), antidepressants (Gill *et al.,* 2013), anticonvulsants (Nikalje & Vyawahare, 2011) *etc.* Benzothiazepine moiety is present in variety of drugs like Diltiazem (Calcium Channel Blockers), Thiazesim (Antidepressant), Clentiazem (Calcium Channel Blockers), GW - 577 (Bile Acid Protein Inhibitor) *etc.* **(Figure 2)**

Figure 2. Drugs containing Benzothiazepine moiety

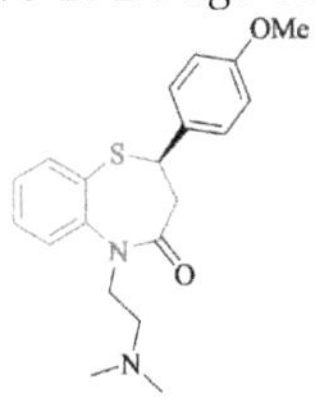

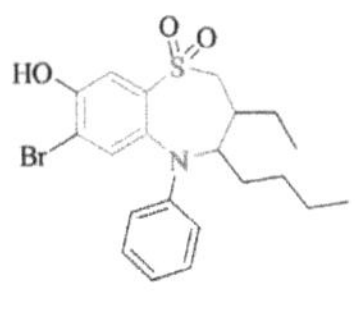

Thiazesim

(Antidepressant)

GW - 577

(Bile Acid Protein Inhibitor)

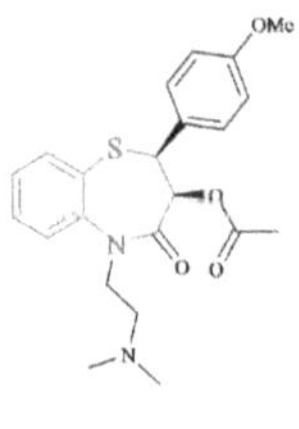

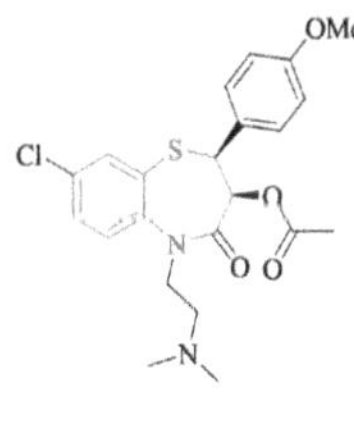

Diltiazem

Clentiazem

o-Aminothiophenol and its derivatives acts as versatile synthons for the preparation of fused 1,5-benzothiazepines, (Sekhar, 2014).

1.2 Recent Advances in Synthesis of Benzothiazepines

Yin *et al* (2020) have designed highly enantioselective synthesis of N-unprotected 2,3-dihydro-1,5-benzothiazepinones **(Figure 3)** by Rh-catalyzed asymmetric hydrogenation of unsaturated medium-ring NH lactams.

Figure 3. Enantioselective synthesis of N-unprotected 2,3-dihydro-1,5-benzothiazepinones

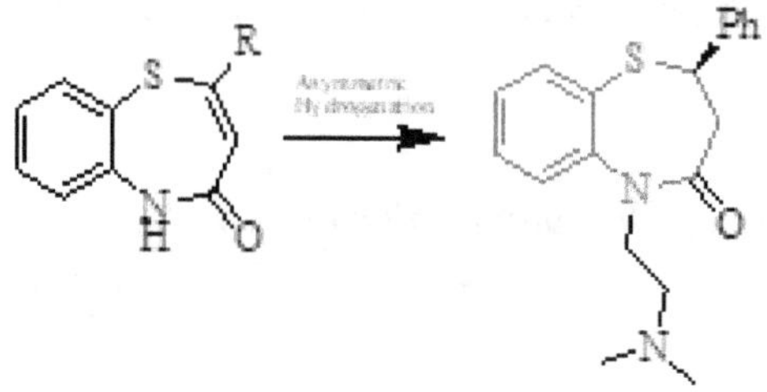

Fang *et al* (2017) reported an unprecedented formal [3 + 4] annulation of α, β-unsaturated acyl azoliums with 2- amino benzene thiols had been utilized to synthesize enantioselective N-H-free 1,5- benzothiazepines. **(Figure 4)**

Figure 4. Synthesis of synthesize enantioselective N-H-free 1,5- benzothiazepines

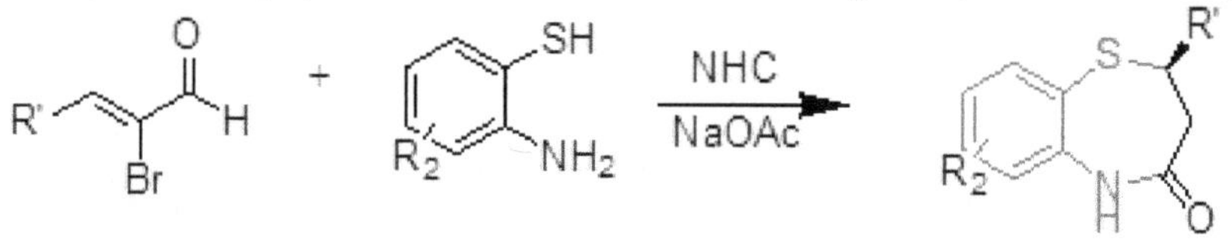

Synthesis and *in vitro* anti-breast cancer activity of some novel 1,5- benzothiazepine derivatives **(Figure 5)** by the condensation of variously substituted chalcones and 2-aminothiophenol via conventional as well as non-conventional inorganic solid-support microwave irradiation methods was described by Ameta and coworkers (2012).

Figure 5. Synthesis of 1,5- benzothiazepine using solid-support microwave irradiation methods

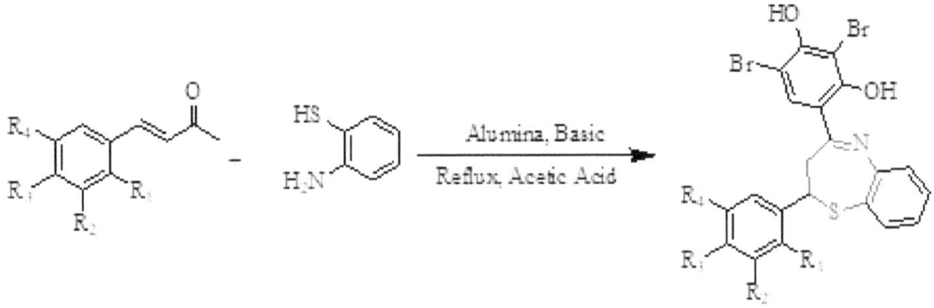

Enantioselective net [4+3] cycloaddition approach to afford optically active 1,5- benzothiazepines **(Figure 6)** by treatment of α,β-unsaturated acyl ammonium intermediates generated by chiral isothiourea catalysts, which undergo two sequential chemo-selective nucleophilic attacks by 2-aminothiophenols was reported by Fukata and coworkers (2015).

Figure 6. Synthesis of optically active 1,5- benzothiazepines

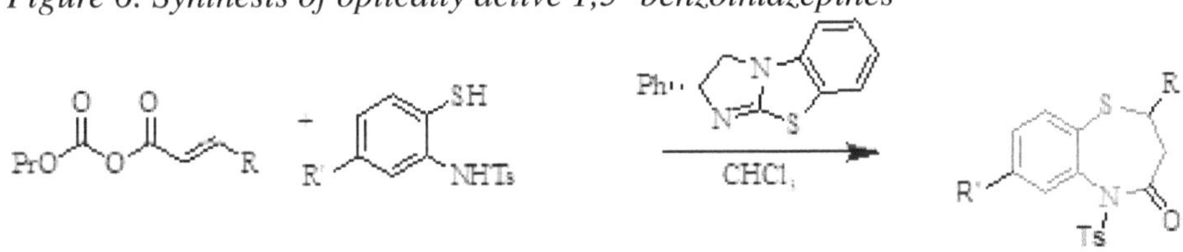

Gutteridge *et al* (2020) have reported synthesis of substituted benzothiazepine compounds **(Figure 7)** with medicinal potential that allows incorporation of structural variation at the three aromatic regions of the structure and at the sulfur atom was developed. Knoevenagel condensation of indan-1,3-diones with aryl aldehydes, yielded 2- arylideneindan-1,3-diones, which underwent thio Michael addition and intramolecular imine formation upon reaction with 2-aminothiophenols.

Figure 7. Synthesis of substituted benzothiazepine compounds

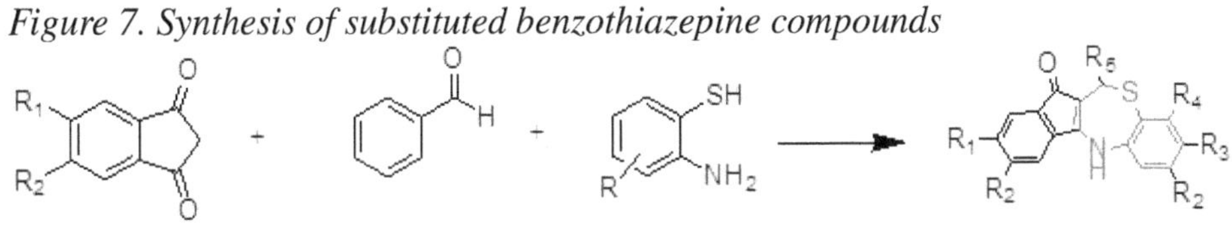

Channegowda and coworkers (2015) has reported novel 1, 5-benzothiazepine derivatives **(Figure 8)** by 2-aminothiophenols on aliphatic electrophones and cyclisation.

Figure 8. Synthesis of 1, 5-benzothiazepine reaction by 2-aminothiophenols

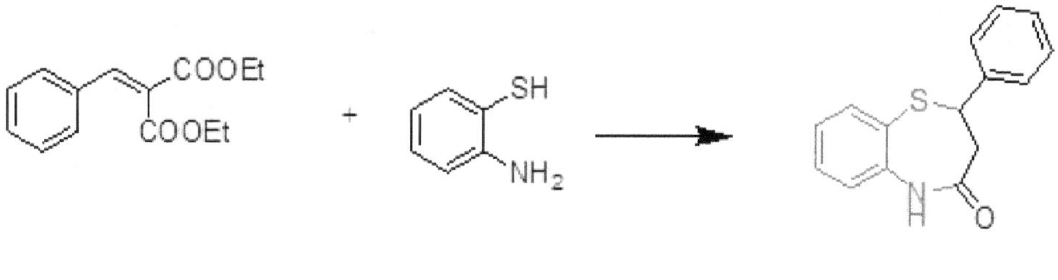

Chobeand coworkers (2019) reported Microwave assisted greener synthesis of biologically active 1,5-benzothiazepine derivatives **(Figure 9)** by using poly(ethylene glycol-400) as an green reaction media and synthesized compounds showed antifungal activity against several pathogens.

Figure 9. Microwave assisted synthesis of 1, 5-benzothiazepine derivatives

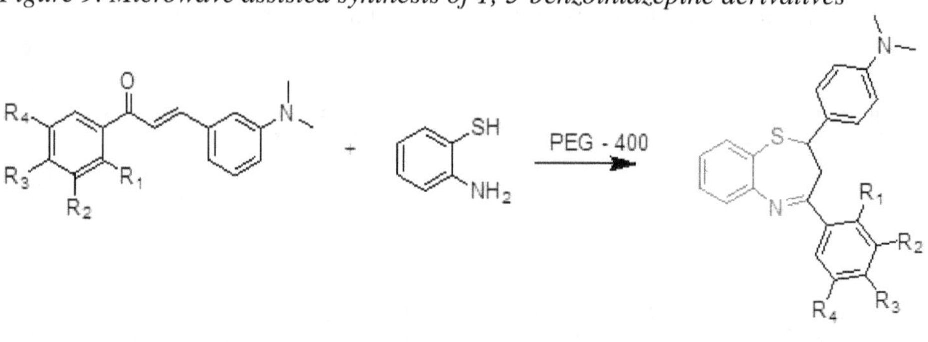

2,3-Dihydro-1,5- benzothiazepines **(Figure 10)** were obtained through a domino process involving a Michael addition of 2-aminothiophenols to chalcones, followed by in-situ cyclization was described by Albanese *et al* (2017).

Figure 10. Synthesis of2,3-Dihydro-1,5- benzothiazepines through a domino process

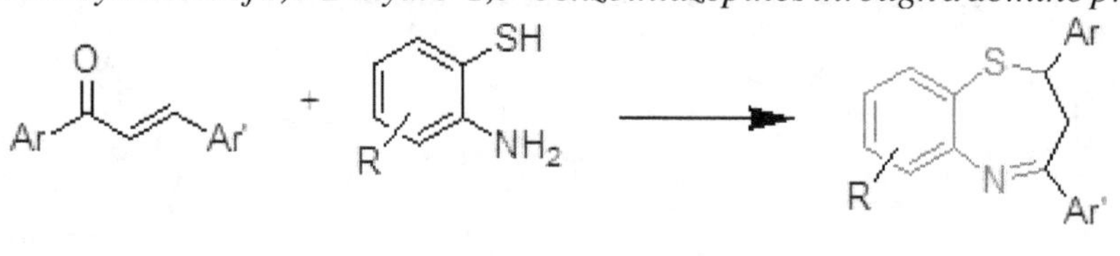

Haroun and coworkers (2022) have reported synthesis and anticancer activity of novel 1,5-benzothiazepine derivatives **(Figure 11)** using polyethylene glycol-400 (PEG-400)-mediated and bleaching clay as green pathways by treatment of different chalcones and o-aminothiophenols to a cyclization step with benzothiazepine derivatives.

Figure 11. Synthesis and anticancer activity of novel 1,5-benzothiazepine derivatives

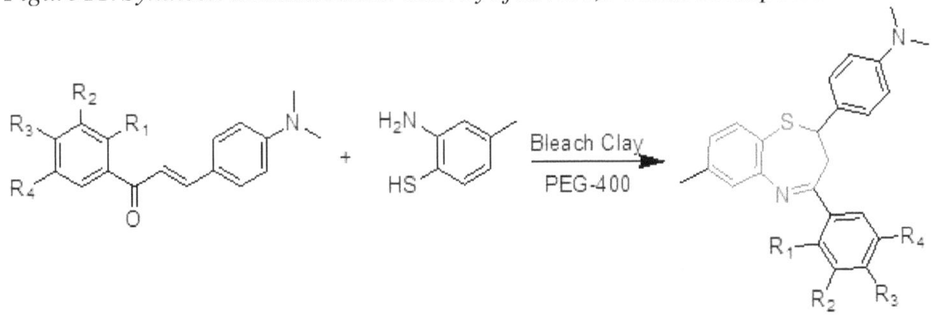

Vyawahare *et al* (2010) have reported microwave assisted synthesis of 2,3-dihydro-2-substituted-4-(naphthalen-2-ol)-yl-1,5-benzothiazepines **(Figure 12)** from 1,3-substituted-prop-2-en-1-one via cyclocondensation of with 2-aminothiophenol in presence of ecofriendly catalyst zinc acetate.

Figure 12. Microwave assisted synthesis of 2,3-dihydro-2-substituted-4-(naphthalen-2-ol)-yl-1,5-benzothiazepines

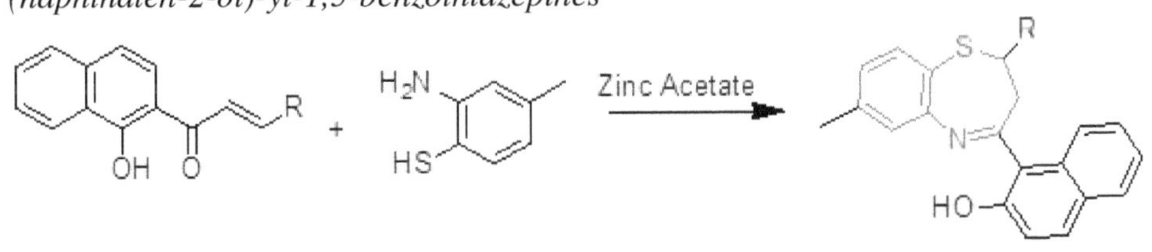

Mhaske *et al* (2014) in year 2014 had been reported Novel 1, 5-benzothiazepine **(Figure 13)** by reaction of 1-(4- chlorophenyl)-1H-pyrazole-4-carbaldehyde, o-hydroxyacetophenones and chalcone.

Figure 13. Synthesis of Novel 1, 5-benzothiazepine derivatives

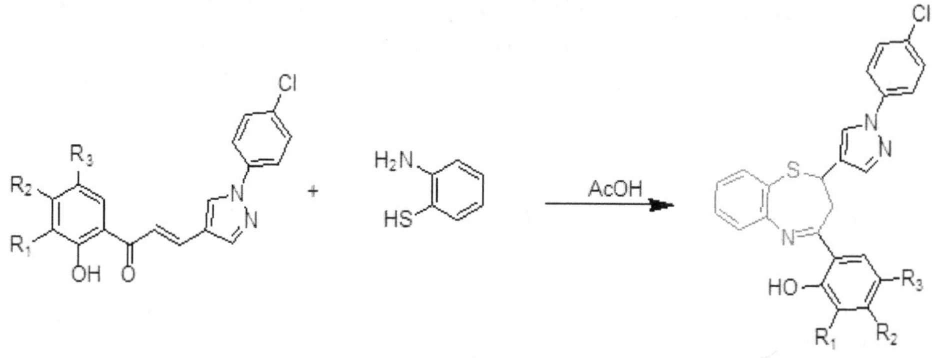

1.3 Biological Applications

1.3.1 Benzothiazepine as Anticancer Agent

Shaik and coworkers (2021) developed a series of chloropyrazine conjugated benzothiazepines I **(Figure 14)** and were evaluated for antimicrobial, antitubercular and cytotoxic activities. These compounds could be considered as potential leads for the development of drugs.

Figure 14. Synthesis of chloropyrazine conjugated benzothiazepines

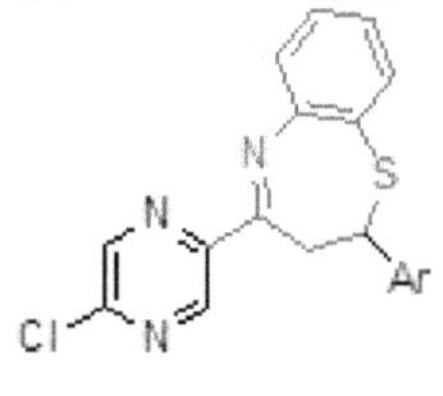

Kotireddy and Ramana (2017) have reported synthesis of benzothiazepines **(Figure 15)** by conventional synthesis method by the reactions between 1,3-diarylprop-2-enones with ortho-amino thiophenol. Synthesized products were tested for anticancer activity against breast colon prostate cancer.

Figure 15. Synthesis of chloropyrazine conjugated benzothiazepines

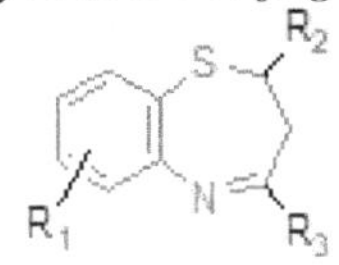

Synthesis of series of 1,5- benzothiazepine derivatives **(Figure 16)** by the reaction of 1,5- benzothiazepine containing 2-phenoxy quinoline moiety with benzohydroximinoyl chlorides via 1,3-dipolar cycloaddition reaction was reported by Dong and coworkers (2011).

Figure 16. Synthesis of series of 1,5- benzothiazepine derivatives

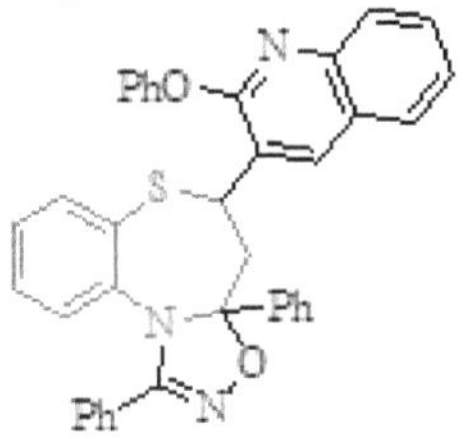

Bhabal and coworkers (2022) have reported synthesis of novel fluorinated 1,5-benzothiazepine derivatives **(Figure 17)** and their biological evaluation as anticancer and antibacterial agents against four human cancer cell lines namely lung (A549), breast (MCF-7), liver (HEPG2) and prostate (P C-3).

Figure 17. Synthesis of novel fluorinated 1,5- benzothiazepine derivatives

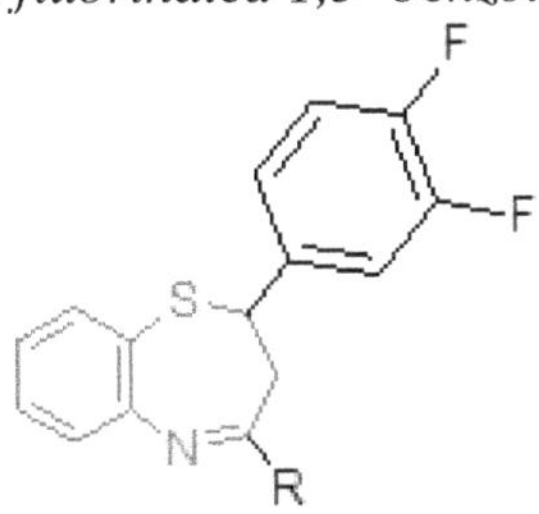

Ngoc and coworkers (2020) reported synthesis, cytotoxic activity, ADMET and molecular docking study of quinoline-based hybrid compounds of 1,5-benzothiazepines **(Figure 18)** by its reaction with aryl aldehydes with yields of 61-87% using piperidine as a catalyst.

Figure 18. Synthesis, cytotoxic activity, ADMET and molecular docking study of quinoline-based hybrid compounds of 1,5-benzothiazepines

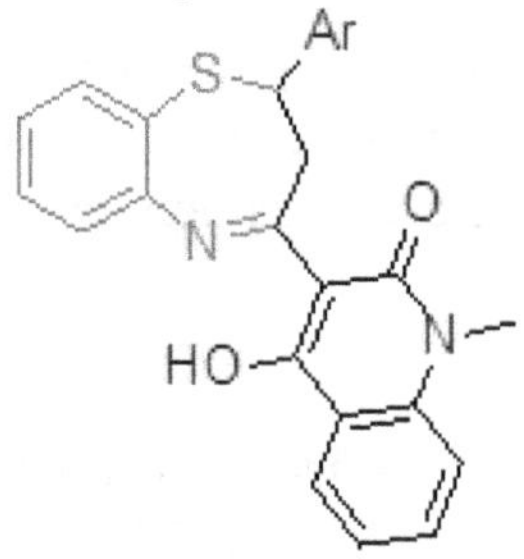

Sarhan and coworkers (2022) have reported synthesis, biological evaluation, and molecular docking of pyrazoline-linked benzene sulfonamides and diaryl 1,5-benzothiazepines **(Figure 19)** prepared from new chalcones and were studied for their in- vitro anticancer profiles on HepG2, HEK-293, MCF-7, and MDA-MB-231 cancer cell lines.

Figure 19. Synthesis of diaryl 1,5- benzothiazepines

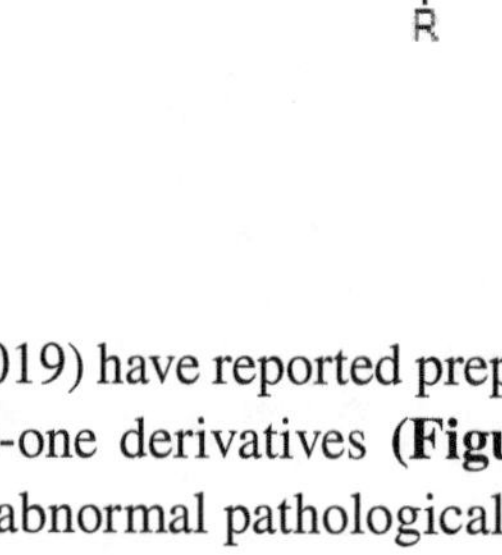

Yong and coworkers (2019) have reported preparation of 2-phenyl-3,4-dihydro-1,4- benzothiazepin-5(2H)-one derivatives **(Figure 20)** useful for prevention or treatment of diseases with abnormal pathological characteristics of GSK3β.

Figure 20. Synthesis of 2-phenyl-3,4-dihydro-1,4- benzothiazepin-5(2H)-one derivatives

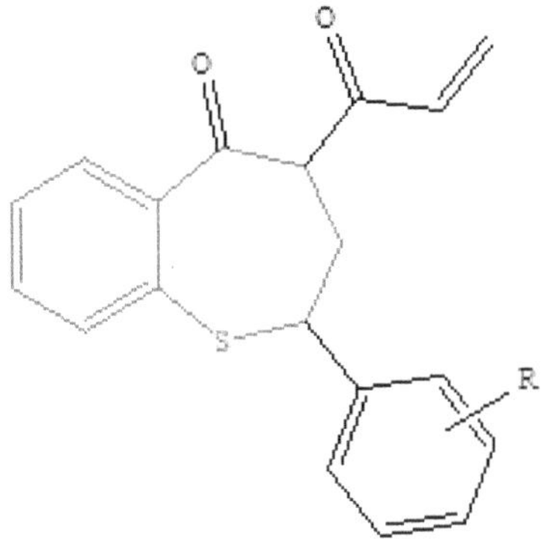

1.3.2 Benzothiazepine as Antimicrobial Agent

Syntheses and antimicrobial studies of 8-substituted-2,5-dihydro-2-(4-methoxyphenyl/3,4-dimethoxyphenyl)-4-(2-thienyl)-1,5-benzothiazepines **(Figure 21)** by reactions of the 5-substituted-2-aminobenzenethiols with α,β-unsaturated thienyl ketones in dry ethanol saturated with dry HCl gas was reported by Pant and coworkers (2008).

Figure 21. Syntheses and antimicrobial studies of 8-substituted-2,5-dihydro-2-(4-methoxyphenyl/3,4-dimethoxyphenyl)-4-(2-thienyl)-1,5-benzothiazepines

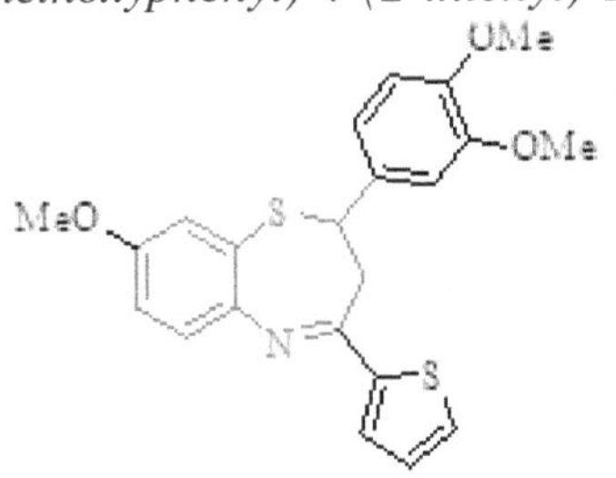

Syntheses and antimicrobial evaluation of 8-substituted-2-(2-chlorophenyl/3-chlorophenyl)-4-(4-hydroxyphenyl/phenyl)-2,3/2,5-dihydro-1,5-benzothiazepines **(Figure 22)** by the reactions of 5- substituted 2-aminobenzenethiols 2-NH -5-X-C H SH, with α,β-unsaturated ketones, 3-(2-chlorophenyl)-1-(4-hydroxyphenyl)-

2-propenone/3-(3-chloro-phenyl)-1-phenyl-2-propenone in dry ethanol in acidic medium was described by Pant *et al* (2021).

Figure 22. Synthesis of 8-substituted-2-(2-chlorophenyl/3-chlorophenyl)-4-(4-hydroxyphenyl/phenyl)-2,3/2,5-dihydro-1,5-benzothiazepines

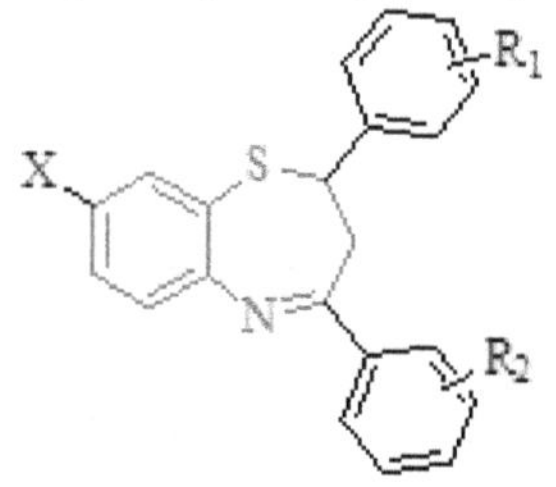

A series of 2,5-dihydro-4-[1-(2-chlorophenyl)-3-methyl-5-pyrazolylidenamino]-2-aryl-1,5 benzothiazepines I **(Figure 23)** was synthesized by microwave enhanced heterocyclization an environmentally benign procedure and they were evaluated for their antimicrobial activity by Desai and coworkers (2007).

Figure 23. Synthesis of 2,5-dihydro-4-[1-(2-chlorophenyl)-3-methyl-5-pyrazolylidenamino]-2-aryl-1,5 benzothiazepines

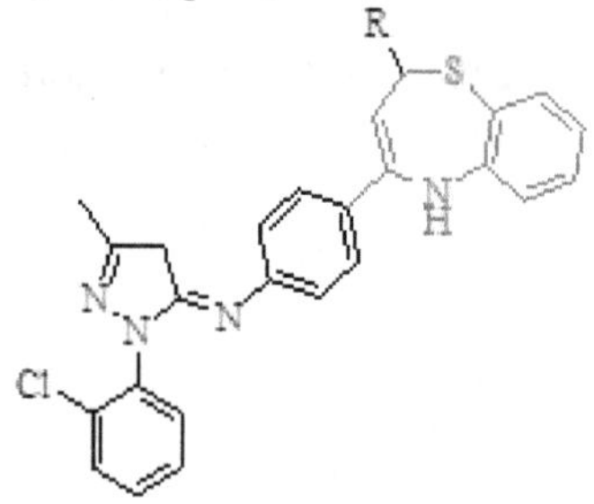

Pant *et al* (2008) have reported syntheses and antimicrobial evaluation of 2-(2-chlorophenyl)-4-(4-chlorophenyl/2-thienyl)-2,5-dihydro-8-substituted-1,5-benzothiazepines **(Figure 24)** via cyclization of 5-substituted-2-aminobenzenethiols with 3-(2-chlorophenyl)-1-(4-chlorophenyl)-2-propenone.

Figure 24. Synthesis of 2-(2-chlorophenyl)-4-(4-chlorophenyl/2-thienyl)-2,5-dihydro-8-substituted-1,5-benzothiazepines

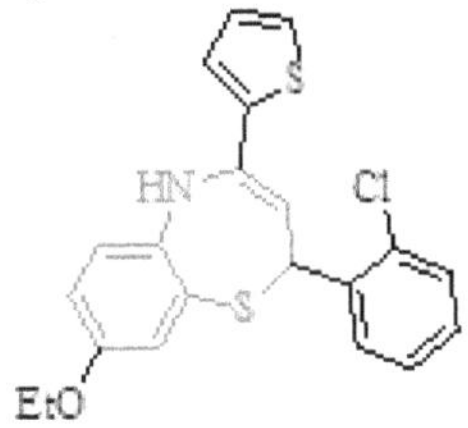

Synthesis of a novel series of 1,5- benzothiazepine derivatives **(Figure 25)** as potential antimicrobial agents was reported by Wang and coworkers (2009).

Figure 25. Synthesis of series of 1,5- benzothiazepine derivatives

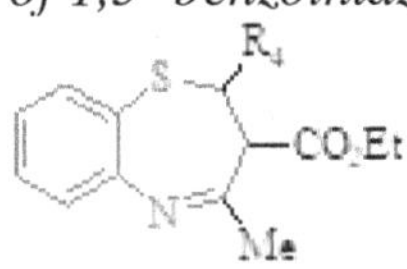

Synthesis and antimicrobial evaluation of some novel 2,3-dihydro-1,5-benzothiazepine derivatives **(Figure 26)** by the treatment of 1-(substituted-2-hydroxy-phenyl)-3-(4'-dimethylamino-phenyl)-prop-2-en-1-ones (chalcones) II with 2 aminothiophenol using ethanol as a solvent in the presence of catalytic amount of lanthanum nitrate was reported by Kendre *et al* (2021).

Figure 26. Synthesis of novel 2,3-dihydro-1,5-benzothiazepine derivatives

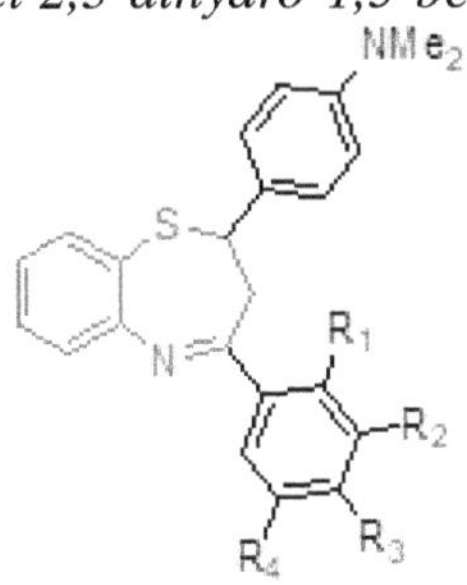

Synthesis and antimicrobial activity of series of 2,5-dihydro-4-[1-(2-chlorophenyl)-3-methyl-5-pyrazolylidenamino]-2-aryl-1,5 benzothiazepines **(Figure 27)** by Microwave irradiation (MWI) under solvent-free conditions was reported by Desai and coworkers (2007).

Figure 27. Synthesis of 2,5-dihydro-4-[1-(2-chlorophenyl)-3-methyl-5-pyrazolylidenamino]-2-aryl-1,5 benzothiazepines

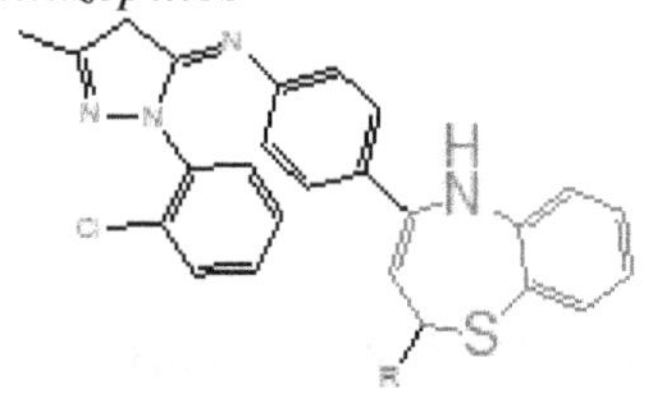

Design, synthesis, and antiviral activities of 1,5- benzothiazepine derivatives containing pyridine moiety **(Figure 28)** have been synthesized by 1,3-dipolar cycloaddition of chalcone by Li and coworker (2017).

Figure 28. Synthesis of 1,5- benzothiazepine derivatives containing pyridine moiety

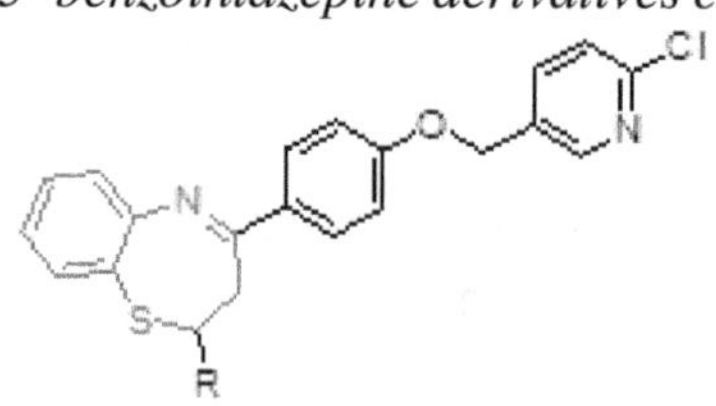

Syntheses and antibacterial, antifungal activity of 8-substituted-2-(2-chlorophenyl/3-chlorophenyl)-4-(4-hydroxyphenyl/phenyl)-2,3/2,5-dihydro-1,5-benzothiazepines **(Figure 29)** from the reactions of 5-substituted-2-aminobenzenethiols 2-NH -5-X-C H SH, with α,β-unsaturated ketones, 3-(2-chlorophenyl)-1-(4-hydroxyphenyl)-2-propenone/3-(3-chloro-phenyl)-1-phenyl-2-propenone by Pant *et al* (2021).

Figure 29. Synthesis of 8-substituted-2-(2-chlorophenyl/3-chlorophenyl)-4-(4-hydroxyphenyl/phenyl)-2,3/2,5-dihydro-1,5-benzothiazepines

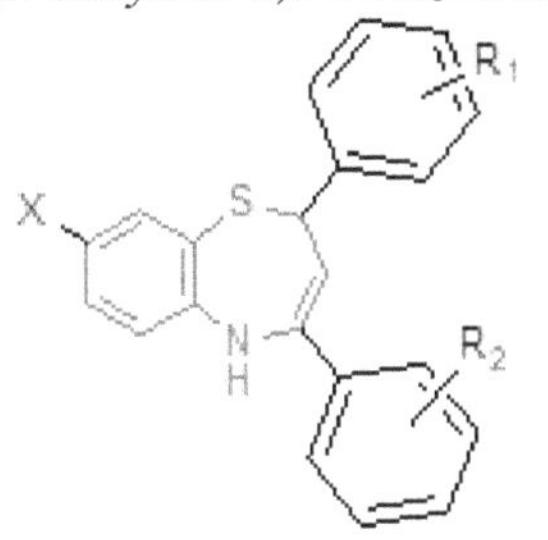

1.3.3 Benzothiazepine as Antitubercular Agent

Wang and coworkers (2021) have reported synthesis of series of diaryl benzo[*b*][1,4]thiazepine derivatives **(Figure 30)** and screened as tubulin polymerization inhibitors with anti-tumor potency.

Figure 30. Synthesis of diaryl benzo[b][1,4]thiazepine derivatives

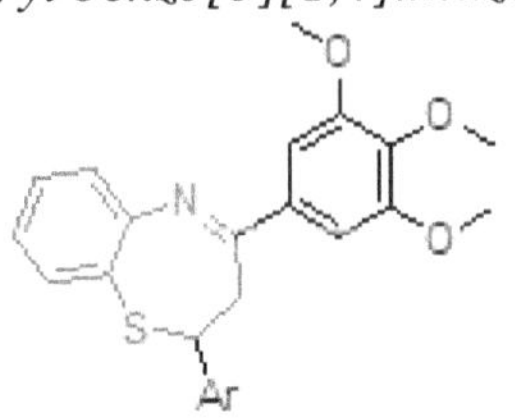

1.3.4 Benzothiazepine as Anticonvulsant Agent

Parjane and coworkers (2020) have reported synthesis of some novel 1,5-benzothiazepine derivatives **(Figure 31)** by the reaction of α,β-unsaturated carbonyl compounds with mercapto aniline and evaluated them for anticonvulsant activity.

Figure 31. Synthesis of novel 1,5-benzothiazepine derivatives

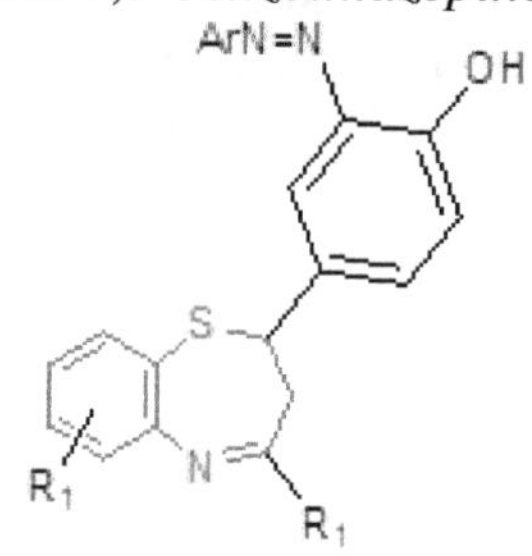

Garg and coworkers (2010) have reported synthesis and anticonvulsant activity of new series of 4-(4'-hydroxyphenyl)-2-(3-substituted phenyl)-3-(4-substituted phenylaminomethylene)-2,3-dihydro-1,5- benzothiazepines and 4-(4'-hydroxyphenyl)-2-(3-substituted phenyl)-3-(4-substituted phenylaminomethylene)-2,3-dihydro-1,5-benzoxazepins. **(Figure 32)**

Figure 32. Synthesis of 4-(4'-hydroxyphenyl)-2-(3-substituted phenyl)-3-(4-substituted phenylaminomethylene)-2,3-dihydro-1,5- benzothiazepines

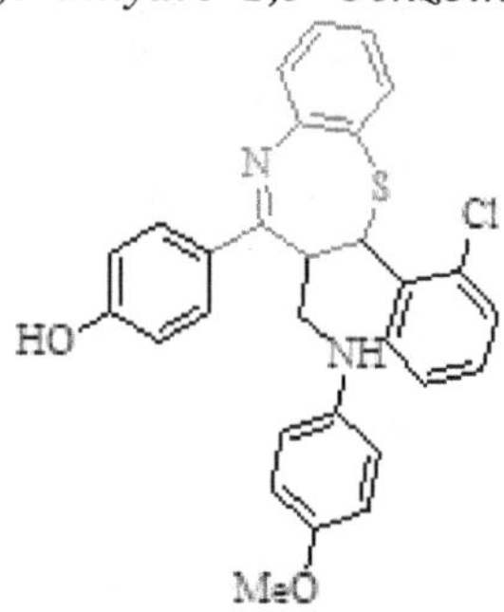

1.3.5 Benzothiazepine as Antioxidant Agent

Pradhan and coworkers (2023) have reported synthesis of a novel series of [1,5]-benzothiazepine-C-β-D-glycoside derivatives **(Figure 33)** in a facile one-pot method and exhibited their anti-oxidant properties.

Figure 33. Synthesis of series of [1,5]-benzothiazepine-C-β-D-glycoside derivatives

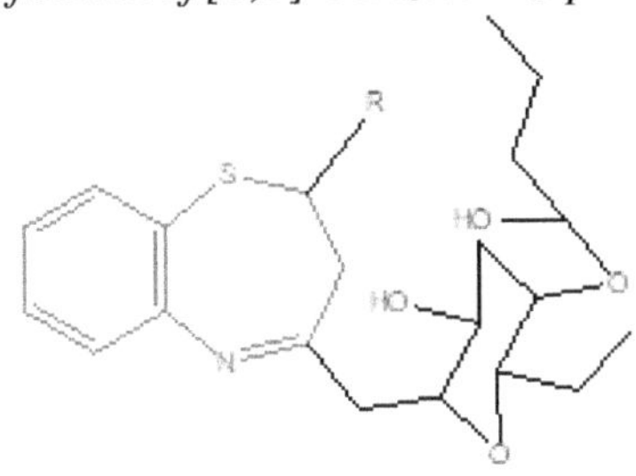

1.3.6 Benzothiazepine as Miscellaneous Agents

Synthesis and evaluation of antihyperglycemic activity of 1,4- benzothiazepine -2-one derivatives **(Figure 34)** in alloxan induced diabetic rats was reported by Srivastav and coworkers (2021).

Figure 34. Synthesis of 1,4- benzothiazepine -2-one derivatives

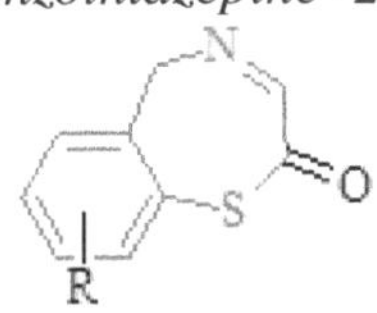

Enantioselective synthesis of N-H-Free 1,5-Benzothiazepines **(Figure 35)** by sulfa-Michael-cyclization reaction between 2-aminothiophenol and α,β-unsaturated pyrazole amides was developed by Wang *et al* (2017).

Figure 35. Synthesis of N-H-Free 1,5-Benzothiazepines

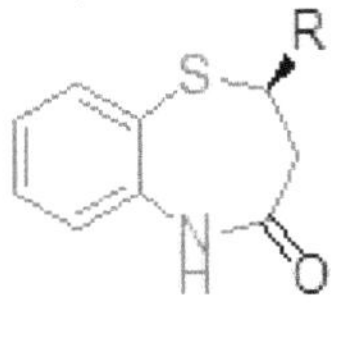

Renuka *et al* (2014) have reported synthesis and antioxidant activity studies of 1,4- benzothiazepine analogues **(Figure 36)** from α,β-unsaturated ketones by a simple and accessible approach.

Figure 36. Synthesis of 1,4- benzothiazepine analogues

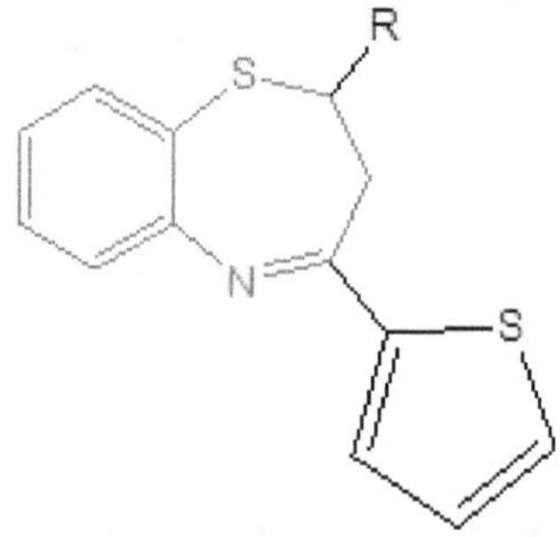

Synthesis and psychotropic study of isoxazolylmethyl substituted 1,5-benzothiazepin-4-one derivatives **(Figure 37)** via 1,3-Dipolar cycloaddition of a nitrile oxide derivative generated in- situ from an aldoxime with substituted benzothiazepinones was reported by Rahaoui and coworkers (2007).

Figure 37. Synthesis of isoxazolylmethyl substituted 1,5-benzothiazepin-4-one derivatives

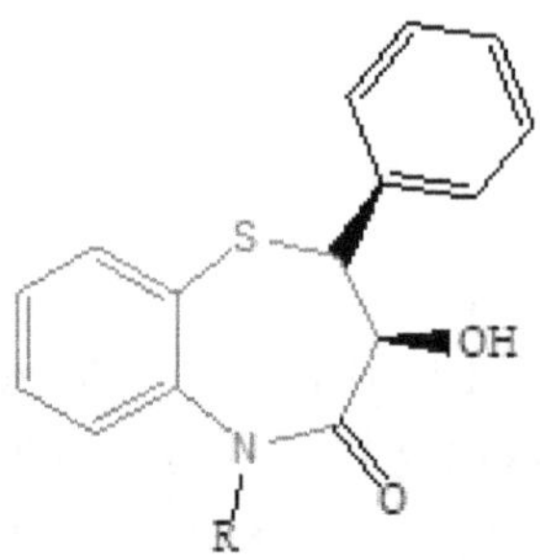

Zhang and coworkers (2017) have reported synthesis of novel crystalline forms of N-(3-amino-3-oxetanyl)methyl]-2-(2,3-dihydro-1,1-dioxido-1,4-benzothiazepin-4(5H-yl)-6-methyl-4 quinazolinamine **(Figure 38)** and evaluation for the treatment of respiratory syncytial virus (RSV) infections.

Figure 38. Synthesis ofN-(3-amino-3-oxetanyl)methyl]-2-(2,3-dihydro-1,1-dioxido-1,4-benzothiazepin-4(5H-yl)-6-methyl-4 quinazolinamine

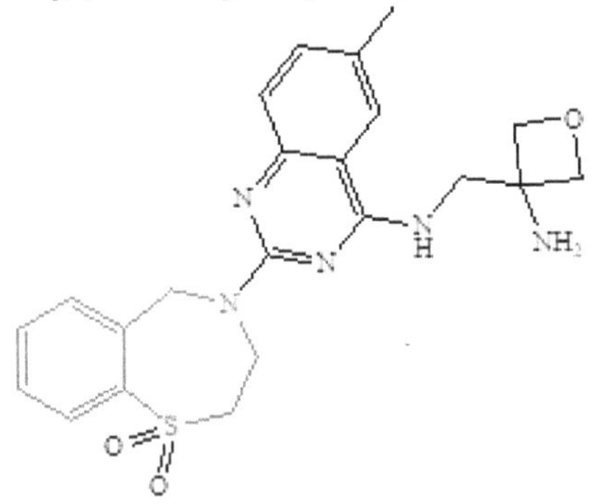

Channegowda and coworkers (2015) have reported synthesis, characterization, anti-angiogenic and anti-oxidant activities of 1,5- benzothiazepin-4-(5H)-one derivatives **(Figure 39)** by the reaction of solution of bromo indazole and [1,1'-bis(diphenylphosphino)ferrocene]palladium(II) dichloride [Pd(dppf)Cl_2] in anhydrous DME.

Figure 39. Synthesis of 1,5- benzothiazepin-4-(5H)-one derivatives

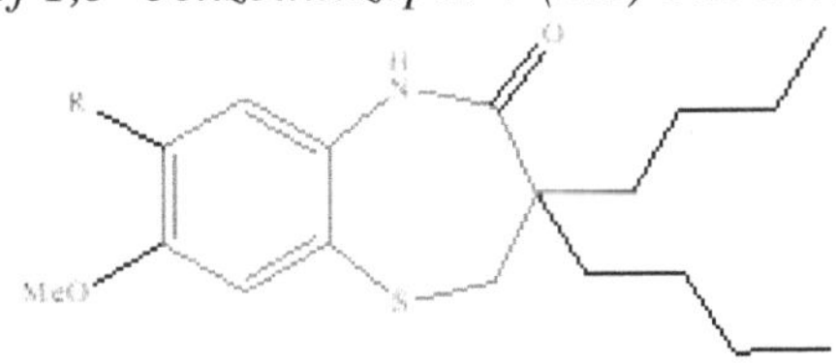

1.4 Conclusion

Literature survey reveals that benzothiazepine containing derivatives are having various biological activities so these derivatives can be considered as good lead compounds for drug design and medicinal chemistry. This synthetic methods can be useful for synthesis of some novel drug molecules which can exhibit important pharmacological activities.

REFERENCES

Albanese, D. C. M., Gaggero, N., & Fei, M. (2017). A practical synthesis of 2,3-dihydro-1,5-benzothiazepines. *Green Chemistry*, 19(23), 5703–5707. DOI: 10.1039/C7GC02097J

Ameta, K. L., Rathore, N. S., & Kumar, B. (2012). Synthesis and in vitro anti breast cancer activity of some novel 1,5-benzothiazepine derivatives. *Journal of the Serbian Chemical Society*, 77(6), 725–731. DOI: 10.2298/JSC110715219A

Bajaj, K., Srivastava, V., Lata, S., & Chandra, R. (2003). *Indian Journal of Chemistry*, 42B, 1723.

Bhabal, S. R., Shaikh, S. F., Yellapurkar, I. P., Pavale, G. S., & Ramana, M. M. V. (2022). Synthesis of novel fluorinated 1,5-benzothiazepine derivatives and their biological evaluation as anticancer and antibacterial agent. *Journal of the Serbian Chemical Society*, 87(10), 1109–1116. DOI: 10.2298/JSC210428041B

Campiani, G., Fiorini, I., De Filippis, M. P., Ciani, S. M., Garofalo, A., Nacci, V., Giorgi, G., Sega, A., Botta, M., Chiarini, A., Budriesi, R., Bruni, G., Romeo, M. R., Manzoni, C., & Mennini, T. (1996). Cardiovascular Characterization of Pyrrolo[2,1-*d*][1,5]benzothiazepine Derivatives Binding Selectively to the Peripheral-Type Benzodiazepine Receptor (PBR): From Dual PBR Affinity and Calcium Antagonist Activity to Novel and Selective Calcium Entry Blockers. *Journal of Medicinal Chemistry*, 39(15), 2922–2938. DOI: 10.1021/jm960162z PMID: 8709127

Channegowda, V. D., Goravanahalli, M., Raghavendrab, N. D., Rekhac, K., Mantelingub, K. S., Rangappab, D., & Bhadregowdaa, G. (2015). Curr. *Chemistry Letters*, 4, 133.

Channegowda, V. D., Nanjappagowda, D. R., Goravanahalli, M. R., & Doddamedur, G. B. (2015). *Journal of Chemical and Pharmaceutical Research*, 7(12), 733.

Chobe, S. S., & Kadam, V. V. (2019). *IJRAR*, 6(1), 492.

Crossley, C. A., Rajani, V., & Yuan, Q. (2023). Modulation of L-type calcium channels in Alzheimer's disease: A potential therapeutic target. *Computational and Structural Biotechnology Journal*, 21, 11–20. DOI: 10.1016/j.csbj.2022.11.049 PMID: 36514335

Desai, K. G., & Desai, K. R. (2007). *Indian Journal of Chemistry*, 46B(7), 1179.

Di Santo, R., Costi, R., Artico, M., Miele, G., Lavecchia, A., Novellino, E., Bergamini, A., Cancio, R., & Maga, G. (2006). Arylthiopyrrole (AThP) Derivatives as Non-Nucleoside HIV-1 Reverse Transcriptase Inhibitors: Synthesis, Structure–Activity Relationships, and Docking Studies (Part 1). *ChemMedChem*, 1(12), 1367–1378. DOI: 10.1002/cmdc.200600119 PMID: 17089433

Dong, C. K., Urgaonkar, S., Cortese, J. F., Gamo, F. J., Garcia-Bustos, J. F., Lafuente, M. J., Patel, V., Ross, L., Coleman, B. I., Derbyshire, E. R., Clish, C. B., Serrano, A. E., Cromwell, M., Barker, R. H.Jr, Dvorin, J. D., Duraisingh, M. T., Wirth, D. F., Clardy, J., & Mazitschek, R. (2011). Identification and Validation of Tetracyclic Benzothiazepines as Plasmodium falciparum Cytochrome bc1 Inhibitors. *Chemistry & Biology*, 18(12), 1602–1610. DOI: 10.1016/j.chembiol.2011.09.016 PMID: 22195562

Dong, Z.-Q., Liu, F.-M., Xu, F., & Yuan, Z.-L. (2011). Synthesis of 1,5-benzothiazepine derivatives bearing 2-phenoxy-quinoline moiety via 1,3-diplolar cycloaddition reaction. *Molecular Diversity*, 15(4), 963–970. DOI: 10.1007/s11030-011-9328-z PMID: 21837389

Evans, B. E., Rittle, K. E., Bock, M. G., DiPardo, R. M., Freidinger, R. M., Whitter, W. L., Lundell, G. F., Veber, D. F., Anderson, P. S., Chang, R. S., Lotti, V. J., Cerino, D. J., Chen, T. B., Kling, P. J., Kunkel, K. A., Springer, J. P., & Hirshfield, J. (1988). Methods for drug discovery: Development of potent, selective, orally effective cholecystokinin antagonists. *Journal of Medicinal Chemistry*, 31(12), 2235–2246. DOI: 10.1021/jm00120a002 PMID: 2848124

Falco, J. L., Pique, M., Gonzalez, M., Buira, I., Mendez, E., Terencio, J., Perez, C., Príncep, M., Palomer, A., & Guglietta, A. (2006). Synthesis, pharmacology and molecular modeling of N-substituted 2-phenyl-indoles and benzimidazoles as potent GABAA agonists. *European Journal of Medicinal Chemistry*, 41(8), 985–990. DOI: 10.1016/j.ejmech.2006.03.031 PMID: 16764969

Fang, C., Lu, T., Zhu, J., Sun, K., & Du, D. (2017). Formal [3 + 4] Annulation of α,β-Unsaturated Acyl Azoliums: Access to Enantioenriched N– *H* -Free 1,5-Benzothiazepines. *Organic Letters*, 19(13), 3470–3473. DOI: 10.1021/acs.orglett.7b01457 PMID: 28598636

Fukata, Y., Asano, K., & Matsubara, S. (2015). Facile Net Cycloaddition Approach to Optically Active 1,5-Benzothiazepines. *Journal of the American Chemical Society*, 137(16), 5320–5323. DOI: 10.1021/jacs.5b02537 PMID: 25856510

Garg, N., & Chandra, T. (2010). Archana, Jain A B & Kumar A. *European Journal of Medicinal Chemistry*, 45(4), 1529. DOI: 10.1016/j.ejmech.2010.01.001 PMID: 20163892

Genton, E. D. W. A. R. D., Barnett, H. J., Fields, W. S., Gent, M., & Hoak, J. C. (1977). XIV. Cerebral ischemia: The role of thrombosis and of antithrombotic therapy. Study group on antithrombotic therapy. *Stroke*, 8(1), 150–175.

Gill, R. K., Aggarwal, N., Kumari, J., Kumari, M., Kaur, P., Kaur, M., Rani, A., Bansal, A., Shah, A., & Bariwal, J. (2013)... *Chemico-Biological Interactions*, 3(3), 146.

Gutteridge, C. E., Sadowski, B. W., Hughes, S. M., Friedlander, J. A., Rodrigo, L. G., Baxter, M. C., Lorei, N. C., Harrell, J. E., & O'Neil, M. T. (2020). Synthesis of Substituted Benzothiazepine Compounds with Medicinal Potential. *International Journal of Organic Chemistry*, 10(3), 123–134. DOI: 10.4236/ijoc.2020.103009

Haroun, M., Chobe, S. S., Alavala, R. R., Mathure, S. M., Jamullamudi, R. N., Nerkar, C. K., Gugulothu, V. K., Tratrat, C., Islam, M. M., Venugopala, K. N., Habeebuddin, M., Telsang, M., Sreeharsha, N., & Anwer, M. K. (2022). 1,5-Benzothiazepine Derivatives: Green Synthesis, In Silico and In Vitro Evaluation as Anticancer Agents. *Molecules (Basel, Switzerland)*, 27(12), 3757. DOI: 10.3390/molecules27123757 PMID: 35744881

Hitoshi, M., Cara, C. Y., & Jasper, A. S. (2022).*Neuropharmacology*, 221, 109290. DOI: 10.1016/j.neuropharm.2022.109290 PMID: 36241085

Kendre, M. M., & Baseer, M. A. (2021).*International Journal of Pharmaceutical Sciences and Research*, 12(2), 1014.

Kotireddy, V., & Ramana, K. V. (2017).*International Journal of Pharmaceutical, Chemical and Biological Sciences*, 7(3), 219.

Kumar, A. (2013).*American Journal of Pharmacological Sciences*, 1(6), 116. DOI: 10.12691/ajps-1-6-2

Kumar, A. (2013).*American Journal of Pharmacological Sciences*, 1(6), 116. DOI: 10.12691/ajps-1-6-2

Li, T., Zhang, J., Pan, J., Wu, Z., Hu, D., & Song, B. (2017). Design, synthesis, and antiviral activities of 1,5-benzothiazepine derivatives containing pyridine moiety. *European Journal of Medicinal Chemistry*, 125, 657–662. DOI: 10.1016/j.ejmech.2016.09.069 PMID: 27721151

Li, W., Liu, X., Zhang, B., & Li, Y. (2013). Synthesis and Antifungal Activity of 2-Carbalkoxyalkyl/(*n* -butyl)-4-aryl-1,5-benzothiazepines. *Youji Huaxue*, 33(07), 1503. DOI: 10.6023/cjoc201210044

Mhaske, G. R., Bajod, B. S., Ambhore, D. M., & Shelke, S. N. (2014).*International Journal of Innovative Research in Science, Engineering and Technology*, 3, 2319.

Mjos, K. D., & Orvig, C. (2014). Metallodrugs in Medicinal Inorganic Chemistry. *Chemical Reviews*, 114(8), 4540–4563. DOI: 10.1021/cr400460s PMID: 24456146

Morton, G. C., Salvino, J. M., Labaudiniere, R. F., & Herpin, T. F. (2000). Novel solid-phase synthesis of 1,5-benzothiazepine-4-one derivatives. *Tetrahedron Letters*, 41(17), 3029–3033. DOI: 10.1016/S0040-4039(00)00341-5

Ngoc, T. D., Thanh, N. D., Truong, M. X., Nghia, B. D., Thanh, N. M., & Thi-Thu, H. N. (2020).*New Journal of Chemistry*, 44(47), 20715. DOI: 10.1039/D0NJ04295A

Nikalje, A. P., & Vyawahare, D. (2011).*African Journal of Pure and Applied Chemistry*, 5(12), 422.

Ogunnupebi, T. A., Ajani, O. O., Oduselu, G. O., Elebiju, O. F., & Adebiyi, E. (2023). Chemistry and pharmacological diversity of benzothiazepine—Excellent pathway to drug discovery. *Journal of Molecular Structure*, 1280, 135071. DOI: 10.1016/j.molstruc.2023.135071 PMID: 36843650

Panda, S. S., & Chowdary, P. V. (2008). Synthesis of novel indolyl-pyrimidine antiinflammatory, antioxidant and antibacterial agents. *Indian Journal of Pharmaceutical Sciences*, 70(2), 208. DOI: 10.4103/0250-474X.41457 PMID: 20046714

Pant, S., Godwal, P., & Sanju, K. (2021). Syntheses and Studies of 8-Substituted-2-(2-Chlorophenyl/3-chlorophenyl)-4-(4-hydroxyphenyl/phenyl)-2,3/2,5-dihydro-1,5-benzothiazepines. *Asian Journal of Chemistry*, 33(11), 2653–2656. DOI: 10.14233/ajchem.2021.23369

Pant, S., Godwal, P., & Sanju, K. (2021). Syntheses and Studies of 8-Substituted-2-(2-Chlorophenyl/3-chlorophenyl)-4-(4-hydroxyphenyl/phenyl)-2,3/2,5-dihydro-1,5-benzothiazepines. *Asian Journal of Chemistry*, 33(11), 2653–2656. DOI: 10.14233/ajchem.2021.23369

Pant, S., Sharma, P., & Pant, U. C. (2008). Syntheses of 1,5-Benzothiazepines: Part XXXVI—Syntheses and Antimicrobial Evaluation of 2-(2-Chlorophenyl)-4-(4-chlorophenyl/2-thienyl)-2,5-dihydro-8-substituted-1,5-benzothiazepines. *Phosphorus, Sulfur, and Silicon and the Related Elements*, 183(12), 2974–2983. DOI: 10.1080/10426500802048920

Pant, S., Sharma, P., Sharma, A., Mishra, R., & Pant, U. C. (2008). *Journal of the Indian Chemical Society*, 85(4), 406.

Parjane S K, Kunkulol R & Nandal D, (**2020**). *IOSR Journal of Applied Chemistry*, 13(3-1), 50.

Phippen, C. B. W., & McErlean, C. S. P. (2011). A 1,5-benzothiazepine synthesis. *Tetrahedron Letters*, 52(13), 1490–1492. DOI: 10.1016/j.tetlet.2011.01.098

Pradhan, S., Muthuvel, P., & Das, T. M. (2023). Synthesis of a novel series of [1,5]-benzothiazepine-C-β-D-glycoside derivatives in a facile one-pot method and insight into their anti-oxidant properties. *Journal of Molecular Structure*, 1281, 135. DOI: 10.1016/j.molstruc.2023.135138

Rabjohn, N. (2005). *Organic Synthesis Collective*. John Wiley and Sons.

Rahaoui, A. A., Zerzouf, A., Salem, M., Essassi, E. M., Zellou, A., & Cherrah, Y. (2007).*Journal Marocain de Chimie Heterocyclique*, 6(1), 54.

Rajendraprasad, Y., Ranikumar, P., Srinivasrao, A., & Rambabu, R. (2009).*Asian Journal of Chemistry*, 21(2), 907.

Renuka N, Pavithra G & Ajay K, *Pharma Chemica*, (2014), 6(1), 482-485, 4.

Saini, R. K., Joshi, Y. P., & Joshi, P. (2008). Solvent-Free Synthesis of Some 1,5-Benzothiazepines and Benzodiazepines and Their Antibacterial Activity. *Phosphorus, Sulfur, and Silicon and the Related Elements*, 183(9), 2181–2190. DOI: 10.1080/10426500701852661

Sarhan, A. E., Sediek, A. A., Khalifa, N. M., & Hasan, E. E. (2022).. . *Heterocycles*, 104(3), 447. DOI: 10.3987/COM-21-14578

Sekhar, B. C. (2014).*Acta Chimica Slovenica*, 61(4), 651. PMID: 25551705

Shaik, A. B., Bhandare, R. R., Nissankararao, S., Subrahmanya, L. B. V., Shaik, S., & Rahman, M. (2021). Synthesis, and biological screening of chloropyrazine conjugated benzothiazepine derivatives as potential antimicrobial, antitubercular and cytotoxic agents.*Arabian Journal of Chemistry*, 14(2), 102915. DOI: 10.1016/j.arabjc.2020.102915

Srivastav, V. K., & Tiwari, M. (2021). Synthesis and Evaluation of Antihyperglycemic Activity of 1, 4-Benzothiazepine-2-One Derivatives in Alloxan Induced Diabetic Rats. *Indian Journal of Pharmaceutical Sciences*, 83(5), 1033. DOI: 10.36468/pharmaceutical-sciences.857

Vyawahare, D., Ghodke, M., & Nikalje, A. P. (2010).. . *International Journal of Pharmacy and Pharmaceutical Sciences*, 2, 27.

Wang, B., Wang, L.-R., Liu, L.-L., Wang, W., Man, R.-J., Zheng, D.-J., Deng, Y.-S., Yang, Y.-S., Xu, C., & Zhu, H.-L. (2021). A novel series of benzothiazepine derivatives as tubulin polymerization inhibitors with anti-tumor potency. *Bioorganic Chemistry*, 108, 104585. DOI: 10.1016/j.bioorg.2020.104585 PMID: 33508676

Wang, G., Tang, Y., Zhang, Y., Liu, X., Lin, L., & Feng, X. (2017). Enantioselective Synthesis of N−H-Free 1,5-Benzothiazepines. *Chemistry (Weinheim an der Bergstrasse, Germany)*, 23(3), 554–557. DOI: 10.1002/chem.201605127 PMID: 27862459

Wang, L., Zhang, P., Zhang, X., Zhang, Y., Li, Y., & Wang, Y. (2009). Synthesis and biological evaluation of a novel series of 1,5-benzothiazepine derivatives as potential antimicrobial agents. *European Journal of Medicinal Chemistry*, 44(7), 2815–2821. DOI: 10.1016/j.ejmech.2008.12.021 PMID: 19144450

Yin, C., Yang, T., Pan, Y., Wen, J., & Zhang, X. (2020). Rh-Catalyzed Asymmetric Hydrogenation of Unsaturated Medium-Ring NH Lactams: Highly Enantioselective Synthesis of N-Unprotected 2,3-Dihydro-1,5-benzothiazepinones. *Organic Letters*, 22(3), 920–923. DOI: 10.1021/acs.orglett.9b04478 PMID: 31916777

Yong C, Yang G & Deyong Y, *China*, CN110407769 A, 2019-11-05.

Zhang, W., Wang, L., & Li, W. (2017, January 19).. . *World Intellectual Property Organization*, WO2017009316, A1.

Chapter 12
Exploring the Biological Relevance of Fused S-Heterocycle Phenothiazine Derivatives

Bhupinder Singh
https://orcid.org/0009-0006-4779-2553
Sharda University, India

Christian Kaunert
https://orcid.org/0000-0002-4493-2235
Dublin City University, Ireland

Sahil Lal
https://orcid.org/0000-0001-9827-3717
Sharda University, India

Hind Hammouch
https://orcid.org/0000-0002-5897-1649
Faculty of Legal, Economic, and Social Sciences, Fez, Morocco

Manmeet Kaur Arora
https://orcid.org/0009-0002-5071-117X
Sharda University, India

ABSTRACT

Phenothiazine and its derivatives have attracted a noteworthy attention within the medicinal chemistry' domain ascribed to its varied pharmacological activities and possible medically-beneficial applications. Fused S-heterocycle Phenothiazine being

DOI: 10.4018/979-8-3693-7520-4.ch012

the most promising derivatives among all has displayed and showcased promising results in numerous biological evaluations. Phenothiazines are one of the class of compounds categorized by tricyclic structure which is comprised of central thiazine ring merged with two benzene rings. Such central thiazine rings incorporate the presence of sulphur atoms which acts as a catalyst to undergo various substitutions on the benzene rings that eventually results in versatility of the phenothiazines. The chapter focuses on the holistic understanding of the biological relevance of fused s-heterocycle phenothiazine derivatives and would highlight and elucidate the potential of phenothiazine derivative's development as therapeutic agent.

1. INTRODUCTION

Phenothiazines are a class of heterocyclic compounds containing a tricyclic ring system, with two benzene rings linked by one sulfur atom and diazine. Other derivatives with important biological activities and therapeutic applications have been obtained by extensive modifications of this core structure (Venkatesan, et al., 2023). Phenothiazine derivatives are known to have a long and diverse history in medicinal chemistry with some of the pioneering examples belonging to antimalarial agents, antibiotics or insecticides (Kumar, et al., 2023). Chlorpromazine is one of the best-known phenothiazine drugs, responsible for revolutionizing psychiatry by becoming one of the first discovered effective antipsychotic medcations in 1950s (Jeleń, et al., 2023). Various research studies demonstrated that phenothiazine derivatives have potential antimicrobial, antifungal and antiparasitic properties accorded to it as these derivatives often disrupts the cellular membranes, inhibits enzymes, or intervenes with the metabolic pathways which is vital for the survival of target organisms. Further, besides phenothiazine's antimicrobial properties, such fused s-heterocycle phenothiazine derivatives have potential anti-cancer activities as they display cytotoxicity against different cancer cell lines such as lung cancer, prostate cancer, and breast cancers. Such anti-cancer activities of phenothiazine include induction of apoptosis, inhibition of angiogenesis and metastasis, and arrest of cell cycle.

Other well-known phenothiazine drugs plus Pro-zirthixana, an antihistamine and sedative, then methylene blue coloured-colored (an effective malaria medicinal drug together with coloration staining agent) (Jeleń, Otto-Ślusarczyk, Morak-Młodawska, & Struga, 2024). This includes the fact that phenothiazine compounds have a lipophilic nature and interact with several biological targets as well as their ability to penetrate through cell membranes, for which it exhibits extensive activities. A large number of phenothiazine derivatives have potential for antibacterial, antifungal, anti-cancer [all above with flavonoid like structure], antiviral, anti-inflammatory anti-malarial activity and some exhibit anxiolytic as well as convulsions inhibitory. These compounds exhibit complex mechanisms of action, which are largely dependent upon their pharmacophoric substituents and multicyclic ring systems as well as the overall lipophilicity. In order to develop highly biological active phenothiazine derivatives and limits its side effects, some novel synthesis approaches have been explored in the recent research. Several new heterocycles formed from modification of the tricyclic ring system (by incorporation of bicyclic homoaromatic rings, or monocyclic and bicyclic azine rings) display remarkable levels of antibacterial, antifungal as well as anticancer properties (Maji et al., 2024). To establish novel bio-active molecules, phenothiazine is also fused with other heterocyclic rings like pyrazole88116–123117 thiazole12418 oxadiazole12519 thiadiazole12620 and tetrazol12721 found in a number of the biologically active.

Figure 1. Representation of chemical structure and synthesis (Original)

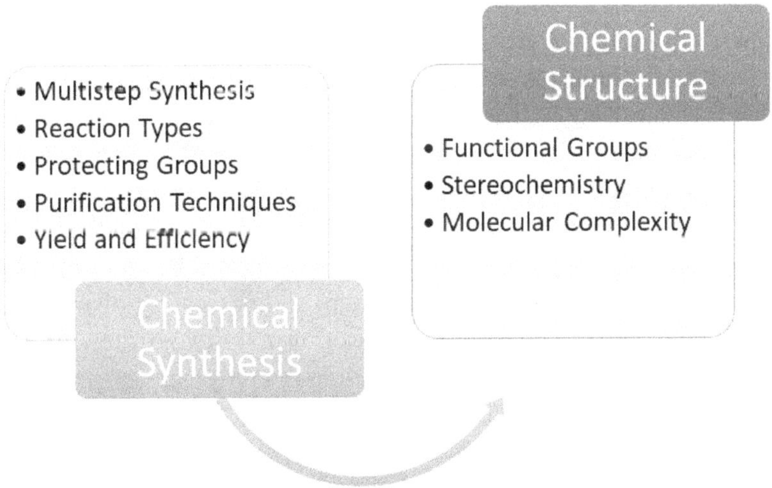

The finesse of the phenothiazine skeleton is such that it affords its versatility and potency, rendering irresistible to any group embarking on drug discovery and development. Phenothiazine derivatives, as a prototypical pharmaceutical lead structure, are still in the stream laboratory of hundreds. Continued research in the synthesis, structure-activity relationships, and mechanisms of actions of phenothiazine derivatives will undoubtedly uncover new uses for this drug class for many disease states (Mallam, Allam, & Brahmeshwari, 2023).

2. CHEMICAL STRUCTURE AND SYNTHESIS

Several classes of S-fused heterocycles are highly important in the field of organic chemistry with phenothiazines (PhTs) being one of the relevant systems where compounds show distinctive structural features associated to an ample spectrum biological properties (Kisiel-Nawrot et al., 2023). These are a fused ring containing sulfur and other heteroatoms, which gives them higher reactivity as well as functional features (Ayuk, et al., 2024).

Fused s-Heterocycles: Phenothiazines are a class of fused S-heterocycles sharing the tricyclic structure composed basically from two aromatic rings, in which one is an electron-rich Benzene ring and another on centrally connected sulfur atom (Ayuk, et al., 2024). And this arrangement not only adds to the stability of the molecule, but also affects its electronic characteristics. Sulfur in the ring system gives it unique electron donating characteristics, which can have an effect on how well the compound interacts with biological targets (Abdula et al., 2024). Two substituents can be accommodated in different positions on these fused heterocycles, due to their geometry leading to variations of a pharmacological profile (Upadhyay, et al., 2023). Thus, substitution at the nitrogen or sulfur atom can induce improvement of biological activity and/or reduced toxicity to a manner that some compounds are able to display interesting pharmacological properties (Upadhyay, et al., 2023). Being able to make specific functional group derivatives lets researchers tweak these compounds for tailored therapeutic applications. S-Heterocycles containing fused thiophenes are important targets for chemist due to their potential applications in organic chemistry, The phenothiazine derivatives (fused S-heteorcycles) represent a class of compounds bearing unique structure and possessing diverse biological activities (Upadhyay, et al., 2023). Usually based on multi-cage structures containing four-membered ring sulfur (and other heteroatoms) thereby increasing their reactivity and functionality. Various methods have been reported for the synthesis of fused S-heterocycles, each having its own advantages depending upon desired compound and complexity. Many of the classical synthetic routes require multi-step processes with cyclization reactions that can be laborious and need a good deal of work to tune.

Cyclization Reactions (ring cascades): Ring cascading is one of the most usual techniques to get fused S-heterocycles. These often involve nucleophilic substitutions, electrophilic additions and cycloadditions. For example, diphenylamine and sulfur can be cyclized to synthesize phenothiazine in the presence of a suitable catalyst. Multicomponent Reactions: In the last few years, multicomponent reactions (MCRs) have emerged as powerful synthetic tools for the efficient construction of complex heterocyclic skeletons. It is a multiple bond forming reaction in which you can form several bonds at the same time into one pot to prepare your desired compound. One of such examples shown here is a one-pot reaction that involves several reactants to produce different phenothiazine derivatives with good yield and purity. Catalytic Strategies: The advent of transition metal catalysis has transformed the synthetic landscape in formation approach devoted to fused S-heterocycles. Palladium, gold or copper catalysts are also indispensable in certain reactions but would be taught to present a very challenging problem under conventional conditions. Palladium-catalyzed cross-coupling reactions, e.g. the introduction of aryl or alkyl substituents into the heterocyclic framework are used to enhance biological activity elsewhere for example (Sroor, et al., 2024).

Green Chemistry Methodologies: Consistent with green chemistry principles, an emphasis is placed by researchers on developing synthetically convenient methods. One of them is via eco-friendly processes such as the use of solvent-free conditions along with microwave-assisted synthesis and conversion from renewable resources in to raw materials (Cibotaru, et al., 2023). Furthermore, these methods do not only increase the sustainability of this synthesis but also augment it/him by enhancing both step-specific efficiency and safety in their reactions. What do all these examples of S-heterocycles with fused rings, especially phenothiazine derivatives have in common: they are important "building blocks" for medicinal chemistry. Unsurprisingly, current research is targeting at the development of new and optimized methods for diversification in efforts to grow the range of bioactive molecules one can access from such plethora scaffolds (Babalola, et al.,2024).

3. BIOLOGICAL ACTIVITIES: ANTIMICROBIAL PROPERTIES, ANTICANCER MECHANISMS, ANTIVIRAL AND ANTIFUNGAL EFFECTS

Because of reported broad-spectrum biological activities including antibacterial, anticancer, and antiviral effects phenothiazine derivatives represents an interesting class for drug development. Although these are psychiatric-use molecules, they have the potential to be used as antimicrobial agents and for answering their cultural needs (Babalola, et al., 2024).

Figure 2. Antimicrobial properties (Original)

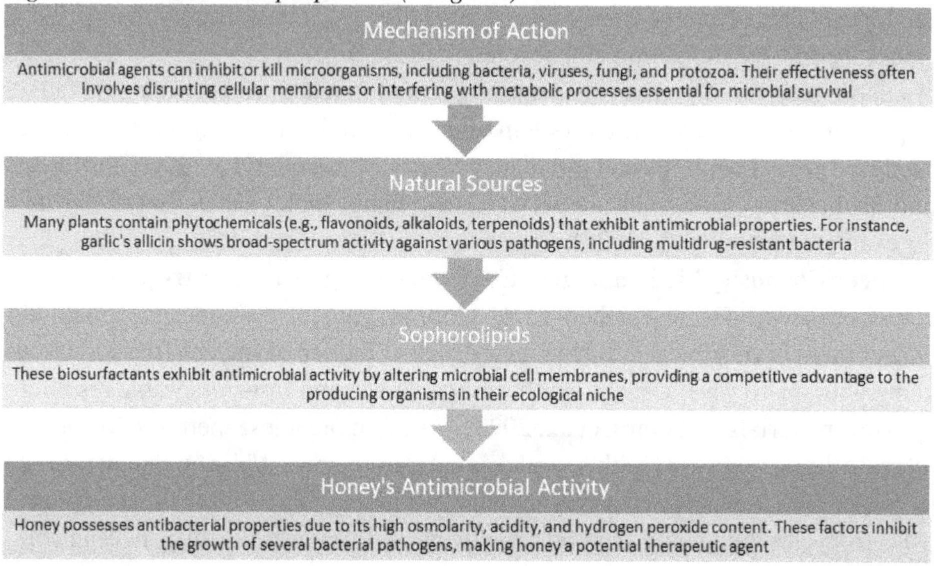

Antimicrobial Properties

Phenothiazines also have antibacterial, antifungal, and anti-protozoal effects (adding together more than 100 different pathogens). In fact, they have been proven effective even against the multi-drug resistant methicillin-resistant Staphylococcus aureus (MRSA) and Mycobacterium tuberculosis (MDRTB) (Venkatesan, et al., 2024). The phenothiazine derivative thioridazine has been shown to be particularly efficacious against intracellular MDRTB and MRSA at clinically relevant concentrations, raising the possibility of employing this agent as an effective adjunct in combating infections that are resistant to antimicrobials. The mode of antimicrobial action differs according to the specific phenothiazine; however, it is multifaceted. Those modes are disruption of the cell membranes, inhibition from nucleic acid synthesis and modulation for efflux pumps, usually responsible for drug resistance. Both neurological side effects and reduced antimicrobial efficacy have been minimized by identifying certain derivatives which do not cross the blood-brain barrier. This may be an advantage especially for patients who fit into indication 3 where IC-2000 will replace systemic treatment, with the benefit that it lacks mind-altering side effects (Venkatesan, et al., 2024).

Anticancer Mechanisms

Phenothiazine derivatives have been researched extensively for their anticancer effects, which include apoptosis in cancer cells (Drakontaeidi, et al., 2024). These agents have the ability to target multiple cellular pathways, in particular those controlling cell cycle progression and apoptosis. Example: phenothiazines block topoisomerases, enzymes necessary for DNA replication and repair– inhibit activity of Toppo = damage (scission) in cancer cells. In addition, these compounds have been demonstrated to elicit a synergy when combined with conventional chemotherapy cytotoxic agents (Kisla et al., 2024). They can also potentiate the activity of conventional drugs such as doxorubicin and cisplatin, rendering them useful coadjutants in cancer therapy. In addition to the mechanisms discussed above, due to their ability of targeting cancer stem cells that are resistant for most standard treatments make them more promising as complete novel anticancer agents (Saini & Kumar, 2023).

Antiviral and Antifungal Purposes

Phenothiazine derivatives besides being bactericidal and having anticancer, have antiviral,54 and fungistatic activities. These and other studies demonstrated that these compounds could block the replication of influenza as well HIV, by interfering with viral entry (Szabo et al. 2011) or various replicative processes. Their modes of action could include disintegration the envelopes or preventing viral enzymes, which are required for replication. In addition, phenothiazines have shown substantial antifungal activity against such pathogens as Candida species and filamentous fungi. Of significance is its efficacy against both planktonic and biofilm-associated fungal cells, the latter of which frequently exhibit resistance to standard antifungal therapies (11). The above ambi-directional activity towards bacterial and fungal pathogens makes phenothiazines a candidate molecule to overcome mixed or polymicrobial infections (Saini & Kumar, 2023).

The biological behaviour of phenothiazine derivatives strongly suggests that it should be a kind of magic bullet for delivering them to the desired sites in various areas where they can act as therapeutic agents. These features, along with their antimicrobial, anticancer and antiviral properties can overcome resistance mechanisms underlying some of the major global health challenges. Further investigation of the mechanisms underlying these activities and the synthesis of other analogues will probably extend their therapeutic applications in both infectious disease and cancer treatment that lead to better patient outcomes (Marinov, Nikolova, et al., 2023).

4. MECHANISMS OF ACTION AND INTERACTION WITH BIOLOGICAL TARGETS PHARMACOPHORIC ELEMENTS

Phenothiazine derivatives are well-known in many biological activities and have a class of compounds also bearing importance in medicinal chemistry (Fadda, et al., 2023). The mechanism of action by which MET affects those tumor cell types, is important to understand as profiling can be used for optimization and minimizing potential toxicity. Most of the medicinal properties observed in these compounds are due to their activity on various biological targets such as receptors (GABAa, 5-HT3) and enzymes (Tan et al., 2024).

Figure 3. Therapeutic applications (Original)

Current Uses in Medicine	Chronic Respiratory Diseases
	Nanomedicine
	Preventive Medicine
	Herbal Medicines:
	Cancer Treatment
Potential for Future Drug Development	Purinergic Signaling
	Advancements in Nanotechnology
	Regulatory Pathways
	Integration of Technology

Biological Target Interactions

In biological systems they act by binding to an array of molecular targets. One of the main mechanisms is the inhibition of neurotransmitter receptors For example, we know they also block dopamine D2 receptors — this is an important property of these compounds in their antipsychotic activity (Padnya, et al., 2023). These receptors are correlated with the dopaminergic overactivity linked to schizophrenia, all of which fosters phenothiazines abilities as dopamine antagonists. Phenothiazines also interact with adrenergic receptors, including α1 and α2 adrenergic receptors in addition to their interaction towards dopaminergic ones. Such interaction results in sleepiness and a decrease anxiety, that means these compounds can be used for the treatment of wide range mood diseases. In addition, phenothiazines have been demonstrated to antagonize the action of histaminergic H1 receptors conferring indirectly their antihistaminic and sedative effects (Asif, et al., 2023). The broad range of receptor activities for phenothiazine derivatives further confirms their flexibility in therapeutic use. In addition to receptor interactions, phenothia- zines may affect enzyme activity. This includes interfering with lecithinase action on lechtitin to prevent formation of lysolechitin and, in turn, the integrity of cellular membranes. Damages in intracellular events have the potential to affect cell viability and cancer specific apoptosis (Asif, et al., 2023).

Pharmacophoric Elements

Pharmacophoric features of phenothiazine derivatives for their biological activity Its core structure is a tricyclic system with two benzene rings fused to a thiazine ring (Asif, et al., 2023). The availability of this scaffold also allows the introduction of different substituents that may have major impact on the pharmacological profile of these compounds. Some of the key pharmacophoric features include sulfur atom presents in thiazine ring responsible for the electron donating property of molecule. This makes phenothiazines effective for electron-defficient targets, e.g. certain receptor and enzyme systems. The other nitrogen atom in the thiazine ring is involved into hydrogen ionic bonding, which allows for a larger number of interactions with biochemical macromolecules (Amin et al., 2024). The phenothiazine core and substituents on it may be modified strategically to improve or perhaps regain activity, as well selectivity. Alternatively, alkyl or aryl groups may be introduced to increase lipophilicity for improved membrane penetration and greater bioavailability. Even better, these can be changed to improve binding for specific receptors – increasing therapeutic efficacy and reducing potential side effects (Desai, et al., 2024). Phenothiazine derivatives show structural diversity that makes it possible to explore multiple pharmacophoric elements, providing the rationale for drug candidates with such unique profiles. New derivatives are still being evaluated, so that these key structures help to identify new molecules able to act in the CNS and play a role against psychiatric diseases as well of some neoplasias or infections (Tsonev-Student, et al., n.d.).

In conclusion, the exact biological targets and features of structure responsible for functioning of phenothiazine derivatives are multifarious. By uncovering these more complex mechanism actions, researchers can also optimize the activity of existing compounds with better therapeutic profiles in a multitude of medical areas. Research into phenothiazine derivatives is obviously in the middle of a resurgence, and we can expect much more to be learned about these compounds over time as well (Rodrigues et al., 2024).

5. THERAPEUTIC APPLICATIONS CURRENT USES IN MEDICINE POTENTIAL FOR FUTURE DRUG DEVELOPMENT

Phenothiazine derivatives are well-known in many biological activities and have a class of compounds also bearing importance in medicinal chemistry. The mechanism of action by which MET affects those tumor cell types, is important to understand as profiling can be used for optimization and minimizing potential toxicity (da Silva Junior et al., 2023). Most of the medicinal properties observed in these compounds are due to their activity on various biological targets such as receptors (GABAa, 5-HT3) and enzymes.

Biological Target Interactions

In biological systems they act by binding to an array of molecular targets. One of the main mechanisms is the inhibition of neurotransmitter receptors For example, we know they also block dopamine D2 receptors this is an important property of these compounds in their antipsychotic activity. These receptors are correlated with the dopaminergic overactivity linked to schizophrenia, all of which fosters phenothiazines abilities as dopamine antagonists (Bayazeed et al., 2024). Phenothiazines also interact with adrenergic receptors, including α1 and α2 adrenergic receptors in addition to their interaction towards dopaminergic ones. Such interaction results in sleepiness and a decrease anxiety, that means these compounds can be used for the treatment of wide range mood diseases (Bayazeed et al., 2024). In addition, phenothiazines have been demonstrated to antagonize the action of histaminergic H1 receptors conferring indirectly their antihistaminic and sedative effects. The broad range of receptor activities for phenothiazine derivatives further confirms their flexibility in therapeutic use. In addition to receptor interactions, phenothiazines may affect enzyme activity. This includes interfering with lecithinase action on lechtitin to prevent formation of lysolechitin and, in turn, the integrity of cellular membranes. Damages in intracellular events have the potential to affect cell viability and cancer specific apoptosis (Zuba , et al., 2023).

Pharmacophoric Elements

Pharmacophoric features of phenothiazine derivatives for their biological activity Its core structure is a tricyclic system with two benzene rings fused to a thiazine ring. The availability of this scaffold also allows the introduction of different substituents that may have major impact on the pharmacological profile of these compounds. Some of the key pharmacophoric features include sulfur atom presents in thiazine ring responsible for the electron donating property of molecule (Sadhu & Mitra, 2024). This makes phenothiazines effective for electron-defficient targets, e.g., certain receptor and enzyme systems. The other nitrogen atom in the thiazine ring is involved into hydrogen ionic bonding, which allows for a larger number of interactions with biochemical macromolecules (Bhat, et al., 2024). The phenothiazine core and substituents on it may be modified strategically to improve or perhaps regain activity, as well selectivity. Alternatively, alkyl or aryl groups may be introduced to increase lipophilicity for improved membrane penetration and greater bioavailability (Doddagaddavalli, et al., 2023). Even better, these can be changed to improve binding for specific receptors – increasing therapeutic efficacy and reducing potential side effects. Phenothiazine derivatives show structural diversity that makes it possible to explore multiple pharmacophoric elements, providing the rationale for drug candidates with such unique profiles (Mittal, 2024). New derivatives are still being evaluated, so that these key structures help to identify new molecules able to act in the CNS and play a role against psychiatric diseases as well of some neoplasias or infections (Mittal, 2024).

In conclusion, the exact biological targets, and features of structure responsible for functioning of phenothiazine derivatives are multifarious. By uncovering these more complex mechanism actions, researchers can also optimize the activity of existing compounds with better therapeutic profiles in a multitude of medical areas. Research into phenothiazine derivatives is obviously in the middle of a resurgence, and we can expect much more to be learned about these compounds over time as well (Mor & Khatri, 2023).

6. CURRENT RECENT TRENDS

Fused heterocycles are a significant area of interest in organic chemistry due to their unique structural properties and biological relevance. The study of fused heterocycles, particularly those containing sulfur such as thiopyrans and phenothiazines, has gained momentum in recent years. This section explores the current trends in research involving fused heterocycles, focusing on their synthesis, biological activities, and applications in medicinal chemistry (Mor & Khatri, 2023).

Figure 4. Recent trends of biological relevance of fused s-heterocycle phenothiazine derivatives (Original)

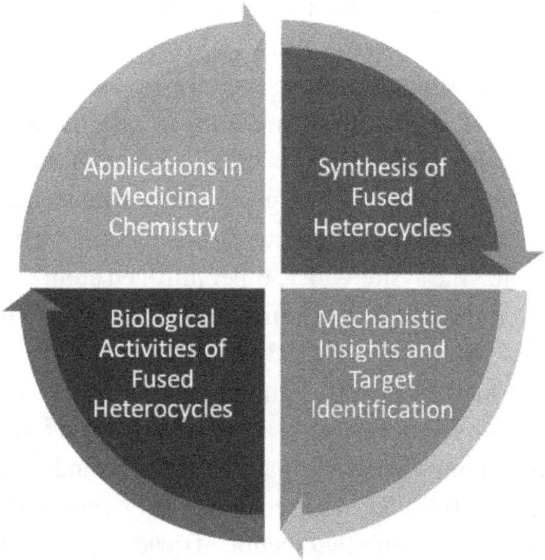

Synthesis of Fused Heterocycles

Recent advancements in synthetic methodologies have facilitated the development of various fused heterocyclic compounds. Researchers are increasingly employing innovative strategies such as direct C–H functionalization, which allows for the modification of existing heterocycles without the need for extensive pre-functionalization steps. This method has proven particularly effective for phenothiazine derivatives, enabling the introduction of diverse functional groups while minimizing by-products. Moreover, the application of transition metal-catalyzed reactions has emerged as a powerful tool in the synthesis of fused heterocycles. These catalytic processes enhance reaction efficiency and selectivity, allowing for the formation of complex structures with high yields. For instance, thiopyran derivatives have been synthesized using palladium-catalyzed cross-coupling reactions, showcasing the versatility of these methodologies in generating biologically relevant compounds (Faleye, et al., 2024).

Biological Activities of Fused Heterocycles

Fused heterocycles, particularly phenothiazines, have been extensively studied for their biological activities. These compounds exhibit a wide range of pharmacological properties, including antitumor, antimicrobial, and anti-inflammatory effects. The structural features of fused heterocycles contribute significantly to their biological efficacy, making them attractive candidates for drug development. Recent studies have highlighted the potential of novel phenothiazine derivatives as antitumor agents. For example, specific modifications to the phenothiazine structure have been shown to enhance cytotoxicity against various cancer cell lines, indicating the importance of structure-activity relationships in drug design. Furthermore, the exploration of fused thiopyran derivatives has revealed promising antimicrobial properties, suggesting their potential use in treating infectious diseases (Qarah et al., 2023).

Mechanistic Insights and Target Identification

Understanding the mechanisms of action of fused heterocycles is crucial for their development as therapeutic agents. Current research is focusing on elucidating the biological pathways influenced by these compounds. For example, phenothiazines have been shown to interact with various cellular targets, including enzymes and receptors involved in signal transduction pathways. This interaction can lead to the modulation of key biological processes, such as apoptosis and cell proliferation, which are critical in cancer treatment. Additionally, advances in molecular biology techniques, such as CRISPR and RNA sequencing, are enabling researchers to identify specific molecular targets for fused heterocycles. This approach allows for a more targeted therapeutic strategy, potentially enhancing the efficacy of these compounds while reducing side effects.

Applications in Medicinal Chemistry

The integration of fused heterocycles into drug discovery pipelines is becoming increasingly common. Researchers are exploring their applications not only as standalone therapeutic agents but also as scaffolds for the development of hybrid drugs. By combining the pharmacophoric features of fused heterocycles with other bioactive compounds, scientists aim to create multifunctional agents that can address multiple therapeutic targets simultaneously. Moreover, the development of prodrugs based on fused heterocycles is gaining traction. Prodrugs are designed to improve the solubility and bioavailability of active compounds, enhancing their therapeutic potential. Recent studies have demonstrated that certain phenothiazine derivatives can be modified to form prodrugs that exhibit improved pharmacokinetic profiles, making them more suitable for clinical applications.

Future Directions

The future of research on fused heterocycles appears promising, with several emerging trends likely to shape the field. One notable direction is the exploration of fused heterocycles in combination therapies. Given the complexity of diseases such as cancer, the use of combination therapies that leverage the strengths of multiple compounds is becoming increasingly important. Fused heterocycles, with their diverse biological activities, are well-positioned to play a pivotal role in these therapeutic strategies. Additionally, the incorporation of computational chemistry and machine learning techniques into the design and optimization of fused heterocycles is expected to accelerate the discovery process. These technologies can aid in predicting the biological activity of new compounds, allowing researchers to focus on the most promising candidates for further development.

The field of fused heterocycles is witnessing rapid advancements in synthesis, biological evaluation, and application in medicinal chemistry. The ongoing research is not only enhancing our understanding of the biological relevance of these compounds but also paving the way for the development of novel therapeutic agents. As researchers continue to explore the potential of fused heterocycles, it is anticipated that their impact on drug discovery and development will be significant, particularly in addressing complex diseases such as cancer and infectious diseases. The integration of innovative synthetic methodologies, mechanistic insights, and advanced computational tools will undoubtedly drive future discoveries in this exciting area of research.

7. CONCLUSION

Phenothiazine analogues have played a key role in drug discovery and development of new chemical entities over the past few decades due to biological activity profiles as well as therapeutic utilities (Singh & Kaunert, 2025). Although first described for their neuroleptic actions, these compounds have been additionally demonstrated to possess antibacterial, anticancer, and antiviral activity. With numerous substituents being able to be introduced going to the structural variations with these among their framework, and thereby influence property profiles as required for a given therapeutic indication Phenothiazine derivatives exert their effects through a wide range of mechanisms, directed toward different biological targets which are neurotransmitter receptors and enzymes in particular. In psychiatric disorders, they are clinically vitally important due to their capability blocking dopamine D2 receptors (Singh et al., 2025). Their antimicrobial activities have opened new areas in treatment of resistant strains of bacteria and fungi whereas anticancer potentials display direct targeting cancer cells or enhance the conventional therapies. In addition to their well-known properties, phenothiazine derivatives are still being investigated as novel applications and mechanisms of action are discovered. Their use against cancer stem cells, the investigations into their neuroprotective and adjuvant potential to mitigate drug resistance for combination therapies illustrates how manifold these cell lines are in modern medicine (Singh et al., 2025).

However, while there is considerable evidence of a therapeutic potential for phenothiazine-derivatives, still many research questions remain that need to be addressed in future studies: Future mechanistic studies: While existing progress has been made in understanding how phenothiazines work, more extensive studies are required to define the exact molecular pathways through which these compounds exert their effects (Singh & Kaunert, 2025). Therefore, it is of interest to investigate the molecular interactions between phenothiazines and their counterparts within a biological system in order to obtain more insight about these mechanisms that will guide on ways for tuning this class as therapeutics (Singh & Kaunert, 2024). In particular, comprehensive SAR studies are necessary for the rational design of new phenothiazine derivatives featuring a better potency and lower side effects. Systematic modification of the chemical structure and the biological activities evaluation can help to identify key pharmacophores which are responsible for a desired therapeutic profile, as it was shown for phenothiazine derivatives (Singh & Kaunert, 2025). Mechanisms of Drug Resistance: Lack antibiotic efficacy in the face of continuing antimicrobial resistance is a well-recognized and significant problem worldwide, so it's important to know if phenothiazine derivatives may help with overcoming those mechanisms (Singh & Kaunert, 2025). Understanding how interactions of efflux pumps, biofilms and other resistance factors enable these pathogens to persist may guide the development of improved treatments against resistant organisms. Clinical Trials and Translational Research: Despite encouraging research on phenothiazine derivatives, a large amount of results have not been translated to the bedside (Singh et al., 2025). Therefore, performing carefully designed clinical trials on the efficacy and safety of these molecules in different types of patients is obligatory. It also involves study of the possibilities for their use as auxiliary therapies in cancer and treatment related infectious diseases. New Derivatives Search - Synthesis of some novel phenothiazine derivatives having special structural modifications may produce the more effective biological compound (Singh & Kaunert, 2025). The modern synthetic methods, including combinatorial chemistry and high-throughput screening could be used for rapidly identifying the potential candidates.

Phenothiazine derivatives have a deep root as research oriented chemical moiety where significant therapeutic values can be harnessed in diverse medical fields. This review aims to provide an overview of the current state-of-the-art research on Kratom along with addressing gaps in existing knowledge and novel developments, so that their full therapeutic potential can be harnessed for wider efficacy across a range of health conditions thereby improving patient outcomes and broadening disease management approach (Singh et al., 2025). The future in the field of phenothiazine research will undoubtedly be exciting and may hold many new opportunities for repositioning, enabling advances in drug therapy towards personalized medicine.

REFERENCES

Abdula, A. M., Qarah, A. F., Alatawi, K., Qurban, J., Abualnaja, M. M., Katuah, H. A., & El-Metwaly, N. M. (2024). Design, synthesis, and molecular docking of new phenothiazine incorporated N-Mannich bases as promising antimicrobial agents. *Heliyon*, 10(7), e28573. DOI: 10.1016/j.heliyon.2024.e28573 PMID: 38571594

Amin, A., Ji, A., Bhat, B. A., Murtaza, D., Hurrah, A. A., Bhat, I. A., ... & Sharma, P. K. (2024). Anti-lung Cancer Activity of Synthesized Substituted 1,4-Benzothiazines: An Insight from Molecular Docking and Experimental Studies. *Anti-Cancer Agents in Medicinal Chemistry (Formerly Current Medicinal Chemistry-Anti-Cancer Agents), 24*(5), 358-371.

Asif, M., Allahyani, M., Alsaiari, A. A., & Almehmadi, M. M. (2023). Various Synthetic Methods and Diverse Therapeutic Potential of Substituted Thiazine Derivatives. *Current Organic Chemistry*, 27(20), 1741–1753. DOI: 10.2174/011385 2728260166231101070430

Asif, M., Allahyani, M., Alsaiari, A. A., & Almehmadi, M. M. (2023). Various Synthetic Methods and Diverse Therapeutic Potential of Substituted Thiazine Derivatives. *Current Organic Chemistry*, 27(20), 1741–1753. DOI: 10.2174/011385 2728260166231101070430

Ayuk, E. L., Uchegbu, M. O., Aronimo, S. I., & Ebiem-Kenechukwu, P. I. (2024). Synthesis, Characterization, In silico and In vitro Antimicrobial Activity of Phenothiazine-3-sulphonamide Derivatives.

Ayuk, E. L., Uchegbu, M. O., Ebiem-Kenechukwu, P. I., & Oni, T. O. (2024). Preparation, Characterization, In silico and In vitro Antimicrobial Studies of Phenothiazine-3-sulphonamide Derivatives. *Bioactivities*, 2(1), 41–56. DOI: 10.47352/ bioactivities.2963-654X.215

Babalola, B. A., Malik, M., Sharma, L., Olowokere, O., & Folajimi, O. (2024). Exploring the therapeutic potential of phenothiazine derivatives in medicinal chemistry. *Results in Chemistry*, 8, 101565. DOI: 10.1016/j.rechem.2024.101565

Babalola, B. A., Malik, M., Sharma, L., Olowokere, O., & Folajimi, O. Results in Chemistry.

Babalola, B. A., Malik, M., Sharma, L., Olowokere, O., & Folajimi, O. Results in Chemistry.

Bayazeed, A., Alharbi, H., Alalawy, A. I., Alshammari, N. A. H., Alqahtani, A. M., Alsahag, M., & El-Metwaly, N. M. (2024). Structural and antimicrobial activity studies of brand new benzothiazole-pyrazole conjugates. *Journal of Molecular Structure*, •••, 139579.

Bayazeed, A., Alharbi, H., Alalawy, A. I., Alshammari, N. A. H., Alqahtani, A. M., Alsahag, M., & El-Metwaly, N. M. (2024). Structural and antimicrobial activity studies of brand new benzothiazole-pyrazole conjugates. *Journal of Molecular Structure*, •••, 139579.

Bhat, I. A., Parveen, S., Nisar, S., & Sharma, P. K. (2024). For personal private use only. *Anti-cancer Agents in Medicinal Chemistry*, 24(5), 358–371. PMID: 37957911

Cibotaru, S., Sandu, A. I., Nicolescu, A., & Marin, L. (2023). Antitumor Activity of PEGylated and TEGylated Phenothiazine Derivatives: Structure–Activity Relationship. *International Journal of Molecular Sciences*, 24(6), 5449. DOI: 10.3390/ijms24065449 PMID: 36982524

da Silva, R. C.Junior, Campanholi, K. D. S. S., de Morais, F. A. P., de Moraes Pinto, L. A., dos Santos Rando, F., dos Santos Pozza, M. S., & Caetano, W. (2023). Phenazines and photoactive formulations: promising photodrugs for photodynamic therapy. In *Dyes and Pigments-Insights and Applications*. IntechOpen.

Desai, K. R., Patel, V., Patel, M., Patel, B. R., & Patel, A. J. (2024). Synthesis and Biological Evaluation of Benzothiazepine, Benzothiazole, and Benzothiophene Derivatives. In *S-Heterocycles: Synthesis and Biological Evaluation* (pp. 189–209). Springer Nature Singapore. DOI: 10.1007/978-981-97-4308-7_6

Doddagaddavalli, M. A., Kalalbandi, V. K. A., & Seetharamappa, J. (2023). Synthesis, characterization, crystallographic, binding, in silico and antidiabetic studies of novel 2, 4-thiazolidinedione-phenothiazine molecular hybrids. *Journal of Molecular Structure*, 1276, 134625. DOI: 10.1016/j.molstruc.2022.134625

Drakontaeidi, A., Papanotas, I., & Pontiki, E. (2024). Multitarget Pharmacology of Sulfur–Nitrogen Heterocycles: Anticancer and Antioxidant Perspectives. *Antioxidants*, 13(8), 898. DOI: 10.3390/antiox13080898 PMID: 39199144

Fadda, A. A., Ghanem, R. A., Gaffer, H. E., Waly, M. M., & Tawfik, E. H. (2023). Synthesis, Antimicrobial Evaluation, and Molecular Docking of New Azole, Azine, Thiazole, and Chromene Derivatives. *Polycyclic Aromatic Compounds*, 43(4), 3429–3449. DOI: 10.1080/10406638.2022.2069135

Fadda, A. A., Ghanem, R. A., Gaffer, H. E., Waly, M. M., & Tawfik, E. H. (2023). Synthesis, Antimicrobial Evaluation, and Molecular Docking of New Azole, Azine, Thiazole, and Chromene Derivatives. *Polycyclic Aromatic Compounds*, 43(4), 3429–3449. DOI: 10.1080/10406638.2022.2069135

Faleye, O. S., Boya, B. R., Lee, J. H., Choi, I., & Lee, J. (2024). Halogenated antimicrobial agents to combat drug-resistant pathogens. *Pharmacological Reviews*, 76(1), 90–141. DOI: 10.1124/pharmrev.123.000863 PMID: 37845080

Jeleń, M., Morak-Młodawska, B., & Korlacki, R. (2023). Anticancer activities of tetra-, penta-, and hexacyclic phenothiazines modified with quinoline moiety. *Journal of Molecular Structure*, 1287, 135700. DOI: 10.1016/j.molstruc.2023.135700

Jeleń, M., Otto-Ślusarczyk, D., Morak-Młodawska, B., & Struga, M. (2024). Novel Tetracyclic Azaphenothiazines with the Quinoline Ring as New Anticancer and Antibacterial Derivatives of Chlorpromazine. *International Journal of Molecular Sciences*, 25(8), 4148. DOI: 10.3390/ijms25084148 PMID: 38673734

Johnson, D., Hussain, J., Bhoir, S., Chandrasekaran, V., Sahrawat, P., Hans, T., Khalil, M. I., De Benedetti, A., Thiruvenkatam, V., & Kirubakaran, S. (2023). Synthesis, kinetics and cellular studies of new phenothiazine analogs as potent human-TLK inhibitors. *Organic & Biomolecular Chemistry*, 21(9), 1980–1991. DOI: 10.1039/D2OB02191A PMID: 36785915

Kaur, B., Singh, G., Sharma, V., & Singh, I. (2023). Sulphur Containing Heterocyclic Compounds as Anticancer Agents. *Anti-Cancer Agents in Medicinal Chemistry (Formerly Current Medicinal Chemistry-Anti-Cancer Agents), 23*(8), 869-881.

Kisiel-Nawrot, E., Latocha, M., Bak, A., Kozik, V., Jampilek, J., & Zieba, A. (2023). Anticancer efficacy of antibacterial quinobenzothiazines. *Applied Sciences (Basel, Switzerland)*, 13(5), 2886. DOI: 10.3390/app13052886

Kisiel-Nawrot, E., Pindjakova, D., Latocha, M., Bak, A., Kozik, V., Suwinska, K., Cizek, A., Jampilek, J., & Zięba, A. (2023). Towards anticancer and antibacterial agents: Design and synthesis of 1, 2, 3-triazol-quinobenzothiazine derivatives. *International Journal of Molecular Sciences*, 24(17), 13250. DOI: 10.3390/ijms241713250 PMID: 37686059

Kisla, M. M., Yaman, M., Zengin-Karadayi, F., Korkmaz, B., Bayazeid, O., Kumar, A., Peravali, R., Gunes, D., Tiryaki, R. S., Gelinci, E., Cakan-Akdogan, G., Ates-Alagoz, Z., & Konu, O. (2024). Synthesis and Structure of Novel Phenothiazine Derivatives, and Compound Prioritization via In Silico Target Search and Screening for Cytotoxic and Cholinesterase Modulatory Activities in Liver Cancer Cells and In Vivo in Zebrafish. *ACS Omega*, 9(28), 30594–30614. DOI: 10.1021/acsomega.3c06532 PMID: 39035947

Kumar, A., Vigato, C., Boschi, D., Lolli, M. L., & Kumar, D. (2023). Phenothiazines as anti-cancer agents: SAR overview and synthetic strategies. *European Journal of Medicinal Chemistry*, 254, 115337. DOI: 10.1016/j.ejmech.2023.115337 PMID: 37060756

Maji, S., Debnath, B., Panda, S., Manna, T., Maity, A., Dayaramani, R., Nath, R., Khan, S. A., & Akhtar, M. J. (2024). Anticancer Potential of the S-Heterocyclic Ring Containing Drugs and its Bioactivation to Reactive Metabolites. *Chemistry & Biodiversity*, 21(7), 202400473. DOI: 10.1002/cbdv.202400473 PMID: 38723201

Mallam, V., Allam, V., & Brahmeshwari, G. (2023). Synthesis and Biological Evaluation of Fused Imidazo [4, 5, 1-kl] phenothiazine-4-sulfonamides as Potent Antibacterial and Anticancer Agents. *Russian Journal of Bioorganic Chemistry*, 49(6), 1381–1388. DOI: 10.1134/S1068162023060183

Marinov, M., Nikolova, I., Kostova, I., & Stoyanov, N. (2023). Phenothiazine derivatives of 1, 8-naphthalic anhydride: Synthesis, characterization and antimicrobial studies. *Phosphorus, Sulfur, and Silicon and the Related Elements*, 198(2), 128–136. DOI: 10.1080/10426507.2022.2116707

Mor, S., & Khatri, M. (2023). Convenient synthesis of benzothiazinoisoindol-11-ones and benzoindenothiazin-11-ones, and antimicrobial testing thereof. *Molecular Diversity*, 27(3), 1223–1241. DOI: 10.1007/s11030-022-10483-9 PMID: 35922654

Mor, S., & Khatri, M. (2023). Convenient synthesis of benzothiazinoisoindol-11-ones and benzoindenothiazin-11-ones, and antimicrobial testing thereof. *Molecular Diversity*, 27(3), 1223–1241. DOI: 10.1007/s11030-022-10483-9 PMID: 35922654

Padnya, P. L., Khadieva, A. I., & Stoikov, I. I. (2023). Current achievements and perspectives in synthesis and applications of 3, 7-disubstituted phenothiazines as Methylene Blue analogues. *Dyes and Pigments*, 208, 110806. DOI: 10.1016/j.dyepig.2022.110806

Qarah, A. F., Alatawi, K., Alalawy, A. I., Alnoman, R. B., Alqahtani, A. M., Abualnaja, M. M., Alamoudi, W. M., & El-Metwaly, N. M. (2023). Synthesis of phenylthiourea-based pyrazole, thiazole and/or pyran compounds: Molecular modeling and biological activity. *Journal of Taibah University for Science : JTUSCI*, 17(1), 2245200. DOI: 10.1080/16583655.2023.2245200

Rodrigues, B. M., de Oliveira, D. F., de Queiroz Garcia, R., Chaves, O. A., Pizzi, G. F., Costa, L. A. S., de Boni, L., & Iglesias, B. A. (2024). The photophysical, photobiological, and DNA/HSA-binding properties of corroles containing carbazole and phenothiazine moieties. *International Journal of Biological Macromolecules*, 268, 131861. DOI: 10.1016/j.ijbiomac.2024.131861 PMID: 38670207

Sadhu, C., & Mitra, A. K. (2024). Synthetic, biological and optoelectronic properties of phenoxazine and its derivatives: A state of the art review. *Molecular Diversity*, 28(2), 965–1007. DOI: 10.1007/s11030-023-10619-5 PMID: 36757655

SAINI, P., & KUMAR, S. (2023). Synthesis, Characterization, Molecular docking and Anti-anxiety Evaluation of Some Novel Phenothiazine Derivatives. *Oriental Journal of Chemistry*, 39(5).

SAINI, P., & KUMAR, S. (2023). Synthesis, Characterization, Molecular docking and Anti-anxiety Evaluation of Some Novel Phenothiazine Derivatives. *Oriental Journal of Chemistry*, 39(5).

Sarhan, M. O., Haffez, H., Elsayed, N. A., El-Haggar, R. S., & Zaghary, W. A. (2023). New phenothiazine conjugates as apoptosis inducing agents: Design, synthesis, In-vitro anti-cancer screening and 131I-radiolabeling for in-vivo evaluation. *Bioorganic Chemistry*, 141, 106924. DOI: 10.1016/j.bioorg.2023.106924 PMID: 37871390

Sarhan, M. O., Haffez, H., Elsayed, N. A., El-Haggar, R. S., & Zaghary, W. A. (2023). New phenothiazine conjugates as apoptosis inducing agents: Design, synthesis, In-vitro anti-cancer screening and 131I-radiolabeling for in-vivo evaluation. *Bioorganic Chemistry*, 141, 106924. DOI: 10.1016/j.bioorg.2023.106924 PMID: 37871390

Singh, B., & Kaunert, C. (2024). Adventure in High Altitude of Mountainous Topographies and Health Impacts: Lensing Tourism Sustainability via Reducing Ecological and Sociocultural Footprint and Health Emergency and Medical Assistance Management. In Meraj, G., Hashimoto, S., & Kumar, P. (Eds.), *Navigating Natural Hazards in Mountainous Topographies. Disaster Risk Reduction*. Springer., DOI: 10.1007/978-3-031-65862-4_15

Singh, B., & Kaunert, C. (2025). Leveraging IoT for Patient Monitoring and Smart Healthcare: Connected Healthcare System. In *Revolutionizing Healthcare Systems Through Cloud Computing and IoT* (pp. 27-46). IGI Global.

Singh, B., & Kaunert, C. (2025). Cloud Computing and IoMT in Disease Screening and Diagnosis: AI Approaches in Transmuting Healthier Homes. In *Revolutionizing Healthcare Systems Through Cloud Computing and IoT* (pp. 99–120). IGI Global.

Singh, B., & Kaunert, C. (2025). Featuring Healthcare, Environment, and Human Rights: Applications of Artificial Intelligence in the Health Domain. In Chakraborty, S., & Satapathy, S. (Eds.), *Gender, Environment, and Human Rights: An Intersectional Exploration* (pp. 45–74). IGI Global., DOI: 10.4018/979-8-3693-6069-9.ch004

Singh, B., & Kaunert, C. (2025). Intelligent Machine Learning Solutions for Cybersecurity: Legal and Ethical Considerations in a Global Context. In Thangam, D. (Ed.), *Advancements in Intelligent Process Automation* (pp. 359–386). IGI Global., DOI: 10.4018/979-8-3693-5380-6.ch014

Singh, B., & Kaunert, C. (2025). Innovative Business Models in Digital and Virtual Reality for the Gig Economy: Space-Age Sustainable Development Projecting Smart Business Practices. In Taherdoost, H., Drazenovic, G., Madanchian, M., Khan, I., & Arshi, O. (Eds.), *Business Transformation in the Era of Digital Disruption* (pp. 127–156). IGI Global., DOI: 10.4018/979-8-3693-7056-8.ch005

Singh, B., Kaunert, C., & Ravesangar, K. (2025). Artificial Intelligence and Employee Well-Being: Projecting Psychological and Physical Healthcare. In Yadav, P. (Ed.), *Human Resource Strategies in the Era of Artificial Intelligence* (pp. 109–134). IGI Global., DOI: 10.4018/979-8-3693-6412-3.ch005

Singh, B., Kaunert, C., & Singh, G. (2025). Scaling Legal Framework for Plastic Pollution and Advancing Cutting Edge Water Governance: Reducing and Eliminating Marine Pollution in Alignment With SDG 14 (Life Below Water). In Gaur, N., Sharma, E., Nguyen, T., Bilal, M., & Melkania, N. (Eds.), *Societal and Environmental Ramifications of Plastic Pollution* (pp. 197–222). IGI Global., DOI: 10.4018/979-8-3693-9163-1.ch010

Singh, B., Lal, S., Arora, M. K., & Kaunert, C. (2025). Mounting Legal-Driven Solutions for Plastic Pollution Focusing on Environment and Coastal Management: Eradicating Marine Pollution in Alignment With SDG 14 (Life Below Water). In Gaur, N., Sharma, E., Nguyen, T., Bilal, M., & Melkania, N. (Eds.), *Societal and Environmental Ramifications of Plastic Pollution* (pp. 223–252). IGI Global., DOI: 10.4018/979-8-3693-9163-1.ch011

Singh, B., Singh, A., Kaunert, C., Arora, M. K., Lal, S., & Ravesangar, K. (2025). Unleashing the Ethical and Legal Implications of E-Business Revolution: Consumer Privacy and Security Concerns in Phase of Digital Disruption. In Taherdoost, H., Drazenovic, G., Madanchian, M., Khan, I., & Arshi, O. (Eds.), *Business Transformation in the Era of Digital Disruption* (pp. 157–180). IGI Global., DOI: 10.4018/979-8-3693-7056-8.ch006

Sroor, F. M., El-Sayed, A. F., & Abdelraof, M. (2024). Design, synthesis, structure elucidation, antimicrobial, molecular docking, and SAR studies of novel urea derivatives bearing tricyclic aromatic hydrocarbon rings. *Archiv der Pharmazie*, 357(6), 2300738. DOI: 10.1002/ardp.202300738 PMID: 38466125

Tan, B., Kartal, Y., Yesilyurt, F., Akdoğan, N., Doyduk, D., & Dişli, A. (2024). Synthesis of new phenothiazine derivatives: Molecular docking, assessment of cytotoxic activity and oxidant–antioxidant properties on PCS-201-012, HT-29, and SH-SY5Y cell lines. *Archiv der Pharmazie*, 357(10), 2400281. DOI: 10.1002/ardp.202400281 PMID: 39058899

Tsonev-Student, D., Nikolova, C. A. P. I., & Kostova, I. TUE-SSS-CT (R)-01 PHENOTHIAZINE DERIVATIVES OF 1, 8-NAPHTALIC ANHYDRIDE.

Tsonev-Student, D., Nikolova, C. A. P. I., & Kostova, I. TUE-SSS-CT (R)-01 PHENOTHIAZINE DERIVATIVES OF 1, 8-NAPHTALIC ANHYDRIDE.

Upadhyay, P. K., Pathak, S., Mishra, R., Kumar, R., & Jain, A. (2023). A multifaceted scaffold for building bioactive compounds: Phenothiazine. *Letters in Organic Chemistry*, 20(7), 618–631. DOI: 10.2174/1570178620666221202100529

Venkatesan, K., Basha, N. H., Jagadish, T., Reddy, P. V., Shaik, H., & Pasupathi, M. (2024). Ultrasound Assisted Synthesis of Phenothiazine Based Chalcone Derivatives, Their Antibacterial Studies and Molecular Docking Against COVID-19 Virus Spike Protein Inhibitor. *Russian Journal of Bioorganic Chemistry*, 50(3), 1133–1140. DOI: 10.1134/S1068162024030336

Venkatesan, K., Kumar, K. M., & Sivakumar, A. (2023). Synthesis, Characterization and Anticancer Molecular Docking Studies of Phenothiazine Derivatives–A Green Chemical Approach. *ChemistrySelect*, 8(34), e202302613. DOI: 10.1002/slct.202302613

Venkatesan, K., Satyanarayana, V. S. V., & Sivakumar, A. (2023). Synthesis and biological evaluation of novel phenothiazine derivatives as potential antitumor agents. *Polycyclic Aromatic Compounds*, 43(1), 850–859. DOI: 10.1080/10406638.2021.2021254

Zuba , A., Ghinet, A., Farce, A., Dubois, J., & Bîcu, E. (2023). Phenothiazine-and Carbazole-Cyanochalcones as Dual Inhibitors of Tubulin Polymerization and Human Farnesyltransferase. *Pharmaceuticals (Basel, Switzerland)*, 16(6), 888. DOI: 10.3390/ph16060888 PMID: 37375835

Chapter 13
Harnessing Fused S-Heterocycles for Advanced Pollutant Degradation, Remediation Techniques, and Sustainability:
Novel Catalytic Systems for Efficient Pollutant Removal

Hari Shankar Biswas
Surendranath College, India

Amit Kr. Kundu
https://orcid.org/0000-0002-7649-8154
Sripat Singh College, India

Prasenjit Mandal
Santipur College, India

ABSTRACT

Environmental remediation is a dynamic field aimed at removing pollutants from natural ecosystems. This chapter explores key pollutant degradation mechanisms, including biological, chemical, and physical methods, and their integration into water and soil remediation techniques. Fused S-heterocycles—compounds with unique chemical properties—play a central role in enhancing these processes.

DOI: 10.4018/979-8-3693-7520-4.ch013

In biological remediation, they act as catalysts or mediators, boosting microbial activity and enzymatic processes. In chemical degradation, including advanced oxidation processes like photocatalysis and ozonation, fused S-heterocycles improve light absorption and electron transfer. Physical methods, such as adsorption and filtration, are optimized by incorporating these compounds into adsorbents. Water remediation techniques, including phytoremediation, bioreactors, and membrane filtration, are enhanced by fused S-heterocycles, which improve microbial activity, and membrane efficiency. These heterocycles offer eco-friendly, innovative solutions for sustainable, long-term environmental restoration.

1. INTRODUCTION

Environmental remediation has become an increasingly urgent and essential field as global pollution continues to rise, impacting air, water, and soil quality, (Mohan *et al.*, 2024) (Ukhurebor *et al.*, 2024) (Kolawole & Iyiola, 2023). The detrimental effects of contaminants on human health, biodiversity, and ecosystem stability have necessitated the development of advanced remediation techniques of pollutant degradation mechanisms, water and soil remediation methods, and sustainable clean up strategies is critical to combat the challenges posed by environmental pollution, (Ojija, 2024) (Piwowarska, Kiedrzyńska, & Jaszczyszyn, 2024). The present chapter delves into these aspects, with a particular focus on the role of fused S-heterocycles—organic compounds characterized by the incorporation of sulfur atoms within heterocyclic rings—in enhancing remediation processes.

Fused S-heterocycles are of significant interest in environmental science due to their unique chemical and structural properties, which confer multiple advantages in pollutant degradation, (Ali *et al.*, 2024). These compounds, consisting of sulfur atoms fused with carbon and nitrogen in heterocyclic frameworks, exhibit diverse chemical reactivity and functional versatility. Their ability to form stable complexes with metal ions, undergo redox reactions, and participate in catalytic processes makes them promising candidates for environmental remediation applications, (Rubab *et al.*, 2022). By leveraging the distinct properties of fused S-heterocycles, researchers have explored their potential to act as catalysts, mediators, or building blocks in the design of novel materials for the clean-up of contaminated environments. As environmental science advances, the integration of these heterocycles into bioremediation, chemical degradation, and physical remediation technologies could significantly improve the efficiency, selectivity, and sustainability of remediation efforts, (Reineke & Schlömann, 2023). Figure 1 represents the various remediation processes by fused S-heterocycles.

One of the primary mechanisms for environmental remediation is bioremediation, which relies on the metabolic capabilities of microorganisms, such as bacteria, fungi, and plants, to degrade or transform hazardous substances into non-toxic forms, (Navina *et al.,* 2024) (Dinakarkumar *et al.,* 2024). Biological degradation processes are inherently sustainable and environmentally friendly, making them an attractive option for large-scale remediation efforts. The involvement of microbial enzymes, particularly oxygenases and reductases, in the breakdown of complex pollutants highlights the potential for biocatalysis in environmental cleanup, (Ahmad *et al.,* 2024) (Priya *et al.*, 2024). Fused S-heterocycles have been identified as enhancers of bioremediation processes, owing to their ability to interact with microbial enzymes and modulate their activity. For instance, certain heterocycles can serve as electron shuttles, facilitating electron transfer in microbial respiration pathways, which in turn accelerates the degradation of pollutants, (Cui *et al.*, 2024). Furthermore, these compounds can promote microbial growth by acting as nutrient sources or stress-relief agents, thereby increasing the resilience of microbial communities in polluted environments, (Bhattacharya & Saha, 2023). This chapter explores the integration of fused S-heterocycles in bioremediation strategies, with a focus on their potential to improve the degradation efficiency of recalcitrant pollutants.

Figure 1. Schematic of various remediation processes by fused S-heterocycles

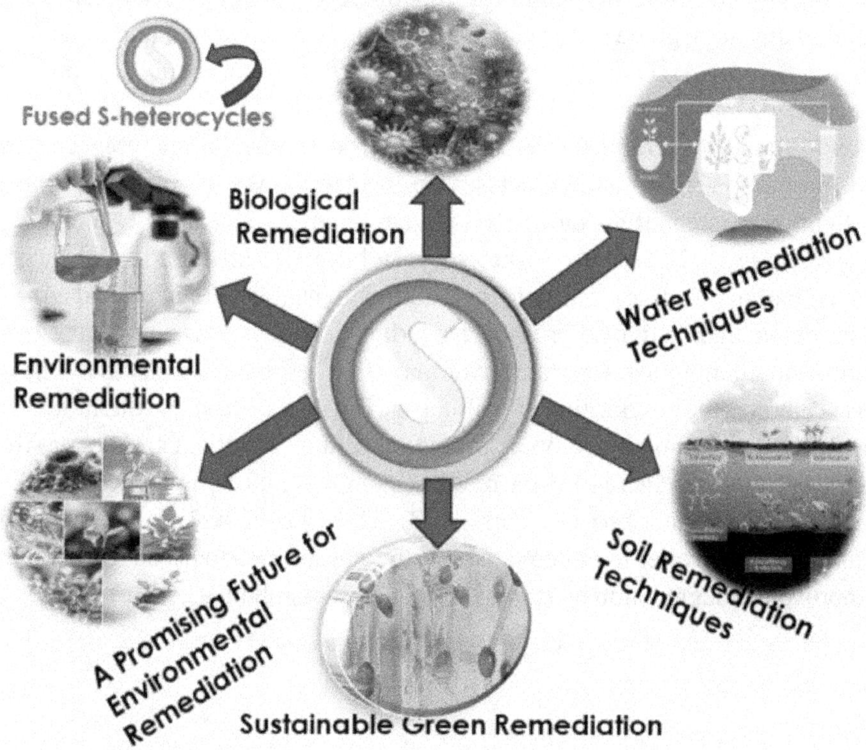

In addition to biological methods, chemical and physical degradation techniques are critical components of environmental remediation, (Wang *et al.,* 2024). Advanced oxidation processes (AOPs) such as photocatalysis, ozonation, and Fenton reactions represent powerful chemical approaches to pollutant degradation, (Rakhi *et al.,* 2024). These processes rely on the generation of reactive oxygen species (ROS), which can oxidize a broad spectrum of organic contaminants, (Okpara *et al.,* 2024). The introduction of fused S-heterocycles into these systems can enhance the performance of chemical degradation methods. For example, when incorporated into photocatalysts, fused S-heterocycles can improve light absorption properties, facilitate charge separation, and promote efficient electron transfer, all of which contribute to enhanced ROS generation and pollutant oxidation, (Ali *et al.*, 2024). Moreover, these heterocycles can be tailored to act as ligands or stabilizers in metal-based catalysts, thereby improving the selectivity and durability of catalytic systems, (Ghorai *et al.,* 2024). Beyond chemical processes, physical remediation techniques such as adsorption, filtration, and volatilization also stand to benefit from the inclusion of fused S-heterocycles. By engineering the structural properties of these compounds, researchers can design advanced adsorbent materials with high surface areas, specific affinities for contaminants, and enhanced pollutant removal capacities, (Zuo *et al.,* 2024). These innovations have the potential to make environmental remediation processes more efficient and targeted, addressing both organic and inorganic pollutants in diverse environmental settings.

As the environmental remediation field continues to evolve, there is a growing emphasis on developing sustainable and eco-friendly approaches. The principles of green chemistry, which prioritize the reduction or elimination of hazardous substances in chemical processes, align with the goals of sustainable remediation, (Szekely, 2024). Fused S-heterocycles, with their tunable chemical properties, can be synthesized using green chemistry techniques, ensuring minimal environmental impact during their production and application, (Borah, Banerjee, & Allam, 2024). Additionally, these heterocycles can be designed to degrade into non-toxic by products, further supporting the sustainability of remediation efforts, (Frühauf, Behringer, & Meyer-Almes, 2023). Beyond the chemical design of individual compounds, integrated remediation strategies that combine biological, chemical, and physical methods offer a holistic approach to addressing complex contamination scenarios. By incorporating fused S-heterocycles into multiple stages of the remediation process, synergistic effects can be achieved, leading to more comprehensive and efficient pollutant removal, (An *et al.*, 2024). This chapter concludes by highlighting the future prospects of fused S-heterocycles in environmental remediation, emphasizing their versatility, effectiveness, and potential to contribute to long-term ecological balance. Through continued research and innovation, fused S-heterocycles may play a pivotal role in shaping the future of sustainable environmental clean up strategies.

2. POLLUTANT DEGRADATION MECHANISMS

Pollutant degradation mechanisms encompass a range of complex biochemical, chemical, and physical processes aimed at breaking down contaminants into less harmful substances, (Santhappan *et al.,* 2024). These processes are integral to environmental remediation, playing a crucial role in transforming hazardous pollutants into forms that pose minimal risk to human health and ecosystems, (Asghar *et al.,* 2024). The effectiveness of degradation mechanisms is contingent upon the nature of the pollutants, environmental conditions, and the specific degradation strategy employed. This section delves into the intricacies of biological, chemical, and physical degradation processes, with a particular emphasis on the role of fused S-heterocycles in enhancing these mechanisms.

2.1. Biological Degradation

Biological degradation, commonly referred to as bioremediation, leverages the natural metabolic processes of microorganisms, plants, and fungi to degrade or transform organic pollutants into less toxic or non-toxic substances, (Kuppan *et al.,* 2024). This process is inherently sustainable, as it relies on naturally occurring organisms and their biochemical pathways to break down contaminants. The versatility of biological degradation makes it a valuable tool for addressing a wide range of pollutants, from hydrocarbons and pesticides to industrial chemicals and heavy metals, (de Carvalho *et al.*, 2024). Microorganisms, particularly bacteria and fungi, are at the forefront of bioremediation efforts. These organisms utilize their metabolic machinery to convert hazardous compounds into energy and biomass, (Alabssawy & Hashem, 2024). Key enzymes involved in microbial degradation include oxygenases and reductases, which catalyze oxidation-reduction reactions essential for breaking down complex organic molecules, (Khaliq, 2023). Oxygenases, for instance, introduce oxygen atoms into organic molecules, rendering them more water-soluble and amenable to further breakdown, (Beškoski *et al.*, 2024). Reductases, on the other hand, facilitate the transfer of electrons, enabling the reduction of pollutants to less harmful forms, (Bajaj *et al.,* 2024). One of the major challenges in bioremediation is the efficiency of microbial degradation, especially in environments with high pollutant concentrations or recalcitrant contaminants, (Bala *et al.,* 2022). This is where fused S-heterocycles come into play. Fused S-heterocycles are organic compounds that incorporate sulfur atoms into heterocyclic rings, resulting in unique structural and chemical properties. These compounds have garnered significant attention in environmental science due to their ability to enhance microbial degradation processes, (Rizzo *et al.,* 2024).

Fused S-heterocycles can act as catalysts or mediators in bioremediation by interacting with microbial enzymes, particularly oxygenases and reductases, to enhance their activity, (Bhandari & Sharma, 2021). For example, these heterocycles can serve as electron shuttles, facilitating electron transfer between microbial cells and pollutants, thereby accelerating the degradation process. In anaerobic environments, where oxygen is limited, fused S-heterocycles can act as alternative electron acceptors, supporting the metabolism of anaerobic microorganisms and enabling the breakdown of pollutants that would otherwise persist in the environment, (Kumar & Verma, 2024). In addition to their role in enzymatic activity, fused S-heterocycles can also promote microbial growth and resilience in polluted environments. Certain heterocycles have been shown to enhance the production of extracellular polymeric substances (EPS) by microorganisms, which are critical for the formation of biofilms, (Fu *et al.,* 2021). Biofilms provide a protective environment for microbial communities, allowing them to thrive in harsh conditions and sustain prolonged degradation activity. Furthermore, fused S-heterocycles can serve as chelating agents, binding to heavy metals and reducing their bioavailability, thus mitigating their toxic effects on microbial populations, (Tabassum *et al.,* 2024). Plants also play a significant role in biological degradation through a process known as phytoremediation, (Wani *et al.,* 2023). Phytoremediation involves the use of plants to absorb, accumulate, and degrade pollutants from soil and water. Plants with high tolerance to contaminants, such as certain species of grasses and trees, can uptake pollutants through their roots and translocate them to their biomass, where they can be metabolized or sequestered. Fused S-heterocycles can enhance phytoremediation by acting as growth promoters or chelating agents, improving the ability of plants to accumulate and degrade pollutants, (Ijoma *et al.,* 2024). Hence, biological degradation is a powerful tool for environmental remediation, particularly when combined with innovative strategies that leverage the unique properties of fused S-heterocycles. These compounds offer the potential to improve microbial and plant-based degradation processes, making bioremediation a more effective and sustainable solution for pollutant removal.

2.2. Chemical Degradation

Chemical degradation refers to the breakdown of pollutants through chemical reactions, often involving the generation of reactive species that can oxidize or reduce contaminants, (Rayaroth *et al.,* 2023). Advanced oxidation processes (AOPs) are among the most widely used chemical degradation techniques due to their ability to generate highly reactive oxygen species (ROS), such as hydroxyl radicals (•OH), superoxide anions (O_2^-), and singlet oxygen ($^1O_2$) (Dong *et al.,* 2022). These ROS have strong oxidative potential, enabling them to break down a wide range of organic pollutants, including persistent organic pollutants (POPs), pesticides, and industrial chemicals. Prominent AOPs include photocatalysis, ozonation, and Fenton reactions, each of which relies on different methods for generating ROS. Photocatalysis, for example, uses light energy to excite a photocatalyst, typically a semiconductor material such as titanium dioxide (TiO_2), leading to the generation of electron-hole pairs that produce ROS, (Chauke *et al.,* 2024). Ozonation involves the use of ozone (O_3) as an oxidizing agent, while Fenton reactions rely on the reaction between hydrogen peroxide (H_2O_2) and iron ions (Fe^{2+}) to produce hydroxyl radicals, (Merchant *et al.,* 2024). Fused S-heterocycles have shown promise in enhancing the efficiency of chemical degradation processes, particularly in AOPs. These compounds can be incorporated into photocatalysts to improve their light absorption properties and facilitate electron transfer, both of which are critical for maximizing ROS generation and pollutant oxidation. For instance, the incorporation of fused S-heterocycles into TiO_2-based photocatalysts has been shown to enhance photocatalytic activity by extending the light absorption range into the visible spectrum and improving charge separation efficiency. This results in higher ROS production and more effective degradation of pollutants under light irradiation.

In addition to photocatalysis, fused S-heterocycles can also enhance ozonation and Fenton reactions. These heterocycles can act as ligands or stabilizers in metal-based catalysts, improving their stability and reactivity. In Fenton reactions, for example, fused S-heterocycles can chelate iron ions, stabilizing them in solution and promoting the continuous generation of hydroxyl radicals. This can lead to more efficient and sustained pollutant degradation, particularly in the presence of complex organic compounds that require prolonged oxidation. Beyond AOPs, chemical degradation also encompasses reductive processes, which involve the reduction of pollutants through electron transfer. Reductive dehalogenation, for instance, is a key process in the degradation of halogenated organic compounds, such as chlorinated solvents and pesticides, (King & Mitch, 2024). Fused S-heterocycles, with their electron-donating properties, can serve as electron donors or mediators in reductive degradation processes, enhancing the reduction of pollutants and facilitating their breakdown into less harmful forms. The versatility of fused S-heterocycles in chemical degradation processes makes them valuable tools for environmental remediation. By enhancing the efficiency of ROS generation, electron transfer, and catalyst stability, these compounds can improve the effectiveness of chemical methods for pollutant removal.

2.3. Physical Degradation

Physical degradation refers to the removal or isolation of pollutants through physical means, such as adsorption, filtration, and volatilization, (Hassen *et al.*, 2024). Unlike biological and chemical degradation, which transform pollutants into less harmful substances, physical degradation primarily focuses on separating contaminants from the environment to prevent further exposure and damage. These techniques are often used in conjunction with other remediation methods to achieve comprehensive cleanup of contaminated sites. Adsorption is one of the most widely used physical degradation techniques, particularly for the removal of organic pollutants and heavy metals from water and soil, (Miyah *et al.*, 2024). Adsorbents such as activated carbon, biochar, and zeolites are commonly employed due to their high surface area and affinity for various contaminants. The effectiveness of adsorption depends on the adsorbent's surface properties, including its porosity, surface charge, and chemical functionality. Fused S-heterocycles can be engineered to create novel adsorbent materials with enhanced adsorption properties. The incorporation of sulfur atoms into heterocyclic frameworks can introduce functional groups that improve the affinity of adsorbents for specific contaminants. For example, fused S-heterocycles with thiol (-SH) or sulfide (-S-) groups can exhibit strong interactions with heavy metal ions, making them effective for the removal of metal pollutants from water and soil, (Fosnacht & Pluth, 2024). Additionally, the unique structural properties of fused S-heterocycles can be exploited to design adsorbents with high porosity and surface area, further improving their adsorption capacity.

Filtration is another important physical degradation technique, particularly for water remediation. Membrane filtration processes such as reverse osmosis, ultrafiltration, and nanofiltration use semi-permeable membranes to separate contaminants from water based on size, charge, and chemical properties, (Singh *et al.,* 2024). The performance of these membranes depends on their selectivity, permeability, and resistance to fouling. Fused S-heterocycles can enhance membrane filtration by improving membrane materials. These heterocycles can be incorporated into polymeric membranes to increase their selectivity and permeability, allowing for more efficient separation of contaminants. Additionally, the incorporation of fused S-heterocycles can improve the fouling resistance of membranes, extending their lifespan and reducing maintenance costs, (Tripathy & Gupta, 2023). This is particularly important in the treatment of industrial wastewater, where membrane fouling by organic and inorganic pollutants is a major challenge. Volatilization, which involves the transfer of pollutants from a liquid or solid phase to a gas phase, is another physical degradation technique. This method is particularly useful for the removal of volatile organic compounds (VOCs) from contaminated soil and water. Fused S-heterocycles can play a role in volatilization processes by acting as mediators or facilitators of pollutant transfer. For instance, certain heterocycles can interact with VOCs to enhance their volatilization rate, making them easier to capture and remove from the environment.

Table 1. Role of fused S-heterocycles in various remediation techniques

Remediation Technique	Role of Fused S-Heterocycles	Examples
Biological Degradation	- Act as electron shuttles to facilitate electron transfer in microbial degradation processes.	- Enhance microbial enzyme activity (e.g., oxygenases, reductases) through electron transfer, improving pollutant breakdown.
	- Serve as alternative electron acceptors in anaerobic environments.	- Enable anaerobic degradation of pollutants by supporting the metabolism of anaerobic microorganisms.
	- Serve as alternative electron acceptors in anaerobic environments.	- Stimulate extracellular polymeric substance (EPS) production in microbial communities, protecting them in harsh environments.
	- Act as chelating agents, reducing the bioavailability of heavy metals, mitigating toxic effects on microbes	- Reduce toxicity of heavy metals in polluted environments, improving microbial degradation efficiency.

continued on following page

Table 1. Continued

Remediation Technique	Role of Fused S-Heterocycles	Examples
Chemical Degradation	- Improve photocatalytic activity by enhancing light absorption and charge separation in photocatalysts.	- Extend the light absorption range in TiO_2-based photocatalysts, boosting the efficiency of advanced oxidation processes (AOPs).
	- Enhance ozonation and Fenton reactions by stabilizing metal ions and promoting ROS generation.	- Chelate iron ions in Fenton reactions, leading to sustained hydroxyl radical production and more effective pollutant degradation.
	- Serve as electron donors or mediators in reductive degradation processes.	- Facilitate the reduction of halogenated organic compounds through enhanced electron transfer
Physical Degradation	- Improve adsorption properties of adsorbents by introducing functional groups that enhance contaminant affinity.	- Incorporate thiol (-SH) or sulfide (-S-) groups in adsorbents, improving the removal of heavy metals from water and soil.
	- Enhance membrane selectivity, permeability, and fouling resistance in filtration processes.	- Integrate fused S-heterocycles into polymeric membranes, boosting performance in reverse osmosis, ultrafiltration, and nanofiltration.
	- Facilitate the volatilization of volatile organic compounds (VOCs).	- Interact with VOCs to enhance volatilization rates, aiding in the removal of these compounds from contaminated environments.

3. WATER REMEDIATION TECHNIQUES

Water contamination is a significant global issue, with pollutants ranging from heavy metals to organic compounds, pesticides, and pharmaceutical residues infiltrating water bodies. Water remediation is vital for ensuring clean and safe water for drinking, agriculture, industrial use, and ecosystem health, (Srivastav *et al.,* 2024). Multiple techniques have been developed to address various types of contaminants, and recent research highlights the potential of fused S-heterocycles in improving these methods. This section explores three major water remediation techniques: phytoremediation, bioreactors, and membrane filtration, emphasizing how fused S-heterocycles can enhance their efficiency and sustainability. Figure 2 represent the various techniques of water and Soil remediation.

3.1. Phytoremediation

Phytoremediation is an eco-friendly and cost-effective water remediation strategy that harnesses the natural ability of plants to absorb, accumulate, and degrade pollutants, (Lavanya, Viswanath, & Sivapullaiah, 2024). Plants are known to take up contaminants through their root systems, translocate them to their biomass, and either store, degrade, or volatilize them. The success of phytoremediation largely depends on the plant species used, as some plants possess higher tolerance levels for specific contaminants and have better pollutant translocation and degradation capabilities. Plants such as water hyacinths, reeds, and certain types of grasses have been extensively studied for their phytoremediation potential, (Gupta & Tyagi, 2022). These plants can absorb a range of pollutants, including heavy metals, pesticides, and organic contaminants, from water. Once taken up, the pollutants can be sequestered in plant tissues, transformed into less toxic compounds, or even released as volatile compounds. This makes phytoremediation particularly effective for cleaning up polluted wetlands, lakes, and rivers. Despite the promise of phytoremediation, challenges such as slow remediation rates, limited plant tolerance to high contaminant levels, and incomplete degradation of certain pollutants persist, (Salifu *et al.*, 2024). Fused S-heterocycles offer a potential solution to these challenges. These compounds, with their unique chemical structures and sulfur-containing functional groups, can enhance the phytoremediation process in several ways.

First, fused S-heterocycles can act as plant growth promoters, stimulating the growth of roots and shoots, which in turn can increase the uptake of pollutants. Certain heterocyclic compounds have been shown to enhance the production of phytohormones like auxins and gibberellins, which regulate plant growth and development, (Hasan *et al.*, 2024). By promoting more robust root systems, fused S-heterocycles can increase the plant's ability to access and absorb contaminants from deeper layers of water and soil. Second, fused S-heterocycles can function as chelating agents, forming complexes with heavy metals and facilitating their uptake by plants. Fused S-heterocycles form stable complexes with heavy metals like cadmium (Cd^{2+}) and lead (Pb^{2+}) through coordination with sulfur atoms, enhancing metal solubility and uptake by plants for improved phytoremediation. Additionally, these heterocycles can interact with and enhance enzyme activity, such as peroxidases and laccases, by inducing conformational changes that increase catalytic efficiency, accelerating the degradation of organic pollutants within plant tissues. However, when complexed with chelating agents, these metals become more soluble and easier for plants to absorb. Fused S-heterocycles, with their sulfur-containing ligands, can form stable chelates with heavy metals, enhancing their uptake by plants and subsequent sequestration or degradation. Lastly, certain fused S-heterocycles may enhance the enzymatic activities within plants that are responsible for degrading organic pollutants. For instance, these compounds can upregulate the production of enzymes like peroxidases and laccases, which are involved in breaking down complex organic molecules such as pesticides and hydrocarbons. This enzymatic enhancement could lead to more efficient degradation of contaminants within plant tissues, reducing the risk of secondary pollution when plant biomass is harvested or decomposes.

These compounds can improve plant growth, increase the bioavailability of heavy metals, and boost enzymatic degradation pathways, making phytoremediation a more effective and comprehensive approach to cleaning polluted water bodies.

3.2. Bioreactors

Bioreactors are engineered systems designed to support and optimize the growth of microorganisms that degrade pollutants in contaminated water, (Narciso-Ortiz *et al.,* 2023). These systems can operate under controlled conditions, such as temperature, pH, and oxygen levels, to maximize the efficiency of microbial degradation processes, (Lutoslawski, Ryznar-Luty, & Cibis, 2024). Bioreactors can be designed to function under aerobic or anaerobic conditions, depending on the nature of the pollutants and the metabolic pathways of the microorganisms involved, (Elsayed *et al.,* 2024). Aerobic bioreactors are commonly used for the treatment of organic pollutants, such as petroleum hydrocarbons, organic acids, and phenolic compounds. In these systems, oxygen is supplied to the microbial communities to support oxidative degradation processes, wherein microorganisms use oxygen as an electron acceptor to break down contaminants into carbon dioxide, water, and biomass. On the other hand, anaerobic bioreactors are employed for the treatment of more recalcitrant pollutants, such as chlorinated solvents and heavy metals, (Bertolini, Zecvchin, & Cavalca, 2023). In bioreactors, fused S-heterocycles function as electron mediators, shuttling electrons between microbial cells and pollutants. In anaerobic systems, they facilitate electron transfer to alternative acceptors like sulfate in sulfate-reducing bacteria, aiding metal-sulfide degradation. In aerobic bioreactors, they enhance electron transfer to oxygen through interactions with redox-active enzymes like oxidases, improving pollutant degradation and overall bioreactor efficiency in treating complex organic contaminants. These compounds can be incorporated into bioreactor materials or introduced as additives to support the microbial communities responsible for pollutant degradation. One of the primary ways fused S-heterocycles can enhance bioreactors is by promoting microbial attachment to surfaces within the reactor. In many bioreactors, microorganisms form biofilms on solid surfaces, which provide a stable environment for microbial growth and activity, (Lago *et al.,* 2024). Fused S-heterocycles can act as biofilm promoters by facilitating the adhesion of microbial cells to reactor surfaces, thereby increasing the surface area available for microbial colonization. This, in turn, can lead to higher biomass concentrations and more efficient degradation of pollutants.

In addition to promoting biofilm formation, fused S-heterocycles can enhance the metabolic activity of microorganisms within the bioreactor. These compounds can serve as electron donors or mediators in microbial degradation pathways, particularly in anaerobic systems where alternative electron acceptors are required. For example, fused S-heterocycles can donate electrons to sulfate-reducing bacteria, facilitating the reduction of sulfate to sulfide and enabling the degradation of sulfide-bound pollutants. Similarly, in aerobic systems, fused S-heterocycles can act as redox mediators, shuttling electrons between microbial cells and contaminants, thereby enhancing the efficiency of electron transfer processes. Stability is another critical factor in bioreactor performance. Microbial communities in bioreactors are often exposed to fluctuating environmental conditions, such as changes in temperature, pH, and pollutant concentrations, which can affect their stability and degradation efficiency, (Wang *et al.*, 2024). Fused S-heterocycles can contribute to microbial stability by acting as buffers or protective agents against environmental stressors. For instance, certain heterocycles can scavenge reactive oxygen species (ROS) that are generated during oxidative degradation processes, protecting microbial cells from oxidative damage and ensuring sustained degradation activity. Moreover, fused S-heterocycles can enhance the resilience of microbial communities by promoting the production of extracellular polymeric substances (EPS), which are critical components of biofilms. EPS provide structural stability to biofilms, protect microorganisms from environmental stress, and facilitate nutrient exchange within the biofilm matrix. By stimulating EPS production, fused S-heterocycles can help maintain the integrity of biofilms in bioreactors, ensuring continuous and efficient pollutant degradation. These compounds can improve microbial attachment, activity, and stability within bioreactors, leading to more efficient and resilient degradation processes. The incorporation of fused S-heterocycles into bioreactor materials and strategies offers exciting possibilities for advancing bioreactor-based water remediation.

3.3. Membrane Filtration

Membrane filtration is a widely used technique for water remediation, particularly in the treatment of wastewater and the purification of drinking water, (Qi *et al.,* 2024). Membrane filtration processes involve the use of semi-permeable membranes that separate contaminants from water based on size, charge, and chemical properties, (Singh *et al.,* 2024). Common membrane filtration techniques include reverse osmosis (RO), ultrafiltration (UF), and nanofiltration (NF). Each of these techniques targets different types of contaminants, from dissolved salts and heavy metals to organic compounds and pathogens. Reverse osmosis is one of the most effective membrane filtration techniques for removing dissolved salts, heavy metals, and other ions from water. RO membranes have extremely small pore sizes, typically on the order of 0.1 to 1 nanometer, which allows them to reject even the smallest ions and molecules, (Chang *et al.,* 2024). Ultrafiltration and nanofiltration, on the other hand, have larger pore sizes, ranging from 1 to 100 nanometers, and are used for the removal of larger organic molecules, pathogens, and particulate matter,

(Zhang *et al.,* 2023). Despite the effectiveness of membrane filtration, challenges such as membrane fouling, low permeability, and limited selectivity persist, (Osman *et al.,* 2024). Membrane fouling occurs when contaminants, particularly organic matter, and suspended solids, accumulate on the surface of the membrane, leading to a decrease in permeability and filtration efficiency. This necessitates frequent cleaning and replacement of membranes, increasing operational costs and reducing the sustainability of the process. Fused S-heterocycles offer potential solutions to these challenges by enhancing the selectivity, permeability, and fouling resistance of membrane materials. Incorporating fused S-heterocycles into membranes creates hydrophilic or hydrophobic domains, enhancing water flux and reducing fouling. The sulfur atoms engage in dipole interactions with water and can selectively remove heavy metals like Hg^{2+} or Pb^{2+} by forming stable metal-thiol complexes, improving membrane selectivity through coordination chemistry.

One of the key ways fused S-heterocycles can enhance membrane filtration is by increasing the selectivity of membranes for specific contaminants. Fused S-heterocycles can be functionalized with various chemical groups that target particular pollutants, such as heavy metals, organic compounds, or pathogens. For example, membranes functionalized with thiol groups (-SH) can exhibit high affinity for heavy metal ions, enabling more efficient removal of metals like lead, mercury, and cadmium from water. Similarly, membranes functionalized with hydrophobic or hydrophilic groups can selectively reject organic compounds or pathogens based on their chemical properties. In addition to improving selectivity, fused S-heterocycles can enhance the permeability of membranes, allowing for higher water flux and more efficient filtration. The incorporation of fused S-heterocycles into membrane materials can create more porous and hydrophilic structures, which facilitate the passage of water molecules while rejecting contaminants, (Dhara *et al.,* 2024). This can lead to higher filtration rates and reduced energy consumption, particularly in pressure-driven processes like reverse osmosis. Fouling resistance is another critical area where fused S-heterocycles can make a significant impact. Membrane fouling is a major challenge in water remediation, as it reduces filtration efficiency and increases maintenance costs, (Osman *et al.*, 2024). Fused S-heterocycles can be designed to resist fouling by incorporating anti-fouling functional groups into the membrane structure. For example, membranes functionalized with zwitter ionic groups, which contain both positive and negative charges, can create a hydrated layer on the membrane surface that repels foulants and prevents their accumulation. Similarly, membranes with antimicrobial or anti-adhesive properties can prevent the growth of biofilms and the attachment of organic matter, further reducing fouling. Moreover, fused S-heterocycles can improve the chemical stability of membranes, making them more resistant to harsh conditions such as extreme pH, temperature, and chemical exposure. This is particularly important in industrial water treatment applications, where membranes are often exposed to aggressive chemicals and high temperatures that can degrade their performance. By incorporating fused S-heterocycles into the membrane material, it is possible to create more durable and long-lasting membranes that can withstand challenging operational conditions without compromising filtration efficiency. These compounds can improve membrane selectivity, permeability, and fouling resistance, leading to more efficient and sustainable water treatment processes. The development of novel membrane materials incorporating fused S-heterocycles represents a promising avenue for advancing water filtration technologies and addressing the growing global demand for clean and safe water.

Figure 2. Schematic of various techniques of water and soil remediation

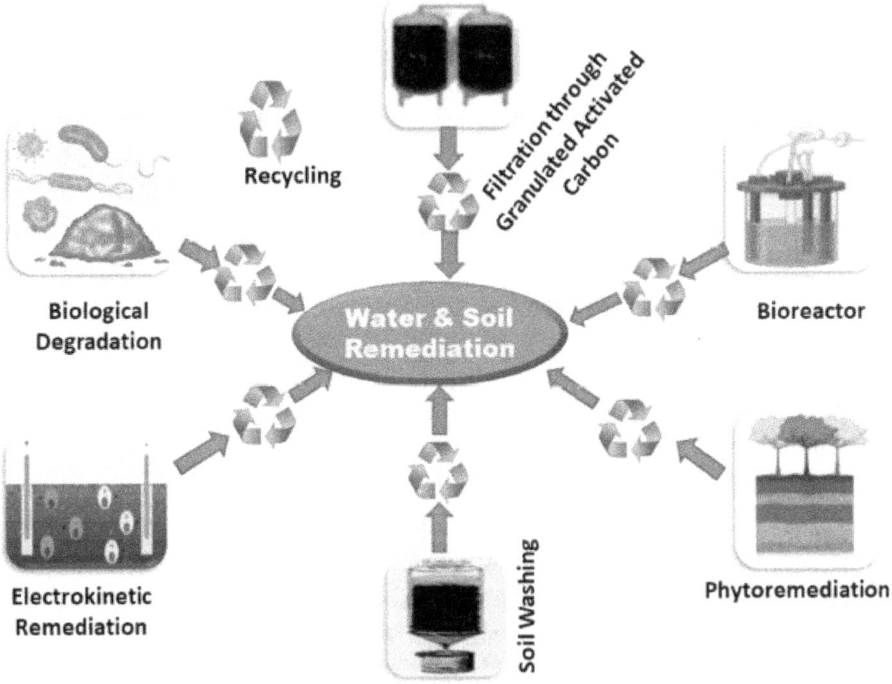

4. SOIL REMEDIATION TECHNIQUES

Soil contamination is a widespread environmental issue, resulting from industrial activities, agricultural practices, and improper waste disposal, (Adepoju *et al.*, 2024). Contaminants such as heavy metals, organic pollutants, and pesticides can accumulate in soil, posing risks to human health, agriculture, and ecosystems. Soil remediation techniques aim to remove or neutralize these contaminants, restoring soil health and fertility. This section explores three major soil remediation techniques: soil washing, bioremediation, and electrokinetic remediation, with a focus on how fused S-heterocycles can enhance these methods.

4.1. Soil Washing

Soil washing is a physical-chemical remediation technique that involves the use of aqueous solutions to extract contaminants from soil, (Dhara *et al.,* 2024). The process typically involves mixing contaminated soil with a washing solution that solubilizes or mobilizes the pollutants, allowing them to be separated from the soil particles. Soil washing is particularly effective for the removal of heavy metals, hydrocarbons, and other inorganic and organic pollutants. The effectiveness of soil washing depends on the choice of washing agents, which can include surfactants, acids, bases, and chelating agents. Surfactants are commonly used to enhance the solubility of hydrophobic organic compounds, while acids and bases can help dissolve heavy metal contaminants. Chelating agents, such as ethylenediaminetetraacetic acid (EDTA), are often employed to bind to metal ions and facilitate their removal from the soil. Fused S-heterocycles offer promising opportunities to enhance soil washing by serving as novel washing agents with high selectivity and efficiency. Fused S-heterocycles form stable rings with heavy metals, enhancing their solubility and preventing re-adsorption onto soil particles. Additionally, they interact with hydrophobic organic pollutants via van der Waals forces and π-π stacking, aiding pollutant extraction.

One of the key advantages of using fused S-heterocycles as washing agents is their ability to form stable complexes with heavy metals. The sulfur atoms in these heterocycles can act as ligands, binding to metal ions and forming soluble complexes that can be easily extracted from the soil. This can improve the efficiency of soil washing, particularly for metals that are tightly bound to soil particles and difficult to mobilize with conventional washing agents. In addition to their metal-chelating properties, fused S-heterocycles can enhance the solubility of hydrophobic organic compounds, such as polycyclic aromatic hydrocarbons (PAHs) and pesticides, by acting as surfactants, (Kanungo *et al.*, 2024). The amphiphilic nature of certain heterocycles allows them to interact with both hydrophobic pollutants and the aqueous washing solution, thereby increasing the solubility and mobility of these contaminants. This can lead to more effective removal of organic pollutants from soil during the washing process. Furthermore, fused S-heterocycles can be engineered to exhibit selective binding properties, allowing them to target specific contaminants while minimizing the extraction of essential soil nutrients. This selectivity can reduce the environmental impact of soil washing and help preserve soil fertility after remediation.

4.2. Bioremediation

Bioremediation is a biological soil remediation technique that employs microorganisms to degrade or transform organic contaminants in soil, (Mokrani *et al.*, 2024). This method leverages the natural metabolic processes of bacteria, fungi, and other microorganisms to break down pollutants into less harmful or non-toxic compounds. Bioremediation is particularly effective for the treatment of organic pollutants, such as hydrocarbons, pesticides, and industrial chemicals, (de Carvalho *et al.,* 2024). Several bioremediation techniques are commonly used, including composting, bioaugmentation, and bioventing. Composting involves the aerobic degradation of organic pollutants through the controlled microbial decomposition of organic matter. Bioaugmentation involves the introduction of specific microbial strains or consortia that are capable of degrading particular pollutants. Bioventing, on the other hand, enhances the natural in situ biodegradation of contaminants by supplying oxygen to the subsurface environment. The efficiency of bioremediation depends on various factors, including the availability of nutrients, electron acceptors, and suitable environmental conditions for microbial growth. Fused S-heterocycles can play a crucial role in enhancing bioremediation by acting as nutrient sources, electron donors or acceptors, and metabolic enhancers, (Sharma *et al.,* 2023).

In bioremediation, fused S-heterocycles serve as redox mediators, shuttling electrons between microbial enzymes and contaminants. Their sulfur atoms undergo reversible oxidation, accelerating organic pollutant degradation, especially in anaerobic environments where microbial activity is slower, enhancing overall bioremediation efficiency. In anaerobic bioremediation processes, such as those involving sulfate-reducing bacteria, fused S-heterocycles can donate electrons to facilitate the reduction of sulfate to sulfide, enabling the degradation of sulfur-bound pollutants. Similarly, in aerobic bioremediation, fused S-heterocycles can act as redox mediators, shuttling electrons between microbial cells and contaminants, thereby enhancing the efficiency of electron transfer and pollutant degradation. Additionally, fused S-heterocycles can serve as nutrient sources for microorganisms, stimulating their growth and activity in contaminated soils. Certain heterocyclic compounds can provide essential nutrients, such as nitrogen or sulfur, which are critical for microbial metabolism and pollutant degradation. By supplementing the soil with fused S-heterocycles, it is possible to enhance microbial activity and accelerate the breakdown of organic contaminants. Moreover, fused S-heterocycles can enhance the production of extracellular enzymes involved in pollutant degradation. These enzymes, such as oxygenases and hydrolases, play a key role in breaking down complex organic molecules into simpler, less toxic compounds. Fused S-heterocycles can upregulate the expression of these enzymes, leading to more efficient degradation of pollutants in the soil.

4.3. Electrokinetic Remediation

Electrokinetic remediation is an emerging soil remediation technique that uses electric fields to mobilize and remove contaminants from soil, (Ngieng *et al.,* 2024). This method is particularly effective for the treatment of heavy metals, organic pollutants, and radionuclides in soils with low permeability, such as clay or silt, (Kumawat *et al.,* 2022). Electrokinetic remediation involves the application of a direct current across electrodes inserted into the contaminated soil, which induces the movement of charged particles, such as metal ions and organic molecules, toward the electrodes for extraction. Fused S-heterocycles enhance conductivity by increasing soil solution ionic strength. Sulfur atoms release ions, improving contaminant migration towards electrodes. Additionally, metal-heterocycle complexes increase heavy metal mobility, facilitating movement towards the anode for extraction, making them effective agents in environmental cleanup processes, (Kou *et al.,* 2024). One of the key challenges in electrokinetic remediation is the low conductivity of certain soils, which can limit the effectiveness of the electric field in mobilizing contaminants, (Taneja, Karaca, & Haritash, 2024). Fused S-heterocycles can be incorporated into electrode materials to enhance their conductivity, allowing for more efficient current distribution and contaminant mobilization. These compounds can also act as electrolytes, increasing the ionic strength of the soil solution and facilitating the movement of charged contaminants toward the electrodes. In addition to improving soil conductivity, fused S-heterocycles can enhance the selectivity of electrokinetic remediation by forming complexes with specific contaminants, such as heavy metals. These complexes can increase the mobility of metal ions in the soil, allowing them to be more easily extracted by the electric field. The selective binding properties of fused S-heterocycles can also reduce the extraction of non-target ions, such as essential nutrients, minimizing the impact on soil fertility. Furthermore, fused S-heterocycles can improve the stability and durability of electrode materials used in electrokinetic remediation. These compounds can protect electrodes from corrosion and degradation, ensuring consistent performance over prolonged remediation periods. This can reduce the need for frequent electrode replacement and maintenance, making electrokinetic remediation a more cost-effective and sustainable option for soil remediation.

5. SUSTAINABLE APPROACHES TO ENVIRONMENTAL CLEANUP

Sustainable approaches to environmental cleanup have become increasingly essential in our efforts to address pollution and protect ecosystems while minimizing negative impacts on the environment, (Bradu *et al.*, 2023). These approaches prioritize the use of environmentally friendly technologies and processes that reduce waste, conserve resources, and ensure long-term ecological balance, (Hegab *et al.*, 2023). Central to these efforts is the development and application of green chemistry, eco-friendly materials, and integrated remediation strategies. In this context, the role of fused S-heterocycles, a class of organic compounds characterized by sulfur-containing heterocyclic rings, becomes significant. These compounds offer unique chemical properties that can be harnessed to enhance various remediation processes, from pollutant degradation to the development of sustainable materials.

5.1. Green Chemistry in Environmental Clean Up

Green chemistry is an essential framework for creating sustainable remediation processes. It focuses on designing chemical processes that are not only effective in removing contaminants but also safe for the environment, (Gürses *et al.*, 2024). The goal of green chemistry is to prevent pollution at the source by reducing or eliminating the use and generation of hazardous substances. This aligns with the principles of sustainability, as it encourages the use of renewable resources, minimizes waste, and enhances energy efficiency. One of the key challenges in environmental remediation is the development of chemical agents and catalysts that can effectively degrade pollutants without creating additional environmental hazards. Traditional chemical processes often rely on harsh reagents and produce toxic byproducts, which can be detrimental to both human health and ecosystems. Green chemistry seeks to address these challenges by promoting the use of non-toxic reagents, reducing the number of reaction steps, and employing safer solvents and materials. Fused S-heterocycles, with their tunable chemical properties, offer significant potential in the application of green chemistry principles to environmental cleanup, (Nasri *et al.*, 2022). These compounds can be synthesized using environmentally friendly methods that reduce energy consumption and the generation of waste. For example, the synthesis of fused S-heterocycles can be achieved through catalytic processes that minimize the use of toxic reagents and solvents. Catalysts based on fused S-heterocycles themselves can be designed to be highly selective, reducing the need for excess chemicals and minimizing the production of undesired byproducts.

However, in their synthesis, fused S-heterocycles can play a crucial role as green catalysts in pollutant degradation processes. Their ability to catalyze redox reactions, including oxidation and reduction of contaminants, can be harnessed in advanced oxidation processes (AOPs) that are widely used in water and soil remediation. AOPs such as photocatalysis, ozonation, and Fenton reactions generate reactive oxygen species (ROS) that break down organic pollutants. Fused S-heterocycles can enhance these reactions by acting as electron transfer mediators, improving the efficiency of ROS generation and increasing the rate of pollutant degradation. The incorporation of fused S-heterocycles into photocatalysts, for instance, can improve light absorption and electron transfer, leading to more efficient degradation of organic contaminants with minimal environmental impact. Moreover, green chemistry principles extend to the end-of-life considerations of the chemicals and materials used in environmental remediation. Fused S-heterocycles can be designed to degrade into non-toxic by-products after they have fulfilled their role in the remediation process. This aligns with the goal of creating self-limiting systems that do not persist in the environment after use, thereby reducing the potential for secondary pollution.

5.2. Eco-Friendly Materials for Remediation

Eco-friendly materials play a crucial role in sustainable environmental cleanup by offering biodegradable, non-toxic, and renewable alternatives to conventional materials, (Hussain, Akhter, & Maktedar, 2024). These materials are designed to minimize environmental impact while maintaining high performance in remediation processes. The development of such materials is critical for achieving long-term sustainability in environmental cleanup efforts, as they help reduce reliance on non-renewable resources and prevent the accumulation of persistent pollutants. Fused S-heterocycles are emerging as promising candidates for the development of eco-friendly materials in environmental remediation. Their unique structural properties, including the presence of sulfur atoms and heterocyclic rings, allow for the design of materials with specific functionalities tailored to address various environmental challenges. These materials can be engineered to have high selectivity for particular pollutants, biodegradability, and low toxicity, making them ideal for sustainable remediation applications. One area where fused S-heterocycles can contribute to eco-friendly materials is in the design of adsorbents for pollutant removal. Adsorption is a widely used technique for removing contaminants from water and soil, and the effectiveness of this process depends largely on the properties of the adsorbent material. Traditional adsorbents, such as activated carbon and synthetic resins, are often derived from non-renewable resources and can have negative environmental impacts if not properly managed, (El Messaoudi *et al.*, 2024). Fused S-heterocycles, on the other hand, can be derived from renewable resources and designed to degrade into harmless byproducts after use. Their high surface area and tunable chemical properties make them suitable for capturing a wide range of contaminants, from heavy metals to organic pollutants.

For example, fused S-heterocycles can be incorporated into biochar, a carbon-rich material derived from the pyrolysis of organic waste, (Jin *et al.*, 2024). Biochar is known for its ability to adsorb pollutants due to its porous structure and high surface area. By functionalizing biochar with fused S-heterocycles, it is possible to enhance its adsorption capacity and selectivity for specific contaminants. This combination creates a biodegradable and renewable adsorbent material that can effectively remove pollutants from water and soil while minimizing environmental impact. Another application of eco-friendly materials based on fused S-heterocycles is in the development of biodegradable polymers for environmental cleanup. Polymers are widely used in various remediation technologies, including membrane filtration, as they provide mechanical strength and chemical resistance. However, many conventional polymers are derived from petrochemical sources and are not biodegradable, leading to concerns about plastic pollution and environmental persistence. Fused S-heterocycles can be used to synthesize biodegradable polymers that break down into non-toxic components after use, reducing the environmental burden associated with polymer-based materials. In addition to their biodegradability, fused S-heterocycle-based polymers can be designed to have specific properties, such as hydrophilicity or hydrophobicity, depending on the target application. For instance, in water remediation, hydrophilic polymers are desirable for creating membranes that allow water to pass through while rejecting contaminants. Fused S-heterocycles can be incorporated into the polymer backbone to create hydrophilic structures with high water permeability and selectivity, resulting in more efficient filtration processes.

5.3. Integrated Remediation Strategies

Integrated remediation strategies are essential for addressing complex contamination scenarios that involve multiple types of pollutants and require the application of different remediation techniques, (Lee *et al.*, 2023). These strategies combine various physical, chemical, and biological approaches to achieve comprehensive and efficient cleanup of contaminated sites. The integration of multiple techniques allows for the synergistic effects that can enhance the overall effectiveness of the remediation process. The versatility of fused S-heterocycles makes them valuable components in integrated remediation strategies. These compounds can be incorporated into different remediation technologies, providing complementary benefits that optimize pollutant removal. By leveraging the unique properties of fused S-heterocycles, it is possible to create integrated systems that address a wide range of contaminants and environmental conditions. One example of an integrated remediation strategy that can benefit from fused S-heterocycles is the combination of phytoremediation and chemical degradation processes. Phytoremediation involves the use of plants to absorb, accumulate, and degrade pollutants from soil and water. Certain plants are capable of taking up contaminants through their roots and translocating them to their biomass, where they can be metabolized or sequestered. However, the efficiency of phytoremediation can be limited by the bioavailability of contaminants and the metabolic capabilities of the plants. Fused S-heterocycles can enhance phytoremediation by acting as chelating agents or plant growth promoters. These compounds can increase the bioavailability of heavy metals and other pollutants, making them more accessible to plants. Additionally, fused S-heterocycles can stimulate plant growth and resilience, allowing them to tolerate higher concentrations of contaminants and improve their overall remediation performance. By combining phytoremediation with chemical degradation processes that involve fused S-heterocycles, it is possible to achieve more complete and efficient removal of pollutants from contaminated sites.

Another integrated approach involves the use of bioreactors in conjunction with advanced oxidation processes (AOPs). Bioreactors are engineered systems that support the growth of microorganisms for the treatment of contaminated water or soil. These systems can operate under aerobic or anaerobic conditions, depending on the target pollutants and microbial communities involved. Fused S-heterocycles can be incorporated into bioreactor materials to enhance microbial attachment, activity, and stability, leading to more efficient degradation of organic pollutants. At the same time, AOPs, such as photocatalysis and Fenton reactions, can be used to generate reactive oxygen species that break down pollutants in the contaminated media. Fused S-heterocycles can enhance these processes by improving the efficiency of ROS generation and electron transfer. By integrating bioreactors with AOPs, it is possible to achieve a two-pronged approach to pollutant degradation, where biological and chemical processes work together to maximize contaminant removal. Integrated remediation strategies can also involve the use of multiple physical techniques, such as adsorption and membrane filtration, to achieve comprehensive pollutant removal. Adsorption is effective for capturing contaminants from water or soil, while membrane filtration provides a physical barrier that separates pollutants from the treated media. Fused S-heterocycles can be used to enhance both techniques by improving the selectivity and efficiency of adsorbents and membranes.

6. CASE STUDIES OF FUSED S-HETEROCYCLES IN ENVIRONMENTAL REMEDIATION

In a study by Zhang *et al.* (2018), thiophene-based fused S-heterocycles were successfully employed to remove heavy metals from industrial wastewater. These compounds demonstrated a strong affinity for metal ions like lead (Pb) and cadmium (Cd), forming stable complexes that facilitated their extraction. The researchers applied these heterocycles in soil washing experiments, where they significantly increased the removal efficiency of these toxic metals compared to conventional chelating agents, such as EDTA. A field study in agricultural regions of India tested fused S-heterocyclic compounds as catalysts for degrading persistent organic pollutants (POPs), particularly chlorinated pesticides. The compounds catalyzed hydrolysis reactions, breaking down harmful pesticides into less toxic byproducts. The remediation process led to a substantial reduction in pollutant levels in the soil, enabling the restoration of agricultural productivity while reducing health risks associated with pesticide exposure. In an offshore oil spill remediation effort, fused S-heterocycles were used to enhance the bioremediation process. Sulfur-containing heterocycles were introduced into bioreactors to stimulate the growth of sulfate-reducing bacteria, which degraded hydrocarbons more efficiently under anaerobic conditions. The field application showed a significant reduction in oil pollutants, with microbial activity boosted by the addition of these heterocyclic compounds, proving their effectiveness in large-scale environmental applications. These case studies underscore the versatility of fused S-heterocycles in addressing diverse environmental pollutants through chemical complexation, catalytic degradation, and bioremediation enhancements.

CONCLUSION

The exploration of pollutant degradation mechanisms and remediation techniques highlights the significant potential of fused S-heterocycles in advancing environmental cleanup efforts. Characterized by their sulfur-containing heterocyclic rings, these compounds exhibit exceptional chemical and biological properties that enhance both catalytic and biological processes involved in pollution control. Fused S-heterocycles contribute to improved efficiency in diverse remediation techniques, including phytoremediation, bioreactors, membrane filtration, and soil washing, by enhancing pollutant degradation, promoting microbial activity, and improving material properties. Their ability to act as catalysts, mediators, and eco-friendly materials aligns with sustainable approaches to environmental cleanup, supporting the principles of green chemistry and minimizing ecological impact. Looking ahead, future research should focus on developing novel fused S-heterocycles with optimized properties for specific contaminants and integrating these compounds into hybrid remediation systems that combine multiple techniques for comprehensive pollution management. Advancements in the synthesis of these compounds, particularly through green chemistry methods, and their application in real-world scenarios will be crucial for addressing complex contamination challenges. By continuing to innovate and refine the use of fused S-heterocycles, we can further enhance environmental remediation strategies, ultimately contributing to a cleaner, healthier, and more sustainable environment.

ACKNOWLEDGMENTS

One of the authors, H.S. Biswas, gratefully acknowledges financial support from the University Grant Commission (UGC) under reference number F.PSW-140/15-16. The authors also extend their gratitude to Dr. Indranil Kar, Principal of Surendranath College, as well as Dr. Suchandra Chatterjee, Dr. Lalita Das, Dr. Harisadhan Ghosh, Dr. Apurba Biswas, and Dr. Noor Salam for their invaluable suggestions and contributions.

REFERENCES

Adepoju, A. O., Femi-Adepoju, A., Jalloh, A., & Faeflen, S. (2024). Soil pollution and management practices. In *Environmental Pollution and Public Health* (pp. 187–236). Elsevier. DOI: 10.1016/B978-0-323-95967-4.00011-8

Ahmad, N., Ahmad, H. R., Sajid, Z., Tahir, M., Salma, H., Tamimi, S. A., & Ayub, M. A. (2024). Role of microbial enzymes in wastewater processing and treatment. In *Enzyme Biotechnology for Environmental Sustainability* (pp. 119–160). Academic Press. DOI: 10.1016/B978-0-443-22072-2.00003-6

Alabssawy, A. N., & Hashem, A. H. (2024). Bioremediation of hazardous heavy metals by marine microorganisms: A recent review. *Archives of Microbiology*, 206(3), 103. DOI: 10.1007/s00203-023-03793-5 PMID: 38358529

Ali, I., Hasan, S. Z., Hozaifa, M., Imanova, G., & Kurniawan, T. A. (2024). *Recent Advances in Hydrogen Storage Methods. Green Hydrogen Economy for Environmental Sustainability* (Vol. 2). Applications, Challenges, and Policies.

Ali, I., Hasan, S. Z., Hozaifa, M., Imanova, G., & Kurniawan, T. A. (2024). *Recent Advances in Hydrogen Storage Methods. Green Hydrogen Economy for Environmental Sustainability* (Vol. 2). Applications, Challenges, and Policies.

An, X., Li, N., Zhang, L., Xu, Z., Zhang, S., & Zhang, Q. (2024). New insights into the typical nitrogen-containing heterocyclic compound-quinoline degradation and detoxification by microbial consortium: Integrated pathways, meta-transcriptomic analysis and toxicological evaluation. *Journal of Hazardous Materials*, 465, 133158. DOI: 10.1016/j.jhazmat.2023.133158 PMID: 38061124

Asghar, N., Hussain, A., Nguyen, D. A., Ali, S., Hussain, I., Junejo, A., & Ali, A. (2024). Advancement in nanomaterials for environmental pollutants remediation: A systematic review on bibliometrics analysis, material types, synthesis pathways, and related mechanisms. *Journal of Nanobiotechnology*, 22(1), 26. DOI: 10.1186/s12951-023-02151-3 PMID: 38200605

Bajaj, D., Yadav, V., Dhyani, A., & Wadhwa, N. K. (2024). Exploration of Microbial Enzymes in Bioremediation. In *Microbial Approaches for Sustainable Green Technologies* (pp. 250–272). CRC Press. DOI: 10.1201/9781003407683-13

Bala, S., Garg, D., Thirumalesh, B. V., Sharma, M., Sridhar, K., Inbaraj, B. S., & Tripathi, M. (2022). Recent strategies for bioremediation of emerging pollutants: A review for a green and sustainable environment. *Toxics*, 10(8), 484. DOI: 10.3390/toxics10080484 PMID: 36006163

Bertolini, M., Zecchin, S., & Cavalca, L. (2023). Sequential Anaerobic/Aerobic microbial transformation of chlorinated ethenes: Use of sustainable approaches for aquifer decontamination. *Water (Basel)*, 15(7), 1406. DOI: 10.3390/w15071406

Beškoski, V., Lješević, M., Jiménez, B., Muñoz-Arnanz, J., Colomer-Vidal, P., Inui, H., & Nakano, T. (2024). *Bioremediation of Soils Contaminated with PFAS: An Update on Available Techniques*. Pilot Studies, Challenges, and Future Directions.

Bhandari, G., & Sharma, M. (2021). Microbial enzymes in the bioremediation of pollutants: Emerging potential and challenges. *Mycoremediation and Environmental Sustainability*, 3, 75–94. DOI: 10.1007/978-3-030-54422-5_3

Bhattacharya, S., & Saha, J. (2023). Green Synthesis and Application of Biogenic Nanomaterials as a Blueprint in Mitigation of Abiotic Stress in Crop Plants: A Conceptual Review. *Advancement of GI-Science and Sustainable Agriculture: A Multi-dimensional Approach*, 155-182.

Borah, B., Banerjee, S., & Allam, B. K. (2024). Recent Advances in the Catalytic Applications of Tin Dioxide-Based Materials in the Synthesis of Bioactive Heterocyclic Compounds. *Tetrahedron Green Chem*, 100048.

Bradu, P., Biswas, A., Nair, C., Sreevalsakumar, S., Patil, M., Kannampuzha, S., Mukherjee, A. G., Wanjari, U. R., Renu, K., Vellingiri, B., & Gopalakrishnan, A. V. (2023). Recent advances in green technology and Industrial Revolution 4.0 for a sustainable future. *Environmental Science and Pollution Research International*, 30(60), 124488–124519. DOI: 10.1007/s11356-022-20024-4 PMID: 35397034

Chang, Z. H., Sum, J. Y., Lau, W. J., Ang, W. L., Teow, Y. H., Ooi, B. S., & Yeap, S. P. (2024). Current State-of-the-Art of Non-Reverse Osmosis-like Forward Osmosis Technology. *Journal of Membrane Science*, 711, 123209. DOI: 10.1016/j.memsci.2024.123209

Chauke, N. M., Mohlala, R. L., Ngqoloda, S., & Raphulu, M. C. (2024). Harnessing visible light: Enhancing TiO2 photocatalysis with photosensitizers for sustainable and efficient environmental solutions. *Frontiers in Chemical Engineering*, 6, 1356021. DOI: 10.3389/fceng.2024.1356021

Chiu, J. C., & Gani, P. (2024). Soil Washing Methods for Effective Removal of Heavy Metal Contaminants. *Industrial and Domestic Waste Management*, 4(1), 56–71. DOI: 10.53623/idwm.v4i1.444

Cui, S., Wang, R., Chen, Q., Pugliese, L., & Wu, S. (2024). Geobatteries in environmental biogeochemistry: Electron transfer and utilization. *Environmental Science and Ecotechnology*, 22, 100446. DOI: 10.1016/j.ese.2024.100446 PMID: 39104555

de Carvalho Lima, E. N., Souza, L. H. M., Aguiar, E. M., Octaviano, A. L. M., Justo, J. F., & Piqueira, J. R. C. (2024). Can bacteria and carbon-based nanomaterials revolutionize nanoremediation strategies for industrial effluents? *Frontiers in Nanotechnology*, 6, 1389107. DOI: 10.3389/fnano.2024.1389107

de Carvalho Lima, E. N., Souza, L. H. M., Aguiar, E. M., Octaviano, A. L. M., Justo, J. F., & Piqueira, J. R. C. (2024). Can bacteria and carbon-based nanomaterials revolutionize nanoremediation strategies for industrial effluents? *Frontiers in Nanotechnology*, 6, 1389107. DOI: 10.3389/fnano.2024.1389107

Dhara, S., Sontakke, A. D., Samanta, N. S., Uppaluri, R. V. S., & Purkait, M. K. (2024). Biowaste-alkaline lignin and GO integrated polysulfone ultrafiltration membrane fabrication for Pb2+ and Eosin Y dye removal. *Separation and Purification Technology*, 341, 126894. DOI: 10.1016/j.seppur.2024.126894

Dinakarkumar, Y., Gnanasekaran, R., Reddy, G. K., Vasu, V., Balamurugan, P., & Murali, G. (2024). *Fungal bioremediation: An overview of the mechanisms, applications and future perspectives*. Environmental Chemistry and Ecotoxicology.

Dong, C., Fang, W., Yi, Q., & Zhang, J. (2022). A comprehensive review on reactive oxygen species (ROS) in advanced oxidation processes (AOPs). *Chemosphere*, 308, 136205. DOI: 10.1016/j.chemosphere.2022.136205 PMID: 36049639

El Messaoudi, N., El Khomri, M., El Mouden, A., Bouich, A., Jada, A., Lacherai, A., Iqbal, H. M. N., Mulla, S. I., Kumar, V., & Américo-Pinheiro, J. H. P. (2024). Regeneration and reusability of non-conventional low-cost adsorbents to remove dyes from wastewaters in multiple consecutive adsorption–desorption cycles: A review. *Biomass Conversion and Biorefinery*, 14(11), 11739–11756. DOI: 10.1007/s13399-022-03604-9

Elsayad, R. M., Sharshir, S. W., Khalil, A., & Basha, A. M. (2024). Recent advancements in wastewater treatment via anaerobic fermentation process: A systematic review. *Journal of Environmental Management*, 366, 121724. DOI: 10.1016/j.jenvman.2024.121724 PMID: 38971071

Fosnacht, K. G., & Pluth, M. D. (2024). Activity-Based Fluorescent Probes for Hydrogen Sulfide and Related Reactive Sulfur Species. *Chemical Reviews*, 124(7), 4124–4257. DOI: 10.1021/acs.chemrev.3c00683 PMID: 38512066

Frühauf, A., Behringer, M., & Meyer-Almes, F. J. (2023). Significance of five-membered heterocycles in human histone deacetylase inhibitors. *Molecules (Basel, Switzerland)*, 28(15), 5686. DOI: 10.3390/molecules28155686 PMID: 37570656

Fu, J., Zhang, Y., Lin, S., Zhang, W., Shu, G., Lin, J., Li, H., Xu, F., Tang, H., Peng, G., Zhao, L., Chen, S., & Fu, H. (2021). Strategies for interfering with bacterial early stage biofilms. *Frontiers in Microbiology*, 12, 675843. DOI: 10.3389/fmicb.2021.675843 PMID: 34168632

Ganguly, N., Roy, A., Banerjee, S., & Pandit, S. (2024). Contaminated Soils: Remedies, Hurdles, and Perspectives. In *Advancements in Bio-systems and Technologies for Wastewater Treatment* (pp. 251–268). Springer International Publishing. DOI: 10.1007/978-3-031-58331-5_13

Ghorai, P., Mondal, U., Hazra, A., & Banerjee, P. (2024). Luminescent metal organic frameworks (LMOFs) and allied composites for the unveiling of organic environmental contaminants (explosive NACs, PAHs and EDCs) sensing through 'Molecular Recognition': A chronicle of recent penetration and future modelling. *Coordination Chemistry Reviews*, 518, 216085. DOI: 10.1016/j.ccr.2024.216085

Gupta, A., & Tyagi, T. (2022). Phytoremediation and therapeutic potential of neglected plants: An invasive aquatic weeds and ornamental plant. In *Biotechnological innovations for environmental bioremediation* (pp. 259–290). Springer Nature Singapore. DOI: 10.1007/978-981-16-9001-3_12

Gürses, A., Güneş, K., Şahin, E., & Açıkyıldız, M. (2024). Technological Applications for Green and Sustainable Production and Elimination of Environmental Pollution. In *Green Chemistry, its Role in Achieving Sustainable Development Goals, Volume1* (pp. 81-128). CRC Press.

Hasan, A., Tabassum, B., Hashim, M., & Khan, N. (2024). Role of plant growth promoting rhizobacteria (PGPR) as a plant growth enhancer for sustainable agriculture: A review. *Bacteria*, 3(2), 59–75.

Hassen, W., Hassen, B., El Ouaer, M., & Hassen, A. (2024). Strategies for Removal of Emerging Compounds of Concern Through Algal Niche Adaptation. In *Algae as a Natural Solution for Challenges in Water-Food-Energy Nexus: Toward Carbon Neutrality* (pp. 161–198). Springer Nature Singapore. DOI: 10.1007/978-981-97-2371-3_9

Hegab, H., Shaban, I., Jamil, M., & Khanna, N. (2023). Toward sustainable future: Strategies, indicators, and challenges for implementing sustainable production systems. *Sustainable Materials and Technologies*, 36, e00617. DOI: 10.1016/j.susmat.2023.e00617

Hussain, S., Akhter, R., & Maktedar, S. S. (2024). *Advancements in sustainable food packaging: from eco-friendly materials to innovative technologies*. Sustainable Food Technology.

Ijoma, G. N., Lopes, T., Mannie, T., & Mhlongo, T. N. (2024). Exploring macrophytes' microbial populations dynamics to enhance bioremediation in constructed wetlands for industrial pollutants removal in sustainable wastewater treatment. *Symbiosis*, 92(3), 323–354. DOI: 10.1007/s13199-024-00981-9

Jin, M., Zhou, Q., Fu, L., & Wu, W. (2024). Application of Biochar-Based Catalysts for Soil and Water Pollution Control. *Topics in Catalysis*, •••, 1–24. DOI: 10.1007/s11244-024-01962-4

Kanungo, J., Sahoo, T., Swain, L. P., & Behera, I. D. (2024). Toxicity of Persistent Hydrocarbon Pollutants, Sources and Sustainable Remediation Process. In *Impact of Petroleum Waste on Environmental Pollution and its Sustainable Management Through Circular Economy* (pp. 39–65). Springer Nature Switzerland.

Khaliq, N. (2023). Microbial enzymes as a robust process to mitigate pollutants of environmental concern. In *Microbial Biomolecules* (pp. 241–267). Academic Press. DOI: 10.1016/B978-0-323-99476-7.00022-3

King, J. F., & Mitch, W. A. (2024). Electrochemical reduction of halogenated organic contaminants using carbon-based cathodes: A review. *Critical Reviews in Environmental Science and Technology*, 54(4), 342–367. DOI: 10.1080/10643389.2023.2239130

Kolawole, A. S., & Iyiola, A. O. (2023). Environmental pollution: threats, impact on biodiversity, and protection strategies. In *Sustainable utilization and conservation of Africa's biological resources and environment* (pp. 377–409). Springer Nature Singapore. DOI: 10.1007/978-981-19-6974-4_14

Kou, B., Yuan, Y., Zhu, X., Ke, Y., Wang, H., Yu, T., & Tan, W. (2024). Effect of soil organic matter-mediated electron transfer on heavy metal remediation: Current status and perspectives. *The Science of the Total Environment*, 917, 170451. DOI: 10.1016/j.scitotenv.2024.170451 PMID: 38296063

Kumar, V., & Verma, P. (2024). Pulp-paper industry sludge waste biorefinery for sustainable energy and value-added products development: A systematic valorization towards waste management. *Journal of Environmental Management*, 352, 120052. DOI: 10.1016/j.jenvman.2024.120052 PMID: 38244409

Kumawat, T. K., Sharma, V., Kumawat, V., Biyani, M., Kumari, N., Garg, R., & Kandwani, N. (2022). Electrokinetic-Assisted Bioremediation and Phytoremediation for the Treatment of Polluted Soil. In *Advances in Bioremediation and Phytoremediation for Sustainable Soil Management: Principles, Monitoring and Remediation* (pp. 379–397). Springer International Publishing. DOI: 10.1007/978-3-030-89984-4_24

Kuppan, N., Padman, M., Mahadeva, M., Srinivasan, S., & Devarajan, R. (2024). *A comprehensive review of sustainable bioremediation techniques: Eco friendly solutions for waste and pollution management.* Waste Management Bulletin.

Lago, A., Rocha, V., Barros, O., Silva, B., & Tavares, T. (2024). Bacterial biofilm attachment to sustainable carriers as a clean-up strategy for wastewater treatment: A review. *Journal of Water Process Engineering*, 63, 105368. DOI: 10.1016/j.jwpe.2024.105368

Lavanya, M. B., Viswanath, D. S., & Sivapullaiah, P. V. (2024). Phytoremediation: An eco-friendly approach for remediation of heavy metal-contaminated soils-A comprehensive review. *Environmental Nanotechnology, Monitoring & Management*, 22, 100975. DOI: 10.1016/j.enmm.2024.100975

Lee, H., Sam, K., Coulon, F., De Gisi, S., Notarnicola, M., & Labianca, C. (2023). Recent developments and prospects of sustainable remediation treatments for major contaminants in soil: A review. *The Science of the Total Environment*, •••, 168769. PMID: 38008308

Lutosławski, K., Ryznar-Luty, A., & Cibis, E. (2024). Efficiency of aerobic biodegradation of sugar beet distillery stillage under dissolved oxygen tension-controlled conditions. *PLoS One*, 19(7), e0306330. DOI: 10.1371/journal.pone.0306330 PMID: 38968255

Merchant, A. I., Vakili, A. H., Kocaman, A., & Amr, S. S. A. (2024). New advancement of advanced oxidation processes for the treatment of Petroleum wastewater. *Desalination and Water Treatment*, 319, 100565. DOI: 10.1016/j.dwt.2024.100565

Miyah, Y., El Messaoudi, N., Benjelloun, M., Acikbas, Y., Şenol, Z. M., Ciğeroğlu, Z., & Lopez-Maldonado, E. A. (2024). Advanced applications of hydroxyapatite nanocomposite materials for heavy metals and organic pollutants removal by adsorption and photocatalytic degradation: A review. *Chemosphere*, 358, 142236. DOI: 10.1016/j.chemosphere.2024.142236 PMID: 38705409

Mohan, C., Robinson, J., Vodwal, L., & Kumari, N. (2024). Sustainable Development Goals for addressing environmental challenges. In *Green Chemistry Approaches to Environmental Sustainability* (pp. 357–374). Elsevier. DOI: 10.1016/B978-0-443-18959-3.00007-0

Mokrani, S., Houali, K., Yadav, K. K., Arabi, A. I. A., Eltayeb, L. B., AwjanAlreshidi, M., Benguerba, Y., Cabral-Pinto, M. M. S., & Nabti, E. (2024). Bioremediation techniques for soil organic pollution: Mechanisms, microorganisms, and technologies-A comprehensive review. *Ecological Engineering*, 207, 107338. DOI: 10.1016/j.ecoleng.2024.107338

Narciso-Ortiz, L., Tec-Caamal, E. N., Aguirre-García, G. J., & Lizardi-Jiménez, M. A. (2023). Bioreactors for bioremediation of polluted water. In *Current Status of Marine Water Microbiology* (pp. 345–364). Springer Nature Singapore. DOI: 10.1007/978-981-99-5022-5_15

Nasri, S., Bayat, M., Miankooshki, F. R., & Samet, N. H. (2022). Recent developments in green approaches for sustainable synthesis of indole-derived scaffolds. *Molecular Diversity*, 26(6), 1–35. DOI: 10.1007/s11030-021-10376-3 PMID: 35031935

Navina, B. K., Velmurugan, N. K., Kumar, P. S., Rangasamy, G., Palanivelu, J., Thamarai, P., & Shakoor, A. (2024). Fungal bioremediation approaches for the removal of toxic pollutants: Mechanistic understanding for biorefinery applications. *Chemosphere*, 350, 141123. DOI: 10.1016/j.chemosphere.2024.141123 PMID: 38185426

Ngieng, H. Y., Jusoh, M. N. H., Ahmad, N., Al Masud, M. A., Samaraweera, H., & Mohamed, M. M. (2024). Enhanced Soil Decontamination via Electrokinetic Removal of Organic Pollutants. *Industrial and Domestic Waste Management*, 4(1), 42–55. DOI: 10.53623/idwm.v4i1.442

Ojija, F. (2024). Emerging environmental contaminants: Sources, effects on biodiversity and humans, remediation, and conservation implications. *Science Progress*, 107(2), 00368504241253720. DOI: 10.1177/00368504241253720 PMID: 38715402

Okpara, E. C., Wojuola, O. B., Quadri, T. W., & Banks, C. E. (2024). An overview of advanced oxidation processes using copper-based catalytic degradation of organic pollutants in water. *Applied Materials Today*, 36, 102053. DOI: 10.1016/j.apmt.2023.102053

Osman, A. I., Chen, Z., Elgarahy, A. M., Farghali, M., Mohamed, I. M., Priya, A. K., Hawash, H. B., & Yap, P. S. (2024). Membrane technology for energy saving: Principles, techniques, applications, challenges, and prospects. *Advanced Energy and Sustainability Research*, 5(5), 2400011. DOI: 10.1002/aesr.202400011

Piwowarska, D., Kiedrzyńska, E., & Jaszczyszyn, K. (2024). A global perspective on the nature and fate of heavy metals polluting water ecosystems, and their impact and remediation. *Critical Reviews in Environmental Science and Technology*, 54(19), 1–23. DOI: 10.1080/10643389.2024.2317112

Priya, A. K., Muruganandam, M., Kumar, A., Senthilkumar, N., Shkir, M., Pandit, B., Imran, M., Prakash, C., & Ubaidullah, M. (2024). Recent advances in microbial-assisted degradation and remediation of xenobiotic contaminants; challenges and future prospects. *Journal of Water Process Engineering*, 60, 105106. DOI: 10.1016/j.jwpe.2024.105106

Qi, Y., Li, D., Zhang, S., Li, F., & Hua, T. (2024). Electrochemical filtration for drinking water purification: A review on membrane materials, mechanisms and roles. *Journal of Environmental Sciences (China)*, 141, 102–128. DOI: 10.1016/j.jes.2023.06.033 PMID: 38408813

Rakhi, P. K., Mishra, K., Scaria, J., Suresh Kumar, M., & Nidheesh, P. V. (2024). Advanced oxidation processes. In *Clean Water: Next Generation Technologies* (pp. 107–116). Springer International Publishing. DOI: 10.1007/978-3-031-48228-1_8

Rayaroth, M. P., Boczkaj, G., Aubry, O., Aravind, U. K., & Aravindakumar, C. T. (2023). Advanced oxidation processes for degradation of water pollutants—ambivalent impact of carbonate species: A review. *Water (Basel)*, 15(8), 1615. DOI: 10.3390/w15081615

Reineke, W., & Schlömann, M. (2023). Microbial degradation of pollutants. In *Environmental microbiology* (pp. 161–290). Springer Berlin Heidelberg. DOI: 10.1007/978-3-662-66547-3_6

Rizzo, C., Pace, A., Pibiri, I., Buscemi, S., & Palumbo Piccionello, A. (2024). From Conventional to Sustainable Catalytic Approaches for Heterocycles Synthesis. *ChemSusChem*, 17(8), e202301604. DOI: 10.1002/cssc.202301604 PMID: 38140917

Rubab, L., Anum, A., Al-Hussain, S. A., Irfan, A., Ahmad, S., Ullah, S., Al-Mutairi, A. A., & Zaki, M. E. (2022). Green chemistry in organic synthesis: Recent update on green catalytic approaches in synthesis of 1, 2, 4-thiadiazoles. *Catalysts*, 12(11), 1329. DOI: 10.3390/catal12111329

Salifu, M., John, M. A., Abubakar, M., Bankole, I. A., Ajayi, N. D., & Amusan, O. (2024). Phytoremediation Strategies for Heavy Metal Contamination: A Review on Sustainable Approach for Environmental Restoration. *Journal of Environmental Protection*, 15(4), 450–474. DOI: 10.4236/jep.2024.154026

Santhappan, J. S., Kalaiselvan, N., Assis, S. M., Amjith, L. R., Glivin, G., & Mathimani, T. (2024). Origin, types, and contribution of emerging pollutants to environmental degradation and their remediation by physical and chemical techniques. *Environmental Research*, 257, 119369. DOI: 10.1016/j.envres.2024.119369 PMID: 38848998

Sharma, S., Shaikh, S., Mohana, S., Desai, C., & Madamwar, D. (2023). Current trends in bioremediation and bio-integrated treatment of petroleum hydrocarbons. *Environmental Science and Pollution Research International*, 31(46), 1–20. DOI: 10.1007/s11356-023-30479-8 PMID: 37861831

Singh, G., Yadav, P., Singh, S. K., Ranjan, N., Verma, H., Ferreira, L. F. R., & Kumar, A. (2024). Membrane processes for wastewater treatment. In *Advances in Chemical Pollution, Environmental Management and Protection* (Vol. 10, pp. 35-56). Elsevier. DOI: 10.1016/bs.apmp.2023.08.001

Srivastav, A. L., Patel, N., Rani, L., Kumar, P., Dutt, I., Maddodi, B. S., & Chaudhary, V. K. (2024). Sustainable options for fertilizer management in agriculture to prevent water contamination: A review. *Environment, Development and Sustainability*, 26(4), 8303–8327. DOI: 10.1007/s10668-023-03117-z

Szekely, G. (2024). The 12 principles of green membrane materials and processes for realizing the United Nations' sustainable development goals. *RSC Sustainability*, 2(4), 871–880. DOI: 10.1039/D4SU00027G

Tabassum, G., Kumar, A., Khan, F. N., Verma, A. K., Naz, S., Ali, S. M., & Dev, K. (2024). Phytochemicals as Potent Therapeutic Molecules in Cancer Treatment in Relation to Sustainable Agriculture System. In *Recent Advancements in Sustainable Agricultural Practices: Harnessing Technology for Water Resources, Irrigation and Environmental Management* (pp. 283–312). Springer Nature Singapore. DOI: 10.1007/978-981-97-2155-9_15

Taneja, S., Karaca, O., & Haritash, A. K. (2024). Electrokinetic remediation: Past experiences and future roadmap for sustainable remediation of metal-contaminated soils. *Journal of Geochemical Exploration*, 259, 107437. DOI: 10.1016/j.gexplo.2024.107437

Tripathy, D. B., & Gupta, A. (2023). Nanomembranes-affiliated water remediation: Chronology, properties, classification, challenges and future prospects. *Membranes*, 13(8), 713. DOI: 10.3390/membranes13080713 PMID: 37623773

Ukhurebor, K. E., Hossain, I., Pal, K., Jokthan, G., Osang, F., Ebrima, F., & Katal, D. (2024). Applications and contemporary issues with adsorption for water monitoring and remediation: A facile review. *Topics in Catalysis*, 67(1), 140–155. DOI: 10.1007/s11244-023-01817-4

Wang, L., Yi, Z., Zhang, P., Xiong, Z., Zhang, G., & Zhang, W. (2024). Comprehensive strategies for microcystin degradation: A review of the physical, chemical, and biological methods and genetic engineering. *Journal of Environmental Management*, 365, 121707. DOI: 10.1016/j.jenvman.2024.121707 PMID: 38968883

Wang, X., Cheng, L., Tan, Q., Xing, Y., Wu, H., Tian, Q., Zhang, Y., & Zheng, L. (2024). Lag in pollutant removal efficiency relative to microbial community dynamics in a denitrifying phosphorus removal bioreactor. *Chemical Engineering Journal*, 497, 154498. DOI: 10.1016/j.cej.2024.154498

Wani, Z. A., Ahmad, Z., Asgher, M., Bhat, J. A., Sharma, M., Kumar, A., Sharma, V., Kumar, A., Pant, S., Lukatkin, A. S., & Anjum, N. A. (2023). Phytoremediation of potentially toxic elements: Role, status and concerns. *Plants*, 12(3), 429. DOI: 10.3390/plants12030429 PMID: 36771511

Zhang, J., Li, G., Yuan, X., Li, P., Yu, Y., Yang, W., & Zhao, S. (2023). Reduction of ultrafiltration membrane fouling by the pretreatment removal of emerging pollutants: A review. *Membranes*, 13(1), 77. DOI: 10.3390/membranes13010077 PMID: 36676884

Zhang, X., Liu, Y., Zhang, M., & Wang, H. (2018). Thiophene-based fused S-heterocycles for removal of heavy metals from industrial wastewater. *Journal of Hazardous Materials*, 354, 125–134. DOI: 10.1016/j.jhazmat.2018.04.059

Zuo, Z., Zhang, H., Gao, S., Wang, C., Chen, W. T., & Hu, G. (2024). Recent progress in porous organic frameworks for electrochemical sensing of environmental pollutants. *Journal of Environmental Chemical Engineering*, 12(5), 113263. DOI: 10.1016/j.jece.2024.113263

Compilation of References

Abbasriyaludeen, A. R., Chitra, K., Ramasamy, S., Murugan, P., & Chandrasekar, P. (2021). Molecular engineering of twisted dipolar chromophores for efficiency boosted BHJ solar cells. *Journal of Materials Chemistry. C, Materials for Optical and Electronic Devices*, 9(13), 4562–4575. DOI: 10.1039/D1TC00708D

Abbracchio, M. P., & Cattabeni, F. (1999). Brain adenosine receptors as targets for therapeutic intervention in neurodegenerative diseases. *Annals of the New York Academy of Sciences*, 890(1), 79–92. DOI: 10.1111/j.1749-6632.1999.tb07983.x PMID: 10668415

Abdelhamid, A. A., Salah, H. A., & Marzouk, A. A. (2020). Synthesis of imidazole derivatives: Ester and hydrazide compounds with antioxidant activity using ionic liquid as an efficient catalyst. *Journal of Heterocyclic Chemistry*, 57(2), 676–685. DOI: 10.1002/jhet.3808

Abdel-Maksoud, M. S., Ammar, U. M., & Oh, C. H. (2019). Anticancer profile of newly synthesized BRAF inhibitors possess 5-(pyrimidin-4-yl) imidazo[2,1-b] thiazole scaffold. *Bioorganic & Medicinal Chemistry*, 2019(27), 2041–2051. DOI: 10.1016/j.bmc.2019.03.062 PMID: 30955995

Abdel-Maksoud, M. S., Kim, M. R., El-Gamal, M. I., Gamal El-Din, M. M., Tae, J., Choi, H. S., Lee, K.-T., Yoo, K. H., & Oh, C.-H. (2015, May 5). Gamal, El-Din, M.M., Tae, J., Choi, H.S., Lee, K.T., Yoo, K.H., Oh, C.H. (2015). Design, synthesis, in vitro antiproliferative evaluation, and kinase inhibitory effects of a new series of imidazo[2,1-b] thiazole derivatives. *European Journal of Medicinal Chemistry*, 95, 453–463. DOI: 10.1016/j.ejmech.2015.03.065 PMID: 25841200

Abdel-Megid, M., Elmahdy, K. M., Elkazak, A. M., Seada, M. H., & Mohamed, O. F. (2016). Chemistry of Thienopyrimidines and Their Biological Applications. *Journal of Pharmaceutical and Applied Chemistry*, 2(3), 103–127. DOI: 10.18576/jpac/020301

Abdelnaby, R. M., El-Malah, A. A., FakhrEldeen, R. R., Saeed, M. M., Nadeem, R. I., Younis, N. S., Abdel-Rahman, H. M., & El-Dydamony, N. M. (2022). In vitro anticancer activity screening of novel fused thiophene derivatives as VEGFR-2/AKT dual inhibitors and apoptosis inducers. *Pharmaceuticals (Basel, Switzerland)*, 15(6), 700. DOI: 10.3390/ph15060700 PMID: 35745619

Abdelwahab, A. B., Hanna, A. G., & Kirsch, G. (2016). Synthesis of novel 3-Acetyl-2-aminothiophenes and Investigation of their behaviour in the reaction with Vilsmeier–Haack reagent. *Synthesis*, 2881–2888.

Abdelwahab, A. B., Hanna, A. G., & Kirsch, G. (2017). Targeted synthesis of 4-Chloro-3-formylthieno [2,3-*b*] pyridine and/or 4-chlorothieno [2,3-*b*] pyridine by reaction between N-protected 3-Acetyl-2-aminothiophenes and Vilsmeier–Haack reagent. *Synthesis*, 49(13), 2971–2979. DOI: 10.1055/s-0036-1588992

Abdel-Wahab, B. F., Khidre, R. E., & Awad, G. E. (2017). Design and Synthesis of Novel 6-(5-Methyl-1 H-1, 2, 3-triazol-4-yl) -5-[(2-(thiazol-2-yl) hydrazono) methyl] imidazo [2, 1-b] thiazoles as Antimicrobial Agents. *Journal of Heterocyclic Chemistry*, 54(1), 489–494. DOI: 10.1002/jhet.2610

Abdula, A. M., Qarah, A. F., Alatawi, K., Qurban, J., Abualnaja, M. M., Katuah, H. A., & El-Metwaly, N. M. (2024). Design, synthesis, and molecular docking of new phenothiazine incorporated N-Mannich bases as promising antimicrobial agents. *Heliyon*, 10(7), e28573. DOI: 10.1016/j.heliyon.2024.e28573 PMID: 38571594

Abo-Bakr, A., & Abbas-Temirek, H. H. (2015)... *International Journal of Current Research*, 7, 14434.

Aboelez, M. O., Kamel, M. S., Belal, A., El Badry Abdel-Aziz, A., Abourehab, M. A., Abdel-Ghany, H., & El Hamd, A. (2022). M.; El-Remaily, M. A. E. A. A. Microwave-assisted synthesis, spectroscopic characterization, and biological evaluation of fused thieno [2, 3-*d*] pyrimidines as potential anti-cancer agents targeting EGFRWT and EGFRT790M. *Molecular Diversity*, •••, 1–17. PMID: 35780205

Abongwa, M., Martin, R. J., & Robertson, A. P. (2017). A brief review on the mode of action of antinematodal drugs. *Acta Veterinaria*, 67(2), 137–152. DOI: 10.1515/acve-2017-0013 PMID: 29416226

Abo-salema, H. M., El-kadyb, D. S., Abd-Elhalimb, M. M., Solimanc, A. A. F., Ebaida, M. S., & El-Sawya, E. R. (2021)... *Egyptian Journal of Chemistry*, 64(2), 561–571.

Addu, N., Miriyala, H., Kapavarapu, R., Kolli, S. K., & Pal, M. (2023). Wang-OSO3H catalyzed one-pot sonochemical synthesis of 1,2,4-benzothiadiazine-1,1-dioxide derivatives: Their in silico / in vitro assessments against MtbCM. *Journal of Molecular Structure*, 1283, 135313. DOI: 10.1016/j.molstruc.2023.135313

Adepoju, A. O., Femi-Adepoju, A., Jalloh, A., & Faeflen, S. (2024). Soil pollution and management practices. In *Environmental Pollution and Public Health* (pp. 187–236). Elsevier. DOI: 10.1016/B978-0-323-95967-4.00011-8

Adeyemi, O. S., Eseola, A. O., Plass, W., Atolani, O., Sugi, T., Han, Y., Batiha, G. E. S., Kato, K., Awakan, O. J., Olaolu, T. D., Nwonuma, C. O., Alejolowo, O., Owolabi, A., Rotimi, D., & Kayode, O. T. (2020). Imidazole derivatives as anti-parasitic agents and use of molecular modelling to investigate the structure activity relationship. *Parasitology Research*, 119(6), 1925–1941. DOI: 10.1007/s00436-020-06668-6 PMID: 32279093

aEl-Remaily, M. A. E. A. A. A., Soliman, A. M., Khalifa, M. E., El-Metwaly, N. M., Alsoliemy, A., El-Dabea, T., & Abu-Dief, A. M. (2022). Rapidly, highly yielded and green synthesis of dihydrotetrazolo [1, 5-a] pyrimidine derivatives in aqueous media using recoverable Pd (II) thiazole catalyst accelerated by ultrasonic: Computational studies. *Applied Organometallic Chemistry*, 36(2, No. e6320), e6320. DOI: 10.1002/aoc.6320bEl Remaily, M. A. E. A. A. A., Naidu, V. R., Ni, S., & Franzén, J. (2015). Carbocation Catalysis: Oxa-Diels-Alder Reactions of Unactivated Aldehydes and Simple Dienes. *European Journal of Organic Chemistry*, 2015(30), 6610–6614. DOI: 10.1002/ejoc.201501112

Afsina Abdulla, C., Neetha, M., Aneeja, T., & Anilkumar, G. (2020). Synthesis and applications of imidazothiazoles: An Overview. *ChemistrySelect*, 5(33), 10374–10386. DOI: 10.1002/slct.202002842

Ahmad, N., Ahmad, H. R., Sajid, Z., Tahir, M., Salma, H., Tamimi, S. A., & Ayub, M. A. (2024). Role of microbial enzymes in wastewater processing and treatment. In *Enzyme Biotechnology for Environmental Sustainability* (pp. 119–160). Academic Press. DOI: 10.1016/B978-0-443-22072-2.00003-6

aKim, O. K., Woo, H. Y., Heuer, W. B., Kim, K. S., & Lee, K. S. (2000). Electronic/photonic property modulations of D-π-D (A) chromophores by DTT as π-center. *American Chemical Society, Polymer* Preprints, *Division of Polymer Chemistry, 41*(1), 795-796.

Alabssawy, A. N., & Hashem, A. H. (2024). Bioremediation of hazardous heavy metals by marine microorganisms: A recent review. *Archives of Microbiology*, 206(3), 103. DOI: 10.1007/s00203-023-03793-5 PMID: 38358529

Alamery, S., AlAjmi, A., Wani, T. A., & Zargar, S. (2023). In silico and in vitro exploration of poziotinib and olmutinib synergy in lung cancer: Role of hsa-miR-7-5p in regulating apoptotic pathway marker genes. *Medicina*, 59(11), 1923. DOI: 10.3390/medicina59111923 PMID: 38003971

Albanese, D. C. M., Gaggero, N., & Fei, M. (2017). A practical synthesis of 2,3-dihydro-1,5-benzothiazepines. *Green Chemistry*, 19(23), 5703–5707. DOI: 10.1039/C7GC02097J

Alexey S. chiriapkin, Ivan P. Kodonidi, dmitry I. pozdnyakov, A. A. G. (2022). Synthesis, in vitro and docking studies of 2-substituted 5,6,7,8-tetrahydrobenzo[4,5]thieno[2,3-d]pyrimidine-4(3H)-one derivatives as agents for the treatment of Alzheimers disease. *Chimica Techno Acta, 9(2).*

Ali, A. R., El-Bendary, E. R., Ghaly, M. A., & Shehata, I. A. (2014). Synthesis, in vitro anticancer evaluation and in silico studies of novel imidazo[2,1-b] thiazole derivatives bearing pyrazole moieties. *European Journal of Medicinal Chemistry*, 75, 492–500. DOI: 10.1016/j.ejmech.2013.12.010 PMID: 24576591

Ali, E. M. H., Abdel-maksoud, M. S., & Oh, C. (2019). Bioorganic & Medicinal Chemistry Thieno [2, 3- d] pyrimidine as a promising scaffold in medicinal chemistry : Recent advances. *Bioorganic & Medicinal Chemistry*, 27(7), 1159–1194. DOI: 10.1016/j.bmc.2019.02.044 PMID: 30826188

Ali, I., Hasan, S. Z., Hozaifa, M., Imanova, G., & Kurniawan, T. A. (2024). *Recent Advances in Hydrogen Storage Methods. Green Hydrogen Economy for Environmental Sustainability* (Vol. 2). Applications, Challenges, and Policies.

Alinaghizadeh, F., Zahedifar, M., Seifi, M., & Sheibani, H. (2016). Cascade synthesis of thieno[2,3-*b*]pyridines by using intramolecular cyclization reactions of 3-cyano-2-(organylmethylthio) pyridines. *Journal of the Brazilian Chemical Society*, 27, 663–669. DOI: 10.5935/0103-5053.20150309

Al-Mulla, A. (2017). A review: Biological importance of heterocyclic compounds. *Der Pharma Chemica*, 9(13), 141–147.

Al-Rashood, S. T., Elshahawy, S. S., El-Qaias, A. M., El-Behedy, D. S., Hassanin, A. A., El-Sayed, S. M., El-Messery, S. M., Shaldam, M. A., & Hassan, G. S. (2020). New thiazolopyrimidine as anti-cancer agents: Synthesis, biological evaluation, DNA binding, molecular modeling and ADMET study. *Bioorganic & Medicinal Chemistry Letters*, 30(23), 127611. DOI: 10.1016/j.bmcl.2020.127611 PMID: 33068712

Altowyan, M. S., Haukka, M., Soliman, S. M., Barakat, A., Alaswad, S. O., Boraei, A. T. A., Gad, E. M., & Youssef, M. F. (2023). Synthesis, Characterization and Single Crystal X-ray Diffraction Analysis of Fused Triazolo/Thiadiazole Clubbed with Indole Scaffold. *Crystals*, 13(3), 423. DOI: 10.3390/cryst13030423

Al-Wahaibi, L. H., Abu-Melha, H. M., & Ibrahim, D. A. (2018). Synthesis of Novel 1,2,4-Triazolyl Coumarin Derivatives as Potential Anticancer Agents. *Journal of Chemistry*, 2018, 1–8. DOI: 10.1155/2018/5201374

Al-Wahaibi, L. H., Akilandeswari, G., Anusha, R., Al-Shaalan, N. H., Alkmali, O. M., El-Emam, A. A., Percino, J. M., & Thamotharan, S. (2019). Insights into the nature of weak noncovalent interactions in 3-(4-fluorophenyl)-6-(2-fluorophenyl)-1,2,4-triazolo[3,4-b][1,3,4]thiadiazole, a potential bioactive agent: X-ray, QTAIM and molecular docking analysis. *Journal of Molecular Structure*, 1183, 331–341. DOI: 10.1016/j.molstruc.2019.01.106

Al-Wahaibi, L. H., Karthikeyan, S., Blacque, O., El-Masry, A. A., Hassan, H. M., Percino, M. J., El-Emam, A. A., & Thamotharan, S. (2022). Structural and Energetic Properties of Weak Noncovalent Interactions in Two Closely Related 3,6-Disubstituted-[1,2,4]triazolo[3,4- *b*][1,3,4]thiadiazole Derivatives: In Vitro Cyclooxygenase Activity, Crystallography, and Computational Investigations. *ACS Omega*, 7(38), 34506–34520. DOI: 10.1021/acsomega.2c04252 PMID: 36188268

Aly, A. A., Brown, A. B., Ramadan, M., Gamal-Eldeen, A. M., Abdel-Aziz, M., Abuo-Rahma, G. E. D. A. A., & Radwan, M. F. (2010). Thieno[2,3-d]pyrimidines in the synthesis of antitumor and antioxidant agents. *Archiv der Pharmazie*, 343(5), 301–309. DOI: 10.1002/ardp.200900245 PMID: 20232374

Amery, W. K., & Bruynseels, J. P. (1992). Levamisole, the story and the lessons. *International Journal of Immunopharmacology*, 14(3), 481–486. DOI: 10.1016/0192-0561(92)90179-O PMID: 1618599

Ameta, K. L., Rathore, N. S., & Kumar, B. (2012). Synthesis and in vitro anti breast cancer activity of some novel 1,5-benzothiazepine derivatives. *Journal of the Serbian Chemical Society*, 77(6), 725–731. DOI: 10.2298/JSC110715219A

Amin, A., Ji, A., Bhat, B. A., Murtaza, D., Hurrah, A. A., Bhat, I. A., ... & Sharma, P. K. (2024). Anti-lung Cancer Activity of Synthesized Substituted 1,4-Benzothiazines: An Insight from Molecular Docking and Experimental Studies. *Anti-Cancer Agents in Medicinal Chemistry (Formerly Current Medicinal Chemistry-Anti-Cancer Agents), 24*(5), 358-371.

Amira T. A. Mohamed. (2019). *Synthesis and antitumor activity evaluation of some thienopyrimidine derivatives*. DOI: 10.1080/00397911.2019.1697822

Ammar, U. M., Abdel-Maksoud, M. S., Ali, E. M. H., Mersal, K. I., Ho Yoo, K., & Oh, C. H. (2020). Structural optimization of imidazothiazole derivatives affords a new promising series as B-Raf V600E inhibitors, synthesis, in vitro assay and in silico screening. *Bioorganic Chemistry*, 100, 103967. DOI: 10.1016/j.bioorg.2020.103967 PMID: 32470760

Amorim, R., de Meneses, M. D., Borges, J. C., da Silva Pinheiro, L. C., Caldas, L. A., Cirne-Santos, C. C., de Mello, M. V., de Souza, A. M., Castro, H. C., de Palmer Paixao, I. C., Campos, R. D., Bergmann, I. E., Malirat, V., Bernardino, A. M. R., Rebello, M. A., & Ferreira, D. F. (2017). Thieno[2,3-*b*]pyridine derivatives: A new class of antiviral drugs against Mayaro virus. *Archives of Virology*, 162(6), 1577–1587. DOI: 10.1007/s00705-017-3261-0 PMID: 28213871

Amr, A. E. G. E., Sherif, M. H., Assy, M. G., Al-Omar, M. A., & Ragab, I. (2010). Antiarrhythmic, serotonin antagonist and antianxiety activities of novel substituted thiophene derivatives synthesized from 2-amino-4,5,6,7-tetrahydro-N- phenylbenzo[b]thiophene-3-carboxamide. *European Journal of Medicinal Chemistry*, 45(12), 5935–5942. DOI: 10.1016/j.ejmech.2010.09.059 PMID: 20950897

Anbar, H. S., El-Gamal, M. I., Tarazi, H., Lee, B. S., Jeon, H. R., Kwon, D., & Oh, C. H. (2010). Imidazothiazole-based potent inhibitors of V600E-B-RAF kinase with promising anti-melanoma activity: Biological and computational studies. *Journal of Enzyme Inhibition and Medicinal Chemistry*, 35(1), 1712–1726. DOI: 10.1080/14756366.2020.1819260 PMID: 32962435

Ando, S., Nishida, J. I., Inoue, Y., Tokito, S., & Yamashita, Y. (2004). Synthesis, physical properties, and field-effect transistors of novel thiophene/thiazolothiazole co-oligomers. *Journal of Materials Chemistry*, 14(12), 1787. DOI: 10.1039/b403699a

Andreani, A., Burnelli, S., Granaiola, M., Leoni, A., Locatelli, A., Morigi, R., Rambaldi, M., Varoli, L., Calonghi, N., Cappadone, C., Voltattorni, M., Zini, M., Stefanelli, C., Masotti, L., & Shoemaker, R. H. (2008). Antitumor activity of new substituted 3-(5-imidazo[2,1-b]thiazolylmethylene)-2-indolinones and 3-(5-imidazo[2,1-b] thiadiazolylmethylene)-2-indolinones: Selectivity against colon tumor cells and effect on cell cycle-related events. *Journal of Medicinal Chemistry*, 51, 809–816. DOI: 10.1021/jm701246g PMID: 19006285

Andreani, A., Granadilla, M., Leoni, A., Locatelli, A., Morigi, R., Rambaldi, M., Varoli, L., Lannigan, D., Smith, J., Scudiero, D., Kondapaka, S., & Shoemaker, R. H. (2011). Imidazo[2,1-b] thiazole guanylhydrazones as RSK2 inhibitors. *European Journal of Medicinal Chemistry*, 46(9), 4311–4323. DOI: 10.1016/j.ejmech.2011.07.001 PMID: 21794960

Andreani, A., Granaiola, M., Locatelli, A., Morigi, R., Rambaldi, M., Varoli, L., Calonghi, N., Cappadone, C., Farruggia, G., Stefanelli, C., Masotti, L., Nguyen, T. L., Hamel, E., & Shoemaker, R. H. (2012). Substituted 3-(5-imidazo [2, 1-b] thiazolylmethylene)-2-indolinones and analogues: Synthesis, cytotoxic activity, and study of the mechanism of action. *Journal of Medicinal Chemistry*, 55(5), 2078–2088. DOI: 10.1021/jm2012694 PMID: 22283430

Andreani, A., Leoni, A., Locatelli, A., Morigi, R., Rambaldi, M., Cervellati, R., Greco, E., Kondratyuk, T. P., Park, E. J., Huang, K., VanBreemen, R. B., & Pezzuto, J. M. (2013). Chemopreventive and antioxidant activity of 6-substituted imidazo[2,1-b] thiazoles. *European Journal of Medicinal Chemistry*, 68, 412–421. DOI: 10.1016/j.ejmech.2013.07.052 PMID: 23994869

Andreani, A., Rambaldi, M., Locatelli, A., Bossa, R., Galatulas, I., & Salvatore, G. (1994). Antitumor and cardiotonic activity of imidazo[2,1-b] thiazole guanylhydrazones. *In Vivo (Athens, Greece)*, 8, 1031–1032. PMID: 7772733

Andrew, S. Bell, Zhiyoung Yu, jennie A. hutton, Megan H. Wright, James A. Brannigan, daniel Paape, shieley M. Roberts, Charlotte L. sutherell, Markus Ritzefeld, Anathony J. Wilkinson, Deborah F. Smith, Robin J. Leatherbarrow, and E. W. T. (2020). Novel Thienopyrimidine Inhibitors of Leishmania N-Myrostoyltransferase with On-Target Activity in Intracellular Amastigotes. *Journal of Medicinal Chemistry*.

An, X., Li, N., Zhang, L., Xu, Z., Zhang, S., & Zhang, Q. (2024). New insights into the typical nitrogen-containing heterocyclic compound-quinoline degradation and detoxification by microbial consortium: Integrated pathways, meta-transcriptomic analysis and toxicological evaluation. *Journal of Hazardous Materials*, 465, 133158. DOI: 10.1016/j.jhazmat.2023.133158 PMID: 38061124

Armour, M., Davies, A. G., Upadhyay, J., & Wassermann, A. (1967). Colored electrically conducting polymers from furan, pyrrole, and thiophene. *Journal of Polymer Science. Part A-1, Polymer Chemistry*, 5(7), 1527–1538. DOI: 10.1002/pol.1967.150050704

Arya, N., Munde, M. K., Dwivedi, J., Jagdale, A. Y., & Jain, K. S. (2013). Design, synthesis, and antihyperlipidemic evaluation of novel 2-[1-(Substitutedphenyl)-4-oxo-azetidin-2-yl]-5,6-disubstitutedthieno[2,3-d] pyrimidin-4(3H)-ones. *Archiv der Pharmazie*, 346(8), 588–595. DOI: 10.1002/ardp.201300001 PMID: 23881664

Asghar, N., Hussain, A., Nguyen, D. A., Ali, S., Hussain, I., Junejo, A., & Ali, A. (2024). Advancement in nanomaterials for environmental pollutants remediation: A systematic review on bibliometrics analysis, material types, synthesis pathways, and related mechanisms. *Journal of Nanobiotechnology*, 22(1), 26. DOI: 10.1186/s12951-023-02151-3 PMID: 38200605

Ashe, A. J., Fang, X., Schiesher, M., Richardson, A. D., & Hedberg, K. (2000). Aromatic Heterocycles Containing Arsenic and Sulfur: The Synthesis of 1,2- and 1,3-Thiaarsole and the Gas-Phase Molecular Structure of 1,2-Thiaarsole. *Journal of the American Chemical Society*, 122(29), 7012–7016. DOI: 10.1021/ja993924z

Asif, M., Allahyani, M., Alsaiari, A. A., & Almehmadi, M. M. (2023). Various Synthetic Methods and Diverse Therapeutic Potential of Substituted Thiazine Derivatives. *Current Organic Chemistry*, 27(20), 1741–1753. DOI: 10.2174/0113852728260166231101070430

Attenni, B., Hernando, J. I. M., Malancona, S., Narjes, F., Ontoria, J. M. O., & Rowley, M. (2010). *U.S. Patent No. 7,781,431*. Washington, DC: U.S. Patent and Trademark Office.

Attenni, B., Hernando, J. I. M., Malancona, S., Narjes, F., Ontoria, J. M. O., & Rowley, M. (2005).

Ayuk, E. L., Uchegbu, M. O., Aronimo, S. I., & Ebiem-Kenechukwu, P. I. (2024). Synthesis, Characterization, In silico and In vitro Antimicrobial Activity of Phenothiazine-3-sulphonamide Derivatives.

Ayuk, E. L., Uchegbu, M. O., Ebiem-Kenechukwu, P. I., & Oni, T. O. (2024). Preparation, Characterization, In silico and In vitro Antimicrobial Studies of Phenothiazine-3-sulphonamide Derivatives. *Bioactivities*, 2(1), 41–56. DOI: 10.47352/bioactivities.2963-654X.215

Azam, F., El-gnidi, B. A., Alkskas, I. A., Ahmed, M. A., Azam, F., El-gnidi, B. A., Alkskas, I. A., & Ahmed, M. A. (2010). *oxidative stress in mice neuroleptic-induced catalepsy and oxidative stress in mice. 6366*. DOI: 10.3109/14756361003671052

Babalola, B. A., Malik, M., Sharma, L., Olowokere, O., & Folajimi, O. Results in Chemistry.

Babalola, B. A., Malik, M., Sharma, L., Olowokere, O., & Folajimi, O. (2024). Exploring the therapeutic potential of phenothiazine derivatives in medicinal chemistry. *Results in Chemistry*, 8, 101565. DOI: 10.1016/j.rechem.2024.101565

Baburajeev, C. P., Mohan, C. D., Rangappa, S., Mason, D. J., Fuchs, J. E., Bender, A., Barash, U., Vlodavsky, I., Basappa, , & Rangappa, K. S. (2017). Identification of Novel Class of Triazolo-Thiadiazoles as Potent Inhibitors of Human Heparanase and their Anticancer Activity. *BMC Cancer*, 17(1), 235. DOI: 10.1186/s12885-017-3214-8 PMID: 28359266

Badr, S. M. I., & Barwa, R. M. (2011). Synthesis of some new [1,2,4]triazolo[3,4-b][1,3,4]thiadiazines and [1,2,4]triazolo[3,4-b][1,3,4] thiadiazoles starting from 5-nitro-2-furoic acid and evaluation of their antimicrobial activity. *Bioorganic & Medicinal Chemistry*, 19(15), 4506–4512. DOI: 10.1016/j.bmc.2011.06.024 PMID: 21723735

Bae, S. H., Park, J. H., Choi, H. G., Kim, H., & Kim, S. H. (2018). Imidazole Antifungal Drugs Inhibit the Cell Proliferation and Invasion of Human Breast Cancer Cells. *Biomolecules & Therapeutics*, 26(5), 494–502. DOI: 10.4062/biomolther.2018.042 PMID: 30092625

Baig, M. F., Nayak, V. L., Budaganaboyina, P., Mullagiri, K., Sunkari, S., Gour, J., & Kamal, A. (2018). Synthesis and biological evaluation of imidazo [2, 1-b] thiazole-benzimidazole conjugates as microtubule-targeting agents. *Bioorganic Chemistry*, 77, 515–526.

Bajaj, D., Yadav, V., Dhyani, A., & Wadhwa, N. K. (2024). Exploration of Microbial Enzymes in Bioremediation. In *Microbial Approaches for Sustainable Green Technologies* (pp. 250–272). CRC Press. DOI: 10.1201/9781003407683-13

Bajaj, K., Srivastava, V., Lata, S., & Chandra, R. (2003).*Indian Journal of Chemistry*, 42B, 1723.

Baker, R. K., Rupprecht, K. M., Pessolano, A. A., & Durette, P. L. (1988) US4767766 A, United States.

Baker, D. J., Greenhaff, P. L., & Timmons, J. A. (2006). Glycogen phosphorylase inhibition as a therapeutic target: A review of the recent patent literature. *Expert Opinion on Therapeutic Patents*, 16(4), 459–466. DOI: 10.1517/13543776.16.4.459

Bakr F Abdel-Wahab, S. S. (2014)..*Synthesis*, 46, 1709–1716. DOI: 10.1055/s-0033-1338627

Balaji, G., Della Pelle, A. M., Popere, B. C., Chandrasekaran, A., & Thayumanavan, S. (2012). Synthesis and properties of thienopyrrole based heteroacenes–indolodibenzothienopyrrole and dicarbazolodithienopyrrole. *Organic & Biomolecular Chemistry*, 10(17), 3455–3462. DOI: 10.1039/c2ob25087j PMID: 22421802

Balaji, G., Phua, D. I., Shim, W. L., & Valiyaveettil, S. (2010). Synthesis and characterization of unsymmetric indolodithienopyrrole and extended diindolodithienopyrrole. *Organic Letters*, 12(2), 232–235. DOI: 10.1021/ol902528b PMID: 20000317

Balaji, G., & Valiyaveettil, S. (2009). Synthesis and properties of symmetric and unsymmetric dibenzothienopyrroles. *Organic Letters*, 11(15), 3358–3361. DOI: 10.1021/ol901133m PMID: 19572751

Bala, S., Garg, D., Thirumalesh, B. V., Sharma, M., Sridhar, K., Inbaraj, B. S., & Tripathi, M. (2022). Recent strategies for bioremediation of emerging pollutants: A review for a green and sustainable environment. *Toxics*, 10(8), 484. DOI: 10.3390/toxics10080484 PMID: 36006163

Banik, S., Bandyopadhyay, S., & Ganguly, S. (2003). Bioeffects of microwave–a brief review. *Bioresource Technology*, 87(2), 155–159. DOI: 10.1016/S0960-8524(02)00169-4 PMID: 12765354

Barker, A. J., Faull, A. W., & Kettle, J. G. (2002). *U.S. Patent No. 6,479,527*. Washington, DC: U.S. Patent and Trademark Office.

Barradas, J. S., Errea, M. I., D'Accorso, N. B., Sepúlveda, C. S., & Damonte, E. B. (2011). Imidazo [2, 1-b] thiazole carbohydrate derivatives: Synthesis and antiviral activity against Junin virus, agent of Argentine hemorrhagic fever. *European Journal of Medicinal Chemistry*, 46(1), 259–264. DOI: 10.1016/j.ejmech.2010.11.012 PMID: 21115214

Barrows, R. D., Hammill, J. T., Tran, M. C., Falade, M. O., Rice, A. L., Davis, C. W., Emge, T. J., Rablen, P. R., Kiplin Guy, R., & Knapp, S. (2020). Evaluation of 1,1-cyclopropylidene as a thioether isostere in the 4-thio-thienopyrimidine (TTP) series of antimalarials. *Bioorganic & Medicinal Chemistry*, 28(22), 115758. DOI: 10.1016/j.bmc.2020.115758 PMID: 33007559

Bartlett, J. B., Freeman, S., Kenny, P., Morley, A., & Whittamore, P. (2003).

Batchu, H., Bhattacharyya, S., & Batra, S. (2012). Iodine-mediated intramolecular electrophilic aromatic cyclization in allylamines: A general route to synthesis of quinolines, pyrazolo[4,3-*b*] pyridines, and thieno[3,2-*b*]pyridines. *Organic Letters*, 14(24), 6330–6333. DOI: 10.1021/ol303109m PMID: 23230893

Battula, S. R. K., Rama Kishore Putta, V. P., Subbareddy, G. V., Chakravarthy, I. E., & Saravanan, V. A. (2017). Divergent and metal free synthesis of sulfoximine tethered imidazoles, imidazopyridines, imidazothiazoles, imidazobenzothiazines, thiazoles and selenazoles. *Organic & Biomolecular Chemistry*, 15(17), 3742–3755. DOI: 10.1039/C7OB00601B PMID: 28406509

Bayazeed, A., Alharbi, H., Alalawy, A. I., Alshammari, N. A. H., Alqahtani, A. M., Alsahag, M., & El-Metwaly, N. M. (2024). Structural and antimicrobial activity studies of brand new benzothiazole-pyrazole conjugates. *Journal of Molecular Structure*, •••, 139579.

Bayoumy, B. E. (1991). *Acta Poloniae Pharmaceutica*, 48, 13. PMID: 1669421

Benzodifuranyl, N. (2020). *Anti-Inflammatory and Analgesic Agents*. 1–25.

Bergman, J., & Janosik, T. (2002). Recent progress in the chemistry of sulfur-containing indoles. In *Progress in Heterocyclic Chemistry* (Vol. 14, pp. 1–18). Elsevier., DOI: 10.1016/S0959-6380(02)80003-4

Bernardino, A. M., Pinheiro, L. C., Ferreira, V. F., & Azevedo, A. R. (2004). Synthesis and antiviral activity of new 4-(phenylamino)thieno[2,3-*b*]pyridine derivatives. *Heterocyclic Communications*, 10(6), 407–410. DOI: 10.1515/HC.2004.10.6.407

Bertolini, M., Zecchin, S., & Cavalca, L. (2023). Sequential Anaerobic/Aerobic microbial transformation of chlorinated ethenes: Use of sustainable approaches for aquifer decontamination. *Water (Basel)*, 15(7), 1406. DOI: 10.3390/w15071406

Beškoski, V., Lješević, M., Jiménez, B., Muñoz-Arnanz, J., Colomer-Vidal, P., Inui, H., & Nakano, T. (2024). *Bioremediation of Soils Contaminated with PFAS: An Update on Available Techniques*. Pilot Studies, Challenges, and Future Directions.

Bettendorff, L. (2021). Update on Thiamine Triphosphorylated Derivatives and Metabolizing Enzymatic Complexes. *Biomolecules*, 11(11), 1645. DOI: 10.3390/biom11111645 PMID: 34827643

Bevk, D., Marin, L., Lutsen, L., & Vanderzande, D. (2013). Maes, W. *RSC Advances*, 3, 11418. DOI: 10.1039/c3ra40851e

Bhabal, S. R., Shaikh, S. F., Yellapurkar, I. P., Pavale, G. S., & Ramana, M. M. V. (2022). Synthesis of novel fluorinated 1,5-benzothiazepine derivatives and their biological evaluation as anticancer and antibacterial agent. *Journal of the Serbian Chemical Society*, 87(10), 1109–1116. DOI: 10.2298/JSC210428041B

Bhandari, G., & Sharma, M. (2021). Microbial enzymes in the bioremediation of pollutants: Emerging potential and challenges. *Mycoremediation and Environmental Sustainability*, 3, 75–94. DOI: 10.1007/978-3-030-54422-5_3

Bhat, I. A., Parveen, S., Nisar, S., & Sharma, P. K. (2024). For personal private use only. *Anti-cancer Agents in Medicinal Chemistry*, 24(5), 358–371. PMID: 37957911

Bhatia, R., & Kumar, B. (2023). p53 as a potential target for treatment of cancer: A perspective on recent advancements in small molecules with structural insights and SAR studies. *European Journal of Medicinal Chemistry*, 247, 115020. DOI: 10.1016/j.ejmech.2022.115020 PMID: 36543034

Bhattacharya, S., & Saha, J. (2023). Green Synthesis and Application of Biogenic Nanomaterials as a Blueprint in Mitigation of Abiotic Stress in Crop Plants: A Conceptual Review. *Advancement of GI-Science and Sustainable Agriculture: A Multi-dimensional Approach*, 155-182.

Bhatt, N., Vyas, K., Nimavat, K., Govender, T., Kruger, H. G., Maguire, G. E. M., & Bhatt, P. (2013). Synthesis of novel N-alkyl-2-(3,5-dimethyl-1,1-dioxido-2H-1,2,6-thiadiazin-4-yl)benzamides as potential antimicrobial agents. *Journal of Pharmacy Research*, 6(3), 355–360. DOI: 10.1016/j.jopr.2013.03.008

Bhavsar, Z. A., Acharya, P. T., Jethava, D. J., & Patel, H. D. (2020). Recent advances in development of anthelmintic agents: Synthesis and biological screening. *Synthetic Communications*, 50(7), 917–946. DOI: 10.1080/00397911.2019.1695276

Bhongade, B. A., Talath, S., Gadad, R. A., & Gadad, A. K. (2016). Biological activities of imidazo [2, 1-b][1, 3, 4] thiadiazole derivatives: A review. *Journal of Saudi Chemical Society*, 20, S463–S475. DOI: 10.1016/j.jscs.2013.01.010

Bhowmik, D., Kamandar, P., Gupta, J., Mete, S., & Hatvate, N. (2024). An efficient, green and micellar catalyzed preparative-scale synthesis of 3,4-Disubstituted isoxazole-5(4H)-ones in the water. *Sustainable Chemistry for the Environment*, 5, 100070. DOI: 10.1016/j.scenv.2024.100070

Biehl, E. R. (2011) "Five-Membered Ring Systems: Thiophenes and Se/Te Derivatives." In *Progress in Heterocyclic Chemistry*, vol. 22, pp. 109-141. Elsevier.

Biradar, M. S., Nargund, S. L., & Thapa, S. (2023). Synthesis and Cytotoxicity Assay of Aniline Substituted Thienopyrimidines for Anti-Colorectal Cancer Activity. *Results in Chemistry*, 5(February), 100926. DOI: 10.1016/j.rechem.2023.100926

Bishop, R. (1996). Designing new lattice inclusion hosts. *Chemical Society Reviews*, 25(5), 311. DOI: 10.1039/cs9962500311

Blackburn, C., Achab, A., Elder, A., Ghosh, S., Guo, J., Harriman, G., & Jones, M. (2005). Synthesis of 3-Amino-1,2,4-benzothiadi- azine 1,1-Dioxides via a Tandem Aza-Wittig/Heterocumulene Annulation. *The Journal of Organic Chemistry*, 70(24), 10206–10209. DOI: 10.1021/jo051843h PMID: 16292875

Black, J. L., Richelson, E., & Richardson, J. W.BLACK. (1985). Antipsychotic Agents: A Clinical Update. *Mayo Clinic Proceedings*, 60(11), 777–789. DOI: 10.1016/S0025-6196(12)60421-2 PMID: 2865413

Blair, J. B., Marona-Lewicka, D., Kanthasamy, A., Lucaites, V. L., Nelson, D. L., & Nichols, D. E. (1999). Thieno [3, 2-b]-and thieno [2, 3-b] pyrrole bioisosteric analogues of the hallucinogen and serotonin agonist N, N-dimethyltryptamine. *Journal of Medicinal Chemistry*, 42(6), 1106–1111. DOI: 10.1021/jm980692q PMID: 10090793

Blenkle, M., Boldt, P., Bräuchle, C., Grahn, W., Ledoux, I., Nerenz, H., Stadler, S., Wichern, J., & Zyss, J. (1996). Chalcogens as electron donors for selected nonlinear optic phores. *Journal of the Chemical Society, Perkin Transactions 2: Physical Organic Chemistry*, 2(7), 1377–1384. DOI: 10.1039/P29960001377

Bodige, V., Ravula, P., Gulipalli, K. C., Endoori, S., Cherukumalli, P., Jn, N. S., & Seelam, V. (2019). Design, synthesis, antitubercular and antibacterial activities of pyrrolo[3,2-*b*]pyridine-3-carboxamide linked 2-methoxypyridine derivatives and in silico docking studies. *Synthetic Communications*, 49(17), 2219–2234. DOI: 10.1080/00397911.2019.1618874

Borah, B., Banerjee, S., & Allam, B. K. (2024). Recent Advances in the Catalytic Applications of Tin Dioxide-Based Materials in the Synthesis of Bioactive Heterocyclic Compounds. *Tetrahedron Green Chem*, 100048.

Borcea, A. M., Ionu , I., Cri an, O., & Oniga, O. (2021). An Overview of the Synthesis and Antimicrobial, Antiprotozoal, and Antitumor Activity of Thiazole and Bisthiazole Derivatives. *Molecules (Basel, Switzerland)*, 26(3), 624. DOI: 10.3390/molecules26030624 PMID: 33504100

Boschelli, D. H., Wu, B., Sosa, A. C. B., Durutlic, H., Ye, F., Raifeld, Y., Golas, J. M., & Boschelli, F. (2004). Identification of 7-phenylaminothieno-[3,2-*b*]pyridine-6-carbonitriles as a new class of Src kinase inhibitors. *Journal of Medicinal Chemistry*, 47(27), 6666–6668. DOI: 10.1021/jm049237m PMID: 15615514

Bosson-Vanga, H., Primas, N., Franetich, J. F., Lavazec, C., Gomez, L., Ashraf, K., & Mazier, D. (2021). A new thienopyrimidinone chemotype shows multistage activity against Plasmodium falciparum, including artemisinin-resistant parasites. *Microbiology Spectrum*, 9(2), e00274–e21.

Bouherrou, H., Saidoun, A., Abderrahmani, A., Abdellaziz, L., Rachedi, Y., Dumas, F., & Demenceau, A. (2017). Synthesis and Biological Evaluation of New Substituted Hantzsch Thiazole Derivatives from Environmentally Benign One-Pot Synthesis Using Silica Supported Tungstosilisic Acid as Reusable Catalyst. *Molecules (Basel, Switzerland)*, 22(5), 757. DOI: 10.3390/molecules22050757 PMID: 28481282

Bradu, P., Biswas, A., Nair, C., Sreevalsakumar, S., Patil, M., Kannampuzha, S., Mukherjee, A. G., Wanjari, U. R., Renu, K., Vellingiri, B., & Gopalakrishnan, A. V. (2023). Recent advances in green technology and Industrial Revolution 4.0 for a sustainable future. *Environmental Science and Pollution Research International*, 30(60), 124488–124519. DOI: 10.1007/s11356-022-20024-4 PMID: 35397034

Braverman, S., & Freund, M. (1990). Rearrangements of bridged diallenes. A facile synthesis of novel condensed heterocycles by tandem [3,3]-sigmatropic rearrangement and double intramolecular michael addition of diallenyl disulfides and diselenides. Isolation of stable diallenyl diselenides. *Tetrahedron*, 46(16), 5759–5776. DOI: 10.1016/S0040-4020(01)87773-5

Bremner, D. H., Dunn, A. D., Wilson, K. A., Sturrock, K. R., & Wishart, G. (1998). The synthesis of thienopyridines from ortho-halogenated pyridine derivatives. Part 3. *Synthesis*, 1998(8), 1095–1097. DOI: 10.1055/s-1998-2124

Budriesi, R., Ioan, P., Locatelli, A., Cosconati, S., Leoni, A., Ugenti, M. P., Andreani, A., di Toro, R., Bedini, A., Spampinato, S., Marinelli, L., Novellino, E., & Chiarini, A. (2008). Imidazo[2,1-b] thiazole system: A scaffold endowing dihydropyridines with selective cardiodepressant activity. *Journal of Medicinal Chemistry*, 51(6), 1592–1600. DOI: 10.1021/jm070681+ PMID: 18303827

Buttar, D., Edge, M., Emery, S. C., Fitzek, M., Forder, C., Griffen, A., Hayter, B., Hayward, C. F., Hopcroft, P. J., Luke, R. W., Page, K., Stawpert, J., & Wright, A. (2008). Discovery of imidazole vinyl pyrimidines as a novel class of kinase inhibitors which inhibit Tie-2 and are orally bioavailable. *Bioorganic & Medicinal Chemistry Letters*, 18(16), 4723–4726. DOI: 10.1016/j.bmcl.2008.06.106 PMID: 18676144

Cabaj, W., Stankiewicz, M., Jonas, W., & Moore, L. (1995). Levamisole and its influence on the immune response of lambs. *Veterinary Research Communications*, 19(1), 17–26. DOI: 10.1007/BF01839248 PMID: 7762136

Cai, J., & Goodnow, R. A., Jr. (2008). *U.S. Patent No. 7,396,940*. Washington, DC: U.S. Patent and Trademark Office.

Campiani, G., Fiorini, I., De Filippis, M. P., Ciani, S. M., Garofalo, A., Nacci, V., Giorgi, G., Sega, A., Botta, M., Chiarini, A., Budriesi, R., Bruni, G., Romeo, M. R., Manzoni, C., & Mennini, T. (1996). Cardiovascular Characterization of Pyrrolo[2,1-*d*][1,5]benzothiazepine Derivatives Binding Selectively to the Peripheral-Type Benzodiazepine Receptor (PBR): From Dual PBR Affinity and Calcium Antagonist Activity to Novel and Selective Calcium Entry Blockers. *Journal of Medicinal Chemistry*, 39(15), 2922–2938. DOI: 10.1021/jm960162z PMID: 8709127

Cao, J., Wang, L., You, G., Li, L., Yao, L., Zhou, Z., Yang, E., Cai, W., Ling, Q., & Zhen, H. (2022). Dopant-free, hole-transporting polymers containing benzotriazole acceptor unit for perovskite solar cells. *Dyes and Pigments*, 200, 110170. DOI: 10.1016/j.dyepig.2022.110170

Castanedo, G. M., & Sutherlin, D. P. (2001). Synthesis of tetrasubstituted thiophenes on solid-support using the Gewald reaction. *Tetrahedron Letters*, 42(41), 7181–7184. DOI: 10.1016/S0040-4039(01)01470-8

Castanedo, G., Liu, Y., Crawford, J. J., & Braun, M.-G. (2016). Synthesis of fused imidazole-containing ring systems via dual oxidative amination of C (sp3)–H bonds. *The Journal of Organic Chemistry*, 81(18), 8617–8624. DOI: 10.1021/acs.joc.6b01517 PMID: 27529722

Castanheira, E. M., Abreu, A. S., Queiroz, M.-J. R., Ferreira, P. M., Coutinho, P. J., Nazareth, N., & Nascimento, M. S.-J. (2009). Fluorescence properties of a potential antitumoral benzothieno [3, 2-b] pyrrole in solution and lipid membranes. *Journal of Photochemistry and Photobiology A Chemistry*, 206(2-3), 220–226. DOI: 10.1016/j.jphotochem.2009.07.007

Cen, J., & Zhongguo Yiyao Gongye Zazhi, (1997). *Chemical Abstract*, 127, 331417.

Chambhare, R. V., Khadse, B. G., Bobde, A. S., & Bahekar, R. H. (2003). Synthesis and preliminary evaluation of some N-[5-(2-furanyl)-2-methyl-4-oxo-4H-thieno[2,3-d]pyrimidin-3-yl]-carboxamide and 3-substituted-5-(2-furanyl)-2-methyl-3H-thieno[2,3-d]pyrimidin-4-ones as antimicrobial agents. *European Journal of Medicinal Chemistry*, 38(1), 89–100. DOI: 10.1016/S0223-5234(02)01442-3 PMID: 12593919

Chang, Z. H., Sum, J. Y., Lau, W. J., Ang, W. L., Teow, Y. H., Ooi, B. S., & Yeap, S. P. (2024). Current State-of-the-Art of Non-Reverse Osmosis-like Forward Osmosis Technology. *Journal of Membrane Science*, 711, 123209. DOI: 10.1016/j.memsci.2024.123209

Channegowda, V. D., Goravanahalli, M., Raghavendrab, N. D., Rekhac, K., Mantelingub, K. S., Rangappab, D., & Bhadregowdaa, G. (2015). Curr. *Chemistry Letters*, 4, 133.

Channegowda, V. D., Nanjappagowda, D. R., Goravanahalli, M. R., & Doddamedur, G. B. (2015).*Journal of Chemical and Pharmaceutical Research*, 7(12), 733.

Chauke, N. M., Mohlala, R. L., Ngqoloda, S., & Raphulu, M. C. (2024). Harnessing visible light: Enhancing TiO2 photocatalysis with photosensitizers for sustainable and efficient environmental solutions. *Frontiers in Chemical Engineering*, 6, 1356021. DOI: 10.3389/fceng.2024.1356021

Chen, H. Y., Nikolka, M., Wadsworth, A., Yue, W., Onwubiko, A., Xiao, M., White, A. J., Baran, D., Sirringhaus, H., & McCulloch, I. (2018). A thieno[2,3-*b*] pyridine-flanked diketopyrrolopyrrole polymer as a n-type polymer semiconductor for allpolymer solar cells and organic field-effect transistors. *Macromolecules*, 51(1), 71–79. DOI: 10.1021/acs.macromol.7b00934

Chen, X., Zhu, C., Guo, F., Qiu, X., Yang, Y., Zhang, S., He, M., Parveen, S., Jing, C., Li, Y., & Ma, B. (2010). Acetic acid derivatives of 3,4-dihydro-2H-1,2,4-benzothiadiazine 1,1-dioxide as a novel class of potent aldose reductase inhibitors. *Journal of Medicinal Chemistry*, 53(23), 8330–8344. DOI: 10.1021/jm100962a PMID: 21062005

Chhabra, S., & Shah, K. (2023). Synthesis and anticancer activity of 1,2,4-Benzothiadiazine-1,1-dioxides. *Results in Chemistry*, 5, 100873. DOI: 10.1016/j.rechem.2023.100873

Chiarelli, L. R., Salina, E. G., Mori, G., Azhikina, T., Riabova, O., Lepioshkin, A., Grigorov, A., Forbak, M., Madacki, J., Orena, B. S., Manfredi, M., Gosetti, F., Buzzi, A., Degiacomi, G., Sammartino, J. C., Marengo, E., Korduláková, J., Riccardi, G., Mikušová, K., & Pasca, M. R. (2020). New Insights into the Mechanism of Action of the Thienopyrimidine Antitubercular Prodrug TP053. *ACS Infectious Diseases*, 6(2), 313–323. DOI: 10.1021/acsinfecdis.9b00388 PMID: 31729215

Chinsembu, K. C. (2016). Tuberculosis and nature's pharmacy of putative anti-tuberculosis agents. *Acta Tropica*, 153, 46–56. DOI: 10.1016/j.actatropica.2015.10.004 PMID: 26464047

Chirra, N., Shanigarapu, V., Abburi, N. P., Sinegubova, E. O., Pedapati, R. K., Esaulkova, Y. L., Zarubaev, V. V., & Kantevari, S. (2024). Synthesis and antiviral activity of novel imidazo [2, 1-b] thiazoles coupled with morpholine and thiomorpholines. *Journal of Heterocyclic Chemistry*, 61(3), 496–505. DOI: 10.1002/jhet.4778

Chiu, J. C., & Gani, P. (2024). Soil Washing Methods for Effective Removal of Heavy Metal Contaminants. *Industrial and Domestic Waste Management*, 4(1), 56–71. DOI: 10.53623/idwm.v4i1.444

Chobe, S. S., & Kadam, V. V. (2019).*IJRAR*, 6(1), 492.

Choudhary, D., Kumar, B., & Thakur, A. (2024). Cancer and insulin-like growth factor inhibitors: Recent advancements and SAR analysis. In *Current Molecular Targets of Heterocyclic Compounds for Cancer Therapy* (pp. 161–181). Academic Press. DOI: 10.1016/B978-0-323-96121-9.00008-5

Cibotaru, S., Sandu, A. I., Nicolescu, A., & Marin, L. (2023). Antitumor Activity of PEGylated and TEGylated Phenothiazine Derivatives: Structure–Activity Relationship. *International Journal of Molecular Sciences*, 24(6), 5449. DOI: 10.3390/ijms24065449 PMID: 36982524

Comoy, C., Banaszak, E., & Fort, Y. (2006). New efficient access to thieno[3,2-*b*] pyridine derivatives via regioselective lithiation of 3-methylthiopyridine. *Tetrahedron*, 62(25), 6036–6041. DOI: 10.1016/j.tet.2006.04.008

Contributor, O., & Masurier, N. (n.d.). *Thienopyrimidine. 4*(Scheme 2).

Cox, T. R. (2021). The matrix in cancer. *Nature Reviews. Cancer*, 21(4), 217–238. DOI: 10.1038/s41568-020-00329-7 PMID: 33589810

Crossley, C. A., Rajani, V., & Yuan, Q. (2023). Modulation of L-type calcium channels in Alzheimer's disease: A potential therapeutic target. *Computational and Structural Biotechnology Journal*, 21, 11–20. DOI: 10.1016/j.csbj.2022.11.049 PMID: 36514335

Cui, S., Wang, R., Chen, Q., Pugliese, L., & Wu, S. (2024). Geobatteries in environmental biogeochemistry: Electron transfer and utilization. *Environmental Science and Ecotechnology*, 22, 100446. DOI: 10.1016/j.ese.2024.100446 PMID: 39104555

Da Pozzo, E., La Pietra, V., Cosimelli, B., Da Settimo, F., Giacomelli, C., Marinelli, L., Martini, C., Novellino, E., Taliani, S., & Greco, G. (2014). p53 functional inhibitors behaving like pifithrin-β counteract the Alzheimer peptide non-β-amyloid component effects in human SH-SY5Y cells. *ACS Chemical Neuroscience*, 5(5), 390–399. DOI: 10.1021/cn4002208 PMID: 24646317

da Silva, R. C.Junior, Campanholi, K. D. S. S., de Morais, F. A. P., de Moraes Pinto, L. A., dos Santos Rando, F., dos Santos Pozza, M. S., & Caetano, W. (2023). Phenazines and photoactive formulations: promising photodrugs for photodynamic therapy. In *Dyes and Pigments-Insights and Applications*. IntechOpen.

Dabaeva, V. V., Bagdasaryan, M. R., Paronikyan, E. G., Dashyan, S. S., Paronikyan, R. G., Nazaryan, I. M., & Akopyan, A. G. (2020). Synthesis and Neurotropic Activity of Novel Condensed Cyclopentanopyrido[3′,2′:4,5]-Thieno[3,2-D]Pyrimidine Derivatives. *Pharmaceutical Chemistry Journal*, 54(7), 707–713. DOI: 10.1007/s11094-020-02259-y

Dahlmann, U., & Neidlein, R. (1997). The Diyne Reaction of 3, 3′-Bis (phenylethynyl)-2, 2′-bithiophene Derivatives via Rhodium Complexes: A novel approach to condensed benzo [2, 1-b: 3, 4-b′] dithiophenes. *Helvetica Chimica Acta*, 80(1), 111–120. DOI: 10.1002/hlca.19970800111

Dangi, R. R., Hussain, N., & Talesara, G. L. (2011). Synthesis characterization and biological evaluation of some alkoxyphthalimide derivatives of 3-(4-substituted phenyl)-6, 6-diphenyl-3,3a-dihydro-2H-imidazo[2,1-b] pyrazolo[3,4-d] [1,3] thiazol-7(6H)-one. *Medicinal Chemistry Research*, 20(9), 1490–1498. DOI: 10.1007/s00044-010-9392-4

Davis, L. E., Shalin, S. C., & Tackett, A. J. (2019). Current state of melanoma diagnosis and treatment. *Cancer Biology & Therapy*, 20(11), 1366–1379. DOI: 10.1080/15384047.2019.1640032 PMID: 31366280

Dawar, R., Karan, S., Bhardwaj, S., Meena, D. K., Padhan, S. R., Reddy, K. S., & Bana, R. S. (2023). Role of Sulphur Fertilization in Legume Crops: A Comprehensive Review. *International Journal of Plant and Soil Science*, 35(21), 718–727. DOI: 10.9734/ijpss/2023/v35i214033

de Carvalho Lima, E. N., Souza, L. H. M., Aguiar, E. M., Octaviano, A. L. M., Justo, J. F., & Piqueira, J. R. C. (2024). Can bacteria and carbon-based nanomaterials revolutionize nanoremediation strategies for industrial effluents? *Frontiers in Nanotechnology*, 6, 1389107. DOI: 10.3389/fnano.2024.1389107

De La, J., Curtis, C., Lawrence, D., Paul, G., Laurence, S., Ronald, B., Lu, S., Gatoulis, S. C., Brown, B., Migoya, E., & Shore, N. (2023). ORIGINAL RESEARCH ARTICLE A Phase I Clinical Trial Evaluating the Safety and Dosing of Relugolix with Novel Hormonal Therapy for the Treatment of Advanced Prostate Cancer. *Targeted Oncology*, 18(3), 383–390. DOI: 10.1007/s11523-023-00967-5 PMID: 37060432

Demydchuk, B. A., Sokolyuk, P. A., Vasilenko, O. M., Brovarets, V. S., & Drach, B. S. (2011)... *J. Org. Pharm. Chem.*, 1, 47.

Deng, J., Peng, L., Zhang, G., Lan, X., Li, C., Chen, F., Zhou, Y., Lin, Z., Chen, L., Dai, R., Xu, H., Yang, L., Zhang, X., & Hu, W. (2011). European Journal of Medicinal Chemistry The highly potent and selective dipeptidyl peptidase IV inhibitors bearing a thienopyrimidine scaffold effectively treat type 2 diabetes. *European Journal of Medicinal Chemistry*, 46(1), 71–76. DOI: 10.1016/j.ejmech.2010.10.016 PMID: 21106276

Deng, X., Tan, X., An, T., Ma, Q., Jin, Z., Wang, C., Meng, Q., & Hu, C. (2019). Synthesis, Characterization, and Biological Activity of a Novel Series of Benzo [4,5] imidazo[2,1-b] thiazole Derivatives as Potential Epidermal Growth Factor Receptor Inhibitors. *Molecules (Basel, Switzerland)*, 24(4), 682. DOI: 10.3390/molecules24040682 PMID: 30769844

Deprets, S., & Kirsch, G. (2000).*European Journal of Organic Chemistry*, 1353. Advance online publication. DOI: 10.1039/C6RA17315B

Desai, K. R., Patel, V., Patel, M., Patel, B. R., & Patel, A. J. (2024). Synthesis and Biological Evaluation of Benzothiazepine, Benzothiazole, and Benzothiophene Derivatives. In *S-Heterocycles: Synthesis and Biological Evaluation* (pp. 189–209). Springer Nature Singapore. DOI: 10.1007/978-981-97-4308-7_6

Dessi, A., Calamante, M., Mordini, A., Zani, L., Taddei, M., & Reginato, G. (2014). Microwave-activated synthesis of thiazolo[5,4-d]thiazoles by a condensation/oxidation sequence. *RSC Advances*, 4(3), 1322–1328. DOI: 10.1039/C3RA45015E

Dhara, S., Sontakke, A. D., Samanta, N. S., Uppaluri, R. V. S., & Purkait, M. K. (2024). Biowaste-alkaline lignin and GO integrated polysulfone ultrafiltration membrane fabrication for Pb2+ and Eosin Y dye removal. *Separation and Purification Technology*, 341, 126894. DOI: 10.1016/j.seppur.2024.126894

Di Noia, V., D'Aveni, A., D'Argento, E., Rossi, S., Ghirardelli, P., Bortolotti, L., Vavassori, V., Bria, E., & Ceresoli, G. (2021). Treating disease progression with osimertinib in EGFR-mutated non-small-cell lung cancer: Novel targeted agents and combination strategies. *ESMO Open*, 6(6), 100280. DOI: 10.1016/j.esmoop.2021.100280 PMID: 34634633

Di Santo, R., Costi, R., Artico, M., Miele, G., Lavecchia, A., Novellino, E., Bergamini, A., Cancio, R., & Maga, G. (2006). Arylthiopyrrole (AThP) Derivatives as Non-Nucleoside HIV-1 Reverse Transcriptase Inhibitors: Synthesis, Structure–Activity Relationships, and Docking Studies (Part 1). *ChemMedChem*, 1(12), 1367–1378. DOI: 10.1002/cmdc.200600119 PMID: 17089433

Dienes, Y., Eggenstein, M., Kárpáti, T., Sutherland, T. C., Nyulászi, L., & Baumgartner, T. (2008). Phosphorus-Based Heteropentacenes: Efficiently Tunable Materials for Organic n-Type Semiconductors. *Chemistry (Weinheim an der Bergstrasse, Germany)*, 14(32), 9878–9889. DOI: 10.1002/chem.200801549 PMID: 18844201

Dinakarkumar, Y., Gnanasekaran, R., Reddy, G. K., Vasu, V., Balamurugan, P., & Murali, G. (2024). *Fungal bioremediation: An overview of the mechanisms, applications and future perspectives*. Environmental Chemistry and Ecotoxicology.

Dincel, E. D., Gürsoy, E., Yilmaz-Ozden, T., & Ulusoy-Güzeldemirci, N. (2020). Antioxidant activity of novel imidazo [2, 1-b] thiazole derivatives: Design, synthesis, biological evaluation, molecular docking study and in silico ADME prediction. *Bioorganic Chemistry*, 103, 104220. DOI: 10.1016/j.bioorg.2020.104220 PMID: 32896742

Dincel, E. D., Hasbal-Celikok, G., Yilmaz-Ozden, T., & Ulusoy-Güzeldemirci, N. (2022). Design, synthesis, biological evaluation, molecular docking, and dynamic simulation study of novel imidazo [2, 1-b] thiazole derivatives as potent antioxidant agents. *Journal of Molecular Structure*, 1258, 132673. DOI: 10.1016/j.molstruc.2022.132673

Djuidje, E. N., Sciabica, S., Buzzi, R., Dissette, V., Balzarini, J., Liekens, S., Serra, E., Andreotti, E., Manfredini, S., Vertuani, S., & Baldisserotto, A. (2020). Design, synthesis and evaluation of benzothiazole derivatives as multifunctional agents. *Bioorganic Chemistry*, 101, 103960. DOI: 10.1016/j.bioorg.2020.103960 PMID: 32559579

Doddagaddavalli, M. A., Kalalbandi, V. K. A., & Seetharamappa, J. (2023). Synthesis, characterization, crystallographic, binding, in silico and antidiabetic studies of novel 2, 4-thiazolidinedione-phenothiazine molecular hybrids. *Journal of Molecular Structure*, 1276, 134625. DOI: 10.1016/j.molstruc.2022.134625

Dong, C. K., Urgaonkar, S., Cortese, J. F., Gamo, F. J., Garcia-Bustos, J. F., Lafuente, M. J., Patel, V., Ross, L., Coleman, B. I., Derbyshire, E. R., Clish, C. B., Serrano, A. E., Cromwell, M., Barker, R. H.Jr, Dvorin, J. D., Duraisingh, M. T., Wirth, D. F., Clardy, J., & Mazitschek, R. (2011). Identification and Validation of Tetracyclic Benzothiazepines as Plasmodium falciparum Cytochrome bc1 Inhibitors. *Chemistry & Biology*, 18(12), 1602–1610. DOI: 10.1016/j.chembiol.2011.09.016 PMID: 22195562

Dong, C., Fang, W., Yi, Q., & Zhang, J. (2022). A comprehensive review on reactive oxygen species (ROS) in advanced oxidation processes (AOPs). *Chemosphere*, 308, 136205. DOI: 10.1016/j.chemosphere.2022.136205 PMID: 36049639

Dong, Jin Hong, Seung Hyun Jung, Jisook Kim, Danbee Jung, Young Gil Ahn, K. H. S. and K. H. M. (2019). Synthesis and biological evaluation of novel thienopyrimidine derivatives as diacylglycerol acyltransferase 1 (DGAT-1) inhibitors. *Journal of Enzyme Inhibition and Medicinal Chemistry*, •••, 35.

Dong, Z.-Q., Liu, F.-M., Xu, F., & Yuan, Z.-L. (2011). Synthesis of 1,5-benzothiazepine derivatives bearing 2-phenoxy-quinoline moiety via 1,3-diplolar cycloaddition reaction. *Molecular Diversity*, 15(4), 963–970. DOI: 10.1007/s11030-011-9328-z PMID: 21837389

Dorai, S. T., Lakshmikanth, K., Tiwari, P., Saini, S. M., & Chandrashekharappa, S. (2023). One-pot construction of novel trifluoromethyl dihydro-imidazo [1, 2-a] pyridine: A greener approach. *Tetrahedron*, 148, 133691. DOI: 10.1016/j.tet.2023.133691

Dorsey, F. B. D., McDaniel, S. L., Levin, R. B., Vacca, J. P., Darke, P. L., Zugay, J. A., Emini, E. A., Schleif, W. A., Lin, J. H., Chen, I. W., Holloway, M. K., Anderson, P. S., & Huff Bioorg, J. R. (1994). Synthesis and evaluation of pyridyl analogs of L-735,524: Potent HIV-1 protease inhibitors Med. *Bioorganic & Medicinal Chemistry Letters*, 4(23), 2769–2774. DOI: 10.1016/S0960-894X(01)80592-8

Dotsenko, V. V., Buryi, D. S., Lukina, D. Y., & Krivokolysko, S. G. (2020). Recent advances in the chemistry of thieno[2,3-*b*]pyridines 1. Methods of synthesis of thieno[2,3-*b*]pyridines. *Russian Chemical Bulletin*, 69(10), 1829–1858. DOI: 10.1007/s11172-020-2969-2

Dotsenko, V. V., Krivokolysko, S. G., & Litvinov, V. P. (2003). Synthesis of Hexa- and Octahydropyrido[3'2':4,5]thieno[3,2-*d*]pyrimidines. *Chemistry of Heterocyclic Compounds*, 39(1), 110–112. DOI: 10.1023/A:1023033027915

Drakontaeidi, A., Papanotas, I., & Pontiki, E. (2024). Multitarget Pharmacology of Sulfur–Nitrogen Heterocycles: Anticancer and Antioxidant Perspectives. *Antioxidants*, 13(8), 898. DOI: 10.3390/antiox13080898 PMID: 39199144

Drapak, I., Perekhoda, L., Demchenko, N., Suleiman, M., Rakhimova, M., Demchuk, I., Taran, S., Seredynska, N., & Gerashchenko, I. (2019). Cardioprotective Activity of Some 2-Arylimino-1,3-Thiazole Derivatives. *Scientia Pharmaceutica*, 87(1), 1–8. DOI: 10.3390/scipharm87010007

Du Bois, D. J. (2003). *U.S. Patent No. 6,576,653*. Washington, DC: U.S. Patent and Trademark Office.

Du Bois, D. J. (2004). *U.S. Patent No. 6,828,343*. Washington, DC: U.S. Patent and Trademark Office.

Dunn, A. D., & Norrie, R. (1987). Nucleophilic displacements in pyridine rings. *Journal of Heterocyclic Chemistry*, 24(1), 85–89. DOI: 10.1002/jhet.5570240118

Du, P., Zhou, H., Sui, Y., Liu, Q., & Zou, K. (2016). Asymmetric synthesis of 3,4-dihydro-2H-1,2,4-benzothiadiazine-1,1- dioxides catalyzed by scandium(III)-inda-Pybox. *Tetrahedron*, 72(12), 1573–1578. DOI: 10.1016/j.tet.2016.02.009

Durso, M., Gentili, D., Bettini, C., Zanelli, A., Cavallini, M., De Angelis, F., Grazia Lobello, M., Biondo, V., Muccini, M., Capelli, R., & Melucci, M. (2013). π-Core tailoring for new high performance thieno(bis)imide based n-type molecular semiconductors. *Chemical Communications (Cambridge)*, 49(39), 4298–4300. DOI: 10.1039/C2CC37053K PMID: 23165167

Duwuri, M., Konkar, S., Hong, K. H., Blagg, B. S. J., & Krise, J. P. (2006). *ACS Chemical Biology*, 1(5), 309–315. DOI: 10.1021/cb6001202 PMID: 17163760

Edwin, G.Tse, M. K. and matthew H. T. (2019). The past, present and future of anti-malarial medicines. *Malaria Journal*. PMID: 30902052

Eissa, K. I., Kamel, M. M., Mohamed, L. W., Doghish, A. S., Alnajjar, R., Al-Karmalawy, A. A., & Kassab, A. E. (2023). Design, synthesis, and biological evaluation of thienopyrimidine derivatives as multifunctional agents against Alzheimer's disease. *Drug Development Research*, 84(5), 937–961.

El Alami, A., El Maraghi, A., & Sdassi, H. (2024). Review of synthesis process of benzoxazole and benzothiazole derivatives. *Synthetic Communications*, 54(10), 769–801. DOI: 10.1080/00397911.2024.2337089

El Messaoudi, N., El Khomri, M., El Mouden, A., Bouich, A., Jada, A., Lacherai, A., Iqbal, H. M. N., Mulla, S. I., Kumar, V., & Américo-Pinheiro, J. H. P. (2024). Regeneration and reusability of non-conventional low-cost adsorbents to remove dyes from wastewaters in multiple consecutive adsorption–desorption cycles: A review. *Biomass Conversion and Biorefinery*, 14(11), 11739–11756. DOI: 10.1007/s13399-022-03604-9

El Shehry, M. F., Abu-Hashem, A. A., & El-Telbani, E. M. (2010). Synthesis of 3-((2,4-dichlorophenoxy)methyl)-1,2,4-triazolo(thiadiazoles and thiadiazines) as anti-inflammatory and molluscicidal agents. *European Journal of Medicinal Chemistry*, 45(5), 1906–1911. DOI: 10.1016/j.ejmech.2010.01.030 PMID: 20153090

El-Atawy, M. A., Ferretti, F., & Ragaini, F. (2017). Palladium-Catalyzed Intramolecular Cyclization of Nitroalkenes: Synthesis of Thienopyrroles. *European Journal of Organic Chemistry*, 2017(14), 1902–1910. DOI: 10.1002/ejoc.201700165

Elgiushy, H. R., Abou-Taleb, N. A., Holz, G. G., Chepurny, O. G., Pirmettis, I., Kakabakos, S., Karageorgos, V., Liapakis, G., Albohy, A., Abouzid, K. A. M., & Hammad, S. F. (2021). Synthesis, in vitro biological investigation, and molecular dynamics simulations of thiazolopyrimidine based compounds as corticotrophin releasing factor receptor-1 antagonists. *Bioorganic Chemistry*, 114, 105079. DOI: 10.1016/j.bioorg.2021.105079 PMID: 34174633

Elmongy, E., kedr, mohammed, abotaleb, nageh, & Abbas, S. (2021). Design and synthesis of new thienopyrimidine derivatives along with their antioxidant activity. *Egyptian Journal of Chemistry, 0*(0), 0–0. DOI: 10.21608/ejchem.2021.23721.2412

Elsayad, R. M., Sharshir, S. W., Khalil, A., & Basha, A. M. (2024). Recent advancements in wastewater treatment via anaerobic fermentation process: A systematic review. *Journal of Environmental Management*, 366, 121724. DOI: 10.1016/j.jenvman.2024.121724 PMID: 38971071

El-Sayed, H. A., Assy, M. G., & Mohamed, S., A. (. (2020). An efficient synthesis and antimicrobial activity of *N* -bridged triazolo[3,4- *b*]thiadiazine and triazolo[3,4- *b*]thiadiazole derivatives under microwave irradiation. *Synthetic Communications*, 50(7), 997–1007. DOI: 10.1080/00397911.2020.1726397

El-Shershaby, M. H., Ghiaty, A., Bayoumi, A. H., Ahmed, H. E. A., El-Zoghbi, M. S., El-Adl, K., & Abulkhair, H. S. (2021). 1,2,4-Triazolo[4,3- *c*]quinazolines: A bioisosterism-guided approach towards the development of novel PCAF inhibitors with potential anticancer activity. *New Journal of Chemistry*, 45(25), 11136–11152. DOI: 10.1039/D1NJ00710F

El-Shoukrofy, M. S., Abd El Razik, H. A., AboulWafa, O. M., Bayad, A. E., & El-Ashmawy, I. M. (2019). Pyrazoles containing thiophene, thienopyrimidine and thienotriazolopyrimidine as COX-2 selective inhibitors: Design, synthesis, in vivo anti-inflammatory activity, docking and in silico chemo-informatic studies. *Bioorganic Chemistry*, 85(February), 541–557. DOI: 10.1016/j.bioorg.2019.02.036 PMID: 30807897

El-Taweel, F. M. A., Elagamey, A. A., El-Kenawy, A., & Waly, M. A. (2001). Novel synthesis of thiazole, coumarin,pyridine,thiophene and thieno[2,3-*b*]pyridine derivatives. *Phosphorus, Sulfur, and Silicon and the Related Elements*, 176(1), 215–225. DOI: 10.1080/10426500108055120

Engstrom, K. M., Baize, A. L., Franczyk, T. S., Kallemeyn, J. M., Mulhern, M. M., Rickert, R. C., & Wagaw, S. (2009). Improved synthesis of 3-Substituted-4-amino[3,2-*c*]thienopyridines. *The Journal of Organic Chemistry*, 74(10), 3849–3855. DOI: 10.1021/jo9003772 PMID: 19371092

Eras, J., Galvez, C., & Garcta, F. (1984). Reactivity of thienopyrroles. Synthesis of isomeric nitro and bromothienopyrroles. *Journal of Heterocyclic Chemistry*, 21(1), 215–217. DOI: 10.1002/jhet.5570210143

Eslam, M.H. Ali, Mohammed S. Abdel-Maksoud, C.-H. oh. (2019). Thieno[2,3-d] pyrimidine as a promising scaffold in medicinal chemistry: Recent advances. *Bioorganic & Medicinal Chemistry*. PMID: 30826188

Etivand, N., Ahmadi Sabegh, M., & Khalafy, J. (2019). Synthesis of a new series of benzo[d] imidazo [2, 1-b] thiazole-1-ium hydroxides by a one-pot, three-component reaction in water. *Monatshefte für Chemie*, 150(2), 317–325. DOI: 10.1007/s00706-018-2315-7

Evans, B. E., Rittle, K. E., Bock, M. G., DiPardo, R. M., Freidinger, R. M., Whitter, W. L., Lundell, G. F., Veber, D. F., Anderson, P. S., Chang, R. S., Lotti, V. J., Cerino, D. J., Chen, T. B., Kling, P. J., Kunkel, K. A., Springer, J. P., & Hirshfield, J. (1988). Methods for drug discovery: Development of potent, selective, orally effective cholecystokinin antagonists. *Journal of Medicinal Chemistry*, 31(12), 2235–2246. DOI: 10.1021/jm00120a002 PMID: 2848124

Ewida, M. A., Abou El Ella, D. A., Lasheen, D. S., Ewida, H. A., El-Gazzar, Y. I., & El-Subbagh, H. I. (2018). Imidazo [2′, 1′: 2, 3] thiazolo [4, 5-d] pyridazinone as a new scaffold of DHFR inhibitors: Synthesis, biological evaluation and molecular modeling study. *Bioorganic Chemistry*, 80, 11–23. DOI: 10.1016/j.bioorg.2018.05.025 PMID: 29864684

Fadda, A. A., Ghanem, R. A., Gaffer, H. E., Waly, M. M., & Tawfik, E. H. (2023). Synthesis, Antimicrobial Evaluation, and Molecular Docking of New Azole, Azine, Thiazole, and Chromene Derivatives. *Polycyclic Aromatic Compounds*, 43(4), 3429–3449. DOI: 10.1080/10406638.2022.2069135

Falco, J. L., Pique, M., Gonzalez, M., Buira, I., Mendez, E., Terencio, J., Perez, C., Príncep, M., Palomer, A., & Guglietta, A. (2006). Synthesis, pharmacology and molecular modeling of N-substituted 2-phenyl-indoles and benzimidazoles as potent GABAA agonists. *European Journal of Medicinal Chemistry*, 41(8), 985–990. DOI: 10.1016/j.ejmech.2006.03.031 PMID: 16764969

Faleye, O. S., Boya, B. R., Lee, J. H., Choi, I., & Lee, J. (2024). Halogenated antimicrobial agents to combat drug-resistant pathogens. *Pharmacological Reviews*, 76(1), 90–141. DOI: 10.1124/pharmrev.123.000863 PMID: 37845080

Fan, C., Miao, J., Jiang, B., Yang, C., Wu, H., Qin, J., & Cao, Y. (2013). Highly efficient, solution-processed orange–red phosphorescent OLEDs by using new iridium phosphor with thieno[3, 2-*c*]pyridine derivative as cyclometalating ligand. *Organic Electronics*, 14(12), 3392–3398. DOI: 10.1016/j.orgel.2013.10.013

Fang, B., Hu, C., Ding, Y., Qin, H., Luo, Y., Xu, Z., Meng, J., & Chen, Z. (2021). Discovery of 4 H-thieno [3, 2-b] pyrrole derivatives as potential anticancer agents. *Journal of Heterocyclic Chemistry*, 58(8), 1610–1627. DOI: 10.1002/jhet.4285

Fang, C., Lu, T., Zhu, J., Sun, K., & Du, D. (2017). Formal [3 + 4] Annulation of α,β-Unsaturated Acyl Azoliums: Access to Enantioenriched N– *H* -Free 1,5-Benzothiazepines. *Organic Letters*, 19(13), 3470–3473. DOI: 10.1021/acs.orglett.7b01457 PMID: 28598636

Fan, Z., Yang, Z., Zhang, H., Mi, N., Wang, H., Cai, F., Zuo, X., Zheng, Q., & Song, H. (2010). Synthesis, Crystal Structure, and Biological Activity of 4-Methyl-1,2,3-thiadiazole-Containing 1,2,4-Triazolo[3,4- *b*][1,3,4]thiadiazoles. *Journal of Agricultural and Food Chemistry*, 58(5), 2630–2636. DOI: 10.1021/jf9029628 PMID: 20014761

Farid, N. A., Kurihara, A., & Wrighton, S. A. (2010). Metabolism and disposition of the thienopyridine antiplatelet drugs ticlopidine, clopidogrel, and prasugrel in humans. *Journal of Clinical Pharmacology*, 50(2), 126–142. DOI: 10.1177/0091270009343005 PMID: 19948947

Fascio, M. L., Errea, M. I., & D'accorso, N. B. (2015). Imidazothiazole and related heterocyclic systems. Synthesis, chemical and biological properties. *European Journal of Medicinal Chemistry*, 90, 666–683. DOI: 10.1016/j.ejmech.2014.12.012 PMID: 25499987

Fatma E. M. El-baih, H. A. S. A.-B. and H. M. A.-H. (n.d.). *Synthesis of Some Thienopyrimidine Derivatives.*

Fei, Z., Pattanasattayavong, P., Han, Y., Schroeder, B. C., Yan, F., Kline, R. J., Anthopoulos, T. D., & Heeney, M. (2014). Influence of side-chain regiochemistry on the transistor performance of high-mobility, all-donor polymers. *Journal of the American Chemical Society*, 136(43), 15154–15157. DOI: 10.1021/ja508798s PMID: 25302474

Feng, M., Tang, B., Liang, S., & Jiang, X. (2016). Sulfur containing scaffolds in drugs: Synthesis and application in medicinal chemistry. *Current Topics in Medicinal Chemistry*, 16, 1200–1216. DOI: 10.2174/15680266156661509151117 41 PMID: 26369815

Fernandes, S. S. M., Castro, M. C. R., Mesquita, I., Andrade, L., Mendes, A., & Raposo, M. M. M. (2017). Synthesis and characterization of novel thieno [3,2-*b*] thiophene based metal-free organic dyes with different heteroaromatic donor moieties as sensitizers for dye-sensitized solar cells. *Dyes and Pigments*, 136, 46–53. DOI: 10.1016/j.dyepig.2016.08.020

Field, M. C., Horn, D., Fairlamb, A. H., Ferguson, M. A. J., Gray, D. W., Read, K. D., De Rycker, M., Torrie, L. S., Wyatt, P. G., Wyllie, S., & Gilbert, I. H. (2017). Anti-trypanosomatid drug discovery: An ongoing challenge and a continuing need. *Nature Reviews. Microbiology*, 15(4), 217–231. DOI: 10.1038/nrmicro.2016.193 PMID: 28239154

Fitch, D. M., Evans, K. A., Chai, D., & Duffy, K. J. (2005). A Highly Efficient, Asymmetric Synthesis of Benzothiadiazine-Substituted Tetramic Acids: Potent Inhibitors of Hepatitis C Virus RNA-Dependent RNA Polymerase. *Organic Letters*, 7(24), 5521–5524. DOI: 10.1021/ol052371w PMID: 16288546

Flefel, E., El-Sayed, W., Mohamed, A., El-Sofany, W., & Awad, H. (2017). Synthesis and Anti-cancer Activity of New 1-Thia-4-azaspiro[4.5]decane, Their Derived Thiazolopyrimidine and 1,3,4-Thiadiazole Thioglycosides. *Molecules (Basel, Switzerland)*, 22(1), 170. DOI: 10.3390/molecules22010170 PMID: 28117699

Fosnacht, K. G., & Pluth, M. D. (2024). Activity-Based Fluorescent Probes for Hydrogen Sulfide and Related Reactive Sulfur Species. *Chemical Reviews*, 124(7), 4124–4257. DOI: 10.1021/acs.chemrev.3c00683 PMID: 38512066

Fouad, M. M., El-Bendary, E. R., suddek, G. M., Shehata, I. A., & El-Kerdawy, M. M. (2018). Synthesis and in vitro antitumor evaluation of some new thiophenes and thieno[2,3-d]pyrimidine derivatives. *Bioorganic Chemistry*, 81(2), 587–598. DOI: 10.1016/j.bioorg.2018.09.022 PMID: 30248510

Francotte, P., Goffin, E., Fraikin, P., Lestage, P., Van Heugen, J.-C., Gillotin, F., Danober, L., Thomas, J.-Y., Chiap, P., Caignard, D.-H., Pirotte, B., & de Tullio, P. (2010). New Fluorinated 1,2,4-Benzothiadiazine 1,1-Dioxides: Discovery of an Orally Active Cognitive Enhancer Acting through Potentiation of the 2-Amino-3-(3-hydroxy-5-methylisoxazol-4-yl)propionic Acid Receptors. *Journal of Medicinal Chemistry*, 53(4), 1700–1711. DOI: 10.1021/jm901495t PMID: 20108934

from ketone.pdf. (n.d.).

Frühauf, A., Behringer, M., & Meyer-Almes, F. J. (2023). Significance of five-membered heterocycles in human histone deacetylase inhibitors. *Molecules (Basel, Switzerland)*, 28(15), 5686. DOI: 10.3390/molecules28155686 PMID: 37570656

Fu, D. J., Li, J., & Yu, B. (2021). Annual review of LSD1/KDM1A inhibitors in 2020. *European Journal of Medicinal Chemistry*, 214, 113254. DOI: 10.1016/j.ejmech.2021.113254 PMID: 33581557

Fu, J., Zhang, Y., Lin, S., Zhang, W., Shu, G., Lin, J., Li, H., Xu, F., Tang, H., Peng, G., Zhao, L., Chen, S., & Fu, H. (2021). Strategies for interfering with bacterial early stage biofilms. *Frontiers in Microbiology*, 12, 675843. DOI: 10.3389/fmicb.2021.675843 PMID: 34168632

Fukata, Y., Asano, K., & Matsubara, S. (2015). Facile Net Cycloaddition Approach to Optically Active 1,5-Benzothiazepines. *Journal of the American Chemical Society*, 137(16), 5320–5323. DOI: 10.1021/jacs.5b02537 PMID: 25856510

Fuller, L. S., & Iddon, B. (1999). K A Smith J. Thienothiophenes. Part 3.1,2 On the ring-opening reactions of 3-lithiated and 3,6-dilithiated thieno[3,2-*b*]thiophenes; new routes to polyfunctionalized thiophenes and enediynes Chem. Soc. *Journal of the Chemical Society, Perkin Transactions 1: Organic and Bio-Organic Chemistry*, 1(10), 1273–1278. DOI: 10.1039/a901300h

Gadkari, Y. U., Hatvate, N. T., Takale, B. S., & Telvekar, V. N. (2020). Concentrated solar radiation as a renewable heat source for a preparative-scale and solvent-free Biginelli reaction. *New Journal of Chemistry*, 44(20), 8167–8170. DOI: 10.1039/D0NJ01351J

Gadkari, Y. U., Hatvate, N. T., & Telvekar, V. N. (2021a). Concentrated solar radiation-assisted one-pot/multicomponent synthesis of pyranopyrazole derivatives under neat condition. *Research on Chemical Intermediates*, 47(10), 4245–4255. DOI: 10.1007/s11164-021-04530-7

Gadkari, Y. U., Hatvate, N. T., & Telvekar, V. N. (2021b). Solar energy as a renewable energy source for preparative-scale as well as solvent and catalyst-free Hantzsch reaction. *Sustainable Chemistry and Pharmacy*, 21, 100444. DOI: 10.1016/j.scp.2021.100444

Gadkari, Y. U., Jadhav, N. L., Hatvate, N. T., & Telvekar, V. N. (2020). Concentrated Solar Radiation Aided Green Approach for Preparative Scale and Solvent-Free Synthesis of 3-Methyl-4-(hetero)arylmethylene Isoxazole-5(4H)-ones. *ChemistrySelect*, 5(39), 12320–12323. DOI: 10.1002/slct.202003348

Gali, R., Banothu, J., Porika, M., Velpula, R., Bavantula, R., Abbagani, S. (2014). Synthesis and in vitro cytotoxic activity of novel coumarinylimidazo [2, 1-b] thiazole derivatives. Royal Society of Chemistry advance,4, 53812-9.

Gali, R., Banothu, J., Porika, M., Velpula, R., Bavantula, R., & Abbagani, S. (2014). Synthesis and in vitro cytotoxic activity of novel coumarinylimidazo [2, 1-b] thiazole derivatives. *RSC Advances*, 4(96), 53812–53819. DOI: 10.1039/C4RA11428K

Ganguly, N., Roy, A., Banerjee, S., & Pandit, S. (2024). Contaminated Soils: Remedies, Hurdles, and Perspectives. In *Advancements in Bio-systems and Technologies for Wastewater Treatment* (pp. 251–268). Springer International Publishing. DOI: 10.1007/978-3-031-58331-5_13

Gao, J. H., Li, R. J., Li, L. Q., Meng, Q., Jiang, H., Li, H. X., & Hu, W. P. (2007). High-Performance Field-Effect Transistor Based on Dibenzo [d, d′] thieno [3, 2-b; 4, 5-b′] dithiophene, an Easily Synthesized Semiconductor with High Ionization Potential. *Advanced Materials*, 19(19), 3008–3011. DOI: 10.1002/adma.200701167

Gao, P., Cho, D., Yang, X., Enkelmann, V., Baumgarten, M., & Müllen, K. (2010). Heteroheptacenes with fused thiophene and pyrrole rings. *Chemistry (Weinheim an der Bergstrasse, Germany)*, 16(17), 5119–5128. DOI: 10.1002/chem.200903562 PMID: 20349476

Gao, Y., Zhang, J., & Liu, Y. (2013). *European Journal of Medicinal Chemistry*, 66, 49–55.

Garcia, F., & Gálvez, C. A. R. M. E. N. (1985). The synthesis of hienopyrroles. *Synthesis*, 1985(02), 143–156. DOI: 10.1055/s-1985-31136

García-Valverde, M., & Torroba, T. (2005). Sulfur-nitrogen heterocycles. *MDPI*, 10, 318–320.

Gardner, J. O., & Beard, C. C. (1979). Synthesis and Antimicrobial Activity of 3H-l,2,4-Thiadiazolo Fused Heterocycles. *Journal of Pharmaceutical Sciences*, 68(2), 182–185. DOI: 10.1002/jps.2600680216 PMID: 423088

Garg, N., & Chandra, T. (2010). Archana, Jain A B & Kumar A. *European Journal of Medicinal Chemistry*, 45(4), 1529. DOI: 10.1016/j.ejmech.2010.01.001 PMID: 20163892

Geedkar, D., Kumar, A., & Sharma, P. (2020). Multiwalled carbon nanotubes crowned with nickel-ferrite magnetic nanoparticles assisted heterogeneous catalytic strategy for the synthesis of benzo [d] imidazo [2, 1-b] thiazole scaffolds. *Journal of Heterocyclic Chemistry*, 57(12), 4331–4347. DOI: 10.1002/jhet.4140

Genton, E. D. W. A. R. D., Barnett, H. J., Fields, W. S., Gent, M., & Hoak, J. C. (1977). XIV. Cerebral ischemia: The role of thrombosis and of antithrombotic therapy. Study group on antithrombotic therapy. *Stroke*, 8(1), 150–175.

Ger. Pat. 10031389; RZhKhim, 19O93P (2002).

Ghorab, M. M., Mohamed, Y. A., Mohamed, S. A., & Ammar, Y. A. (1996). SYNTHESIS AND RADIATION STABILITY OF NOVEL THIAZOLOPYRIMIDINES WITH EXPECTED ANTI-FUNGAL ACTIVITY. *Phosphorus, Sulfur, and Silicon and the Related Elements*, 108(1–4), 249–256. DOI: 10.1080/10426509608029657

Ghorai, P., Mondal, U., Hazra, A., & Banerjee, P. (2024). Luminescent metal organic frameworks (LMOFs) and allied composites for the unveiling of organic environmental contaminants (explosive NACs, PAHs and EDCs) sensing through 'Molecular Recognition': A chronicle of recent penetration and future modelling. *Coordination Chemistry Reviews*, 518, 216085. DOI: 10.1016/j.ccr.2024.216085

Ghosh, S., Bedi, A., & Zade, S. S. (2015). Thienopyrrole and selenophenopyrrole donor fused with benzotriazole acceptor: Microwave assisted synthesis and electrochemical polymerization. *RSC Advances*, 5(7), 5312–5320. DOI: 10.1039/C4RA14850A

Gill, R. K., Aggarwal, N., Kumari, J., Kumari, M., Kaur, P., Kaur, M., Rani, A., Bansal, A., Shah, A., & Bariwal, J. (2013).. . *Chemico-Biological Interactions*, 3(3), 146.

Gomha, S. M., Kheder, N. A., Abdelaziz, M. R., Mabkhot, Y. N., & Alhajoj, A. M. (2017). A facile synthesis and anticancer activity of some novel thiazoles carrying 1,3,4-thiadiazole moiety. *Chemistry Central Journal*, 11(1), 25. DOI: 10.1186/s13065-017-0255-7 PMID: 29086817

Gomha, S. M., & Riyadh, S. M. (2011). Synthesis under Microwave Irradiation of [1,2,4]Triazolo[3,4-b] [1,3,4]thiadiazoles and Other Diazoles Bearing Indole Moieties and Their Antimicrobial Evaluation. *Molecules (Basel, Switzerland)*, 16(10), 8244–8256. DOI: 10.3390/molecules16108244 PMID: 21959297

GOOD. G. (2015). No Title No Title No Title. *Angewandte Chemie International Edition, 6(11), 951–952., 1*(April), 395–397.

Gour, K. (2023). Organic Chemistry Heterocyclic compounds. *The Diverse World of Ringed Molecules.*, 6, 108–110. DOI: 10.37532/jmoc.2023.6(4).108-110

Greenhough, A., Smartt, H. J., Moore, A. E., Roberts, H. R., Williams, A. C., Paraskeva, C., & Kaidi, A. (2009). The COX-2/PGE 2 pathway: Key roles in the hallmarks of cancer and adaptation to the tumour microenvironment. *Carcinogenesis*, 30(3), 377–386. DOI: 10.1093/carcin/bgp014 PMID: 19136477

Grigor'ev, A. A., Karpov, S. V., Kayukov, Y. S., Nasakin, O. E., Gracheva, I. A., & Tafeenko, V. A. (2017). A new route to highly substituted thieno[2,3-*b*]pyridines via cascade heterocyclization of 2-acyl-1,1,3,3-tetracyanopropenide salts. *Chemistry of Heterocyclic Compounds*, 53(2), 230–235. DOI: 10.1007/s10593-017-2044-6

Gronowitz, S. (1986). *Thiophene and its derivatives*. J. Wiley & sons.

Gronowitz, S. (Ed.). (1991). *Chemistry of Heterocyclic Compounds: Thiophene and Its Derivatives, Part Four* (1st ed., Vol. 44). Wiley., DOI: 10.1002/9780470187265

Gronowitz, S., Hornfeldt, A. B., & Yang, Y. H. (1988).*Chemica Scripta*, 28, 281.

Grover, G., Nath, R., Bhatia, R., & Akhtar, M. J. (2020). Synthetic and therapeutic perspectives of nitrogen containing heterocycles as anti-convulsants. *Bioorganic & Medicinal Chemistry*, 28(15), 115585. DOI: 10.1016/j.bmc.2020.115585 PMID: 32631563

Guo, M., Yu, X., Zhu, Y. Z., & Yu, Y. (2023). From Bench to Bedside: What Do We Know about Imidazothiazole Derivatives So Far? *Molecules (Basel, Switzerland)*, 28(13), 5052. DOI: 10.3390/molecules28135052 PMID: 37446714

Gupta, R., & Chaudhary, R. P. (2013). Synthesis, antimicrobial and DFT studies of novel fused thiazolopyrimidine derivatives. *Hc, 19*(3), 207–214. DOI: 10.1515/hc-2013-0024

Gupta, A., & Flynn, B. L. (2017). Linear and angular heteroacenes from double-electrophilic cyclization (DEC) and DEC-reductive elimination of diynes. *Organic Letters*, 19(8), 1939–1941. DOI: 10.1021/acs.orglett.7b00265 PMID: 28388059

Gupta, A., & Tyagi, T. (2022). Phytoremediation and therapeutic potential of neglected plants: An invasive aquatic weeds and ornamental plant. In *Biotechnological innovations for environmental bioremediation* (pp. 259–290). Springer Nature Singapore. DOI: 10.1007/978-981-16-9001-3_12

Gürses, A., Güneş, K., Şahin, E., & Açıkyıldız, M. (2024). Technological Applications for Green and Sustainable Production and Elimination of Environmental Pollution. In *Green Chemistry, its Role in Achieving Sustainable Development Goals, Volume1* (pp. 81-128). CRC Press.

Gutteridge, C. E., Sadowski, B. W., Hughes, S. M., Friedlander, J. A., Rodrigo, L. G., Baxter, M. C., Lorei, N. C., Harrell, J. E., & O'Neil, M. T. (2020). Synthesis of Substituted Benzothiazepine Compounds with Medicinal Potential. *International Journal of Organic Chemistry*, 10(3), 123–134. DOI: 10.4236/ijoc.2020.103009

Güzeldemirci, N. U., & Küçükbasmaci, O. (2010). Synthesis and antimicrobial activity evaluation of new 1,2,4-triazoles and 1,3,4-thiadiazoles bearing imidazo[2,1-b] thiazole moiety. *European Journal of Medicinal Chemistry*, 45(1), 63–68. DOI: 10.1016/j.ejmech.2009.09.024 PMID: 19939519

Guzeldemirci, N. U., Pehlivan, E., & Naesens, L. (2018). Synthesis and antiviral activity evaluation of new 4-thiazolidinones bearing an imidazo [2, 1-b] thiazole moiety. *Marmara Pharmaceutical Journal*, 22(2), 237–248. DOI: 10.12991/mpj.2018.61

Gyűjtő, I., Porcs-Makkay, M., Lukács, G., Pusztai, G., Garádi, Z., Tóth, G., Nyulasi, B., Simig, G., & Volk, B. (2019). Synthesis of 4-methyl-2H-1,2,3-benzothiadiazine1,1-dioxides and their further transformation via alkylation and reduction steps. *Synthetic Communications*, 49(24), 3475–3485. DOI: 10.1080/00397911.2019.1673777

Gyűjtő, I., Simig, G., Porcs-Makkay, M., & Volk, B. (2020). Synthesis and Chemistry of 1,2,3-Benzothiadiazine 1,1-Dioxide Derivatives: A Comprehensive Overview. *Chemistry (Basel, Switzerland)*, 2(3), 674–690. DOI: 10.3390/chemistry2030043

Hafez, H. N., Hussein, H. A. R., & El-Gazzar, A. R. B. A. (2010). Synthesis of substituted thieno[2,3-d]pyrimidine-2,4-dithiones and their S-glycoside analogues as potential antiviral and antibacterial agents. *European Journal of Medicinal Chemistry*, 45(9), 4026–4034. DOI: 10.1016/j.ejmech.2010.05.060 PMID: 20691339

Hagiwara, K., Saitoh, K., Iihama, T., Kawana, T., & Hosaka, H. (1993). Synthesis and Herbicidal Activity of Fused Δ²-1, 2, 4-Thiadiazolines. *Journal of Pesticide Science*, 18(4), 309–318. DOI: 10.1584/jpestics.18.4_309

Hamama, W. S., Ibrahim, M. E., Raoof, H. A., & Zoorob, H. H. (2017). Synthesis and Antimicrobial Evaluation of Some Novel 5-Phenyl-5H-thiazolo [4, 3-b][1, 3, 4] thiadiazole Systems. *Journal of Heterocyclic Chemistry*, 54(4), 2360–2366.

Hamdy, R., Jones, A. T., El-Sadek, M., Hamoda, A. M., Shakartalla, S. B., Al Shareef, Z. M., Soliman, S. S. M., & Westwell, A. D. (2021). New Bioactive Fused Triazolothiadiazoles as Bcl-2-Targeted Anticancer Agents. *International Journal of Molecular Sciences*, 22(22), 12272. DOI: 10.3390/ijms222212272 PMID: 34830153

Han, D., Lu, J., Fan, B., Lu, W., Xue, Y., Wang, M., Liu, T., Cui, S., Gao, Q., Duan, Y., & Xu, Y. (2024). Lysine-Specific Demethylase 1 Inhibitors: A Comprehensive Review Utilizing Computer-Aided Drug Design Technologies. *Molecules (Basel, Switzerland)*, 29(2), 550. DOI: 10.3390/molecules29020550 PMID: 38276629

Hansen, S. K., Harris, Æ. S. O., Hashash, Æ. A., Hoch, Æ. U., Hogan, Æ. J. N., Howlett, A. R., Jacobs, Æ. J. W., Lam, Æ. J. W., Ritchie, Æ. S. C., Romanowski, Æ. M. J., Silverman, J. A., Stockett, Æ. D. E., Teague, Æ. J. N., Zimmerman, Æ. K. M., & Taverna, Æ. Pietro. (2010). *SNS-314, a pan-Aurora kinase inhibitor, shows potent anti-tumor activity and dosing flexibility in vivo.* 707–717. DOI: 10.1007/s00280-009-1076-8

Haristoy, D., Mery, S., Heinrich, B., Mager, L., Nicoud, J. F., & Guillon, D. (2000). Structure and photoconductive behaviour of a sanidic liquid crystal. *Liquid Crystals*, 27(3), 321–328. DOI: 10.1080/026782900202769

Haroun, M., Chobe, S. S., Alavala, R. R., Mathure, S. M., Jamullamudi, R. N., Nerkar, C. K., Gugulothu, V. K., Tratrat, C., Islam, M. M., Venugopala, K. N., Habeebuddin, M., Telsang, M., Sreeharsha, N., & Anwer, M. K. (2022). 1,5-Benzothiazepine Derivatives: Green Synthesis, In Silico and In Vitro Evaluation as Anticancer Agents. *Molecules (Basel, Switzerland)*, 27(12), 3757. DOI: 10.3390/molecules27123757 PMID: 35744881

Harrouche, K., Lahouel, A., Belghobsi, M., Pirotte, B., & Khelili, S. (2019). Synthesis, characterization, and investigation of the antioxidant activity of some 1,2,4-benzothiadiazine-1,1-dioxides bearing sulfonylthioureas moieties. *Canadian Journal of Chemistry*, 97(12), 824–832. DOI: 10.1139/cjc-2019-0194

Harrow, I. D., & Gration, K. A. (1985). Mode of action of the anthelmintics morantel, pyrantel and levamisole on muscle cell membrane of the nematode Ascaris suum. *Pesticide Science*, 16(6), 662–672. DOI: 10.1002/ps.2780160612

Hart, C. J., Riches, A. G., Tiash, S., Abraham, R., Fayd'Herbe, K., Joch, E., Zulfiqar, B., Sykes, M. L., Avery, V. M., Šlapeta, J., Abraham, S., Ryan, J. H., & Skinner-Adams, T. S. (2023). Thieno [3, 2-b] pyrrole 5-carboxamides as potent and selective inhibitors of Giardia duodenalis. *International Journal for Parasitology. Drugs and Drug Resistance*, 23, 54–62. DOI: 10.1016/j.ijpddr.2023.09.002 PMID: 37776606

Hasan, A., Tabassum, B., Hashim, M., & Khan, N. (2024). Role of plant growth promoting rhizobacteria (PGPR) as a plant growth enhancer for sustainable agriculture: A review. *Bacteria*, 3(2), 59–75.

Hassen, W., Hassen, B., El Ouaer, M., & Hassen, A. (2024). Strategies for Removal of Emerging Compounds of Concern Through Algal Niche Adaptation. In *Algae as a Natural Solution for Challenges in Water-Food-Energy Nexus: Toward Carbon Neutrality* (pp. 161–198). Springer Nature Singapore. DOI: 10.1007/978-981-97-2371-3_9

Hatvate, N. T., & Ghodse, S. M. (2020). One-pot three-component synthesis of isoxazole using ZSM-5 as a heterogeneous catalyst. *Synthetic Communications*, 50(23), 3676–3683. DOI: 10.1080/00397911.2020.1815786

Hatvate, N. T., Ghodse, S. M., & Telvekar, V. N. (2018). Metal-free synthesis of 2-aminothiadiazoles via TBHP-Mediated oxidative C-S bond formation. *Synthetic Communications*, 48(3), 285–290. DOI: 10.1080/00397911.2017.1398330

Hatvate, N., & Gupta, J. (2024). An efficient, green and solvent-free three-component synthesis of 2,4,6-triarylpyridines utilizing ZSM-5 as a heterogeneous catalyst. *Synthetic Communications*, 54(20), 1760–1770. DOI: 10.1080/00397911.2024.2403779

Hegab, H., Shaban, I., Jamil, M., & Khanna, N. (2023). Toward sustainable future: Strategies, indicators, and challenges for implementing sustainable production systems. *Sustainable Materials and Technologies*, 36, e00617. DOI: 10.1016/j.susmat.2023.e00617

Henning, N., Kannigadu, C., & Aucamp, J. (2023). *Probing benzothiadiazine-1, 1-dioxide ethylene glycol derivatives against Leishmania : synthesis and in vitro efficacy evaluation*. 1–14.

Hill, R., & Canals, M. (2022). Experimental considerations for the assessment of in vivo and in vitro opioid pharmacology. *Pharmacology & Therapeutics*, 230, 107961. DOI: 10.1016/j.pharmthera.2021.107961 PMID: 34256067

Himaja, M., Karigar, A. A., Ramana, M. V., & Sikarwar, M. S. (2012). *Tropical Journal of Pharmaceutical Research*, 11(4), 611–620.

Himaja, M., Prathap, K. J., & Mali, S. V. (2012). Facile Synthesis of 3,6-Disubstituted-1,2,4-triazolo-[3,4- *b*]-1,3,4-thiadiazoles via Oxidative Cyclization of *n* -Heteroaryl-Substituted Hydrazones and Their Biological Activity. *Journal of Heterocyclic Chemistry*, 49(4), 823–828. DOI: 10.1002/jhet.870

Hitoshi, M., Cara, C. Y., & Jasper, A. S. (2022).*Neuropharmacology*, 221, 109290. DOI: 10.1016/j.neuropharm.2022.109290 PMID: 36241085

Holderfield, M., Deuker, M., McCormick, F., & McMahon, M. (2014). Targeting RAF kinases for cancer therapy: BRAF-mutated melanoma and beyond. *Nature Reviews. Cancer*, 14(7), 455–467. DOI: 10.1038/nrc3760 PMID: 24957944

Hommelsheim, R., van Nahl, R., Hanek, L. M., Ward, J. S., Rissanen, K., & Bolm, C. (2024). Palladium-Catalyzed Domino Reaction for the Synthesis of 3-Amino 1,2,4-Benzothiadiazine 1,1-Dioxides. *Advanced Synthesis & Catalysis*, 366(4), 717–724. DOI: 10.1002/adsc.202300875

Hrncic, D., Rasic-Markovic, A., Macut, D., Mladenovic, D., Susic, V., Djuric, D., & Stanojlovic, O. (2018). Sulfur – Containing Amino Acids in Seizures: Current State of the Art. *Current Medicinal Chemistry*, 25(3), 378–390. DOI: 10.2174/09 29867324666170609090613 PMID: 28595553

Huang, J., & Yu, G. (2021). Structural engineering in polymer semiconductors with aromatic N-heterocycles. *Chemistry of Materials*, 33(5), 1513–1539. DOI: 10.1021/ acs.chemmater.0c03975

Huang, S., Zuo, L., Zhang, L., Guo, X., Cheng, C., He, Y., Cheng, G., Yu, J., Liu, Y., Chen, R., Tang, G., Fan, Y., & Feng, L. (2024). Design, Synthesis, and Mode of Action of Thioacetamide Derivatives as the Algicide Candidate Based on Active Substructure Splicing Strategy. *Journal of Agricultural and Food Chemistry*, 72(13), 7021–7032. DOI: 10.1021/acs.jafc.4c00912 PMID: 38501582

Hu, L., Fan, M., Shi, S., Song, X., Wang, F., He, H., & Qi, B. (2022). Dual target inhibitors based on EGFR: Promising anticancer agents for the treatment of cancers (2017-). *European Journal of Medicinal Chemistry*, 227, 113963. DOI: 10.1016/j. ejmech.2021.113963 PMID: 34749202

Hussain, Asif, MOHAMMAD A. NASEER, & MOHAMMAD SARAFROZ. (2009). SYNTHESIS AND ANTICONVULSANT ACTIVITY OF SOME NOVEL FUSED HETEROCYCLIC 1,2,4-TRIAZOLO-[3,4-b]-1,3,4-THIADIAZOLE DERIVATIVES. *ACTA POLONIAE PHARMACEUTICA –. Drug Research*, 66(2), 135–140.

Hussain, S., Akhter, R., & Maktedar, S. S. (2024). *Advancements in sustainable food packaging: from eco-friendly materials to innovative technologies*. Sustainable Food Technology.

Huwaimel, B. I., Bhakta, M., Kulkarni, C. A., Milliken, A. S., Wang, F., Peng, A., Brookes, P. S., & Trippier, P. C. (2021). Discovery of Halogenated Benzothiadiazine Derivatives with Anticancer Activity**. *ChemMedChem*, 16(7), 1143–1162. DOI: 10.1002/cmdc.202000729 PMID: 33331124

Hu, Z., Martí, J., & Lu, H. (2021). Structure of benzothiadiazine at zwitterionic phospholipid cell membranes. *The Journal of Chemical Physics*, 155(15), 154303. Advance online publication. DOI: 10.1063/5.0065163 PMID: 34686044

Ibrahim, Y. A., Elwahy, A. H. M., & Kadry, A. M. (1996). Thienopyrimidines: Synthesis, reactions, and biological activity. *Advances in Heterocyclic Chemistry*, 65(65), 235–281. DOI: 10.1016/S0065-2725(08)60297-4

Ijoma, G. N., Lopes, T., Mannie, T., & Mhlongo, T. N. (2024). Exploring macrophytes' microbial populations dynamics to enhance bioremediation in constructed wetlands for industrial pollutants removal in sustainable wastewater treatment. *Symbiosis*, 92(3), 323–354. DOI: 10.1007/s13199-024-00981-9

Iqbal, A., Khan, A., Ahmedi, S., Manzoor, N., & Siddiqui, T. (2023). Synthesis, antifungal evaluation, and molecular docking studies of steroidal thiazolopyrimidines. *Steroids*, 193, 109186. DOI: 10.1016/j.steroids.2023.109186 PMID: 36736803

Isci, R., Rahimi Varzeghani, A., Kaya, K., Sütay, B., Tekin, E., & Ozturk, T. (2022). Triphenylamine/Tetraphenylethylene Substituted 4-Thieno[3,2-b]thiophen-3-ylbenzonitriles: Synthesis, Photophysical-Electronic Properties and Applications. *ACS Sustainable Chemistry & Engineering*, 10(4), 1605–1615. DOI: 10.1021/acssuschemeng.1c07240

Islam, F., & Quadery, T. M. (2021). Therapeutic Potential, Synthesis, Patent Evaluation and SAR Studies of Thieno[3,2-d]pyrimidine Derivatives: Recent Updates. *Current Drug Targets*, 22(17), 1944–1963. DOI: 10.2174/1389450122666210526094047 PMID: 34042033

Islam, F., & Zeng, Q. (2024). Advances in Organosulfur-Based Polymers for Drug Delivery Systems. *Polymers*, 16(9), 1207. DOI: 10.3390/polym16091207 PMID: 38732676

Ittamalla, C., Chary, D. Y., Balasubramanian, S., & Reddy, B. V. S. (2024). Rh(III)-Catalyzed C−H Annulation of 3-Aryl-2 H -benzo[e][1,2,4]thiadiazine-1,1-dioxides with Vinylene Carbonates. *European Journal of Organic Chemistry*, 27(4), e202300983. Advance online publication. DOI: 10.1002/ejoc.202300983

Ivanov, V. L., Artemov, V. A., Shestopalov, A. M., Nesterov, V. N., Struchkov, Yu. T., & Litvinov, V. P. (1996). 2-Bromo-1-Phenylethylidenemalononitrile In Synthesis Of Functionally Substituted Thieno[3,2-*b*]Pyridines And Thiazolo[4,5-*b*]Pyridines. *Khimiya Geterotsiklichesikh Soedinenii*, 413.

Jackson, L. M., & Hawkey, C. J. (2000). *COX-2 Selective Nonsteroidal Anti-Inflammatory Drugs Do They Really Offer Any Advantages ? 59*(6), 1207–1216.

Jadhav, V. B., Kulkarni, M. V., Rasal, V. P., Biradar, S. S., & Vinay, M. D. (2008). Synthesis and anti-inflammatory evaluation of methylene bridged benzofuranyl imidazo[2,1-b][1,3,4]thiadiazoles. *European Journal of Medicinal Chemistry*, 3(8), 1721–1729. DOI: 10.1016/j.ejmech.2007.06.023 PMID: 17845827

Jaiswal, S. & Siddiqui, I. R. (2014). Montmorillonite K-10 Supported Solvent-Free Synthesis of Novel 2-thioxoimidazole-4-one N-nucleosides. *chemistry*, Int. J. Pharm. Sci. Rev. Res.,26(1),*16*, 18.

Jaiswal, P., Kawathekar, N., Jain, G., & Jain, S. (2021). *In Silico Design, Synthesis, Anti-Microbial and Anti-Tubercular Evaluation of New Series of Benzothiazinone Derivatives Containing Triazolo Thiadiazole Moiety*. DOI: 10.21203/rs.3.rs-688359/v1

Jaiswal, S. (2019). Five and Six Membered Heterocyclic Compound with Antimicrobial Activity. *Journal for Modern Trends in Science and Technology*, 5(06), 36–39.

Jaiswal, S., & Dwivedi, S. (2017). Alumina Supported Solvent-free Synthesis and Structural Elucidation of New 3-aryl-5-benzylidene-2-thioxohydantoin Derivatives. *Indian Journal of Heterocyclic Chemistry*, 27(3), 265–268.

Jaiswal, S., & Dwivedi, S. (2017). Microwave-assisted eco-friendly synthesis and antimicrobial evaluation of aryl-triazole-1, 3, 4-thiadiazols. *Journal of Medical Research and Innovation*, 1(3), 17–22. DOI: 10.15419/jmri.51

Jaiswal, S., & Mojahid, A. (2020). Innovation in Society through Green\Eco-Friendly Technology. *Journal for Modern Trends in Science and Technology*, 6(01), 37–43.

Jaiswal, S., & Sharma, A. (•••). Green Synthesis of Some Novel Substituted and Unsubstituted Benzimidazole Derivatives by Using Microwave Energy. *International Journal of Pharmaceutical Sciences Review and Research*, 27, 153–156.

Jakovleva, E. E., Foksha, S. P., Brusina, M. A., Kubarskaja, L. G., Piotrovskij, L. B., Bychkov, E. R., & Shabanov, P. D. (2020). Studying the anticonvulsive activity of new ligands of NDMA-receptor complex—Imidazole-4,5-dicarbonic acid derivatives. *Rev. Clin. Pharm. Drug Ther.*, 18(2), 149–154. DOI: 10.17816/RCF182149-154

Jana, A., Bhaumick, P., Panday, A. K., Mishra, R., & Choudhury, L. H. (2019). I_2/DMSO mediated multicomponent reaction for the synthesis of 2-arylbenzo[d] imidazo[2,1-b] thiazole derivatives. *Organic & Biomolecular Chemistry*, 17(21), 5316–5330. DOI: 10.1039/C9OB00515C PMID: 31095156

Janku, F. (2017). Phosphoinositide 3-kinase (PI3K) pathway inhibitors in solid tumors : From laboratory to patients. *Cancer Treatment Reviews*, 59, 93–101. Advance online publication. DOI: 10.1016/j.ctrv.2017.07.005 PMID: 28779636

Jayanthi, M., Venkatesh, P., & Thirunavukkarasu, S. V. (2022). A novel and efficient synthesis of 1, 2, 3,4-tetrahydropyrimidine carboxamide derivatives by Biginelli reaction and their in vitro anti-bacterial activity. *Biomedicine (India)*, 42(2), 280–285. DOI: 10.51248/.v42i2.884

Jeleń, M., Morak-Młodawska, B., & Korlacki, R. (2023). Anticancer activities of tetra-, penta-, and hexacyclic phenothiazines modified with quinoline moiety. *Journal of Molecular Structure*, 1287, 135700. DOI: 10.1016/j.molstruc.2023.135700

Jeleń, M., Otto-Ślusarczyk, D., Morak-Młodawska, B., & Struga, M. (2024). Novel Tetracyclic Azaphenothiazines with the Quinoline Ring as New Anticancer and Antibacterial Derivatives of Chlorpromazine. *International Journal of Molecular Sciences*, 25(8), 4148. DOI: 10.3390/ijms25084148 PMID: 38673734

Jen, A. K.-Y., Rao, V. P., Drost, K. J., Cai, Y., Mininni, R. M., Kenney, J. T., Binkley, E. S., & Dalton, L. R. (1994). The Development of Highly Active, Thermally Stable Chromophores and Polymers for Electrooptic Applications. *Proceedings of SPIE - the International Society for Optical Engineering*, 2285, 49–57. DOI: 10.1117/12.187500

Jiang, H., Oniwa, K., Xu, Z., Bao, M., Yamamoto, Y., & Jin, T. (2017). Synthesis and properties of dicyanomethylene-endcapped thienopyrrole-based quinoidal S, N-heteroacenes. *Bulletin of the Chemical Society of Japan*, 90(7), 789–797. DOI: 10.1246/bcsj.20170083

Jiang, J. K., Boxer, M. B., Vander Heiden, M. G., Shen, M., Skoumbourdis, A. P., Southall, N., Veith, H., Leister, W., Austin, C. P., Park, H. W., Inglese, J., Cantley, L. C., Auld, D. S., & Thomas, C. J. (2010). Evaluation of thieno [3, 2-b] pyrrole [3, 2-d] pyridazinones as activators of the tumor cell specific M2 isoform of pyruvate kinase. *Bioorganic & Medicinal Chemistry Letters*, 20(11), 3387–3393. DOI: 10.1016/j.bmcl.2010.04.015 PMID: 20451379

Jiang, N., Yan, J., Liang, Y., Shi, Y., He, Z., Wu, Y., Zeng, Q., Liu, X., & Peng, J. (2020). Resistance Genes and their Interactions with Bacterial Blight/Leaf Streak Pathogens (Xanthomonas oryzae) in Rice (Oryza sativa L.)—An Updated Review. *Rice (New York, N.Y.)*, 13(1), 3. DOI: 10.1186/s12284-019-0358-y PMID: 31915945

Jin, M., Zhou, Q., Fu, L., & Wu, W. (2024). Application of Biochar-Based Catalysts for Soil and Water Pollution Control. *Topics in Catalysis*, •••, 1–24. DOI: 10.1007/s11244-024-01962-4

Johnson, D., Hussain, J., Bhoir, S., Chandrasekaran, V., Sahrawat, P., Hans, T., Khalil, M. I., De Benedetti, A., Thiruvenkatam, V., & Kirubakaran, S. (2023). Synthesis, kinetics and cellular studies of new phenothiazine analogs as potent human-TLK inhibitors. *Organic & Biomolecular Chemistry*, 21(9), 1980–1991. DOI: 10.1039/D2OB02191A PMID: 36785915

Johnson, J. R., & Ketcham, R. (1960). Thiazolothiazoles. I. The Reaction of Aromatic Aldehydes with Dithioöxamide. *Journal of the American Chemical Society*, 82(11), 2719–2724. DOI: 10.1021/ja01496a017

Johnson, J. R., Rotenberg, D. H., & Ketcham, R. (1970). Thiazolothiazoles. II. Parent heterocycle and its carboxylic and amino derivatives. *Journal of the American Chemical Society*, 92(13), 4046–4050. DOI: 10.1021/ja00716a036

Joshi, K. A., Dhalani, J. M., Bhatt, H. B., & Kapadiya, K. M. (2021). Synthesis of some new 6-aryl-3-(4-isopropylphenyl)[1,2,4]triazolo[3,4-b][1,3,4] thiadiazoles and its anti-microbial study. *Current Chemistry Letters*, 10(4), 479–488. DOI: 10.5267/j.ccl.2021.4.003

Jubeen, F., Iqbal, S. Z., Shafiq, N., Khan, M., Parveen, S., Iqbal, M., & Nazir, A. (2018). Eco-friendly synthesis of pyrimidines and its derivatives: A review on broad spectrum bioactive moiety with huge therapeutic profile. *Synthetic Communications*, 48(6), 601–625. DOI: 10.1080/00397911.2017.1408840

Jung, I. H., Kim, J. H., Nam, S. Y., Lee, C., Hwang, D. H., & Yoon, S. C. (2015). Development of new photovoltaic conjugated polymers based on di (1-benzothieno) [3, 2-b: 2′, 3′-d] pyrrole: benzene ring extension strategy for improving open-circuit voltage. *Macromolecules*, 48(15), 5213–5221. DOI: 10.1021/acs.macromol.5b01129

Jung, I. H., Kim, J. H., Nam, S. Y., Lee, C., Hwang, D. H., & Yoon, S. C. (2016). A di (1-benzothieno)[3, 2-b: 2′, 3′-d] pyrrole and isoindigo-based electron donating conjugated polymer for efficient organic photovoltaics. *Journal of Materials Chemistry. C, Materials for Optical and Electronic Devices*, 4(4), 663–667. DOI: 10.1039/C5TC04220H

Kabir, E., & Uzzaman, M. (2022). A review on biological and medicinal impact of heterocyclic compounds. *Results in Chemistry*, 4, 100606. DOI: 10.1016/j.rechem.2022.100606

Kajita, Y., Sakai, M., Atsuta, S., & Kobayashi, M.KAJITA. (1990). Y.; SAKAI, M.; ATSUTA, S.; KOBAYASHI, M., The immunomodulatory effects of levamisole on rainbow trout, Oncorhynchus mykiss. *Fish Pathology*, 25(2), 93–98. DOI: 10.3147/jsfp.25.93

Kamal, A., Dastagiri, D., Ramaiah, M. J., Reddy, J. S., Bharathi, E. V., Srinivas, C., Pushpavalli, S. N. C. V. L., Pal, D., & Pal-Bhadra, M. (2010). Synthesis of imidazothiazole–chalcone derivatives as anticancer and apoptosis inducing agents. *ChemMedChem*, 5(11), 1937–1947. DOI: 10.1002/cmdc.201000346 PMID: 20836120

Kamal, A., Khan, M. N. A., Reddy, K. S., Rohini, K., Sastry, G. N., Sateesh, B., & Sridhar, B. (2007). Synthesis, structure analysis, and antibacterial activity of some novel 10-substituted 2-(4-piperidyl/phenyl)-5,5-dioxo[1,2,4]triazolo[1,5-b][1,2,4] benzothiadiazine derivatives. *Bioorganic & Medicinal Chemistry Letters*, 17(19), 5400–5405. DOI: 10.1016/j.bmcl.2007.07.043 PMID: 17719222

Kamal, A., Khanna, G. R., Krishnaji, T., & Ramu, R. (2005). A new facile chemoenzymatic synthesis of levamisole. *Bioorganic & Medicinal Chemistry Letters*, 15(3), 613–615. DOI: 10.1016/j.bmcl.2004.11.054 PMID: 15664823

Kamal, A., Shetti, R. V. C. R. N. C., Azeeza, S., Ahmed, S. K., Swapna, P., Reddy, A. M., Khan, I. A., Sharma, S., & Abdullah, S. T. (2010). Anti-tubercular agents. Part 5: Synthesis and biological evaluation of benzothiadiazine 1,1-dioxide based congeners. *European Journal of Medicinal Chemistry*, 45(10), 4545–4553. DOI: 10.1016/j.ejmech.2010.07.015 PMID: 20705368

Kamal, A., Sultana, F., Ramaiah, M. J., Srikanth, Y., Viswanath, A., Kishor, C., Sharma, P., Pushpavalli, S., Addlagatta, A., & Pal-Bhadra, M. (2012). 3-Substituted 2-Phenylimidazo [2, 1-b] benzothiazoles: Synthesis, Anticancer Activity, and Inhibition of Tubulin Polymerization. *ChemMedChem*, 7(2), 292–300. DOI: 10.1002/cmdc.201100511 PMID: 22241597

Kamboj, P., Anjali, , Imtiyaz, K., Rizvi, M. A., Nath, V., Kumar, V., Husain, A., & Amir, M. (2024). Design, synthesis, biological assessment and molecular modelling studies of novel imidazothiazole-thiazolidinone hybrids as potential anticancer and anti-inflammatory agents. *Scientific Reports*, 14(1), 8457. DOI: 10.1038/s41598-024-59063-x PMID: 38605072

Kamboj, P., Mahore, A., Husain, A., Shaquiquzzaman, M., Alam, M. M., & Amir, M. (2023). A review on the synthesis and anticancer potentials of imidazothiazole derivatives. *Mini-Reviews in Medicinal Chemistry*, 23(17), 1733–1759. DOI: 10.2174/1389557523666230201113237 PMID: 36722483

Kamiński, K., Zagaja, M., Łuszczki, J. J., Rapacz, A., Andres-Mach, M., Latacz, G., & Kieć-Kononowicz, K. (2015). Design, Synthesis, and Anticonvulsant Activity of New Hybrid Compounds Derived from 2-(2,5-Dioxopyrrolidin-1-yl)propanamides and 2-(2,5-Dioxopyrrolidin-1-yl)butanamides. *Journal of Medicinal Chemistry*, 58(13), 5274–5286. DOI: 10.1021/acs.jmedchem.5b00578 PMID: 26052884

Kamoutsis, C., Fesatidou, M., Petrou, A., Geronikaki, A., Poroikov, V., Ivanov, M., Soković, M., Ćirić, A., Carazo, A., & Mladěnka, P. (2021). Triazolo Based-Thiadiazole Derivatives. Synthesis, Biological Evaluation and Molecular Docking Studies. *Antibiotics (Basel, Switzerland)*, 10(7), 804. DOI: 10.3390/antibiotics10070804 PMID: 34356726

Kanungo, J., Sahoo, T., Swain, L. P., & Behera, I. D. (2024). Toxicity of Persistent Hydrocarbon Pollutants, Sources and Sustainable Remediation Process. In *Impact of Petroleum Waste on Environmental Pollution and its Sustainable Management Through Circular Economy* (pp. 39–65). Springer Nature Switzerland.

Karaman, B., & Ulusoy Güzeldemirci, N. (2016). Synthesis and biological evaluation of new imidazo [2, 1-b] thiazole derivatives as anticancer agents. *Medicinal Chemistry Research*, 25(11), 2471–2484. DOI: 10.1007/s00044-016-1684-x

Karande, V., Darandale, S. N., & Shinde, D. B. (2013). Microwave assisted nano (ZnO–TiO2) catalyzed synthesis of some new 4,5,6,7-tetrahydro-6-((5-substituted-1, 3, 4- oxadiazol-2-yl) methyl) thieno[2,3-*c*]pyridine as antimicrobial agents. *Bioorganic & Medicinal Chemistry Letters*, 23(7), 2250–2253. DOI: 10.1016/j.bmcl.2013.01.041 PMID: 23434418

Karegoudar, P., Prasad, D. J., Ashok, M., Mahalinga, M., Poojary, B., & Holla, B. S. (2008). Synthesis, antimicrobial and anti-inflammatory activities of some 1,2,4-triazolo[3,4-b][1,3,4]thiadiazoles and 1,2,4-triazolo[3,4-b][1,3,4]thiadiazines bearing trichlorophenyl moiety. *European Journal of Medicinal Chemistry*, 43(4), 808–815. DOI: 10.1016/j.ejmech.2007.06.026 PMID: 17804121

Kassab, A. E., Gedawy, E. M., & Sayed, A. S. (2024). Fused thiophene as a privileged scaffold: A review on anti-Alzheimer's disease potentials via targeting cholinesterases, monoamine oxidases, glycogen synthase kinase-3, and Aβ aggregation. *International Journal of Biological Macromolecules*, 265, 131018. DOI: 10.1016/j.ijbiomac.2024.131018 PMID: 38518928

Katritzky, A. R. (Ed.). (1995). *Comprehensive organic functional group transformations: 7-volume-set*. Pergamon Press.

Kaur, B., Singh, G., Sharma, V., & Singh, I. (2023). Sulphur Containing Heterocyclic Compounds as Anticancer Agents. *Anti-Cancer Agents in Medicinal Chemistry (Formerly Current Medicinal Chemistry-Anti-Cancer Agents), 23*(8), 869-881.

Kavitha, K., Srikrishna, D., Dubey, P. K., & Aparna, P. (2019). An efficient one-pot four-component Gewald reaction: Synthesis of substituted 2-aminothiophenes with coumarin–thiazole scaffolds under environmentally benign conditions. *Journal of Sulfur Chemistry*, 40(2), 195–208. DOI: 10.1080/17415993.2018.1556275

Kawakubo, H., Takagi, S., Yamaura, Y., Katoh, S., Ishimoto, Y., Nagatani, T., Mochizuki, D., Kamata, T., & Sasaki, Y. (1993). (R)-1,2,3,4-tetrahydro[1]benzothieno[2,3-c]pyridines: Novel optically active compounds with strong 5-HT1A receptor binding ability exhibiting anticonflict activity and lessening of memory impairment. *Journal of Medicinal Chemistry*, 36(23), 3526–3532. DOI: 10.1021/jm00075a006 PMID: 7902439

Kazushige Sogawa, and M. D. P. (1970). NII-Electronic Library Service. *Chemical Pharmaceutical Bulletin, 43*, 2091. Retrieved from https://www.mendeley.com/research/geology-volcanic-history-eruptive-style-yakedake-volcano-group-central-japan/

Kendre, B. V, Bhusare, S. R., & Hardas, P. S. (2020). *Antimicrobial and Anti-Inflammatory Activity Studies of Novel Thiazolopyrimidines. 10*(January).

Kendre, M. M., & Baseer, M. A. (2021).*International Journal of Pharmaceutical Sciences and Research*, 12(2), 1014.

Kerassa, A., Belaidi, S., Harkati, D., Lanez, T., Prasad, O., & Sinha, L. (2016). Investigations on Molecular Structure, Electronic Properties, NLO Properties and Comparison of Drug-Likeness of Triazolothiadiazole Derivatives by Quantum Methods and QSAR Analysis. *Reviews in Theoretical Science*, 4(1), 85–96. DOI: 10.1166/rits.2016.1050

Khalifa, N. M., Al, M. A., El, A., Amr, G. E., Baiuomy, A. R., & Abdel, R. F. (2015). *Synthesis and Biological Evaluation of Some Novel Fused Thiazolo [3, 2 a] Pyrimidines as Potential Analgesic and Anti Inflammatory Agents 1. 41*(2), 192–200. DOI: 10.1134/S1068162015020090

Khalili Ghomi, M., Noori, M., Nazari Montazer, M., Zomorodian, K., Dastyafteh, N., Yazdanpanah, S., Sayahi, M. H., Javanshir, S., Nouri, A., Asadi, M., Badali, H., Larijani, B., Irajie, C., Iraji, A., & Mahdavi, M. (2023). [1,2,4]triazolo[3,4-b][1,3,4]thiadiazole derivatives as new therapeutic candidates against urease positive microorganisms: Design, synthesis, pharmacological evaluations, and in silico studies. *Scientific Reports*, 13(1), 10136. DOI: 10.1038/s41598-023-37203-z PMID: 37349372

Khaliq, N. (2023). Microbial enzymes as a robust process to mitigate pollutants of environmental concern. In *Microbial Biomolecules* (pp. 241–267). Academic Press. DOI: 10.1016/B978-0-323-99476-7.00022-3

Khan, M. N. (•••).. . *Journal of Electronic Materials*, 201746(5), 2772–2783.

Khan, R. H., & Rastogi, R. C. (1988). Indian J. Chem. *Sect. B: Org. Chem. Incl. Med. Chem*, 27, 501.

Khan, R. H., & Rastogi, R. C. (1988). Indian J. Chem. *Sect. B: Org. Chem. Incl. Med. Chem.*, 27, 501.

Khedkar, V. M., Arya, N., Coutinho, E. C., Shishoo, C. J., & Jain, K. S. (2014). Docking study of novel antihyperlipidemic thieno[2,3-d]pyrimidine; LM-1554, with some molecular targets related to hyperlipidemia - an investigation into its mechanism of action. *SpringerPlus*, 3(1), 628. Advance online publication. DOI: 10.1186/2193-1801-3-628 PMID: 25392798

Khelili, S., Kihal, N., Yekhlef, M., de Tullio, P., Lebrun, P., & Pirotte, B. (2012). Synthesis and pharmacological activity of N-(2,2-dimethyl-3,4-dihydro-2H-1-benzopyran-4-yl)-4H-1,2,4-benzothiadiazine-3-carboxamides 1,1-dioxides on rat uterus, rat aorta and rat pancreatic β-cells. *European Journal of Medicinal Chemistry*, 54, 873–878. DOI: 10.1016/j.ejmech.2012.05.011 PMID: 22647221

Kidwai, M.* and Mishra, A. D. (n.d.). *A Novel Synthesis of Thieno-pyrimidines Using Inorganic Solid Support. 24.*

Kidwai, M., Venkataramanan, R., & Dave, B. (2001). Solventless synthesis of thiohydantoins over K2CO3. *Green Chemistry*, 3(6), 278–279. DOI: 10.1039/b106034c

Kim, O. K., Woo, H., Kim, J. K., & Huang, Z. (2000, March). New class of light-emitting polymers/oligomers. In *Liquid Crystal Materials, Devices, and Flat Panel Displays* (Vol. 3955, pp. 134-140). SPIE.

Kim, J., Lee, S. Y., Lee, J., Do, Y., & Chang, S. (2008). Synthetic Utility of Ammonium Salts in a Cu-Catalyzed Three-Component Reaction as a Facile Coupling Partner. *The Journal of Organic Chemistry*, 73(23), 9454–9457. DOI: 10.1021/jo802014g PMID: 18956843

Kinases, T., Potential, T., & Activities, A. (2017). *Review Article*. 1–24. DOI: 10.1002/ardp.201700242

King, J. F., & Mitch, W. A. (2024). Electrochemical reduction of halogenated organic contaminants using carbon-based cathodes: A review. *Critical Reviews in Environmental Science and Technology*, 54(4), 342–367. DOI: 10.1080/10643389.2023.2239130

Kishore, K. G., Jácome, A. I., Gómez, A. I., Conejo, A. S., Basavanag, U. M. V., Wrobel, K., & Montaño, R. G. (2016). Synthesis of unsymmetrical bis-heterocycles containing the imidazo[2,1-b] thiazole framework and their benzo[d]fused analogues by an acid-free Groebke–Blackburn–Bienaymé reaction. *Tetrahedron Letters*, 57(31), 3556–3560. DOI: 10.1016/j.tetlet.2016.06.120

Kisiel-Nawrot, E., Latocha, M., Bak, A., Kozik, V., Jampilek, J., & Zieba, A. (2023). Anticancer efficacy of antibacterial quinobenzothiazines. *Applied Sciences (Basel, Switzerland)*, 13(5), 2886. DOI: 10.3390/app13052886

Kisiel-Nawrot, E., Pindjakova, D., Latocha, M., Bak, A., Kozik, V., Suwinska, K., Cizek, A., Jampilek, J., & Zięba, A. (2023). Towards anticancer and antibacterial agents: Design and synthesis of 1, 2, 3-triazol-quinobenzothiazine derivatives. *International Journal of Molecular Sciences*, 24(17), 13250. DOI: 10.3390/ijms241713250 PMID: 37686059

Kisla, M. M., Yaman, M., Zengin-Karadayi, F., Korkmaz, B., Bayazeid, O., Kumar, A., Peravali, R., Gunes, D., Tiryaki, R. S., Gelinci, E., Cakan-Akdogan, G., Ates-Alagoz, Z., & Konu, O. (2024). Synthesis and Structure of Novel Phenothiazine Derivatives, and Compound Prioritization via In Silico Target Search and Screening for Cytotoxic and Cholinesterase Modulatory Activities in Liver Cancer Cells and In Vivo in Zebrafish. *ACS Omega*, 9(28), 30594–30614. DOI: 10.1021/acsomega.3c06532 PMID: 39035947

Klotz, K. N. (2000). Adenosine receptors and their ligands. *Naunyn-Schmiedeberg's Archives of Pharmacology*, 362(4-5), 382–391. DOI: 10.1007/s002100000315 PMID: 11111832

Kochhar, M. M., Salahi-Asbahi, M., & Williams, B. B. (1973). Bicyclic Thiadiazoles I: 2-(2-Thienyl)-5-phenylthiazolo[2,3-b]-1,3,4-thiadiazole. *Journal of Pharmaceutical Sciences*, 62(2), 336–337. DOI: 10.1002/jps.2600620239 PMID: 4686420

Kolawole, A. S., & Iyiola, A. O. (2023). Environmental pollution: threats, impact on biodiversity, and protection strategies. In *Sustainable utilization and conservation of Africa's biological resources and environment* (pp. 377–409). Springer Nature Singapore. DOI: 10.1007/978-981-19-6974-4_14

Koshelev, V. N., Primerova, O., Stepan, V. V., & Stupnikova Anna, S. (2023). Synthesis and Antioxidant Activity of Novel Thiazole and Thiazolidinone Derivatives with Phenolic Fragments. *Applied Sciences (Basel, Switzerland)*, 13(24), 13112. DOI: 10.3390/app132413112

Kotaiah, Y., Harikrishna, N., Nagaraju, K., & Venkata Rao, C. (2012). Synthesis and antioxidant activity of 1,3,4-oxadiazole tagged thieno[2,3-d]pyrimidine derivatives. *European Journal of Medicinal Chemistry*, 58, 340–345. DOI: 10.1016/j.ejmech.2012.10.007 PMID: 23149297

Kotireddy, V., & Ramana, K. V. (2017).*International Journal of Pharmaceutical, Chemical and Biological Sciences*, 7(3), 219.

Kou, B., Yuan, Y., Zhu, X., Ke, Y., Wang, H., Yu, T., & Tan, W. (2024). Effect of soil organic matter-mediated electron transfer on heavy metal remediation: Current status and perspectives. *The Science of the Total Environment*, 917, 170451. DOI: 10.1016/j.scitotenv.2024.170451 PMID: 38296063

Koudad, M., El Hamouti, C., Elaatiaoui, A., Dadou, S., Oussaid, A., Abrigach, F., Pilet, G., Benchat, N., & Allali, M. (2020). Synthesis, crystal structure, antimicrobial activity and docking studies of new imidazothiazole derivatives. *Journal of the Indian Chemical Society*, 17, 297–306.

Krivokolysko, S. G., Dyachenko, V. D., Nesterov, V. N., Sharanin, Yu. A., & Struchkov, Yu. T. (1999).*Russian Journal of Organic Chemistry*, 35, 966.

Kumar, R. S., Praveen, S., Shridharshini, K., Maruthamuthu, M., Mohanapriya, K., & Mythili, A. (2022). Biological applications of imidazothiazole scaffolds: A current review. *Journal of Advanced Chemical Sciences*, 756-769.

Kumar, A. (2013).*American Journal of Pharmacological Sciences*, 1(6), 116. DOI: 10.12691/ajps-1-6-2

Kumar, A., Kumar, B., & Bhatia, R. (2023). Design, synthesis, molecular docking, and biological evaluation of isatin-based fused heterocycles as epidermal growth factor receptor inhibitors. *Assay and Drug Development Technologies*, 21(5), 222–233. DOI: 10.1089/adt.2022.120 PMID: 37439798

Kumar, A., Vigato, C., Boschi, D., Lolli, M. L., & Kumar, D. (2023). Phenothiazines as anti-cancer agents: SAR overview and synthetic strategies. *European Journal of Medicinal Chemistry*, 254, 115337. DOI: 10.1016/j.ejmech.2023.115337 PMID: 37060756

Kumar, M. P., Ravi, T. K., Chawla, R., Bhuvana, S., Sonia, G., & Gopalakrishnan, S. (2010). Microwave assisted synthesis and biological activity of novel coumarinyltriazolothiadiazoles. *Indian Journal of Pharmaceutical Sciences*, 72(3), 357. DOI: 10.4103/0250-474X.70483 PMID: 21188046

Kumar, P. R., Raju, S., Goud, P. S., Sailaja, M., Sarma, M. R., Reddy, G. O., Kumar, M. P., Reddy, V. K., Suresh, T., & Hegde, P. (2004). Synthesis and biological evaluation of thiophene [3, 2-b] pyrrole derivatives as potential anti-inflammatory agents. *Bioorganic & Medicinal Chemistry*, 12(5), 1221–1230. DOI: 10.1016/j.bmc.2003.11.003 PMID: 14980634

Kumar, P., & Kumar, V. (2017)... *Journal of Heterocyclic Chemistry*, 54(2), 465–471.

Kumar, R., & Pandey, M. (2014). A comprehensive review on synthetic approach for antimalarial agents. *European Journal of Medicinal Chemistry*, 85, 315–325. DOI: 10.1016/j.ejmech.2014.07.084

Kumar, V., & Verma, P. (2024). Pulp-paper industry sludge waste biorefinery for sustainable energy and value-added products development: A systematic valorization towards waste management. *Journal of Environmental Management*, 352, 120052. DOI: 10.1016/j.jenvman.2024.120052 PMID: 38244409

Kumawat, T. K., Sharma, V., Kumawat, V., Biyani, M., Kumari, N., Garg, R., & Kandwani, N. (2022). Electrokinetic-Assisted Bioremediation and Phytoremediation for the Treatment of Polluted Soil. In *Advances in Bioremediation and Phytoremediation for Sustainable Soil Management: Principles, Monitoring and Remediation* (pp. 379–397). Springer International Publishing. DOI: 10.1007/978-3-030-89984-4_24

Kuppan, N., Padman, M., Mahadeva, M., Srinivasan, S., & Devarajan, R. (2024). *A comprehensive review of sustainable bioremediation techniques: Eco friendly solutions for waste and pollution management.* Waste Management Bulletin.

Labbe, G., Buelens, J., & Dehaen, W. A. (1993). A new approach to azapentalenes by an addition–rearrangement sequence: Synthesis of fused 1,2,4-thiadiazoles. *Journal of the Chemical Society, Perkin Transactions 1: Organic and Bio-Organic Chemistry*, 1(16), 1825–1826. DOI: 10.1039/P19930001825

Lagardère, P., Mustière, R., Amanzougaghene, N., Hutter, S., Casanova, M., Franetich, J. F., Tajeri, S., Malzert-Fréon, A., Corvaisier, S., Azas, N., Vanelle, P., Verhaeghe, P., Primas, N., Mazier, D., Masurier, N., & Lisowski, V. (2023). New antiplasmodial 4-amino-thieno[3,2-d]pyrimidines with improved intestinal permeability and microsomal stability. *European Journal of Medicinal Chemistry*, 249(January), 115115. Advance online publication. DOI: 10.1016/j.ejmech.2023.115115 PMID: 36680984

Lago, A., Rocha, V., Barros, O., Silva, B., & Tavares, T. (2024). Bacterial biofilm attachment to sustainable carriers as a clean-up strategy for wastewater treatment: A review. *Journal of Water Process Engineering*, 63, 105368. DOI: 10.1016/j.jwpe.2024.105368

Lal, K., Bermeo, R., Cramer, J., Vasile, F., Ernst, B., Imberty, A., Bernardi, A., Varrot, A., & Belvisi, L. (2021). Prediction and Validation of a Druggable Site on Virulence Factor of Drug Resistant Burkholderia cenocepacia. *Chemistry (Basel, Switzerland)*, 27(40), 10341–10348. DOI: 10.1002/chem.202100252 PMID: 33769626

Larsen, A. K., Escargueil, A. E., & Skladanowski, A. (2003). Catalytic topoisomerase II inhibitors in cancer therapy. *Pharmacology & Therapeutics*, 99(2), 167–181. DOI: 10.1016/S0163-7258(03)00058-5 PMID: 12888111

Larsen, A. P., Francotte, P., Frydenvang, K., Tapken, D., Goffin, E., Fraikin, P., Caignard, D.-H., Lestage, P., Danober, L., Pirotte, B., & Kastrup, J. S. (2016). Synthesis and Pharmacology of Mono, Di, and Trialkyl-Substituted 7 Chloro-3,4-dihydro 2H 1,2,4-benzothiadiazine 1,1-Dioxides Combined with X-ray Structure Analysis to Understand the Unexpected Structure–Activity Relationship at AMPA Receptors. *ACS Chemical Neuroscience*, 7(3), 378–390. DOI: 10.1021/acschemneuro.5b00318 PMID: 26771108

Lavanya, M. B., Viswanath, D. S., & Sivapullaiah, P. V. (2024). Phytoremediation: An eco-friendly approach for remediation of heavy metal-contaminated soils-A comprehensive review. *Environmental Nanotechnology, Monitoring & Management*, 22, 100975. DOI: 10.1016/j.enmm.2024.100975

Le, D. T., Truong, N. T. T., Luu, T. H. T., Nguyen, L. T., Hoang, M. H., Huynh, H. P. K., Cu, S. T., Nguyen, Q. T., & Nguyen, H. T. (2022). Donor–acceptor and donor–donor alternating conjugated polymers based on dithieno [3, 2-b: 2′, 3′-d] pyrrole: synthesis, optical properties and organic solar cells applications. *Journal of Polymer Research*, 29(4), 111. DOI: 10.1007/s10965-022-02969-9

Lee, D. J., Kim, K., & Park, Y. J. (2002). Novel Synthesis of 5, 6-Dihydro-4 H-thieno [3, 2-b] pyrrol-5-ones via the Rhodium (II)-Mediated Wolff Rearrangement of 3-(Thieno-2-yl)-3-oxo-2-diazopropanoates. *Organic Letters*, 4(6), 873–876. DOI: 10.1021/ol016995n PMID: 11893174

Lee, H. S., Lee, J. S., Jung, A. R., Cha, W., Kim, H., Son, H. J., Cho, J. H., & Kim, B. S. (2016). Processing temperature control of a diketopyrrolopyrrole-alt-thieno [2,3-*b*]thiophene polymer for high-mobility thin-film transistors and polymer solar cells with high open-circuit voltages. *Polymer*, 105, 79–8. DOI: 10.1016/j.polymer.2016.10.024

Lee, H., Sam, K., Coulon, F., De Gisi, S., Notarnicola, M., & Labianca, C. (2023). Recent developments and prospects of sustainable remediation treatments for major contaminants in soil: A review. *The Science of the Total Environment*, •••, 168769. PMID: 38008308

Lee, J. H. (2015)... *Organic Electronics*, 20, 108–115.

Lee, S. B., & Ha, Y. (2019). Synthesis and characterization of phosphorescent iridium(III) complexes containing thienopyridine ligands. *Molecular Crystals and Liquid Crystals (Philadelphia, Pa.)*, 679(1), 65–70. DOI: 10.1080/15421406.2019.1597547

Lewis, V. A. (2017). Psychopharmacology. In *Pharmacology and Therapeutics for Dentistry* (pp. 133–155). Elsevier., DOI: 10.1016/B978-0-323-39307-2.00010-2

Liao, B., Lin, C., Lee, J., & Yang, J. C. (2017). Lung Cancer Optimal management of EGFR -mutant non-small cell lung cancer with disease progression on fi rst-line tyrosine kinase inhibitor therapy. *Lung Cancer (Amsterdam, Netherlands)*, 110(7), 7–13. DOI: 10.1016/j.lungcan.2017.05.009 PMID: 28676222

Liao, W., Yang, W., Xu, J., Yan, Z., Pan, M., Xu, X., Zhou, S., Zhu, Y., Lan, J., Zeng, M., Han, X., Li, S., Li, Y., Liang, K., Gao, Y., & Peng, Q. (2021). Therapeutic Potential of CUDC-907 (Fimepinostat) for Hepatocarcinoma Treatment Revealed by Tumor Spheroids-Based Drug Screening. *Frontiers in Pharmacology*, 12, 658197. Advance online publication. DOI: 10.3389/fphar.2021.658197 PMID: 34776939

Liaras, K., Geronikaki, A., Glamočlija, J., Ćirić, A., & Soković, M. (2011). Thiazole-based chalcones as potent antimicrobial agents. Synthesis and biological evaluation. *Bioorganic & Medicinal Chemistry*, 19(10), 3135–3140. DOI: 10.1016/j.bmc.2011.04.007 PMID: 21524583

Li, C., Zhang, J., Ren, Z., Xie, R., Yin, C., Ma, W., Zhou, F., Chen, H., & Lin, Y. (2021). Development of 'multiresistance rice' by an assembly of herbicide, insect and disease resistance genes with a transgene stacking system. *Pest Management Science*, 77(3), 1536–1547. DOI: 10.1002/ps.6178 PMID: 33201594

Li, F., Gong, J., Liu, J., Li, Y., & Li, Z. (2020). Design and Synthesis of Novel 3,4-Dihydro-2H-1,2,4-benzothiadiazine 1,1-Dioxides-based Strobilurins as Potent Fungicide Candidates. *Chemical Research in Chinese Universities*, 36(6), 1190–1195. DOI: 10.1007/s40242-020-0160-9

Li, H. (2019)... *Chemical Reviews*, 119(10), 5421–5478.

Li, J., An, B., Song, X., Zhang, Q., Chen, C., Wei, S., Fan, R., Li, X., & Zou, Y. (2021). Design, synthesis and biological evaluation of novel 2,4-diaryl pyrimidine derivatives as selective $EGFR^{L858R/T790M}$ inhibitors. *European Journal of Medicinal Chemistry*, 212, 113019. DOI: 10.1016/j.ejmech.2020.113019 PMID: 33429247

Lio, J.-Ch., He, H.-W., Tan, X.-S., & Ding, M.-W. (2005). *Russian Chemical Bulletin*, Kam, P.C. & Nethery, C.M. (2003). The thienopyridine derivatives (platelet adenosine diphosphate receptor antagonists), pharmacology and clinical developments. *Anaesthesia*, 58(1), 28–35. DOI: 10.1046/j.1365-2044.2003.02960.x PMID: 15601269

Li, P., Tian, P., Chen, Y., Song, X., Xue, W., Jin, L., Hu, D., Yang, S., & Song, B. (2018). Novel bisthioether derivatives containing a 1,3,4-oxadiazole moiety: Design, synthesis, antibacterial and nematocidal activities. *Pest Management Science*, 74(4), 844–852. DOI: 10.1002/ps.4762 PMID: 29024290

Li, T., Zhang, J., Pan, J., Wu, Z., Hu, D., & Song, B. (2017). Design, synthesis, and antiviral activities of 1,5-benzothiazepine derivatives containing pyridine moiety. *European Journal of Medicinal Chemistry*, 125, 657–662. DOI: 10.1016/j.ejmech.2016.09.069 PMID: 27721151

Litvinov, V. P. (2006). The Chemistry of Thienopyrimidines. *Advances in Heterocyclic Chemistry*, 92(06), 83–143. DOI: 10.1016/S0065-2725(06)92003-0

Litvinov, V. P., Dotsenko, V. V., & Krivokolysko, S. G. (2005). Thienopyridines: Synthesis, properties, and biological activity. *Russian Chemical Bulletin*, 54(4), 864–904. DOI: 10.1007/s11172-005-0333-1

Litvinov, V. P., Dotsenko, V. V., & Krivokolysko, S. G. (2007). The chemistry of thienopyridines. *Advances in Heterocyclic Chemistry*, 93, 117–178. DOI: 10.1016/S0065-2725(06)93003-7

Litvinov, V. P., Rodinovskaya, L. A., Sharanin, Yu. A., Shestopalov, A. M., & Senning, A. (1992). Advances in the Chemistry of 3-Cyanopyridin-2(1*H*)-Ones, -Thiones, And -Selenones. *Sulfur Reports*, 13(1), 1–142. DOI: 10.1080/01961779208048951

Litvinov, V. P., Roman, S. V., & Dyachenko, V. D. (2000). Naphthyridines. Structure, physicochemical properties and general methods of synthesis. *Russian Chemical Reviews*, 69(3), 218. DOI: 10.1070/RC2000v069n03ABEH000553

Liu, J., Seibold, S. A., Rieke, C. J., Song, I., Cukier, R. I., & Smith, W. L. (2007). Prostaglandin endoperoxide H synthases: Peroxidase hydroperoxide specificity and cyclooxygenase activation. *The Journal of Biological Chemistry*, 282(25), 18233–18244. DOI: 10.1074/jbc.M701235200 PMID: 17462992

Liu, J., Zhou, W., & Wu, F. (2021). Biological Evaluation and Synthesis of Thiazole Schiff Base Derivatives. *Heterocycles*, 102(7), 1337. DOI: 10.3987/COM-21-14467

Liu, K. G., Robichaud, A. J., Bernotas, R. C., Yan, Y., Lo, J. R., Zhang, M.-Y., Hughes, Z. A., Huselton, C., Zhang, G. M., Zhang, J. Y., Kowal, D. M., Smith, D. L., Schechter, L. E., & Comery, T. A. (2010). 5-Piperazinyl-3-sulfonylindazoles as potent and selective 5-hydroxytryptamine-6 antagonists. *Journal of Medicinal Chemistry*, 53(21), 7639–7646. DOI: 10.1021/jm1007825 PMID: 20932009

Li, W., Liu, X., Zhang, B., & Li, Y. (2013). Synthesis and Antifungal Activity of 2-Carbalkoxyalkyl/(*n* -butyl)-4-aryl-1,5-benzothiazepines. *Youji Huaxue*, 33(07), 1503. DOI: 10.6023/cjoc201210044

Li, Y., Chang, Y., Fu, J., Ding, R., Zhang, L., Liang, T., Liu, Y., Liu, Y., & Hu, J. (2021). Design, synthesis and biological evaluation of aminopyrimidine derivatives bearing a 4, 5, 6, 7-tetrahydrothieno[3,2-*c*]pyridine as potent EGFR inhibitors. *European Journal of Medicinal Chemistry*, 226, 113845. DOI: 10.1016/j.ejmech.2021.113845 PMID: 34534838

Luhman, U., Wentz, F. G., Knieriem, B., & Luttke, W. (1978). Theoretical and spectroscopical investigations of indigo dyes, xviii. properties of 2,2,5,5-tetramethylthieno(3,2-*b*)thiophene-3(2*H*),6(5*H*)-dione, a basic chromophoric system of thioindigo dyes with trans-s-trans-s-trans-configuration of the carbonyl groups. *Chemische Berichte*, 111, 3233. DOI: 10.1002/cber.19781110927

Lutosławski, K., Ryznar-Luty, A., & Cibis, E. (2024). Efficiency of aerobic biodegradation of sugar beet distillery stillage under dissolved oxygen tension-controlled conditions. *PLoS One*, 19(7), e0306330. DOI: 10.1371/journal.pone.0306330 PMID: 38968255

M'esangeau, C., Yous, S., P'er'es, B., Lesieur, D., & Besson, T. (2005). Pictet–Spengler heterocyclizations via microwave-assisted degradation of DMSO. *Tetrahedron Letters*, 46(14), 2465–2468. DOI: 10.1016/j.tetlet.2005.02.043

Mabkhoot, Y. N. (2009). Synthesis and analysis of some bis-heterocyclic compounds containing sulphur. *Molecules (Basel, Switzerland)*, 14(5), 1904–1914. DOI: 10.3390/molecules14051904 PMID: 19471210

Madhusudana, K., Shireesha, B., Naidu, V. G., Ramakrishna, S., Narsaiah, B., Rao, A. R., & Diwan, P. V. (2012). Anti-inflammatory potential of thienopyridines as possible alternative to NSAIDs. *European Journal of Pharmacology*, 678(1–3), 48–54. DOI: 10.1016/j.ejphar.2011.12.019 PMID: 22209879

Mahfouz, , Aziz, A. A., & Elhabashy, F. M. (1990). Mahfouz; Aziz, A. A.; Elhabashy, F., New synthesis of 2-substituted imidazo [2, 1-b] thiazoles and their antimicrobial activities. *Archives of Pharmacal Research*, 13(1), 9–13. DOI: 10.1007/BF02857826

Mahmoud, H. K., Gomha, S. M., Farghaly, T. A., & Awad, H. M. (2021). Synthesis of thiazole linked imidazo [2, 1-b] thiazoles as anticancer agents. *Polycyclic Aromatic Compounds*, 41(8), 1608–1622. DOI: 10.1080/10406638.2019.1689514

Maji, S., Debnath, B., Panda, S., Manna, T., Maity, A., Dayaramani, R., Nath, R., Khan, S. A., & Akhtar, M. J. (2024). Anticancer Potential of the S-Heterocyclic Ring Containing Drugs and its Bioactivation to Reactive Metabolites. *Chemistry & Biodiversity*, 21(7), 202400473. DOI: 10.1002/cbdv.202400473 PMID: 38723201

Makawana, J. A., Patel, M. P., & Patel, R. G. (2011). Synthesis and in vitro antimicrobial evaluation of penta-substituted pyridine derivatives bearing the quinoline nucleus. *Medicinal Chemistry Research*, 21(5), 616–623. DOI: 10.1007/s00044-011-9568-6

Makhova, N. N., Belen'kii, L. I., Gazieva, G. A., Dalinger, I. L., Konstantinova, L. S., Kuznetsov, V. V., Kravchenko, A. N., Krayushkin, M. M., Rakitin, O. A., Starosotnikov, A. M., Fershtat, L. L., Shevelev, S. A., Shirinian, V. Z., & Yarovenko, V. N. (2020). Progress in the chemistry of nitrogen-, oxygen- and sulfur-containing heterocyclic systems. *Russian Chemical Reviews*, 89(1), 55–124. DOI: 10.1070/RCR4914

Malasala, S., Polomoni, A., Ahmad, M. N., Shukla, M., Kaul, G., Dasgupta, A., Chopra, S., & Nanduri, S. (2021). Structure based design, synthesis and evaluation of new thienopyrimidine derivatives as anti-bacterial agents. *Journal of Molecular Structure*, 1234, 130168. DOI: 10.1016/j.molstruc.2021.130168

Malicorne, G., Bompart, J., Giral, L., & Despaux, E. (1991). Synthèse et activité anti-bactérienne d'acides 4,7-dihydro-4-éthyl-7-oxothiéno[3,2-*b*]pyridine-6-carboxyliquesSynthesis and antibacterial activity of 4,7-dihydro-4-ethyl-7-oxothieno(3,2-b)pyridine-6-carboxylic acids. *European Journal of Medicinal Chemistry*, 26(1), 3–11. DOI: 10.1016/0223-5234(91)90207-4

Mallam, V., Allam, V., & Brahmeshwari, G. (2023). Synthesis and Biological Evaluation of Fused Imidazo [4, 5, 1-kl] phenothiazine-4-sulfonamides as Potent Antibacterial and Anticancer Agents. *Russian Journal of Bioorganic Chemistry*, 49(6), 1381–1388. DOI: 10.1134/S1068162023060183

Manaithiya, A., Alam, O., Sharma, V., Javed, M., & Mittal, S. (2021). Bioorganic Chemistry GPR119 agonists : Novel therapeutic agents for type 2 diabetes mellitus. In Bioorganic Chemistry (Vol. 113, Issue May). Elsevier Inc. DOI: 10.1016/j.bioorg.2021.104998

Manasa, K. L., Pujitha, S., Sethi, A., Mohammed, A., Alvala, M., Angeli, A., & Supuran, C. T. (2020). Synthesis and Biological Evaluation of Imidazo [2, 1-*b*] Thiazole based Sulfonyl Piperazines as Novel Carbonic Anhydrase II Inhibitors. *Metabolites*, 10(4), 136. DOI: 10.3390/metabo10040136 PMID: 32244413

Mangwegape, D. K., Zuma, N. H., Aucamp, J., & N'Da, D. D. (2021). Synthesis and in vitro antileishmanial efficacy of novel benzothiadiazine-1,1-dioxide derivatives. *Archiv der Pharmazie*, 354(5), 2000280. Advance online publication. DOI: 10.1002/ardp.202000280 PMID: 33491807

Marco, A. B., Andreu, R., Franco, S., Garín, J., Orduna, J., Villacampa, B., & Alicante, R. (2013). Efficient second-order nonlinear optical chromophores based on dithienothiophene and thienothiophene bridges. *Tetrahedron*, 69(19), 3919–3926. DOI: 10.1016/j.tet.2013.03.027

Marco, A. B., Gindre, D., Iliopoulos, K., Franco, S., Andreu, R., Canevet, D., & Sallé, M. (2018). (Super)gelators derived from push-pull chromophores: Synthesis, gelling properties and second harmonic generation. *Organic & Biomolecular Chemistry*, 16(14), 2470–2478. DOI: 10.1039/C8OB00251G PMID: 29561017

Margutti, S., & Laufer, S. A. (2007). Are MAP Kinases Drug Targets? Yes, but Difficult Ones. *ChemMedChem*, 2(8), 1116–1140. DOI: 10.1002/cmdc.200600271 PMID: 17541990

Marinov, M., Nikolova, I., Kostova, I., & Stoyanov, N. (2023). Phenothiazine derivatives of 1, 8-naphthalic anhydride: Synthesis, characterization and antimicrobial studies. *Phosphorus, Sulfur, and Silicon and the Related Elements*, 198(2), 128–136. DOI: 10.1080/10426507.2022.2116707

Marqués-López, E., & Herrera, R. P. (2015). Bucherer–Bergs and Strecker Multicomponent Reactions. In Multicomponent Reactions (pp. 331–357). Wiley. DOI: 10.1002/9781118863992.ch10

Martin Hernando, J. I., Ontoria, J. M., Malancona, S., Attenni, B., Fiore, F., Bonelli, F., Koch, U., Di Marco, S., Colarusso, S., Ponzi, S., Gennari, N., Vignetti, S. E., del Rosario Rico Ferreira, M., Habermann, J., Rowley, M., & Narjes, F. (2009). Optimization of thienopyrrole-based finger-loop inhibitors of the hepatitis C virus NS5B polymerase. *ChemMedChem*, 4(10), 1695–1713. DOI: 10.1002/cmdc.200900184 PMID: 19672916

Masaret, G. S. (2021). Convenient synthesis and anticancer evaluation of novel pyrazolyl-thiophene, thieno[3,2-*b*]pyridine, pyrazolo[3,4-*d*]thieno[3,2-*b*]pyridine and pyrano[2,3-*d*]thieno[3,2-*b*]pyridine derivative. *Journal of Heterocyclic Chemistry*, 58(6), 1344–1358. DOI: 10.1002/jhet.4262

Mashraqui, S. H., Ghadigaonkar, S., Ashraf, M., Sri Ranjini, A., Ghosh, S., & Das, P. K. (2007). Optically transparent and thermally stable nonlinear optic chromo. phores featuring a thieno [2,3-b]thiophene donor. *Tetrahedron*, 63(40), 10011–10017. DOI: 10.1016/j.tet.2007.07.059

Mashraqui, S. H., Sangvikar, Y. S., & Meetsma, A. (2006). Synthesis and structures of thieno [2,3-b]thiophene incorporated [3.3]dithiacyclophanes. Enhanced first hyperpolarizability in an unsymmetrically polarized cyclophane. *Tetrahedron Letters*, 47(31), 5599–5602. DOI: 10.1016/j.tetlet.2006.05.098

Mason, C. D., & Nord, F. F. (1951). STUDIES ON THE CHEMISTRY OF HETEROCYCLICS. XIII. TRIARYLMETHANE DYES CONTAINING A THIOPHENE RING[1]. *The Journal of Organic Chemistry*, 16(5), 722–727. DOI: 10.1021/jo01145a011

Massin, J., Ducasse, L., Abbas, M., Hirsch, L., Toupance, T., & Olivier, C. (2015). Molecular engineering of carbazole-fluorene sensitizers for high open-circuit voltage DSSCs: Synthesis and performance comparison with iodine and cobalt electrolytes. *Dyes and Pigments*, 118, 76–87. DOI: 10.1016/j.dyepig.2015.02.020

Mathew, B., Mathew, G. E., Sonia, G., Kumar, A., Charles, N. P., & Kumar, P. (2013). Design of 1-(furan-2-yl)-N-(5-substituted phenyl-1, 3, 4-thiadiazol-2-yl) methanimine derivatives as Enoyl-ACP reductase inhibitors: Synthesis, molecular docking studies and anti-tubercular activity. *Bangladesh Journal of Pharmacology*, 8(3), 242–248. DOI: 10.3329/bjp.v8i3.14778

Mathew, V., Keshavayya, J., & Vaidya, V. P. (2006). Heterocyclic system containing bridgehead nitrogen atom: Synthesis and pharmacological activities of some substituted 1,2,4-triazolo[3,4-b]-1,3,4-thiadiazoles. *European Journal of Medicinal Chemistry*, 41(9), 1048–1058. DOI: 10.1016/j.ejmech.2006.03.018 PMID: 16822595

Mathew, V., Keshavayya, J., Vaidya, V. P., & Giles, D. (2007). Studies on synthesis and pharmacological activities of 3,6-disubstituted-1,2,4-triazolo[3,4-b]-1,3,4-thiadiazoles and their dihydro analogues. *European Journal of Medicinal Chemistry*, 42(6), 823–840. DOI: 10.1016/j.ejmech.2006.12.010 PMID: 17331622

Mavrova, A. T., Dimov, S., Yancheva, D., Rangelov, M., Wesselinova, D., & Tsenov, J. A. (2016). Synthesis, anticancer activity and photostability of novel 3-ethyl-2-mercapto-thieno[2,3-d]pyrimidin-4(3H)-ones. *European Journal of Medicinal Chemistry*, 123, 69–79. DOI: 10.1016/j.ejmech.2016.07.022 PMID: 27474924

Mavrova, A. T., Vuchev, D., Anichina, K., & Vassilev, N. (2010). Synthesis, antitrichinnellosis and antiprotozoal activity of some novel thieno[2,3-d]pyrimidin-4(3H)-ones containing benzimidazole ring. *European Journal of Medicinal Chemistry*, 45(12), 5856–5861. DOI: 10.1016/j.ejmech.2010.09.050 PMID: 20950896

Meenakshi Singh, Sushil K Singh, M. T. C. (2011). A NOVEL QSAR MODEL FOR EVALUATION AND PREDICTING THE INHIBITION ACTIVITY OF H1-RECEPTOR ANTAGONISTS: A SERIES OF THIENOPYRIMIDNES. *Journal of Drug Delivery and Therapeutics*.

Mekky, A. E., Sanad, S. M., Said, A. Y., & Elneairy, M. A. (2020). Synthesis, cytotoxicity, in-vitro antibacterial screening and in-silico study of novel thieno[2,3-*b*]pyridines as potential pim-1 inhibitors. *Synthetic Communications*, 50(15), 2376–2389. DOI: 10.1080/00397911.2020.1778033

Merchant, A. I., Vakili, A. H., Kocaman, A., & Amr, S. S. A. (2024). New advancement of advanced oxidation processes for the treatment of Petroleum wastewater. *Desalination and Water Treatment*, 319, 100565. DOI: 10.1016/j.dwt.2024.100565

Mhaske, G. R., Bajod, B. S., Ambhore, D. M., & Shelke, S. N. (2014).*International Journal of Innovative Research in Science, Engineering and Technology*, 3, 2319.

Miaki, T., Kawamura, T., Sugahara, K., Alcoverro, M., Casas, J., Pardàs, M., Ahmadi, A., Abedin, M. A., Mattausch, H. J., Koide, T., Abadi, M., De Alfaro, L., Chack, S., Sharma, P., Mahesan, K. V., Bhargavi, S., Jayadevappa, D., Appia, V., Yezzi, A., . . . He, C. (2017). Main paper.pdf. In IEEE Transactions on Image Processing (Vol. 24, Issue 1, pp. 1–9). Retrieved from http://dx.doi.org/DOI: 10.1016/j.eswa.2013.05.056%0Ahttps://hal.inria.fr/hal-00912934/document#page=36%0Awww.ijirae.com%0Ahttp://www.springerlink.com/content/9662h2hwq81041g2

Mirhoseini, H., & Taghdiri, M. (2016). Extractive oxidation desulfurization of sulfur-containing model fuel using hexamine–phosphotungstate hybrid as effective heterogeneous catalyst. *Fuel*, 167, 60–67. DOI: 10.1016/j.fuel.2015.11.042

Mishra, A., Srivastava, M., Rai, P., Yadav, S., Tripathi, B. P., Singh, J., & Singh, J. (2016). Visible light triggered, catalyst free approach for the synthesis of thiazoles and imidazo [2, 1-b] thiazoles in EtOH: H 2 O green medium. *RSC Advances*, 6(54), 49164–49172. DOI: 10.1039/C6RA05385H

Mitsudo, K., Shimohara, S., Mizoguchi, J., Mandai, H., & Suga, S. (2012). Synthesis of nitrogen-bridged terthiophenes by tandem Buchwald–Hartwig coupling and their properties. *Organic Letters*, 14(11), 2702–2705. DOI: 10.1021/ol300887t PMID: 22594835

Miwa, K., Hitaka, T., Imada, T., Sasaki, S., Yoshimatsu, M., Kusaka, M., Tanaka, A., Nakata, D., Furuya, S., & Endo, S. (2011). pyrimidin-6-yl] phenyl } -3-methoxyurea (TAK-385) as a Potent, Orally Active. *Non-Peptide Antagonist of the Human Gonadotropin-Releasing Hormone Receptor.*, 1, 4998–5012. PMID: 21657270

Miyah, Y., El Messaoudi, N., Benjelloun, M., Acikbas, Y., Şenol, Z. M., Ciğeroğlu, Z., & Lopez-Maldonado, E. A. (2024). Advanced applications of hydroxyapatite nanocomposite materials for heavy metals and organic pollutants removal by adsorption and photocatalytic degradation: A review. *Chemosphere*, 358, 142236. DOI: 10.1016/j.chemosphere.2024.142236 PMID: 38705409

Mjos, K. D., & Orvig, C. (2014). Metallodrugs in Medicinal Inorganic Chemistry. *Chemical Reviews*, 114(8), 4540–4563. DOI: 10.1021/cr400460s PMID: 24456146

Mohamed, K. O., Nissan, Y. M., El-Malah, A. A., Ahmed, W. A., Ibrahim, D. M., Sakr, T. M., & Motaleb, M. A. (2017). Design, synthesis and biological evaluation of some novel sulfonamide derivatives as apoptosis inducers. *European Journal of Medicinal Chemistry*, 135, 424–433. DOI: 10.1016/j.ejmech.2017.04.069 PMID: 28463785

Mohamed, M. A., Khattab, R. R., Wasfy, A. A. F., Abo-riya, M. A., El-Kalyoubi, S., & Hassan, N. A. (2021). Synthesis and Anticancer Activity Evaluation of Some Thienopyrimidine Derivatives. *Russian Journal of Bioorganic Chemistry*, 47(6), 1312–1323. DOI: 10.1134/S1068162021060194

Mohamed, O. S., Al-Taifi, E. A., El-Emary, T. I., & Bakhite, E. A. (2007). Studies on the synthesis of some new cyanopyridine-thione and thieno[2,3-*b*]pyridine derivatives. *Phosphorus, Sulfur, and Silicon and the Related Elements*, 182(5), 1061–1082. DOI: 10.1080/10426500601096369

Mohamed, S. F., Abd-Elghaffar, H. S., Amr, A. E.-G. E., Elnaggar, D. H., Abou-Amra, E. S., Hosny, H. M., Mohamed, A. M., & Abd El-Shafy, D. N. (2023). New poly heterocyclic compounds based on pyrimidine-2-thiones: Synthesis, evaluation of putative anti-viral agents, DFT calculation, and molecular modeling. *Journal of Molecular Structure*, 1291, 136083. DOI: 10.1016/j.molstruc.2023.136083

Mohamed, S. F., Flefel, E. M., Amr, A. E.-G. E., & Abd El-Shafy, D. N. (2010). Anti-HSV-1 activity and mechanism of action of some new synthesized substituted pyrimidine, thiopyrimidine and thiazolopyrimidine derivatives. *European Journal of Medicinal Chemistry*, 45(4), 1494–1501. DOI: 10.1016/j.ejmech.2009.12.057 PMID: 20110135

Mohan, C., Robinson, J., Vodwal, L., & Kumari, N. (2024). Sustainable Development Goals for addressing environmental challenges. In *Green Chemistry Approaches to Environmental Sustainability* (pp. 357–374). Elsevier. DOI: 10.1016/B978-0-443-18959-3.00007-0

Mohan, J., & Kiran, K. (1991). Novel bridgehead nitrogen heterocycles: Synthesis and antimicrobial activity of 2H-imidazo[2,1-a] pyrazolo[3,4-d] thiazoles. *Indian Journal of Chemistry*, 30B, 898–900.

Mohareb, R. M. (1992). Phenyl isothiocyanate in heterocyclic synthesis: Novel synthesis of thiazoles, thieno[2,3-*b*]pyridine, thiophene and thieno[3,2-*c*]pyridazine derivatives. *Monatshefte für Chemie*, 123(4), 341–347. DOI: 10.1007/BF00810946

Mohareb, R. M., Aziz, S. I., & El-Saka, I. (1991).*Sulfur Letters*, 13, 229.

Mohi El-Deen, E. M., Abd El-Meguid, E. A., Karam, E. A., Nossier, E. S., & Ahmed, M. F. (2020). Synthesis and biological evaluation of new pyridothienopyrimidine derivatives as antibacterial agents and *Escherichia coli* topoisomerase II inhibitors. *Antibiotics (Basel, Switzerland)*, 9(10), 695. DOI: 10.3390/antibiotics9100695 PMID: 33066400

Mokrani, S., Houali, K., Yadav, K. K., Arabi, A. I. A., Eltayeb, L. B., AwjanAl-reshidi, M., Benguerba, Y., Cabral-Pinto, M. M. S., & Nabti, E. (2024). Bioremediation techniques for soil organic pollution: Mechanisms, microorganisms, and technologies-A comprehensive review. *Ecological Engineering*, 207, 107338. DOI: 10.1016/j.ecoleng.2024.107338

Molina, P., Arques, A., & Alias, A. (1992). Aza Wittig-type reaction between the iminophosphorane derived from 3-amino-4-phenylthiazole-2(3H)-thione and iso(thio)cyanates: Preparation of mesoionic thiazolo[2,3-b]-1,3,4-thiadiazoles and N,N-bisheteroarylamines. *Tetrahedron*, 48(7), 1285–1298. DOI: 10.1016/S0040-4020(01)90791-4

Mona, E. (2017). Ibrahim, Wafaa S Hamama, Hadeer A Raoof, Hanafi H Zoorob. *Der Pharma Chemica*, 9(12), 28–33.

Monarch, R. M. (1992).*Italian Chemical Society*, 122, 147.

Moraski, G. C., Deboosère, N., Marshall, K. L., Weaver, H. A., Vandeputte, A., Hastings, C., Woolhiser, L., Lenaerts, A. J., Brodin, P., & Miller, M. J. (2020). Intracellular and in vivo evaluation of imidazo[2,1-b] thiazole-5-carboxamide anti-tuberculosis compounds. *PLoS One*, 5(1), e0227224. Advance online publication. DOI: 10.1371/journal.pone.0227224 PMID: 31905374

Mor, S., & Khatri, M. (2023). Convenient synthesis of benzothiazinoisoindol-11-ones and benzoindenothiazin-11-ones, and antimicrobial testing thereof. *Molecular Diversity*, 27(3), 1223–1241. DOI: 10.1007/s11030-022-10483-9 PMID: 35922654

Morton, G. C., Salvino, J. M., Labaudiniere, R. F., & Herpin, T. F. (2000). Novel solid-phase synthesis of 1,5-benzothiazepine-4-one derivatives. *Tetrahedron Letters*, 41(17), 3029–3033. DOI: 10.1016/S0040-4039(00)00341-5

Mostafa, M. S., Radini, I. A. M., El-Rahman, N. M. A., & Khidre, R. E. (2024). Synthetic Methods and Pharmacological Potentials of Triazolothiadiazines: A Review. *Molecules (Basel, Switzerland)*, 29(6), 1326. DOI: 10.3390/molecules29061326 PMID: 38542962

Mukku, N., & Maiti, B. (2020). On water catalyst-free synthesis of benzo [d] imidazo [2, 1-b] thiazoles and novel N-alkylated 2-aminobenzo [d] oxazoles under microwave irradiation. *RSC Advances*, 10(2), 770–778. DOI: 10.1039/C9RA08929B PMID: 35494448

Mu, S., Oniwa, K., Jin, T., Asao, N., Yamashita, M., & Takaishi, S. (2016). A highly emissive distyrylthieno [3,2-b]thiophene based red luminescent organic single crystal: Aggregation induced emission, optical waveguide edge emission, and balanced ambipolar carrier transport. *Organic Electronics*, 34, 23–27. DOI: 10.1016/j.orgel.2016.04.001

Nagarajaiah, H., Khazi, I. M., & Begum, N. S.NAGARAJAIAH. (2012). Synthesis, characterization and biological evaluation of thiazolopyrimidine derivatives. *Journal of Chemical Sciences*, 124(4), 847–855. DOI: 10.1007/s12039-012-0271-z

Nagarajaiah, H., M Khazi, I. A., & Begum, N. S.NAGARAJAIAH. (2015). Synthesis of some new derivatives of thiazolopyrimidines and hydrolysis of its arylidene derivative. *Journal of Chemical Sciences*, 127(3), 467–479. DOI: 10.1007/s12039-015-0797-y

Nagesh, H., Padmashali, B., Sandeep, C., Yuvaraj, T., Siddesh, M., & Mallikarjuna, S. (2014). Synthesis and antimicrobial activity of benzothiophene substituted coumarins, pyrimidines and pyrazole as new scaffold. *International Journal of Pharmaceutical Sciences Review and Research*, 28(2), 6–10.

Nagireddy, P. K. R., Kommalapati, V. K., Siva Krishna, V., Sriram, D., Tangutur, A. D., & Kantevari, S. (2019). Imidazo[2,1-*b*] thiazole-Coupled Natural Noscapine Derivatives as Anticancer Agents. *ACS Omega*, 4(21), 19382–19388. DOI: 10.1021/acsomega.9b02789 PMID: 31763563

Nammalwar, B., & Bunce, R. A. (2024). Recent Advances in Pyrimidine-Based Drugs. *Pharmaceuticals (Basel, Switzerland)*, 17(1), 104. DOI: 10.3390/ph17010104 PMID: 38256937

Narciso-Ortiz, L., Tec-Caamal, E. N., Aguirre-García, G. J., & Lizardi-Jiménez, M. A. (2023). Bioreactors for bioremediation of polluted water. In *Current Status of Marine Water Microbiology* (pp. 345–364). Springer Nature Singapore. DOI: 10.1007/978-981-99-5022-5_15

Narusaka, M., & Narusaka, Y. (2017). Thienopyrimidine-type compounds protect Arabidopsis plants against the hemibiotrophic fungal pathogen Colletotrichum higginsianum and bacterial pathogen Pseudomonas syringae pv. maculicola. *Plant Signaling & Behavior*, 12(3), e1293222.

Nasri, S., Bayat, M., Miankooshki, F. R., & Samet, N. H. (2022). Recent developments in green approaches for sustainable synthesis of indole-derived scaffolds. *Molecular Diversity*, 26(6), 1–35. DOI: 10.1007/s11030-021-10376-3 PMID: 35031935

Navina, B. K., Velmurugan, N. K., Kumar, P. S., Rangasamy, G., Palanivelu, J., Thamarai, P., & Shakoor, A. (2024). Fungal bioremediation approaches for the removal of toxic pollutants: Mechanistic understanding for biorefinery applications. *Chemosphere*, 350, 141123. DOI: 10.1016/j.chemosphere.2024.141123 PMID: 38185426

Ngieng, H. Y., Jusoh, M. N. H., Ahmad, N., Al Masud, M. A., Samaraweera, H., & Mohamed, M. M. (2024). Enhanced Soil Decontamination via Electrokinetic Removal of Organic Pollutants. *Industrial and Domestic Waste Management*, 4(1), 42–55. DOI: 10.53623/idwm.v4i1.442

Ngoc, T. D., Thanh, N. D., Truong, M. X., Nghia, B. D., Thanh, N. M., & Thi-Thu, H. N. (2020).*New Journal of Chemistry*, 44(47), 20715. DOI: 10.1039/D0NJ04295A

Nielsen, O. B. T., Nielsen, C. K., & Feit, P. W. (1973). Aminobenzoic Acid Diuretics. 5.1 3-Amino-4-arylmethyl-5-sulfamylbenzoic Acid Derivatives and Related Compounds. *Journal of Medicinal Chemistry*, 16(10), 1170–1177. DOI: 10.1021/jm00268a022 PMID: 4749473

Nikalje, A. P., & Vyawahare, D. (2011).*African Journal of Pure and Applied Chemistry*, 5(12), 422.

Ningegowda, R., Chandrashekharappa, S., Singh, V., Mohanlall, V., & Venugopala, K. N. (2020). Design, synthesis and characterization of novel 2-(2, 3-dichlorophenyl)-5-aryl-1, 3, 4-oxadiazole derivatives for their anti-tubercular activity against Mycobacterium tuberculosis. *Chemical Data Collections*, 28, 100431. DOI: 10.1016/j.cdc.2020.100431

Nkomba, G., Terre'Blanche, G., Janse van Rensburg, H. D., & Legoabe, L. J. (2022). Design, synthesis and evaluation of amino-3,5-dicyanopyridines and thieno[2,3-*b*]pyridines as ligands of adenosine A_1 receptors for the potential treatment of epilepsy. *Medicinal Chemistry Research*, 31(8), 1277–1297. DOI: 10.1007/s00044-022-02908-9 PMID: 35634433

Noh, J. H., Jeon, N. J., Choi, Y. C., Nazeeruddin, M. K., Grätzel, M., & Seok, S. (2013). Il Nanostructured TiO2/CH3NH3PbI3 heterojunction solar cells employing spiro-OMeTAD/Co-complex as hole-transporting material. *Journal of Materials Chemistry. A, Materials for Energy and Sustainability*, 1(38), 11842–11847. DOI: 10.1039/c3ta12681a

Ogunnupebi, T. A., Ajani, O. O., Oduselu, G. O., Elebiju, O. F., & Adebiyi, E. (2023). Chemistry and pharmacological diversity of benzothiazepine—Excellent pathway to drug discovery. *Journal of Molecular Structure*, 1280, 135071. DOI: 10.1016/j.molstruc.2023.135071 PMID: 36843650

Ojija, F. (2024). Emerging environmental contaminants: Sources, effects on biodiversity and humans, remediation, and conservation implications. *Science Progress*, 107(2), 00368504241253720. DOI: 10.1177/00368504241253720 PMID: 38715402

Okpara, E. C., Wojuola, O. B., Quadri, T. W., & Banks, C. E. (2024). An overview of advanced oxidation processes using copper-based catalytic degradation of organic pollutants in water. *Applied Materials Today*, 36, 102053. DOI: 10.1016/j.apmt.2023.102053

Osaka, I., Sauve, G., Zhang, R., Kowalewski, T., & McCullough, R. D. (2007). Novel Thiophene-Thiazolothiazole Copolymers for Organic Field-Effect Transistors. *Advanced Materials*, 19(23), 4160–4165. DOI: 10.1002/adma.200701058

Osman, A. I., Chen, Z., Elgarahy, A. M., Farghali, M., Mohamed, I. M., Priya, A. K., Hawash, H. B., & Yap, P. S. (2024). Membrane technology for energy saving: Principles, techniques, applications, challenges, and prospects. *Advanced Energy and Sustainability Research*, 5(5), 2400011. DOI: 10.1002/aesr.202400011

Padam, K., Kumar, L. D., & Nisheeth, R. (2015).. . *Indian Journal of Heterocyclic Chemistry*, 25, 91–96.

Padmaja, R. D., Balamurali, M. M., & Chanda, K. (2019). One-Pot, Telescopic Approach for the Chemoselective Synthesis of Substituted Benzo[e]pyrido/pyrazino/pyridazino[1,2- b][1,2,4]thiadiazine dioxides and Their Significance in Biological Systems. *The Journal of Organic Chemistry*, 84(18), 11382–11390. DOI: 10.1021/acs.joc.9b00869 PMID: 31436096

Padnya, P. L., Khadieva, A. I., & Stoikov, I. I. (2023). Current achievements and perspectives in synthesis and applications of 3, 7-disubstituted phenothiazines as Methylene Blue analogues. *Dyes and Pigments*, 208, 110806. DOI: 10.1016/j.dyepig.2022.110806

Padwa, A., Heidelbaugh, T. M., & Kuethe, J. T. (1999). Synthesis of substituted 2-pyridones via the pummerer Cyclization-Deprotonation-Cycloaddition cascade of imidosulfoxides. *The Journal of Organic Chemistry*, 64(6), 2038–2049. DOI: 10.1021/jo982315r PMID: 11674298

Paek, S., Zimmermann, I., Gao, P., Gratia, P., Rakstys, K., Grancini, G., Nazeeruddin, M. K., Malik, A. R., Samia, A. K., & Khalid, A. A.. (2016). Donor-π-donor type hole transporting materials: Marked π-bridge effects on optoelectronic properties, solid-state structure, and perovskite solar cell efficiency. *Chemical Science (Cambridge)*, 7(9), 6068–6075. DOI: 10.1039/C6SC01478J PMID: 30034747

Panda, S. S., & Chowdary, P. V. (2008). Synthesis of novel indolyl-pyrimidine antiinflammatory, antioxidant and antibacterial agents. *Indian Journal of Pharmaceutical Sciences*, 70(2), 208. DOI: 10.4103/0250-474X.41457 PMID: 20046714

Panneer Selvam, T., Richa James, C., Vijaysarathy Dniandev, P., & Karyn Valzita, S. (2012). A mini review of pyrimidine and fused pyrimidine marketed drugs. *Research in Pharmacy, 2*(4), 1–9. Retrieved from www.researchinpharmacy.com

Pant, S., Godwal, P., & Sanju, K. (2021). Syntheses and Studies of 8-Substituted-2-(2-Chlorophenyl/3-chlorophenyl)-4-(4-hydroxyphenyl/phenyl)-2,3/2,5-dihydro-1,5-benzothiazepines. *Asian Journal of Chemistry*, 33(11), 2653–2656. DOI: 10.14233/ajchem.2021.23369

Pant, S., Sharma, P., & Pant, U. C. (2008). Syntheses of 1,5-Benzothiazepines: Part XXXVI—Syntheses and Antimicrobial Evaluation of 2-(2-Chlorophenyl)-4-(4-chlorophenyl/2-thienyl)-2,5-dihydro-8-substituted-1,5-benzothiazepines. *Phosphorus, Sulfur, and Silicon and the Related Elements*, 183(12), 2974–2983. DOI: 10.1080/10426500802048920

Pant, S., Sharma, P., Sharma, A., Mishra, R., & Pant, U. C. (2008). *Journal of the Indian Chemical Society*, 85(4), 406.

Papernaya, L. K., Shatrova, A. A., Kletskov, A. V., Petkevich, S. K., Sterkhova, I. V., Klyba, L. V., & Levkovskaya, G. G. (2017). Microwave synthesis of new azolyl-substituted thiazolo[5,4-d]thiazoles. *Russian Journal of Organic Chemistry*, 53(4), 550–556. DOI: 10.1134/S107042801704008X

Parit, S., Manchare, A., Ghodse, S., & Hatvate, N. (2024). Comparative review on homogeneous and heterogeneous catalyzed synthesis of 1,3-thiazole. *Synthetic Communications*, 54(23), 1–21. DOI: 10.1080/00397911.2024.2399187

Parjane S K, Kunkulol R & Nandal D, (**2020**). *IOSR Journal of Applied Chemistry*, 13(3-1), 50.

Park, J.-H., El-Gamal, M. I., Lee, Y. S., & Oh, C.-H. (2011). New imidazo [2, 1-b] thiazole derivatives: Synthesis, in vitro anticancer evaluation, and in silico studies. *European Journal of Medicinal Chemistry*, 46(12), 5769–5777. DOI: 10.1016/j.ejmech.2011.08.024 PMID: 22033063

Parmentier, J. G., Portevin, B., Golsteyn, R. M., Pierré, A., Hickman, J., Gloanec, P., & De Nanteuil, G. (2009). Synthesis and CHK1 inhibitory potency of hymenialdisine analogues. *Bioorganic & Medicinal Chemistry Letters*, 19(3), 841–844. DOI: 10.1016/j.bmcl.2008.12.001 PMID: 19111462

Paschoalino, M., Marinho, M. D., Santos, I. A., Grosche, V. R., Martins, D. O., Rosa, R. B., & Jardim, A. C. (2023). An update on the development of antiviral against Mayaro virus: From molecules to potential viral targets. *Archives of Microbiology*, 205(4), 106. DOI: 10.1007/s00203-023-03441-y PMID: 36881172

Patel, P. S., Akbari, V. K., Modi, S. D., Belim, M. A., Tailor, R. B., Patel, H. D., & Dewani, B. (2019). Synthesis and Biological Evaluation of Schiff Base Involving Thieno [2, 3-D] Pyrimidine Moiety As Antimicrobial Agents. *Research Journal of Life Sciences, Bioinformatics, Pharmaceutical and Chemical Sciences*, 5(31), 31–41. DOI: 10.26479/2019.0505.03

Patel, S. (2016)... *Journal of Materials Chemistry*, 4(21), 4856–4866.

Patel, S. A., Nilsson, M. B., Le, X., Cascone, T., Jain, R. K., & Heymach, J. V. (2023). Molecular mechanisms and future implications of VEGF/VEGFR in cancer therapy. *Clinical Cancer Research*, 29(1), 30–39. DOI: 10.1158/1078-0432.CCR-22-1366 PMID: 35969170

Pathania, S., Narang, R. K., & Rawal, R. K. (2019). Role of sulphur-heterocycles in medicinal chemistry: An update. *European Journal of Medicinal Chemistry*, 180, 486–508. DOI: 10.1016/j.ejmech.2019.07.043 PMID: 31330449

Peixoto, D., Begouin, A., & Queiroz, M. J. (2012). Synthesis of 2-(hetero)arylthieno[2,3-*b*] or [3, 2-*b*]pyridines from 2,3-dihalopyridines,(hetero)arylalkynes, and Na_2S Further functionalizations. *Tetrahedron*, 68(35), 7082–7094. DOI: 10.1016/j.tet.2012.06.057

Peng, Q., Peng, J.-B., Kang, E. T., Neoh, K. G., & Cao, Y. (2005). Synthesis and Electroluminescent Properties of Copolymers Based on Fluorene and 2,5-Di(2-hexyloxyphenyl)thiazolothiazole. *Macromolecules*, 38(17), 7292–7298. DOI: 10.1021/ma050378n

Peng, S. H., Huang, T. W., Gollavelli, G., & Hsu, C. S. (2017). Thiophene and diketopyrrolopyrrole based conjugated polymers as efficient alternatives to spiro-OMeTAD in perovskite solar cells as hole transporting layers. *Journal of Materials Chemistry. C, Materials for Optical and Electronic Devices*, 5(21), 5193–5198. DOI: 10.1039/C7TC00966F

Peng, X., Qin, F., Xu, M., Zhu, S., Pan, Y., Tang, H., Menga, X., & Wanga, H. (2019). Synthesis of imidazo[1,2-c] thiazoles through Pd-catalyzed bicyclization of tert-butyl isonitrile with thioamides. *Organic & Biomolecular Chemistry*, 17(36), 8403–8407. DOI: 10.1039/C9OB01664C PMID: 31482915

Pereira, J. A., Pessoa, A. M., Cordeiro, M. N. D. S., Fernandes, R., Prudêncio, C., Noronha, J. P., & Vieira, M. (2015). Quinoxaline, its derivatives and applications: A State of the Art review. *European Journal of Medicinal Chemistry*, 97, 664–672. DOI: 10.1016/j.ejmech.2014.06.058 PMID: 25011559

Pereira, N. L., Rihal, C. S., So, D. Y. F., Rosenberg, Y., Lennon, R. J., Mathew, V., Goodman, S. G., Weinshilboum, R. M., Wang, L., Baudhuin, L. M., Lerman, A., Hasan, A., Iturriaga, E., Fu, Y. P., Geller, N., Bailey, K., & Farkouh, M. E. (2019). Clopidogrel pharmacogenetics state-of-the-art review and the TAILOR-PCI study. *Circulation: Cardiovascular Interventions*, 12(4), 1–11. DOI: 10.1161/CIRCINTERVENTIONS.119.007811 PMID: 30998396

Phippen, C. B. W., & McErlean, C. S. P. (2011). A 1,5-benzothiazepine synthesis. *Tetrahedron Letters*, 52(13), 1490–1492. DOI: 10.1016/j.tetlet.2011.01.098

Pisano, C., Vlodavsky, I., Ilan, N., & Zunino, F. (2014). The potential of heparanase as a therapeutic target in cancer. *Biochemical Pharmacology*, 89(1), 12–19. DOI: 10.1016/j.bcp.2014.02.010 PMID: 24565907

Piwowarska, D., Kiedrzyńska, E., & Jaszczyszyn, K. (2024). A global perspective on the nature and fate of heavy metals polluting water ecosystems, and their impact and remediation. *Critical Reviews in Environmental Science and Technology*, 54(19), 1–23. DOI: 10.1080/10643389.2024.2317112

Porter, M. (1977). Vulcanization of Rubber. In Oae, S. (Ed.), *Organic Chemistry of Sulfur* (pp. 71–118). Springer US., DOI: 10.1007/978-1-4684-2049-4_3

Potts, K. T., & Kane, J. (1975). Bridgehead nitrogen heterocycles. IX. Fused-ring systems derived from fusion of the 1,2,4-thiadiazole system with the isoxazole, 1,3,4-oxadiazole, thiazole, 1,2,4-thiadiazole, and 1,3,4-thiadiazole systems. *The Journal of Organic Chemistry*, 40(18), 2600–2604. DOI: 10.1021/jo00906a006

Poudel, M., Kim, G., Bhattarai, P. Y., Shin, S., Zaraei, S. O., Oh, C. H., & Choi, H. S. (2011). Potent Imidazothiazole-based Inhibitor of BRAF V600E Overcomes Acquired Resistance via Inhibition of RAF Dimerization in PLX4032-resistant Melanoma. *Anticancer Research*, 42(6), 2911–2921. DOI: 10.21873/anticanres.15773 PMID: 35641256

Prabhakar, V., Babu Kondra, S., & Maddula, S. R. (2016). Synthesis, Structural Elucidation of Novel Thieno [2,3-d] Pyrimidine Core Unit Containing 1,2,4-Triazoles and Thiophenes as Potent Antimicrobial Activity. *Organic Chemistry: Current Research*, 05(03). Advance online publication. DOI: 10.4172/2161-0401.1000169

Pradhan, S., Muthuvel, P., & Das, T. M. (2023). Synthesis of a novel series of [1,5]-benzothiazepine-C-β-D-glycoside derivatives in a facile one-pot method and insight into their anti-oxidant properties. *Journal of Molecular Structure*, 1281, 135. DOI: 10.1016/j.molstruc.2023.135138

Pricopie, A. I., Foc an, M., Ionu , I., Marc, G., Vlase, L., Găină, L. I., Vodnar, D. C., Simon, E., Barta, G., Pîrnău, A., & Oniga, O. (2020). Novel 2,4-Disubstituted-1,3-Thiazole Derivatives: Synthesis, Anti-Candida Activity Evaluation and Interaction with Bovine Serum Albumine. *Molecules (Basel, Switzerland)*, 25s(5), 1079. DOI: 10.3390/molecules25051079 PMID: 32121062

Prisca Lagardere, Cyril Fersing, Nicolas Masurier, and V. L. (2022). *Thienopyrimidine: A Promising Sacffold to Access Anti-Infective Agents. 15.*

Priya, A. K., Muruganandam, M., Kumar, A., Senthilkumar, N., Shkir, M., Pandit, B., Imran, M., Prakash, C., & Ubaidullah, M. (2024). Recent advances in microbial-assisted degradation and remediation of xenobiotic contaminants; challenges and future prospects. *Journal of Water Process Engineering*, 60, 105106. DOI: 10.1016/j.jwpe.2024.105106

Purohit, D. H., Dodiya, B. L., Ghetiya, R. M., Vekariya, P. B., & Joshi, H. S. (2011). Synthesis and antimicrobial activity of some new 1,3,4-thiadiazoles and 1,3,4-thiadiazines containing 1,2,4-Triazolo nucleus. *Acta Chimica Slovenica*, 58(1), 53–59. PMID: 24061943

Qadir, T., Amin, A., Sharma, P. K., Jeelani, I., & Abe, H. (2022). A Review on Medicinally Important Heterocyclic Compounds. In The Open Medicinal Chemistry Journal (Vol. 16, Issue 1). DOI: 10.2174/18741045-v16-e2202280

Qarah, A. F., Alatawi, K., Alalawy, A. I., Alnoman, R. B., Alqahtani, A. M., Abualnaja, M. M., Alamoudi, W. M., & El-Metwaly, N. M. (2023). Synthesis of phenylthiourea-based pyrazole, thiazole and/or pyran compounds: Molecular modeling and biological activity. *Journal of Taibah University for Science : JTUSCI*, 17(1), 2245200. DOI: 10.1080/16583655.2023.2245200

Qin, J., Xie, P., Ventocilla, C., Zhou, G., Vultur, A., Chen, Q., Liu, Q., Herlyn, M., Winkler, J., & Marmorstein, R. (2012). Identification of a Novel Family of BRAFV600E Inhibitors. *Journal of Medicinal Chemistry*, 55(11), 5220–5230. DOI: 10.1021/jm3004416 PMID: 22537109

Qin, P., Kast, H., Nazeeruddin, M. K., Zakeeruddin, S. M., Mishra, A., Bäuerle, P., & Grätzel, M. (2014). Low band gap S, N-heteroacene-based oligothiophenes as hole-transporting and light absorbing materials for efficient perovskite-based solar cells. *Energy & Environmental Science*, 7(9), 2981–2985. DOI: 10.1039/C4EE01220H

Qi, T., Guo, Y., Liu, Y., Xi, H., Zhang, H., Gao, X., Liu, Y., Lu, K., Du, C., Yu, G., & Zhu, D. (2008). Synthesis and properties of the anti and syn isomers of dibenzothieno [b, d] pyrrole. *Chemical Communications*, (46), 6227–6229. DOI: 10.1039/b813683a PMID: 19082128

Qi, Y., Li, D., Zhang, S., Li, F., & Hua, T. (2024). Electrochemical filtration for drinking water purification: A review on membrane materials, mechanisms and roles. *Journal of Environmental Sciences (China)*, 141, 102–128. DOI: 10.1016/j.jes.2023.06.033 PMID: 38408813

Quelques, E. T. D. E. (1971). *Thi~nopyrimidine~ii. 27.*

Rabjohn, N. (2005). *Organic Synthesis Collective*. John Wiley and Sons.

Rachi, Y., Yamakado, R., & Okada, S. (2018). Synthesis of dyes with a sulfur-containing heterocyclic donor and a pyrroline-type acceptor. *Dyes and Pigments*, 159, 345–351. DOI: 10.1016/j.dyepig.2018.06.032

Rahaoui, A. A., Zerzouf, A., Salem, M., Essassi, E. M., Zellou, A., & Cherrah, Y. (2007).*Journal Marocain de Chimie Heterocyclique*, 6(1), 54.

Rahman, M. A., Salajegheh, A., Smith, R. A., & Lam, A. K.-Y. (2015). Multiple proliferation-survival signalling pathways are simultaneously active in BRAF V600E mutated thyroid carcinomas. *Experimental and Molecular Pathology*, 99(3), 492–497. DOI: 10.1016/j.yexmp.2015.09.006 PMID: 26403329

Rajendraprasad, Y., Ranikumar, P., Srinivasrao, A., & Rambabu, R. (2009).*Asian Journal of Chemistry*, 21(2), 907.

Rakhi, P. K., Mishra, K., Scaria, J., Suresh Kumar, M., & Nidheesh, P. V. (2024). Advanced oxidation processes. In *Clean Water: Next Generation Technologies* (pp. 107–116). Springer International Publishing. DOI: 10.1007/978-3-031-48228-1_8

Ramprasad J., Nayak N., Dalimba U., Yogeeswari P., Sriram D., *Med. Chem. Commun.* 2016, 7, 338 (2016).

Ram, V. J., Sethi, A., Nath, M., & Pratap, R. (2019). *The chemistry of heterocycles: Nomenclature and chemistry of three- to- five- membered heterocycles*. Elsevier.

Rashad, A. E., Shamroukh, A. H., Abdel-Megeid, R. E., Mostafa, A., El-Shesheny, R., Kandeil, A., Ali, M. A., & Banert, K. (2010). Synthesis and screening of some novel fused thiophene and thienopyrimidine derivatives for anti-avian influenza virus (H5N1) activity. *European Journal of Medicinal Chemistry*, 45(11), 5251–5257. DOI: 10.1016/j.ejmech.2010.08.044 PMID: 20828882

Rashdan, H. R. M., Abdelmonsef, A. H., Shehadi, I. A., Gomha, S. M., Soliman, A. M. M., & Mahmoud, H. K. (2020). Synthesis, Molecular Docking Screening and Anti-Proliferative Potency Evaluation of Some New Imidazo[2,1-*b*] Thiazole Linked Thiadiazole Conjugates. *Molecules (Basel, Switzerland)*, 25(21), 4997. DOI: 10.3390/molecules25214997 PMID: 33126630

Rasmussen, S. C. (2016). The nomenclature of fused-ring arenes and heterocycles: A guide to an increasingly important dialect of organic chemistry. *Chemtexts*, 2(4), 1–13. DOI: 10.1007/s40828-016-0035-3

Rassokhina, I. V., Tikhonova, T. A., Kobylskoy, S. G., Babkin, I. Y., Shirinian, V. Z., Gevorgyan, V., Zavarzin, I. V., & Volkova, Y. A. (2017). Synthesis of imidazo [2, 1-b] thiazoles via copper-catalyzed A3-coupling in batch and continuous flow. *The Journal of Organic Chemistry*, 82(18), 9682–9692. DOI: 10.1021/acs.joc.7b01762 PMID: 28799762

Ravichandran, S., & Karthikeyan, E. (2011). Microwave synthesis-a potential tool for green chemistry. *International Journal of Chemtech Research*, 3, 466–470.

Rawcliffe, R., Shkunov, M., Heeney, M., Tierney, S., McCulloch, I., & Campbell, A. (2008). Organic field-effect transistors of poly(2,5-bis(3- dodecylthiophen-2-yl) thieno [2,3-*b*]thiophene) deposited on five different silane self-assembled monolayers. *Chemical Communications (Cambridge)*, (7), 871–873. DOI: 10.1039/B715536K

Rayaroth, M. P., Boczkaj, G., Aubry, O., Aravind, U. K., & Aravindakumar, C. T. (2023). Advanced oxidation processes for degradation of water pollutants—ambivalent impact of carbonate species: A review. *Water (Basel)*, 15(8), 1615. DOI: 10.3390/w15081615

Reddy, M., & Karthik, P. (2018).. . *Journal of Applied Pharmaceutical Science*, 8(3), 19–24.

Redecker, M., Bradley, D. D. C., Inbasekaran, M., & Woo, E. P. (1999). Mobility enhancement through homogeneous nematic alignment of a liquid-crystalline polyfluorene. *Applied Physics Letters*, 74(10), 1400–1402. DOI: 10.1063/1.123563

Reginato, G., Mordini, A., Zani, L., Calamante, M., & Dessi, A. (2016). Photoactive Compounds Based on the Thiazolo[5,4- *d*]thiazole Core and Their Application in Organic and Hybrid Photovoltaics. *European Journal of Organic Chemistry*, 2016(2), 233–251. DOI: 10.1002/ejoc.201501237

Reineke, W., & Schlömann, M. (2023). Microbial degradation of pollutants. In *Environmental microbiology* (pp. 161–290). Springer Berlin Heidelberg. DOI: 10.1007/978-3-662-66547-3_6

Renuka N, Pavithra G & Ajay K, *Pharma Chemica*, (2014), 6(1), 482-485, 4.

Ribeiro, J. A., Sebastiao, A. M., & De Mendonça, A. (2002). Adenosine receptors in the nervous system: Pathophysiological implications. *Progress in Neurobiology*, 68(6), 377–392. DOI: 10.1016/S0301-0082(02)00155-7 PMID: 12576292

Riffat, A., Ahmad, N., Sajid Aqeel Ahmad, M., & Alvi, A. K. (2023). Sulfur: A macronutrient having potential to improve salinity tolerance in plants. *Pakistan Journal of Botany*, 55(4). Advance online publication. DOI: 10.30848/PJB2023-4(33)

Rizzo, C., Pace, A., Pibiri, I., Buscemi, S., & Palumbo Piccionello, A. (2024). From Conventional to Sustainable Catalytic Approaches for Heterocycles Synthesis. *ChemSusChem*, 17(8), e202301604. DOI: 10.1002/cssc.202301604 PMID: 38140917

Robert, J. F., Boukraa, S., Panouse, J. J., Loppinet, V., & Chaumont, J. P. (1990). Derivatives of imidazo[2,1-b]thiazoles X. Anthelmintic activity of imidazo[2,1-b]thiazoles. *European Journal of Medicinal Chemistry*, 25, 731–736. DOI: 10.1016/0223-5234(90)90192-6

Rodrigues, B. M., de Oliveira, D. F., de Queiroz Garcia, R., Chaves, O. A., Pizzi, G. F., Costa, L. A. S., de Boni, L., & Iglesias, B. A. (2024). The photophysical, photobiological, and DNA/HSA-binding properties of corroles containing carbazole and phenothiazine moieties. *International Journal of Biological Macromolecules*, 268, 131861. DOI: 10.1016/j.ijbiomac.2024.131861 PMID: 38670207

Rodrigues, I., Barcellos, A. M., Belladona, A. L., Roehrs, J. A., Cargnelutti, R., Alves, D., Perin, G., & Schumacher, R. F. (2018). Oxone®-mediated direct arylselenylation of imidazo [2, 1-b] thiazoles, imidazo [1, 2-a] pyridines and 1H-pyrazoles. *Tetrahedron*, 74(31), 4242–4246. DOI: 10.1016/j.tet.2018.06.046

Rombo, D. M., Ribeiro, J. A., & Sebastião, A. M. (2018). Role of adenosine receptors in epileptic seizures. In Borea, P., Varani, K., Gessi, S., Merighi, S., & Vincenzi, F. (Eds.), *The Adenosine Receptors* (pp. 309–350). Humana Press. DOI: 10.1007/978-3-319-90808-3_13

Romussi, A., Cappa, A., Vianello, P., Brambillasca, S., Cera, M. R., Dal Zuffo, R., Fagà, G., Fattori, R., Moretti, L., Trifirò, P., Villa, M., Vultaggio, S., Cecatiello, V., Pasqualato, S., Dondio, G., So, C. W. E., Minucci, S., Sartori, L., Varasi, M., & Mercurio, C. (2020). Discovery of reversible inhibitors of KDM1A efficacious in acute myeloid leukemia models. *ACS Medicinal Chemistry Letters*, 11(5), 754–759. DOI: 10.1021/acsmedchemlett.9b00604 PMID: 32435381

Ronald, K. RussellJeffery B. Press. (1988). Thiophene systems. 9. Thienopyrimidinedione derivatives as potential antihypertensive agents. *American Chemical Society, 31*(9).

Rosell, R., Carcereny, E., Gervais, R., Vergnenegre, A., Massuti, B., Felip, E., Palmero, R., Garcia-Gomez, R., Pallares, C., Sanchez, J. M., Porta, R., Cobo, M., Garrido, P., Longo, F., Moran, T., Insa, A., De Marinis, F., Corre, R., Bover, I., & Paz-Ares, L. (2012). Erlotinib versus standard chemotherapy as first-line treatment for European patients with advanced EGFR mutation-positive non-small-cell lung cancer (EURTAC): A multicentre, open-label, randomised phase 3 trial. *The Lancet. Oncology*, 13(3), 239–246. DOI: 10.1016/S1470-2045(11)70393-X PMID: 22285168

Rubab, L., Anum, A., Al-Hussain, S. A., Irfan, A., Ahmad, S., Ullah, S., Al-Mutairi, A. A., & Zaki, M. E. (2022). Green chemistry in organic synthesis: Recent update on green catalytic approaches in synthesis of 1, 2, 4-thiadiazoles. *Catalysts*, 12(11), 1329. DOI: 10.3390/catal12111329

Russell, R. K., Press, J. B., Rampulla, R. A., McNally, J. J., Falotico, R., Reiser, J. A., Bright, D. A., & Tobia, A. (1988). Thiophene Systems. 9. Thienopyrimidinedione Derivatives as Potential Antihypertensive Agents. *Journal of Medicinal Chemistry*, 31(9), 1786–1793. DOI: 10.1021/jm00117a019 PMID: 2842504

Ryu, H., Kim, H., Park, I., Lee, M., Park, Y. S., Jin, D. H., Hur, S. C., Park, J., & Lee, H. (2023). Discovery of novel, thienopyridine-based tyrosine kinase inhibitors targeting tumorigenic RON splice variants. *ACS Medicinal Chemistry Letters*, 14(9), 1198–1207. DOI: 10.1021/acsmedchemlett.3c00206 PMID: 37736180

Sabnis, R. W. (2023). Novel Thienopyrrole Compounds for Treating Autoimmune Diseases and Inflammatory Conditions. *ACS Medicinal Chemistry Letters*, 14(5), 549–550. DOI: 10.1021/acsmedchemlett.3c00131 PMID: 37197455

Sabnis, R. W., Rangnekar, D. W., & Sonawane, N. D. (1999). 2-aminothiophenes by the gewald reaction. *Journal of Heterocyclic Chemistry*, 36(2), 333–345. DOI: 10.1002/jhet.5570360203

Sadhu, C., & Mitra, A. K. (2024). Synthetic, biological and optoelectronic properties of phenoxazine and its derivatives: A state of the art review. *Molecular Diversity*, 28(2), 965–1007. DOI: 10.1007/s11030-023-10619-5 PMID: 36757655

Sagredou, S., Dalezis, P., Nikoleousakos, N., Nikolaou, M., Voura, M., Almpanakis, K., Panayiotidis, M. I., Sarli, V., & Trafalis, D. T. (2020). 3,6-Disubstituted 1,2,4-Triazolo[3,4-b]Thiadiazoles with Anticancer Activity Targeting Topoisomerase II Alpha. *OncoTargets and Therapy*, 13, 7369–7386. DOI: 10.2147/OTT.S254856 PMID: 32801761

SAINI, P., & KUMAR, S. (2023). Synthesis, Characterization, Molecular docking and Anti-anxiety Evaluation of Some Novel Phenothiazine Derivatives. *Oriental Journal of Chemistry*, 39(5).

Saini, R. K., Joshi, Y. P., & Joshi, P. (2008). Solvent-Free Synthesis of Some 1,5-Benzothiazepines and Benzodiazepines and Their Antibacterial Activity. *Phosphorus, Sulfur, and Silicon and the Related Elements*, 183(9), 2181–2190. DOI: 10.1080/10426500701852661

Salah, A. M. B., Fendri, L. B., Bataille, T., Herrera, R. P., & Naïli, H. (2018). Synthesis, structural determination and antimicrobial evaluation of two novel CoII and ZnII halogenometallates as efficient catalysts for the acetalization reaction of aldehydes. *Chemistry Central Journal*, 12(1), 24. DOI: 10.1186/s13065-018-0393-6 PMID: 29497955

Saleh, N., & Yousaf, Z. (2018). Tools and techniques for the optimized synthesis, reproducibility and scale up of desired nanoparticles from plant derived material and their role in pharmaceutical properties. In *Nanoscale Fabrication, Optimization, Scale-Up and Biological Aspects of Pharmaceutical Nanotechnology* (pp. 85–131). Elsevier. DOI: 10.1016/B978-0-12-813629-4.00003-6

Salib, S. B., Khalil, O. M., Kamel, M. M., & El-Dash, Y. (2016). Synthesis and Antitumor Activity of Novel Thienopyrimidine Derivatives Containing Thiosemicarbazide Moiety. *OAlib*, 03(07), 1–7. DOI: 10.4236/oalib.1102876

Salifu, M., John, M. A., Abubakar, M., Bankole, I. A., Ajayi, N. D., & Amusan, O. (2024). Phytoremediation Strategies for Heavy Metal Contamination: A Review on Sustainable Approach for Environmental Restoration. *Journal of Environmental Protection*, 15(4), 450–474. DOI: 10.4236/jep.2024.154026

Saliyeva, L. N., Diachenko, I. V., Vas'kevich, R. I., Slyvka, N. Y., & Vovk, M. V. (2020). Vas' kevich, R. I.; Slyvka, N. Y.; Vovk, M. V., Imidazothiazoles and their hydrogenated analogs: Methods of synthesis and biomedical potential. *Chemistry of Heterocyclic Compounds*, 56(11), 1394–1407. DOI: 10.1007/s10593-020-02827-w

Saliyeva, L., Holota, S., Grozav, A., Yakovychuk, N., Litvinchuk, M., Slyvka, N., & Vovk, M. (2021). Synthesis and Evaluation of Antimicrobial and Anti-inflammatory Activity of 6-aryliden- 2-methyl-2, 3-dihydroimidazo [2, 1-b] [1, 3] thiazoles. *Biointerface Research in Applied Chemistry*, 12(1), 292–303. DOI: 10.33263/BRIAC121.292303

Saliyeva, L., Holota, S., Grozav, A., Yakovychuk, N., Litvinchuk, M., Slyvka, N., & Vovk, M. (2021). Synthesis and Evaluation of Antimicrobial and Anti-inflammatory Activity of 6-aryliden-2-methyl-2, 3-dihydroimidazo [2, 1-b][1, 3] thiazoles. *Appl. Chem*, 12, 292–303.

Saliyeva, L., Vas'kevich, R., Slyvka, N. Y., & Vovk, M. (2018). The Synthesis and Structural Functionalization of 6-Substituted 2, 3-dihydroimidazo [2, 1-b][1, 3] thiazol-5-ones. *J. Org. Pharm. Chem*, 16(2(62)), 31–41. DOI: 10.24959/ophcj.18.940

Samala, G., Devi, P. B., Saxena, S., Meda, N., Yogeeswari, P., & Sriram, D. (2016). Design, synthesis and biological evaluation of imidazo [2, 1-b] thiazole and benzo [d] imidazo [2, 1-b] thiazole derivatives as Mycobacterium tuberculosis pantothenate synthetase inhibitors. *Bioorganic & Medicinal Chemistry*, 24(6), 1298–1307. DOI: 10.1016/j.bmc.2016.01.059 PMID: 26867485

Samsonenko, A. L., Kostyuchenko, A. S., Zheleznova, T. Y., Shuvalov, V. Y., Vlasov, I. S., & Fisyuk, A. S. (2022). Synthesis of new Fused 4H-Thieno [3, 2-b] pyrrole derivatives via decomposition of Methyl 4-Azido-5-arylthiophene-2-carboxylates. *Synthesis*, 54(14), 3227–3238. DOI: 10.1055/a-1799-9339

Sandeep, C. (2013). B. P., Rashmi S. Kulkarni, Synthesis of Isomeric Subtituted 6-acetyl-3-benzoylindolizine-1-carboxylate and 8-acetyl-3-benzoylindolizine-1-carboxylate from subtituteded 3-acetyl pyridinium bromides and their antimicrobial activity. *Journal of Applicable Chemistry*, 2(5), 1049.

Sandeep, C., Padmashali, B., Kulkarni, R. S., Mallikarjuna, S., Siddesh, M., Nagesh, H., & Thriveni, K. (2014). Synthesis of substituted 5-acetyl-3-benzoylindolizine-1-carboxylate from substituted 2-acetyl pyridinium bromides. *Heterocyclic Letters*, 4(03), 371–376.

Sandeep, C., Venugopala, K. N., Khedr, M. A., Padmashali, B., Kulkarni, R. S., Rashmi, V., & Odhav, B. (2017). Design and synthesis of novel indolizine analogues as COX-2 inhibitors: Computational perspective and in vitro screening. *Indian J. Pharm. Educ. Res*, 51, 452–460. DOI: 10.5530/ijper.51.3.73

Santhappan, J. S., Kalaiselvan, N., Assis, S. M., Amjith, L. R., Glivin, G., & Mathimani, T. (2024). Origin, types, and contribution of emerging pollutants to environmental degradation and their remediation by physical and chemical techniques. *Environmental Research*, 257, 119369. DOI: 10.1016/j.envres.2024.119369 PMID: 38848998

Sarhan, A. E., Sediek, A. A., Khalifa, N. M., & Hasan, E. E. (2022)... *Heterocycles*, 104(3), 447. DOI: 10.3987/COM-21-14578

Sarhan, M. O., Haffez, H., Elsayed, N. A., El-Haggar, R. S., & Zaghary, W. A. (2023). New phenothiazine conjugates as apoptosis inducing agents: Design, synthesis, In-vitro anti-cancer screening and 131I-radiolabeling for in-vivo evaluation. *Bioorganic Chemistry*, 141, 106924. DOI: 10.1016/j.bioorg.2023.106924 PMID: 37871390

Sarkar, R., Singh, S., & Dey, S. (2018)... *Bioorganic & Medicinal Chemistry Letters*, 28(16), 2613–2618.

Saundane, A. R., Manjunatha, Y., & Walmik, P. (2009). Synthesis and antimicrobial activities of 5-(2′, 5′-disubstituted 1H- indol-3′-yl)-5h-thiazolo[4,3-b]1,3,4-oxadiazole,-1,3,4- thiadiazole-1,2,4-triazole and their derivatives. *Heterocyclic Communications*, 15(4), 303. DOI: 10.1515/HC.2009.15.4.303

Sawant, R. L. (2014). Synthesis, QSAR and docking studies of 5HT2A receptor antagonizing thiazolo[3,2-a]pyrimidines as anti-psychotic agents. *Marmara Pharmaceutical Journal*, 3(18), 109–109. DOI: 10.12991/mpj.2014187237

Sayed, M. T. M. (2023). *CHEMISTRY Recent updates on thienopyrimidine derivatives as anticancer agents*. 659–681. DOI: 10.1007/s00044-023-03040-y

Sayresmith, N. A., Saminathan, A., Sailer, J. K., Patberg, S. M., Sandor, K., Krishnan, Y., & Walter, M. G. (2019). Photostable Voltage-Sensitive Dyes Based on Simple, Solvatofluorochromic, Asymmetric Thiazolothiazoles. *Journal of the American Chemical Society*, 141(47), 18780–18790. DOI: 10.1021/jacs.9b08959 PMID: 31660737

Schafer, G., Ahmetovic, M., & Abele, S. (2017). Scalable synthesis of trifluoromethylated imidazo-fused N-heterocycles using TFAA and trifluoroacetamide as CF3-reagents. *Organic Letters*, 19(24), 6578–6581. DOI: 10.1021/acs.orglett.7b03291 PMID: 29161044

Schechter, L. E., Lin, Q., Smith, D. L., Zhang, G., Shan, Q., Platt, B., Brandt, M. R., Dawson, L. A., Cole, D., Bernotas, R., Robichaud, A., Rosenzweig-Lipson, S., & Beyer, C. E. (2008). Neuropharmacological profile of novel and selective 5-HT6 receptor agonists: WAY-181187 and WAY-208466. *Neuropsychopharmacology*, 33(6), 1323–1335. DOI: 10.1038/sj.npp.1301503 PMID: 17625499

Scribner, A., Meitz, S., Fisher, M., Wyvratt, M., Leavitt, P., Liberator, P., Gurnett, A. C., Brown, C., Mathew, J., Thompson, D., Schmatz, D., & Biftu, T. (2008). Synthesis and biological activity of anticoccidial agents: 5,6-Diarylimidazo[2,1-b][1,3] thiazoles. *Bioorganic & Medicinal Chemistry Letters*, 18(19), 5263–5267. DOI: 10.1016/j.bmcl.2008.08.063 PMID: 18789686

Seelolla, G., & Ponneri, V. (2016). Synthesis, Antimicrobial, and Antioxidant Activities of Some Fused Heterocyclic [1,2,4]Triazolo[3,4-b][1,3,4]thiadiazole Derivatives. *Journal of Heterocyclic Chemistry*, 53(3), 929–936. DOI: 10.1002/jhet.2348

Sekhar, B. C. (2014).*Acta Chimica Slovenica*, 61(4), 651. PMID: 25551705

Sekhar, T., Thriveni, P., Venkateswarlu, A., Daveedu, T., Peddanna, K., & Sainath, S. B. (2020). One-pot synthesis of thiazolo[3,2-a]pyrimidine derivatives, their cytotoxic evaluation and molecular docking studies. *Spectrochimica Acta. Part A: Molecular and Biomolecular Spectroscopy*, 231, 118056. DOI: 10.1016/j.saa.2020.118056 PMID: 32006911

Selvam, T. P., Karthik, V., Kumar, P. V., & Ali, M. A. (2012). Design, synthesis, antinociceptive, and anti-inflammatory properties of thiazolopyrimidine derivatives. *Toxicological and Environmental Chemistry*, 94(7), 1247–1258. DOI: 10.1080/02772248.2012.703204

Şenkardeş, S., Kulabaş, N., Bi̇ngöl Özakpinar, Ö., Kalayci, S., Şahi̇n, F., Küçükgüzel, İ., & Küçükgüzel, Ş. G. (2020). Synthesis and Anticancer and Antimicrobial Evaluation of Novel Ether-linked Derivatives of Ornidazole. *Turkish Journal of Pharmaceutical Sciences*, 17(1), 81–93. DOI: 10.4274/tjps.galenos.2018.59389 PMID: 32454765

Serafini, M., Torre, E., Aprile, S., Massarotti, A., Fallarini, S., & Pirali, T. (2019). Synthesis, Docking and Biological Evaluation of a Novel Class of Imidazothiazoles as IDO1 Inhibitors. *Molecules (Basel, Switzerland)*, 24(10), 1874. DOI: 10.3390/molecules24101874 PMID: 31096672

Serykh, V. Y., Kaliev, A. R., Ushakov, I. A., Borodina, T. N., Smirnov, V. I., & Rozentsveiga, I. B. (2018). Regioselective reaction of imidazole-2-thiols with N-sulfonylphenyldichloroacetaldimines: En route to novel sulfonylamino-substituted imidazo[2,1-b] thiazoles and thiazolo[3,2-a] benzimidazoles. *ARKIVOC*, 3, 62–75.

Settypalli, T., Chunduri, V. R., Kerru, N., Nallapaneni, H. K., Chintha, V. R., Daggupati, T., Yeguvapalli, S., & Wudayagiri, R. (2019). Design, Synthesis, Neuroprotective and Antibacterial Activities of 1,2,4-Triazolo[3,4-b]1,3,4-thiadiazole Linked Thieno[2,3-d]pyrimidine Derivatives and In Silico Docking Studies. *ChemistrySelect*, 4(5), 1627–1634. DOI: 10.1002/slct.201803917

Severina, H. I., Skupa, O. O., Voloshchuk, N. I., Suleiman, M. M., & Georgiyants, V. A. (2019). Synthesis and anticonvulsant activity of 6-methyl-2-(2-oxo-2-arylethyl) thio) pyrimidin-4(3H)-one derivatives and products of their cyclization. *Farmatsiia (Sofia)*, 66(3), 141–146. DOI: 10.3897/pharmacia.66.e38137

Shahin, A. I., Zaib, S., Zaraei, S.-O., Kedia, R. A., Anbar, H. S., Younas, M. T., Al-Tel, T. H., Khoder, G., & El-Gamal, M. I. (2023). Design and synthesis of novel anti-urease imidazothiazole derivatives with promising antibacterial activity against Helicobacter pylori. *PLoS One*, 18(6), e0286684. DOI: 10.1371/journal.pone.0286684 PMID: 37267378

Shahrasbi, M., AzamiMovahed, M., Ghorban Dadras, O., Daraei, B., Zarghi, A. (2018). Design, Synthesis and Biological Evaluation of New Imidazo[2,1-b] Thiazole Derivatives as Selective COX-2 Inhibitors. *Iranian Journal of Pharmaceutical Research*, 17, 1288–1296. PMID: 30568687

Shahrasbi, M., Movahed, M. A., Dadras, O. G., Daraei, B., & Zarghi, A. (2018). Design, synthesis and biological evaluation of new Imidazo [2, 1-B] thiazole derivatives as selective COX 2 inhibitors. *Iranian Journal of Pharmaceutical Research: IJPR*, 17(4), 1288. PMID: 30568687

Shaik, A. B., Bhandare, R. R., Nissankararao, S., Subrahmanya, L. B. V., Shaik, S., & Rahman, M. (2021). Synthesis, and biological screening of chloropyrazine conjugated benzothiazepine derivatives as potential antimicrobial, antitubercular and cytotoxic agents. *Arabian Journal of Chemistry*, 14(2), 102915. DOI: 10.1016/j.arabjc.2020.102915

Shaik, S. P., Vishnuvardhan, M., Sultana, F., Rao, A. S., Bagul, C., Bhattacharjee, D., Kapure, J. S., Jain, N., & Kamal, A. (2017). Design and synthesis of 1, 2, 3-triazolo linked benzo [d] imidazo [2, 1-b] thiazole conjugates as tubulin polymerization inhibitors. *Bioorganic & Medicinal Chemistry*, 25(13), 3285–3297. DOI: 10.1016/j.bmc.2017.04.013 PMID: 28462842

Shaik, T. B., Malik, M. S., Routhu, S. R., Seddigi, Z. S., Althagafi, I. I., & Kamal, A. (2020). Evaluation of Anticancer and Anti-Mitotic Properties of Quinazoline and Quinazolino-Benzothiadiazine Derivatives. *Anti-cancer Agents in Medicinal Chemistry*, 20(5), 599–611. DOI: 10.2174/1871520620666191224122204 PMID: 31884931

Shan, J., & Sun, H. (2013). The discovery and development of prasugrel. *Expert Opinion on Drug Discovery*, 8(7), 897–905. DOI: 10.1517/17460441.2013.793668 PMID: 23614463

Shankerrao, S., Bodke, Y. D., & Santoshkumar, S. (2017). Synthesis and antimicrobial activity of some imidazothiazole derivatives of benzofuran. *Arabian Journal of Chemistry*, 10, S554–S558. DOI: 10.1016/j.arabjc.2012.10.018

Shareef, M. A., Devi, G. P., Rani Routhu, S., Kumar, C. G., Kamal, A., & Babu, B. N. (2020). New imidazo[2,1-*b*] thiazole-based aryl hydrazones: Unravelling their synthesis and antiproliferative and apoptosis-inducing potential. *RSC Medicinal Chemistry*, 11(10), 1178–1184. DOI: 10.1039/D0MD00188K PMID: 33479622

Shareef, M. A., Sirisha, K., Sayeed, I. B., Khan, I., Ganapathi, T., Akbar, S., Kumar, C. G., Kamal, A., & Babu, B. N. (2019). Synthesis of new triazole fused imidazo[2,1-b] thiazole hybrids with emphasis on Staphylococcus aureus virulence factors. *Bioorganic & Medicinal Chemistry Letters*, 29(19), 126621. DOI: 10.1016/j.bmcl.2019.08.025 PMID: 31431360

Sharma, S., Shaikh, S., Mohana, S., Desai, C., & Madamwar, D. (2023). Current trends in bioremediation and bio-integrated treatment of petroleum hydrocarbons. *Environmental Science and Pollution Research International*, 31(46), 1–20. DOI: 10.1007/s11356-023-30479-8 PMID: 37861831

Shaukat Tamboli, Riyaj; Devidas Amrutkar, Rakesh; Sanchlal Jain, Kishor; Kumaradoss Kathiravan, M. (2013). *Synthesis and in vivo Antihyperlipidemic Potential of Novel Substituted Thieno[3,2-d]pyrimidines. 10.*

Shaw, R., Tewari, R., Yadav, M., Pandey, E., Tripathi, K., Rani, J., Althagafi, I., & Pratap, R. (2024). Recent advancements in the synthesis of fused thienopyridines and their therapeutic applications. *European Journal of Medicinal Chemistry Reports*, 12, 100185. DOI: 10.1016/j.ejmcr.2024.100185

Shehata, M. K., Uzair, M., Zaraei, S. O., Shahin, A. I., Shah, S. J. A., Ullah, S., Iqbal, J., & El–Gamal, M. I. (2023). Synthesis, biological evaluation, and molecular modeling studies of a new series of imidazothiazole or imidazooxazole derivatives as inhibitors of ectonucleoside triphosphate diphosphohydrolases (NTPDases). *Medicinal Chemistry Research*, 32(2), 314–325. DOI: 10.1007/s00044-022-03000-y

Sheth, S., Brito, R., Mukherjea, D., Rybak, L. P., & Ramkumar, V. (2014). Adenosine receptors: Expression, function and regulation. *International Journal of Molecular Sciences*, 15(2), 2024–2052. DOI: 10.3390/ijms15022024 PMID: 24477263

Shetty, N. S. (2014). A Facile Microwave-Assisted Synthesis of Novel Thienopyrimidines and Triazolothienopyrimidines and their Antimicrobial Activities. *International Journal of Advances in Chemical Engineering and Biological Sciences*, 1(1). Advance online publication. DOI: 10.15242/IJACEBS.C0114037

Shetty, N. S., Khazi, I. A. M., & Ahn, C.-J. (2010). Synthesis, anthelmintic and anti-inflammatory activities of some novel imidazothiazole sulfides and sulfones. *Bulletin of the Korean Chemical Society*, 31(8), 2337–2340. DOI: 10.5012/bkcs.2010.31.8.2337

Shi, B., Zhu, Z., Zhu, Y., Zhou, D., Wang, J., Zhou, P., & Jing, H. (2016). Approach for 2-(arylthio)imidazoles and imidazo[2,1-b]thiazoles from imidazo[2,1-b][1,3,4]thiadiazoles by ringopening and -reconstruction. *Organic & Biomolecular Chemistry*, 14(10), 2978–2984. DOI: 10.1039/C6OB00102E PMID: 26881829

Shi, P., Tu, Y., Wang, C., Kong, D., Ma, D., & Bolm, C. (2020). Synthesis of Benzothiadiazine-1-oxides by Rhodium-Catalyzed C−H Amidation/Cyclization. *Organic Letters*, 22(22), 8842–8845. DOI: 10.1021/acs.orglett.0c03212 PMID: 33170727

Shirinian, V. Z., Nikal, D. M., & Krayushk, M. M. (2003). Synthesis of 4-aryl-2, 6, 6-trimethyl-5-oxo-5, 6-dihydro-4 H-thieno [3, 2-b] pyrroles. *Russian Chemical Bulletin*, 52(8), 1873–1876. DOI: 10.1023/A:1026041728967

Shivhare, K. N., Jaiswal, M. K., Srivastava, A., Tiwaria, S. K., Siddiqui, I. R. New (2018). light-activated C–C and C–N bond formation in the synthesis of imidazo [1,2- a] pyridines and imidazo [2,1-b] thiazoles under catalyst and solvent-free conditions. *New Journal of Chemistry, 42*, 16591.

Shreen, S., Uddin, M. B., Ali, M. S., & Sultana, Z. (2021). The role of antiplatelet agents in ischemic events. *Journal of Drug Delivery and Therapeutics*, 11(4-S), 187–194. DOI: 10.22270/jddt.v11i4-S.4926

Shuai, Y., Haoyi, T., Chen, Z., & Hanhong, X. (2024). Design, synthesis and evaluation of insecticidal activity of phenylthiazole derivative containing primary amide and cyanogroup. *Journal of South China Agricultural University*, 45(3), 354–363. DOI: 10.7671/j.issn.1001-411X.202308001

Shushanik, Sh. Dashyan, Eugene v. babaev, ervand g. paronikyan, Armen G. Ayvazyan, R. G. P. and L. S. H. (2022). Evaluation of Neurotropic Activity and Molecular Docking study of New Derivatives of pyrano[4",3":4′,5′]pyrido[3′,2′:4,5]thieno[3,2-d]pyrimidines on the basis of pyrano[3,4-c]pyridines. *Molecules (Basel, Switzerland).* PMID: 35684318

Shyyka, O., Pokhodylo, N., Finiuk, N., Matiychuk, V., Stoika, R., & Obushak, M. (2018). Anticancer activity evaluation of new thieno[2,3-d]pyrimidin-4(3h)-ones and thieno[3,2-d]pyrimidin-4(3h)-one derivatives. *Scientia Pharmaceutica*, 86(3), 28. Advance online publication. DOI: 10.3390/scipharm86030028 PMID: 30012942

Siddesh, M., Padmashali, B., Thriveni, K., & Sandeep, C. (2014). Synthesis of thiophene-linked pyrimidopyrimidines as pharmaceutical leads. *Journal of Chemical Sciences*, 126(3), 821–826. DOI: 10.1007/s12039-014-0614-z

Siddiqui, R., El-Gamal, M. I., Boghossian, A., Saeed, B. Q., Oh, C.-H., Abdel-Maksoud, M. S., Alharbi, A. M., Alfahemi, H., & Khan, N. A. (2022). Imidazothiazole derivatives exhibited potent effects against brain-eating amoebae. *Antibiotics (Basel, Switzerland)*, 11(11), 1515. DOI: 10.3390/antibiotics11111515 PMID: 36358170

Sigismund, S., Avanzato, D., & Lanzetti, L. (2018). Emerging functions of the EGFR in cancer. *Molecular Oncology*, 12(1), 3–20. DOI: 10.1002/1878-0261.12155 PMID: 29124875

Silva, R. N., Monteiro, V. N., Steindorff, A. S., Gomes, E. V., Noronha, E. F., & Ulhoa, C. J. (2019). Trichoderma/pathogen/plant interaction in pre-harvest food security. *Fungal Biology*, 123(8), 565–583. DOI: 10.1016/j.funbio.2019.06.010 PMID: 31345411

Singh, B., & Kaunert, C. (2025). Leveraging IoT for Patient Monitoring and Smart Healthcare: Connected Healthcare System. In *Revolutionizing Healthcare Systems Through Cloud Computing and IoT* (pp. 27-46). IGI Global.

Singh, G., Yadav, P., Singh, S. K., Ranjan, N., Verma, H., Ferreira, L. F. R., & Kumar, A. (2024). Membrane processes for wastewater treatment. In *Advances in Chemical Pollution, Environmental Management and Protection* (Vol. 10, pp. 35-56). Elsevier. DOI: 10.1016/bs.apmp.2023.08.001

Singh, B., & Kaunert, C. (2024). Adventure in High Altitude of Mountainous Topographies and Health Impacts: Lensing Tourism Sustainability via Reducing Ecological and Sociocultural Footprint and Health Emergency and Medical Assistance Management. In Meraj, G., Hashimoto, S., & Kumar, P. (Eds.), *Navigating Natural Hazards in Mountainous Topographies. Disaster Risk Reduction*. Springer., DOI: 10.1007/978-3-031-65862-4_15

Singh, B., & Kaunert, C. (2025). Cloud Computing and IoMT in Disease Screening and Diagnosis: AI Approaches in Transmuting Healthier Homes. In *Revolutionizing Healthcare Systems Through Cloud Computing and IoT* (pp. 99–120). IGI Global.

Singh, B., & Kaunert, C. (2025). Featuring Healthcare, Environment, and Human Rights: Applications of Artificial Intelligence in the Health Domain. In Chakraborty, S., & Satapathy, S. (Eds.), *Gender, Environment, and Human Rights: An Intersectional Exploration* (pp. 45–74). IGI Global., DOI: 10.4018/979-8-3693-6069-9.ch004

Singh, B., & Kaunert, C. (2025). Innovative Business Models in Digital and Virtual Reality for the Gig Economy: Space-Age Sustainable Development Projecting Smart Business Practices. In Taherdoost, H., Drazenovic, G., Madanchian, M., Khan, I., & Arshi, O. (Eds.), *Business Transformation in the Era of Digital Disruption* (pp. 127–156). IGI Global., DOI: 10.4018/979-8-3693-7056-8.ch005

Singh, B., & Kaunert, C. (2025). Intelligent Machine Learning Solutions for Cybersecurity: Legal and Ethical Considerations in a Global Context. In Thangam, D. (Ed.), *Advancements in Intelligent Process Automation* (pp. 359–386). IGI Global., DOI: 10.4018/979-8-3693-5380-6.ch014

Singh, B., Kaunert, C., & Ravesangar, K. (2025). Artificial Intelligence and Employee Well-Being: Projecting Psychological and Physical Healthcare. In Yadav, P. (Ed.), *Human Resource Strategies in the Era of Artificial Intelligence* (pp. 109–134). IGI Global., DOI: 10.4018/979-8-3693-6412-3.ch005

Singh, B., Kaunert, C., & Singh, G. (2025). Scaling Legal Framework for Plastic Pollution and Advancing Cutting Edge Water Governance: Reducing and Eliminating Marine Pollution in Alignment With SDG 14 (Life Below Water). In Gaur, N., Sharma, E., Nguyen, T., Bilal, M., & Melkania, N. (Eds.), *Societal and Environmental Ramifications of Plastic Pollution* (pp. 197–222). IGI Global., DOI: 10.4018/979-8-3693-9163-1.ch010

Singh, M., Singh, S. K., & Chhabria, M. T. (2012). A novel QSAR model for evaluating and predicting the inhibition activity of H1-receptor antagonists: A series of thienopyrimidine derivatives. *Journal of Drug Delivery and Therapeutics*, 2(1).

Sinitsa, A. D., & Onys, P. P. (2016). *Heterocyclization of N - (chlorosulfonyl) imidoyl chlorides with anilines, a new method of synthesis of 1, 2, 4-benzothiadiazine 1, 1-dioxides. 52*(4), 267–274. DOI: 10.1007/s10593-016-1873-z

Sipyagin, K. A. M. (1996).*Geterotsikl. Soedin.*, 3, 364.

Slassi, S., Zaki, H., Amine, A., Yamni, K., & Bouachrine, M. (2020). Quantum Chemical and Molecular Docking Studies of Imidazole and Its Derivatives as the Active Antifungal Components against C. Albicans. *Physical Chemistry Research.*, 8, 457–469. DOI: 10.22036/PCR.2020.211251.1706

Sonawane, K., Said, R., & Hatvate, N. (2024). Synthesis, Properties, and Biological Applications of 1,3,4-Thiadiazoles. In S-Heterocycles (pp. 163–223). Royal Society of Chemistry. DOI: 10.1039/9781837674015-00163

Soyer Can, O., Ünlü, S., Ocak, H., Çöldür, E. Ç., Sipahi, H., & Bilgin Eran, B. (2022). Synthesis of new imidazothiazole derivatives and investigation of their anti-inflammatory and analgesic activities. *Journal of the Iranian Chemical Society*, 1-9.

Srebrenkoska, V., Bogoeva-Gaceva, G., & Dimeski, D. (2013). Bio-based composites produced by conventional techniques using recycled polymer matrix.

Srinath, R., Venkataranganna, M., & Saravanan, J. (2011). Anti-hyperlipidemic Activity of Thieno - [2, 3-d] - Pyrimidin-4-(3H)-ones. *Macedonian Journal of Medical Sciences*, 4(4), 358–366. DOI: 10.3889/MJMS.1857-5773.2011.0189

Srivastav, A. L., Patel, N., Rani, L., Kumar, P., Dutt, I., Maddodi, B. S., & Chaudhary, V. K. (2024). Sustainable options for fertilizer management in agriculture to prevent water contamination: A review. *Environment, Development and Sustainability*, 26(4), 8303–8327. DOI: 10.1007/s10668-023-03117-z

Srivastav, V. K., & Tiwari, M. (2021). Synthesis and Evaluation of Antihyperglycemic Activity of 1, 4-Benzothiazepine-2-One Derivatives in Alloxan Induced Diabetic Rats. *Indian Journal of Pharmaceutical Sciences*, 83(5), 1033. DOI: 10.36468/pharmaceutical-sciences.857

Sroor, F. M., El-Sayed, A. F., & Abdelraof, M. (2024). Design, synthesis, structure elucidation, antimicrobial, molecular docking, and SAR studies of novel urea derivatives bearing tricyclic aromatic hydrocarbon rings. *Archiv der Pharmazie*, 357(6), 2300738. DOI: 10.1002/ardp.202300738 PMID: 38466125

Steinkopf, W., & Lutzkendorf, G. (1913).*Chem. Ztg.*, 36, 379.

Stockmann, V., & Fiksdahl, A. (2008). Preparation of new pyrido [3, 4-b] thienopyrroles and pyrido [4, 3-e] thienopyridazines. *Tetrahedron*, 64(32), 7626–7632. DOI: 10.1016/j.tet.2008.05.062

Stuerga, D., & Gaillard, P. (1996). Microwave heating as a new way to induce localized enhancements of reaction rate. Non-isothermal and heterogeneous kinetics. *Tetrahedron*, 52(15), 5505–5510. DOI: 10.1016/0040-4020(96)00241-4

Sultana, F., Reddy Bonam, S., Reddy, V. G., Nayak, V. L., Akunuri, R., Rani Routhu, S., Alarifi, A., Halmuthur, M. S. K., & Kamal, A. (2018). Synthesis of benzo[d] imidazo[2,1-b] thiazole-chalcone conjugates as microtubule targeting and apoptosis inducing agents. *Bioorganic Chemistry*, 76, 1–12. DOI: 10.1016/j.bioorg.2017.10.019 PMID: 29102724

Sumanth, G., Lakshmikanth, K., Saini, S. M., Mundhe, P., Shivaprasad, K., & Chandrashekharappa, S. (2023). Phenyl pyrrolo [1, 2-a] quinolines-finding of a key by-product during quinolinium salt preparation. *Journal of Molecular Structure*, 1273, 134350. DOI: 10.1016/j.molstruc.2022.134350

Sundaram, R. M., Chatterjee, S., Oliva, R., Laha, G. S., Cruz, C. V., Leach, J. E., & Sonti, R. V. (2014). Update on Bacterial Blight of Rice: Fourth International Conference on Bacterial Blight. *Rice, 7*(1), 12, s12284-014-0012–0017. DOI: 10.1186/s12284-014-0012-7

Suresh Kumar, G. V., Rajendraprasad, Y., Mallikarjuna, B. P., Chandrashekar, S. M., & Kistayya, C. (2010). Synthesis of some novel 2-substituted-5-[isopropylthiazole] clubbed 1,2,4-triazole and 1,3,4-oxadiazoles as potential antimicrobial and antitubercular agents. *European Journal of Medicinal Chemistry*, 45(5), 2063–2074. DOI: 10.1016/j.ejmech.2010.01.045 PMID: 20149496

Szasz, M., Casoni, D., & Cimpoiu, C. (2023). Study on binding affinity profile, bioactivity, and lipophilicity of selected antipsychotic drugs. *Journal of Liquid Chromatography & Related Technologies*, •••, 1–6. DOI: 10.1080/10826076.2023.2216780

Szekely, G. (2024). The 12 principles of green membrane materials and processes for realizing the United Nations' sustainable development goals. *RSC Sustainability*, 2(4), 871–880. DOI: 10.1039/D4SU00027G

Szkudlinski, M. W. (2007).. . *Challenges and Opportunities of Trapping Ligands*, 231–234. Advance online publication. DOI: 10.1124/mol.107.038208.Please

Tabassum, G., Kumar, A., Khan, F. N., Verma, A. K., Naz, S., Ali, S. M., & Dev, K. (2024). Phytochemicals as Potent Therapeutic Molecules in Cancer Treatment in Relation to Sustainable Agriculture System. In *Recent Advancements in Sustainable Agricultural Practices: Harnessing Technology for Water Resources, Irrigation and Environmental Management* (pp. 283–312). Springer Nature Singapore. DOI: 10.1007/978-981-97-2155-9_15

Tamilavan, V., Kim, S., Sung, J., Lee, D. Y., Cho, S., Jin, Y., Jeong, J., Park, S. H., & Hyun, M. H. (2017). Enhanced photovoltaic performances of bis(pyrrolo [3,4-*c*]pyrrole-1,3-dione)-based wide band gap polymer via the incorporation of an appropriate spacer unit between pyrrolo [3,4-*c*]pyrrole-1,3-dione units. *Organic Electronics*, 42, 34–41. DOI: 10.1016/j.orgel.2016.12.012

Tan, B., Kartal, Y., Yesilyurt, F., Akdoğan, N., Doyduk, D., & Dişli, A. (2024). Synthesis of new phenothiazine derivatives: Molecular docking, assessment of cytotoxic activity and oxidant–antioxidant properties on PCS-201-012, HT-29, and SH-SY5Y cell lines. *Archiv der Pharmazie*, 357(10), 2400281. DOI: 10.1002/ardp.202400281 PMID: 39058899

Taneja, S., Karaca, O., & Haritash, A. K. (2024). Electrokinetic remediation: Past experiences and future roadmap for sustainable remediation of metal-contaminated soils. *Journal of Geochemical Exploration*, 259, 107437. DOI: 10.1016/j.gexplo.2024.107437

Tedesco, R., Shaw, A. N., Bambal, R., Chai, D., Concha, N. O., Darcy, M. G., Dhanak, D., Fitch, D. M., Gates, A., Gerhardt, W. G., Halegoua, D. L., Han, C., Hofmann, G. A., Johnston, V. K., Kaura, A. C., Liu, N., Keenan, R. M., Lin-Goerke, J., Sarisky, R. T., & Duffy, K. J. (2006). 3-(1,1-dioxo-2H-(1,2,4)-benzothiadiazin-3-yl)-4-hydroxy-2(1H)-quinolinones, potent inhibitors of hepatitis C virus RNA-dependent RNA polymerase. *Journal of Medicinal Chemistry*, 49(3), 971–983. DOI: 10.1021/jm050855s PMID: 16451063

Testa, L., Biondi Zoccai, G. G., Valgimigli, M., Latini, R. A., Pizzocri, S., Lanotte, S., Laudisa, M. L., Brambilla, N., Ward, M. R., Figtree, G. A., Bedogni, F., & Bhindi, R. (2010). Current concepts on antiplatelet therapy: Focus on the novel thienopyridine and nonthienopyridine agents. *Advances in Hematology*, 2010(1), 595934. DOI: 10.1155/2010/595934 PMID: 21151515

Thakur, S., Arora, S., Katiyar, M. K., Joshi, G., & Kumar, R. (2024). Fused Thienopyrimidines as Versatile Pharmacophores for the Development of Cyclooxygenase-2 Inhibitors. *ChemistrySelect*, 9(22), e202400514. Advance online publication. DOI: 10.1002/slct.202400514

Thiazoles | DrugBank Online. (n.d.). Retrieved July 28, 2024, from https://go.drugbank.com/categories/DBCAT000348

THOMSON, A. E., NICKERSON, M., GASKELL, P., & GRAHAME, G. R. (1962). Clinical observations on an antihypertensive chlorothiazide analogue devoid of diuretic activity. *Canadian Medical Association Journal*, 87, 1306–1310. PMID: 13981225

Thore, S. N., Gupta, S. V., & Baheti, K. G. (2013). Docking, synthesis, and pharmacological investigation of novel substituted thiazole derivatives as non-carboxylic, anti-inflammatory, and analgesic agents. *Medicinal Chemistry Research*, 22(8), 3802–3811. DOI: 10.1007/s00044-012-0382-6

Toda, F. (1995). Solid state organic chemistry: Efficient reactions, remarkable yields, and stereoselectivity. *Accounts of Chemical Research*, 28(12), 480–486. DOI: 10.1021/ar00060a003

Tojo, S., Kohno, T., Tanaka, T., Kamioka, S., Ota, Y., Ishii, T., Kamimoto, K., Asano, S., & Isobe, Y. (2014). Crystal Structures and Structure-Activity Relationships of Imidazothiazole Derivatives as IDO1 Inhibitors. *ACS Medicinal Chemistry Letters*, 5(10), 1119–1123. DOI: 10.1021/ml500247w PMID: 25313323

Tolba, M. S., Kamal, A. M., Ahmed, E. M., & Hassanien, R. (2018). *Synthesis, reactions, and biological study of some new thienopyrimidine derivatives as antimicrobial and anti- inflammatory agents. July*, 1–10. DOI: 10.1002/jccs.201800292

Tolba, M. S., Sayed, A. M., Sayed, M., & Ahmed, M. (2021). Design, synthesis, biological evaluation, and molecular docking of some new Thieno[2,3-d] pyrimidine derivatives. *Journal of Molecular Structure*, 1246, 131179. DOI: 10.1016/j.molstruc.2021.131179

Tolba, M. S., Sayed, M., El-Dean, A. M. K., Hassanien, R., Abdel-Raheem, S. A. A., & Ahmed, M. (2021). Design, synthesis and antimicrobial screening of some new thienopyrimidines. *Organic Communications*, 14(4), 334–345. DOI: 10.25135/acg.oc.114.2109.2214

Tomašič, T., Nabergoj, D., Vrbek, S., Zidar, N., Jakopin, Ž., Žula, A., Hodnik, Ž., Jukič, M., Anderluh, M., Ilaš, J., Dolenc, M. S., Peluso, J., Ubeaud-Séquier, G., Muller, C. D., Mašič, L. P., & Kikelj, D. (2015). Analogues of the marine alkaloids oroidin, clathrodin, and hymenidin induce apoptosis in human HepG2 and THP-1 cancer cells. *MedChemComm*, 6(1), 105–110. DOI: 10.1039/C4MD00286E

Tong, W., Li, W.-H., He, Y., Mo, Z.-Y., Tang, H.-T., Wang, H.-S., & Pan, Y.-M. (2018). Palladium-metalated porous organic polymers as recyclable catalysts for the chemioselective synthesis of thiazoles from thiobenzamides and isonitriles. *Organic Letters*, 20(8), 2494–2498. DOI: 10.1021/acs.orglett.8b00886 PMID: 29620903

Torosyan, S. A., Gimalova, F. A., Zagitov, V. V., Erastov, A. S., & Miftakhov, M. S. (2018). Synthesis of N-Substituted Methyl 4 H-Thieno [3, 2-b] pyrrole-5-carboxylates. *Russian Journal of Organic Chemistry*, 54(6), 912–917. DOI: 10.1134/S1070428018060131

Torosyan, S. A., Nuriakhmetova, Z. F., Gimalova, F. A., & Miftakhov, M. S. (2019). 4H-Thieno [3, 2-b] pyrrole-5-carboxylate Conjugates with Taurine and Its Tetrabutylammonium Salt. *Russian Journal of Organic Chemistry*, 55(12), 1902–1906. DOI: 10.1134/S1070428019120157

Trafalis, D. T., Sagredou, S., Dalezis, P., Voura, M., Fountoulaki, S., Nikoleousakos, N., Almpanakis, K., Deligiorgi, M. V., & Sarli, V. (2021). Anticancer Activity of Triazolo-Thiadiazole Derivatives and Inhibition of AKT1 and AKT2 Activation. *Pharmaceutics*, 13(4), 493. DOI: 10.3390/pharmaceutics13040493 PMID: 33916378

Triloknadh, S., Venkata Rao, C., Nagaraju, K., Hari Krishna, N., Venkata Ramaiah, C., Rajendra, W., Trinath, D., & Suneetha, Y. (2018). Design, synthesis, neuroprotective, antibacterial activities and docking studies of novel thieno[2,3-d]pyrimidine-alkyne Mannich base and oxadiazole hybrids. *Bioorganic & Medicinal Chemistry Letters*, 28(9), 1663–1669. DOI: 10.1016/j.bmcl.2018.03.030 PMID: 29602681

Tripathy, D. B., & Gupta, A. (2023). Nanomembranes-affiliated water remediation: Chronology, properties, classification, challenges and future prospects. *Membranes*, 13(8), 713. DOI: 10.3390/membranes13080713 PMID: 37623773

Tsivgoulis, G. M., & Lehn Angew, J.-M. (1995). Photonic Molecular Devices: Reversibly Photoswitchable Fluorophores for Nondestructive Readout for Optical Memory Chem. *Angewandte Chemie International Edition in English*, 34(10), 1119–1122. DOI: 10.1002/anie.199511191

Tso, H.-H., Wang, J.-S., Wu, C.-Y., & Lin, H.-C. (1998). H-C Lin Novel Supramolecular Liquid- Crystalline Complexes Derived from 5-n-Decyl Substituted Thieno[3,2-*b*]thiophene-2- and Thiophene-2-Carboxylic Acids. *New Journal of Chemistry*, 22(8), 771–773. DOI: 10.1039/a803315c

Tsonev-Student, D., Nikolova, C. A. P. I., & Kostova, I. TUE-SSS-CT (R)-01 PHENOTHIAZINE DERIVATIVES OF 1, 8-NAPHTALIC ANHYDRIDE.

Tu, Y., Wang, C., Xu, S., Lan, Z., Li, W., Han, J., Zhou, Y., Zheng, P., & Zhu, W. (2017). Design, synthesis, and docking studies of quinazoline analogues bearing aryl semicarbazone scaffolds as potent EGFR inhibitors. *Bioorganic & Medicinal Chemistry*, 25(12), 3148–3157. DOI: 10.1016/j.bmc.2017.04.001 PMID: 28428040

Uddin, J., Ullah, S., Halim, S. A., Waqas, M., Ibrar, A., Khan, I., Bin Muhsinah, A., Khan, A., & Al-Harrasi, A. (2023). Triazolothiadiazoles and Triazolothiadiazines as New and Potent Urease Inhibitors: Insights from In Vitro Assay, Kinetics Data, and In Silico Assessment. *ACS Omega*, 8(35), 31890–31898. DOI: 10.1021/acsomega.3c03546 PMID: 37692208

Ukhurebor, K. E., Hossain, I., Pal, K., Jokthan, G., Osang, F., Ebrima, F., & Katal, D. (2024). Applications and contemporary issues with adsorption for water monitoring and remediation: A facile review. *Topics in Catalysis*, 67(1), 140–155. DOI: 10.1007/s11244-023-01817-4

Ulusoy, N., & Gürsoy, E. (2016). Synthesis and evaluation of new imidazo [2, 1-b] thiazoles as antituberculosis agents. *Marmara Pharmaceutical Journal*, 21(1), 102–109. DOI: 10.12991/marupj.259887

Upadhyay, P. K., Pathak, S., Mishra, R., Kumar, R., & Jain, A. (2023). A multifaceted scaffold for building bioactive compounds: Phenothiazine. *Letters in Organic Chemistry*, 20(7), 618–631. DOI: 10.2174/1570178620666221202100529

Uppar, V., Chandrashekharappa, S., Basarikatti, A. I., Banuprakash, G., Mohan, M. K., Chougala, M., ... & Padmashali, B. (2020). Synthesis, antibacterial, and antioxidant studies of 7-amino-3-(4-fluorobenzoyl) indolizine-1-carboxylate derivatives. *Journal of Applied Pharmaceutical Science, 10*(2), 077-085.

Uppar, V., Mudnakudu-Nagaraju, K. K., Basarikatti, A. I., Chougala, M., Chandrashekharappa, S., Mohan, M. K., & Padmashali, B. (2020). Microwave induced synthesis, and pharmacological properties of novel 1-benzoyl-4-bromopyrrolo [1, 2-a] quinoline-3-carboxylate analogues. *Chemical Data Collections*, 25, 100316.

Valiev, R. R., Fliegl, H., & Sundholm, D. (2014). The aromatic character of thienopyrrole-modified 20π-electron porphyrinoids. *Physical Chemistry Chemical Physics*, 16(22), 11010–11016. DOI: 10.1039/c4cp00883a PMID: 24776774

Varma, P. K. (2018). Therapeutic importance of synthetic thiophene. *Chemistry Central Journal*. PMID: 30564984

Varvounis, G., & Giannopoulos, T. (1996). Synthesis, chemistry, and biological properties of thienopyrimidines. *Advances in Heterocyclic Chemistry*, 66(66), 193–283. DOI: 10.1016/S0065-2725(08)60307-4

Vasilkova, N. O., Nikulin, A. V., & Krivenko, A. P. (2020). Synthesis and Structure of Thiazolopyrimidine Derivatives. *Russian Journal of Organic Chemistry*, 56(6), 990–993. DOI: 10.1134/S1070428020060044

Vekariya, R. H., Patel, K. D., Vekariya, M. K., Prajapati, N. P., Rajani, D. P., Rajani, S. D., & Patel, H. D. (2017). Microwave-assisted green synthesis of new imidazo [2, 1-b] thiazole derivatives and their antimicrobial, antimalarial, and antitubercular activities. *Research on Chemical Intermediates*, 43(11), 6207–6231. DOI: 10.1007/s11164-017-2985-5

Venkatesan, K., Basha, N. H., Jagadish, T., Reddy, P. V., Shaik, H., & Pasupathi, M. (2024). Ultrasound Assisted Synthesis of Phenothiazine Based Chalcone Derivatives, Their Antibacterial Studies and Molecular Docking Against COVID-19 Virus Spike Protein Inhibitor. *Russian Journal of Bioorganic Chemistry*, 50(3), 1133–1140. DOI: 10.1134/S1068162024030336

Venkatesan, K., Kumar, K. M., & Sivakumar, A. (2023). Synthesis, Characterization and Anticancer Molecular Docking Studies of Phenothiazine Derivatives–A Green Chemical Approach. *ChemistrySelect*, 8(34), e202302613. DOI: 10.1002/slct.202302613

Venkatesan, K., Satyanarayana, V. S. V., & Sivakumar, A. (2023). Synthesis and biological evaluation of novel phenothiazine derivatives as potential antitumor agents. *Polycyclic Aromatic Compounds*, 43(1), 850–859. DOI: 10.1080/10406638.2021.2021254

Venugopala, K. N., Al-Attraqchi, O. H. A., Tratrat, C., Nayak, S. K., Morsy, M. A., Aldhubiab, B. E., Attimarad, M., Nair, A. B., Sreeharsha, N., Venugopala, R., Haroun, M., Girish, M. B., Chandrashekharappa, S., Alwassil, O. I., & Odhav, B. (2019). Novel Series of Methyl 3-(Substituted Benzoyl)-7-Substituted-2-Phenylindolizine-1-Carboxylates as Promising Anti-Inflammatory Agents: Molecular Modeling Studies. *Biomolecules*, 9(11), 661. DOI: 10.3390/biom9110661 PMID: 31661893

Venugopala, K. N., Chandrashekharappa, S., Bhandary, S., Chopra, D., Khedr, M. A., Aldhubiab, B. E., Attimarad, M., & Odhav, B. (2018). Efficient Synthesis and Characterization of Novel Substituted 3-Benzoylindolizine Analogues via the Cyclization of Aromatic Cycloimmoniumylides with Electrondeficient Alkenes. *Current Organic Synthesis*, 15(3), 388–395. DOI: 10.2174/1570179414666171024155051

Venugopala, K. N., Chandrashekharappa, S., Deb, P. K., Tratrat, C., Pillay, M., Chopra, D., Al-Shar'i, N. A., Hourani, W., Dahabiyeh, L. A., Borah, P., Nagdeve, R. D., Nayak, S. K., Padmashali, B., Morsy, M. A., Aldhubiab, B. E., Attimarad, M., Nair, A. B., Sreeharsha, N., Haroun, M., & Mailavaram, R. (2021). Anti-tubercular activity and molecular docking studies of indolizine derivatives targeting mycobacterial InhA enzyme. *Journal of Enzyme Inhibition and Medicinal Chemistry*, 36(1), 1471–1486. DOI: 10.1080/14756366.2021.1919889 PMID: 34210233

Venugopala, K. N., Tratrat, C., Pillay, M., Chandrashekharappa, S., Al-Attraqchi, O. H. A., Aldhubiab, B. E., Attimarad, M., Alwassil, O. I., Nair, A. B., Sreeharsha, N., Venugopala, R., Morsy, M. A., Haroun, M., Kumalo, H. M., Odhav, B., & Mlisana, K. (2020). In silico design and synthesis of tetrahydropyrimidinones and tetrahydropyrimidinethiones as potential thymidylate kinase inhibitors exerting anti-TB activity against Mycobacterium tuberculosis. *Drug Design, Development and Therapy*, 14, 1027–1039. DOI: 10.2147/DDDT.S228381 PMID: 32214795

Villuendas, P., & Urriolabeitia, E. P. (2013). Primary amines as directing groups in the Ru-catalyzed synthesis of isoquinolines, benzoisoquinolines, and thienopyridines. *The Journal of Organic Chemistry*, 78(11), 5254–5263. DOI: 10.1021/jo400344m PMID: 23650873

Vince, B., Hill, J. M., Lawitz, E. J., O'Riordan, W., Webster, L. R., Gruener, D. M., Mofsen, R. S., Murillo, A., Donovan, E., Chen, J., McCarville, J. F., Sullivan-Bólyai, J. Z., Mayers, D., & Zhou, X. J. (2014). A randomized, double-blind, multiple-dose study of the pan-genotypic NS5A inhibitor Compound 9 in patients infected with hepatitis C virus genotype 1, 2, 3 or 4. *Journal of Hepatology*, 60(5), 920–927. DOI: 10.1016/j.jhep.2014.01.003 PMID: 24434503

Vu, C. B., Bemis, J. E., Disch, J. S., Ng, P. Y., Nunes, J. J., Milne, J. C., Carney, D. P., Lynch, A. V., Smith, J. J., Lavu, S., Lambert, P. D., Gagne, D. J., Jirousek, M. R., Schenk, S., Olefsky, J. M., & Perni, R. B. (2009). Discovery of imidazo[1,2-b] thiazole derivatives as novel SIRT1 activators. *Journal of Medicinal Chemistry*, 52(5), 1275–1283. DOI: 10.1021/jm8012954 PMID: 19199480

Vyawahare, D., Ghodke, M., & Nikalje, A. P. (2010).. . *International Journal of Pharmacy and Pharmaceutical Sciences*, 2, 27.

Waddar, B., Parne, S. R., Gandi, S., Prasanth, G. R., Yaseen, M., & Kariduraganavar, M. Y. (2024). The second-order nonlinear optical properties of novel triazolo[3,4-b] [1, 3, 4] thiadiazole derivative chromophores using DFT calculations. *Structural Chemistry*, 35(1), 253 264. DOI: 10.1007/s11224-023-02178-0

Wallin, J. J., Guan, J., & Prior, W. W. (2012). *GDC-0941, a Novel Class I Selective PI3K Inhibitor, Enhances the Efficacy of Docetaxel in Human Breast Cancer Models by Increasing Cell Death In Vitro and In Vivo GDC-0941, a Novel Class I Selective PI3K Inhibitor*. Enhances Increasing Cell Death In Vi., DOI: 10.1158/1078-0432. CCR-11-2088

Wang, B., Wang, L.-R., Liu, L.-L., Wang, W., Man, R.-J., Zheng, D.-J., Deng, Y.-S., Yang, Y.-S., Xu, C., & Zhu, H.-L. (2021). A novel series of benzothiazepine derivatives as tubulin polymerization inhibitors with anti-tumor potency. *Bioorganic Chemistry*, 108, 104585. DOI: 10.1016/j.bioorg.2020.104585 PMID: 33508676

Wang, G., Tang, Y., Zhang, Y., Liu, X., Lin, L., & Feng, X. (2017). Enantioselective Synthesis of N−H-Free 1,5-Benzothiazepines. *Chemistry (Weinheim an der Bergstrasse, Germany)*, 23(3), 554–557. DOI: 10.1002/chem.201605127 PMID: 27862459

Wang, L., Yi, Z., Zhang, P., Xiong, Z., Zhang, G., & Zhang, W. (2024). Comprehensive strategies for microcystin degradation: A review of the physical, chemical, and biological methods and genetic engineering. *Journal of Environmental Management*, 365, 121707. DOI: 10.1016/j.jenvman.2024.121707 PMID: 38968883

Wang, L., Zhang, P., Zhang, X., Zhang, Y., Li, Y., & Wang, Y. (2009). Synthesis and biological evaluation of a novel series of 1,5-benzothiazepine derivatives as potential antimicrobial agents. *European Journal of Medicinal Chemistry*, 44(7), 2815–2821. DOI: 10.1016/j.ejmech.2008.12.021 PMID: 19144450

Wang, X., Cheng, L., Tan, Q., Xing, Y., Wu, H., Tian, Q., Zhang, Y., & Zheng, L. (2024). Lag in pollutant removal efficiency relative to microbial community dynamics in a denitrifying phosphorus removal bioreactor. *Chemical Engineering Journal*, 497, 154498. DOI: 10.1016/j.cej.2024.154498

Wang, X., Li, Y., Song, P., Ma, F., & Yang, Y. (2019). Second-Order Nonlinear Optical Switch Manipulation of Photosensitive Layer by an External Electric Field Coupled with Graphene Quantum Dots. *The Journal of Physical Chemistry A*, 123(34), 7401–7407. DOI: 10.1021/acs.jpca.9b05249 PMID: 31369271

Wang, Y., Luo, Y., Hu, D., & Song, B. (2022). Design, Synthesis, Anti-Tomato Spotted Wilt Virus Activity, and Mechanism of Action of Thienopyrimidine-Containing Dithioacetal Derivatives. *Journal of Agricultural and Food Chemistry*, 70(20), 6015–6025. DOI: 10.1021/acs.jafc.2c00773 PMID: 35576166

Wani, T. A., Zargar, S., Alkahtani, H. M., Altwaijry, N., & Al-Rasheed, L. S. (2023). Anticancer potential of sulfonamide moieties via in-vitro and in-silico approaches: Comparative investigations for future drug development. *International Journal of Molecular Sciences*, 24(9), 7953. DOI: 10.3390/ijms24097953 PMID: 37175658

Wani, Z. A., Ahmad, Z., Asgher, M., Bhat, J. A., Sharma, M., Kumar, A., Sharma, V., Kumar, A., Pant, S., Lukatkin, A. S., & Anjum, N. A. (2023). Phytoremediation of potentially toxic elements: Role, status and concerns. *Plants*, 12(3), 429. DOI: 10.3390/plants12030429 PMID: 36771511

Wan, Y., Long, J., Gao, H., & Tang, Z. (2021). 2-Aminothiazole: A privileged scaffold for the discovery of anti-cancer agents. *European Journal of Medicinal Chemistry*, 210, 112953. DOI: 10.1016/j.ejmech.2020.112953 PMID: 33148490

Welch, M., & Phillips, R. S. (1999). Improved Syntheses of [3, 2-b]-and [2, 3-b]-fused Selenolo-and Thienopyrroles, and of Furo [3, 2-b] pyrrole. *Heterocyclic Communications*, 5(4), 305–310. DOI: 10.1515/HC.1999.5.4.305

Wetzel, C., Mishra, A., Mena-Osteritz, E., Liess, A., Stolte, M., Würthner, F., & Bäuerle, P. (2014). Synthesis and structural analysis of thiophene-pyrrole-based S, N-heteroacenes. *Organic Letters*, 16(2), 362–365. DOI: 10.1021/ol403153z PMID: 24359234

Wikel, J. H., Denney, M. L., & Vasileff, R. T. (1993). A convenient synthesis of the thieno[3,2- *c*]pyridine nucleus. *Journal of Heterocyclic Chemistry*, 30(1), 289–290. DOI: 10.1002/jhet.5570300150

Wilding, B., & Klempier, N. (2017). Newest Developments in the Preparation of Thieno[2,3-d]pyrimidines. *Organic Preparations and Procedures International*, 49(3), 183–215. DOI: 10.1080/00304948.2017.1320513

Woodring, J. L., Behera, R., Sharma, A., Wiedeman, J., Patel, G., Singh, B., Guyett, P., Amata, E., Erath, J., Roncal, N., Penn, E., Leed, S. E., Rodriguez, A., Sciotti, R. J., Mensa-Wilmot, K., & Pollastri, M. P. (2018). Series of Alkynyl-Substituted Thienopyrimidines as Inhibitors of Protozoan Parasite Proliferation. *ACS Medicinal Chemistry Letters*, 9(10), 996–1001. DOI: 10.1021/acsmedchemlett.8b00245 PMID: 30344906

Wright, J. B. (1968). The preparation of 2 H -1,2,3-benzothiadiazine-1,1-dioxides, 11 H -11 a -dihydrobenzimidazo[1,2- b] [1,2] benzisothiazole-5,5-dioxides, 6 H -dibenzo[c,g] [1,2,5] thiadiazocine-5,5-dioxides and 5 H -Dibenzo[c,g] [1,2,6] thiadiazocine-6,6-dioxides. *Journal of Heterocyclic Chemistry*, 5(4), 453–459. DOI: 10.1002/jhct.5570050403

Wright, S. W., & Corbett, R. L. (1993). A synthesis of 2-aryl-3-hydroxybenzothiophenes and analogs by the base promoted cyclization of N-phenyl-2-(benzylthio) benzamides. *Tetrahedron Letters*, 34(18), 2875–2878. DOI: 10.1016/S0040-4039(00)60469-0

Wu, J., Li, J., Xu, M.-H., & Liu, D. (2014). Structure–activity relationship and interaction studies of new SIRT1 inhibitors with the scaffold of 3-(furan-2-yl)-[1,2,4] triazolo[3,4-b][1,3,4]thiadiazole. *Bioorganic & Medicinal Chemistry Letters*, 24(14), 3050–3056. DOI: 10.1016/j.bmcl.2014.05.028 PMID: 24880902

Wu, N., Yang, Y., Tian, G., An, L., Liu, S., Yan, T., Yi, M., & Bao, X. (2023). Synthesis, X-ray Crystal Structure, and Antimicrobial Studies of New Quinazolin-4(3 *H*)-one Derivatives Containing the 1,2,4-Triazolo[3,4- *b*][1,3,4]thiadiazole Moiety and 4-Piperidinyl Linker. *Journal of Agricultural and Food Chemistry*, 71(49), 19277–19287. DOI: 10.1021/acs.jafc.3c03670 PMID: 38038681

Wu, S., Shi, J., Chen, J., Hu, D., Zang, L., & Song, B. (2021). Synthesis, Antibacterial Activity, and Mechanisms of Novel 6-Sulfonyl-1,2,4-triazolo[3,4- *b*][1,3,4] thiadiazole Derivatives. *Journal of Agricultural and Food Chemistry*, 69(16), 4645–4654. DOI: 10.1021/acs.jafc.1c01204 PMID: 33871992

Xu, Y., He, Z., Liu, H., Chen, Y., Gao, Y., Zhang, S., Wang, M., Lu, X., Wang, C., Zhao, Z., Liu, Y., Zhao, J., Yu, Y., & Yang, M. (2020). 3D-QSAR, molecular docking, and molecular dynamics simulation study of thieno [3, 2-b] pyrrole-5-carboxamide derivatives as LSD1 inhibitors. *RSC Advances*, 10(12), 6927–6943. DOI: 10.1039/C9RA10085G PMID: 35493862

Yadav, K. P., Rahman, M. A., Nishad, S., Maurya, S. K., Anas, M., & Mujahid, M. (2023). Synthesis and biological activities of benzothiazole derivatives: A review. *Intelligent Pharmacy*, 1(3), 122–132. DOI: 10.1016/j.ipha.2023.06.001

Yadav, L. D. S., Vaish, A., & Sharma, S. (1994). New Fungitoxic Fused-Ring Synthetics Incorporating Azoles and Azines in Different Combinations. *Journal of Agricultural and Food Chemistry*, 42(3), 811–813. DOI: 10.1021/jf00039a043

Yahya, S. M. M., Abdelhamid, A. O., Abd-Elhalim, M. M., Elsayed, G. H., & Eskander, E. F. (2017). The effect of newly synthesized progesterone derivatives on apoptotic and angiogenic pathway in MCF-7 breast cancer cells. *Steroids*, 126, 15–23. DOI: 10.1016/j.steroids.2017.08.002 PMID: 28797724

Yakovlev, M. Yu., Kadushkin, A. V., Solov'eva, N. P., Anisimova, O. S., & Granik, V. G. (1998). ω-Substituted-2-(polyenamino)- or annelated nitropyridines from 1-(3-cyano-5-nitropyridyl-2)-pyridinium salts. *Tetrahedron*, 54(21), 5775–5780. DOI: 10.1016/S0040-4020(98)00265-8

Yamada, S., Flesch, K. N., Murakami, K., & Itami, K. (2020). Rapid Access to Kinase Inhibitor Pharmacophores by Regioselective C-H Arylation of Thieno[2,3-d] pyrimidine. *Organic Letters*, 22(4), 1547–1551. DOI: 10.1021/acs.orglett.0c00143 PMID: 32009412

Yamamoto, T., Sanechika, K., & Yamamoto, A. (1980). Preparation of thermostable and electric-conducting poly(2,5-thienylene). *Journal of Polymer Science. Polymer Letters Edition*, 18(1), 9–12. DOI: 10.1002/pol.1980.130180103

Yang, D., An, B., Wei, W., Tian, L., Huang, B., & Wang, H. (2015). Copper-Catalyzed Domino Synthesis of Nitrogen Heterocycle-Fused Benzoimidazole and 1,2,4-Benzothiadiazine 1,1-Dioxide Derivatives. *ACS Combinatorial Science*, 17(2), 113–119. DOI: 10.1021/co500125n PMID: 25549532

Yang, D., Burugupalli, S., Daniel, D., & Chen, Y. (2012). Microwave-assisted one-pot synthesis of isoquinolines, furopyridines, and thienopyridines by palladium-catalyzed sequential Coupling–Imination–Annulation of 2-bromoarylaldehydes with terminal acetylenes and ammonium acetate. *The Journal of Organic Chemistry*, 77(9), 4466–4472. DOI: 10.1021/jo300494a PMID: 22494334

Yang, H., Higgins, B., Kolinsky, K., Packman, K., Go, Z., Iyer, R., Kolis, S., Zhao, S., Lee, R., Grippo, J., Schostack, K., Simcox, M. E., Heimbrook, D., Bollag, G., & Su, F. (2010). RG7204 (PLX4032), a Selective BRAF(V600E) Inhibitor, Displays Potent Antitumor Activity in Preclinical Melanoma Models. *Cancer Research*, 70(13), 5518–5527. DOI: 10.1158/0008-5472.CAN-10-0646 PMID: 20551065

Yang, X., Xi, J., Sun, Y., Zhang, Y., Zhou, G., & Wong, W. Y. (2019). A dopant-free twisted organic small-molecule hole transport material for inverted planar perovskite solar cells with enhanced efficiency and operational stability. *Nano Energy*, 64, 103946. DOI: 10.1016/j.nanoen.2019.103946

Yin, C., Yang, T., Pan, Y., Wen, J., & Zhang, X. (2020). Rh-Catalyzed Asymmetric Hydrogenation of Unsaturated Medium-Ring NH Lactams: Highly Enantioselective Synthesis of N-Unprotected 2,3-Dihydro-1,5-benzothiazepinones. *Organic Letters*, 22(3), 920–923. DOI: 10.1021/acs.orglett.9b04478 PMID: 31916777

Yin, S. Q., Wang, J. J., Zhang, C. M., & Liu, Z. P. (2012). The development of MetAP-2 inhibitors in cancer treatment. *Current Medicinal Chemistry*, 19(7), 1021–1035. DOI: 10.2174/092986712799320709 PMID: 22229417

Yin, X., Wang, L., & Mao, L. (2024). Comparing the Developmental Toxicity Delay and Neurotoxicity of Benzothiazole and Its Derivatives (BTHs) in Juvenile Zebrafish. *Toxics*, 12(5), 341. DOI: 10.3390/toxics12050341 PMID: 38787120

Yong C, Yang G & Deyong Y, *China*, CN110407769 A, 2019-11-05.

Yousif, M. N. M., El-Sayed, W. A., Abbas, H. A. S., Awad, H. M., & Yousif, N. M. (2017). Anti-cancer activity of new substituted pyrimidines, their thioglycosides and thiazolopyrimidine derivatives. *Journal of Applied Pharmaceutical Science*, 7(11), 21–32. DOI: 10.7324/JAPS.2017.71104

Yuan, H., Liu, Q., Zhang, L., Hu, S., Chen, T., Li, H., Chen, Y., Xu, Y., & Lu, T. (2018). Discovery, optimization and biological evaluation for novel c-Met kinase inhibitors. *European Journal of Medicinal Chemistry*, 143, 491–502. DOI: 10.1016/j.ejmech.2017.11.073 PMID: 29202410

Yuan, J. (2019).. . *Materials Science and Engineering R Reports*, 139, 100537.

Yuan, T., Li, X., Xiao, J., & Wang, S. (2011). Characterization of Xanthomonas oryzae-Responsive cis-Acting Element in the Promoter of Rice Race-Specific Susceptibility Gene Xa13. *Molecular Plant*, 4(2), 300–309. DOI: 10.1093/mp/ssq076 PMID: 21208999

Yue, J., Sun, S., Liang, J., Zhong, W., Lan, L., Ying, L., Huang, F., Yang, W., & Cao, Y. (2016). Effects of pyridyl group orientations on the optoelectronic properties of regio-isomeric diketopyrrolopyrrole based π-conjugated polymers. *Journal of Materials Chemistry. C, Materials for Optical and Electronic Devices*, 4(13), 2470–2479. DOI: 10.1039/C6TC00051G

Yurovskaya, M. A., & Mit'kin, O. D. (1999). Functionalization of pyridines. Reactions with formation of a bond between carbon and group IV, V, or VI heteroatom. *Chemistry of Heterocyclic Compounds*, 4, 437.

Zahradnik, P., Magdolen, P., & Zahradnik, P. (2010). Thiazolo[4,5-d]thiazole—A new domain for potential optoelectronic application. *Tetrahedron Letters*, 51(44), 5819–5821. DOI: 10.1016/j.tetlet.2010.08.110

Zaki, R. M., El-Dean, A. M., Radwan, S. M., & Ammar, M. A. (2020). Synthesis, reactions, and antimicrobial activity of novel heterocyclic compounds containing cyclopenta[*d*]thieno[2,3-*b*]pyridine moiety and related fused heterocycles. *Russian Journal of Bioorganic Chemistry*, 46(1), 85–96. DOI: 10.1134/S1068162020010148

Zaraei, S. O., Sbenati, R. M., Alach, N. N., Anbar, H. S., El-Gamal, R., Tarazi, H., Shehata, M. K., Abdel-Maksoud, M. S., Oh, C. H., & El-Gamal, M. I. (2021). Discovery of first-in-class imidazothiazole-based potent and selective ErbB4 (HER4) kinase inhibitors. *European Journal of Medicinal Chemistry*, 224, 113674. DOI: 10.1016/j.ejmech.2021.113674 PMID: 34237622

Zarenezhad, E., Farjam, M., & Iraji, A. (2021). Synthesis and biological activity of pyrimidines-containing hybrids: Focusing on pharmacological application. *Journal of Molecular Structure*, 1230, 129833. DOI: 10.1016/j.molstruc.2020.129833

Zhang, Y., Luo, L., Han, C., Lv, H., Chen, D., Shen, G., Wu, K., Pan, S., & Ye, F. (2017). *Derivatives as Anti-Inflammatory Agents*. DOI: 10.3390/molecules22111960

Zhang, D., Ruan, D., Li, J., Chen, Z., Zhu, W., Guo, F., Chen, K., Li, Y., & Wang, R. (2020). Four undescribed sulfur-containing indole alkaloids with nitric oxide inhibitory activities from Isatis tinctoria L. roots. *Phytochemistry*, 174, 112337. DOI: 10.1016/j.phytochem.2020.112337 PMID: 32163787

Zhang, H., Hu, Y., Chen, X., Yu, L., Huang, Y., Wang, Z., Wang, S., Lou, Y., Ma, X., Sun, Y., Li, J., Ji, D., Li, L., & Hu, W. (2022). A Centrosymmetric Organic Semiconductor with Donor–Acceptor Interaction for Highly Photostable Organic Transistors. *Advanced Functional Materials*, 32(20), 2111705. DOI: 10.1002/adfm.202111705

Zhang, J., Jiang, V., Jiang, Y., Guo, M., Zhang, S., Li, J., He, J., Liu, J., Wang, J., & Ouyang, L. (2016). Recent advances in the development of dual VEGFR and C-met small molecule inhibitors as anticancer drugs. *European Journal of Medicinal Chemistry*, 108, 495–504. DOI: 10.1016/j.ejmech.2015.12.016 PMID: 26717201

Zhang, J., Li, G., Yuan, X., Li, P., Yu, Y., Yang, W., & Zhao, S. (2023). Reduction of ultrafiltration membrane fouling by the pretreatment removal of emerging pollutants: A review. *Membranes*, 13(1), 77. DOI: 10.3390/membranes13010077 PMID: 36676884

Zhang, Q., Wang, Y., Kan, B., Wan, X., Liu, F., Ni, W., Feng, H., Russell, T. P., & Chen, Y. (2015). A solution-processed high performance organic solar cell using a small molecule with the thieno [3,2-*b*]thiophene central unit. *Chemical Communications (Cambridge)*, 51(83), 15268–15271. DOI: 10.1039/C5CC06009E PMID: 26329677

Zhang, W., Wang, L., & Li, W. (2017, January 19).. . *World Intellectual Property Organization*, WO2017009316, A1.

Zhang, X., Liu, Y., Zhang, M., & Wang, H. (2018). Thiophene-based fused S-heterocycles for removal of heavy metals from industrial wastewater. *Journal of Hazardous Materials*, 354, 125–134. DOI: 10.1016/j.jhazmat.2018.04.059

Zhang, X., Yan, J., Wang, H., Wang, Y., Wang, J., & Zhao, D. (2021). Molecular docking, 3D-QSAR, and molecular dynamics simulations of thieno [3, 2-b] pyrrole derivatives against anticancer targets of KDM1A/LSD1. *Journal of Biomolecular Structure & Dynamics*, 39(4), 1189–1202. DOI: 10.1080/07391102.2020.1726819 PMID: 32036765

Zhan, P., Liu, X., & Clercq, E. (2008). Recent Advances in Antiviral Activity of Benzo/Heterothiadiazine Dioxide Derivatives. *Current Medicinal Chemistry*, 15(15), 1529–1540. DOI: 10.2174/092986708784638898 PMID: 18537628

Zhan, W., Zhou, R., Mao, P., Yuan, C., Zhang, T., Liu, Y., Tian, J., Wang, H., & Xue, W. (2024). Synthesis, antifungal activity and mechanism of action of novel chalcone derivatives containing 1,2,4-triazolo-[3,4-b]-1,3,4-thiadiazole. *Molecular Diversity*, 28(2), 461–474. DOI: 10.1007/s11030-022-10593-4 PMID: 36964852

Zhao, P.-L., Duan, A.-N., Zou, M., Yang, H.-K., You, W.-W., & Wu, S.-G. (2012). Synthesis and cytotoxicity of 3,4-disubstituted-5-(3,4,5-trimethoxyphenyl)-4H-1,2,4-triazoles and novel 5,6-dihydro-[1,2,4]triazolo[3,4-b][1,3,4]thiadiazole derivatives bearing 3,4,5-trimethoxyphenyl moiety. *Bioorganic & Medicinal Chemistry Letters*, 22(13), 4471–4474. DOI: 10.1016/j.bmcl.2012.03.023 PMID: 22647723

Zhong, L., Li, Y., Xiong, L., Wang, W., Wu, M., Yuan, T., Yang, W., Tian, C., Miao, Z., Wang, T., & Yang, S. (2021). Small molecules in targeted cancer therapy: Advances, challenges, and future perspectives. *Signal Transduction and Targeted Therapy*, 6(1), 1–48. DOI: 10.1038/s41392-021-00572-w PMID: 34054126

Zhu, S., An, Z., Chen, X., Chen, P., & Liu, Q. (2015). Cyclic thiourea functionalized dyes with binary π-linkers: Influence of different π-conjugation segments on the performance of dye-sensitized solar cells. *Dyes and Pigments*, 116, 146–154. DOI: 10.1016/j.dyepig.2015.01.022

Zubarev, A. A., Shestopalov, A. M., Larionova, N. A., Rodinovskaya, L. A., & Shestopalov, A. A. (2013). New regio-selective method of combinatorial synthesis of substituted thiophenes, thieno[3,2-*b*]pyridines and other heterocycles via combination of 'domino'-type reactions. *Tetrahedron*, 69(46), 9648–9655. DOI: 10.1016/j.tet.2013.09.025

Zuba , A., Ghinet, A., Farce, A., Dubois, J., & Bîcu, E. (2023). Phenothiazine-and Carbazole-Cyanochalcones as Dual Inhibitors of Tubulin Polymerization and Human Farnesyltransferase. *Pharmaceuticals (Basel, Switzerland)*, 16(6), 888. DOI: 10.3390/ph16060888 PMID: 37375835

Zuo, Z., Zhang, H., Gao, S., Wang, C., Chen, W. T., & Hu, G. (2024). Recent progress in porous organic frameworks for electrochemical sensing of environmental pollutants. *Journal of Environmental Chemical Engineering*, 12(5), 113263. DOI: 10.1016/j.jece.2024.113263

About the Contributors

Shrikaant Kulkarni, Ph.D., is currently working as an Adjunct Professor, Faculty of Business, Victorian Institute of Technology, Melbourne, Australia and Adjunct Professor at Chitkara University, Punjab, India. Dr. Kulkarni has been a senior academician and researcher for more than four decades. He has delivered invited lectures and conducted sessions at national and international conferences as well as faculty development programs. He has guided many major and minor projects in subject areas like engineering chemistry, green chemistry, nanotechnology, analytical chemistry, catalysis, chemical engineering materials, industrial organization, and management. He has published over 100+ research papers in national and international journals and conferences of repute. He has authored 50+ book chapters in books published by CRC, Springer Nature, Apple Academic Press, Elsevier, Wiley, and IET. He has edited 25+ books published by Apple Academic Press/CRC Press and Springer Nature. He authored four textbooks in engineering chemistry. Dr Kulkarni possesses MSc, MPhil, and PhD degrees in Chemistry in addition to master's degrees in Economics, Business Management, and Political Science. He has expertise in the fields of green chemistry and engineering, analytical chemistry, green nanoscience and nanotechnology, wastewater treatment, Green and Analytical Chemistry, and advanced areas in Chemical Engineering and Material Science as well as Advanced AI. Two of his Patents have been published and two more are applied for grant.

Hemantkumar Akolkar, Ph.D., Working as Associate Professor and Head, Department of Chemistry, Rayat Shikshan Sanstha's Abasaheb Marathe Arts and New Commerce, Science College, Rajapur, Dist- Ratnagiri, Maharashtra, India Dr. Hemantkumar N. Akolkar is an academician and researcher for last 14 years at both undergraduate and postgraduate level. He has taught subjects like Heterocyclic Chemistry, Green Chemistry, Organic Spectroscopy, Organic Stereochemistry, Chemistry of Natural Products during his career. He possesses Masters in Organic chemistry and Ph.D. in Chemistry from Savitribai Phule Pune University, Pune. He has 55+ research papers in national and international journals and conferences of repute. His areas of interest are Heterocyclic Chemistry, Synthetic Organic Chemistry and Green Chemistry. He is reviewer of several journals in Chemistry of international repute.

Vijay M Khedkar working as Associate Professor in the Department of Pharmacy at Vishwakarma University, Pune, Maharashtra, India A highly optimistic and proactive computational medicinal chemist with Ph.D. degree in Pharmaceutical Chemistry (Computational Medicinal Chemistry) and Post-doctoral research experience. Successful track record of executing lead identification and optimization projects by applying molecular modeling techniques.

Nirmala R. Darekar, M.Sc., Ph. D., SET, B.Ed. Assistant Professor, Department of Chemistry, Rayat Shikshan Sanstha's Radhabai Kale Mahila Mahavidyalaya, Ahmednagar, Maharashtra, India, 414001 Ms. Nirmala R. Darekar is working as an Assistant Professor in Rayat Shikshan Sanstha's Radhabai Kale Mahila Mahavidyalaya, Ahmednagar. Her specialization is in Analytical Chemistry. She has 11 + years of experience in teaching and 6 years of experience in research. She did her Ph.D. from Savitribai Phule Pune University. She has more than 15 Research publication in reputed International journals. Also She has 03 Patents and 13 book chapters to her credit.

***

Manmeet Kaur Arora is a Research Scholar at the School of Law, Sharda University, Greater Noida. She is a graduate of Law and has completed her Master's in Law from Amity University, Noida. Manmeet Kaur is a qualified Company Secretary Executive Programme. She has over 8+ years of Industry Experience and 1 year of experience as an Associate at a reputed IPR Law Firm. Manmeet Kaur has attended 20+ National & International Conferences, Webinars, Seminars, Symposiums, and Paper Presentations and participated in 5+ Faculty Development programs. She has over 50+ publications in various esteemed Journals and Books. She is active at the Family Dispute Resolution Clinic, Meditation Centre, Greater Noida. Manmeet Kaur has a keen interest in the areas of Corporate Law, Intellectual Property Rights, Insolvency, Artificial Intelligence, Machine Learning, and Blockchain Technology.

Hari Shankar Biswas is Assistant Professor in Chemistry Surendranath College, Kolkata. Dr. Biswas is an esteemed Assistant Professor in Chemistry at

Surendranath College, Kolkata, with a multifaceted academic background and a strong dedication to research and education. He embarked on his academic journey by obtaining his B.Sc. degree in Chemistry Honours from R.K.M.V.C. College in 2005, followed by his M.Sc. degree from Presidency College, Kolkata, in 2008. Dr. Biswas further pursued his scholarly pursuits by earning a Ph.D. in Material Science from the Saha Institute of Nuclear Physics, University of Calcutta, in 2015. His professional experience spans various roles, including a stint as a School teacher at Keshub Academy, Kolkata, from 2008 to 2010, before transitioning into academia as an Assistant Professor at Surendranath College. Dr. Biswas specializes in the research domain of Carbon thin Film, Graphene, and Graphene Oxide-based nano Composites, focusing on their synthesis, characterization, and application across Environmental, Biological, and Technological fields. Dr. Biswas's scholarly contributions are significant, with authorship of 12 book chapters and over 26 papers published in esteemed International Journals. He is actively involved in academic societies, holding life membership in the Indian Science Congress Association and the Indian Chemical Society, along with an affiliated membership with the Royal Society of Chemistry. In addition to his research and academic endeavors, Dr. Biswas plays a vital role in the dissemination of scientific knowledge as a series book Editor and reviewer for reputable international publishers and journals. He serves as an Editorial board member for the American Journal of Chemical Engineering, Reseapro Journal, Lincoln University College, etc., and is affiliated with Rotary International. Dr. Biswas's editorial contributions extend to editing various books, including titles such as "Emerging Concepts in Chemical and Biological Science" and "Modern Approaches in Chemical and Biological Sciences," among others. Through his comprehensive involvement in research, education, and editorial responsibilities, Dr. Hari Shankar Biswas continues to make valuable contributions to the field of Chemistry, fostering advancements and innovations for the betterment of society.

Manoj M. Gadewar currently working as Professor at K. R. Mangalam University, Gurugram. An accomplished researcher, with expertise in nano-biotechnology, pharmacology and pre-clinical study design for execution of research work involved in the process of drug development, he has more than Seventeen years of experience in academia and research. He did his PhD from IIT, Guwahati and published more than 45 research papers in various journals of international repute, to his credit 6 book chapters and 9 patents. He has more than 1100 citations and cumulative impact factor of 100 with H-index of 13. He is GATE qualified in 2003 conducted by IIT Madras with all India rank of 763 and receiver of National Doctoral Fellowship offered by AICTE to pursue doctoral study. He received best researcher award from K R MANGALAM University for his extraordinary contribution to research. He has guided more than 30 PG students and 2 students completed their Ph. D under his supervision.

Mithun Kumar Ghosh currently serves as an Assistant Professor in the Department of Chemistry at Medi-Caps University, located in Indore, Madhya Pradesh, India. Previously, he held similar positions at Govt College Hatta, Damoh, and Pandit S.N Shukla University, Shahdol, both situated in Madhya Pradesh. He completed his doctoral studies in Chemistry at Indira Gandhi National Tribal University, Amarkantak, specializing in the synthesis of transition/rare earth metal clusters with applications in catalysis and biology. Dr. Ghosh's ongoing research endeavors focus on synthesizing homo/heterometallic clusters, coordination polymeric complexes, and investigating their potential in antibiotic degradation and Single Molecule Magnets. Dr. Ghosh has authored 43 publications in renowned international journals and holds a patent. Dr. Ghosh actively contributes to the academic community by serving as a reviewer for various international reputed journals. His diverse research interests encompass Metal Organic Frameworks, Catalysts, Density Functional Theory (DFT), photocatalysis, and an array of chemical properties such as dye adsorption, DNA binding, cytotoxicity, antioxidant activity, and sensing properties.

Hind Hammouch, researcher with a PhD in economics and management, specializing in sustainability, sustainable development, CSR (corporate social responsibility) and management control. Extensive experience in empirical research, quantitative analysis and the development of frameworks for integrating sustainability into organizational practices. Proven record of accomplishment in publishing research results and contributing to the advancement of knowledge in sustainability and management control.

Navnath Hatvate holds a Master of Pharmacy and a PhD from the Department of Pharmaceutical Sciences and Technology at the Institute of Chemical Technology, Mumbai. He currently serves as an Assistant Professor at ICT Mumbai, Marathwada Campus. His areas of expertise and research interests include the development of new chemical entities (NCE) and their analogues, modifications of excipients and their applications in drug delivery, the use of AI and ML in drug discovery and drug delivery, process chemistry of APIs and intermediates, and total synthesis of natural products and bio-active compounds.

Shalini Jaiswal Working as Associate Professor, HOD of Applied Science. She has more than 18 years' experience in academics and research and is the author of more than 55 scientific publications in Journal\Conferences. Dr. Shalini Jaiswal received her D. Phil degree in from University of Allahabad, Allahabad India, in

2009. Dr. Shalini Jaiswal has14 month experience as a project fellow on UGC sponsored project entitled Green synthesis of multi- target nucleosides as a potential anti-viral agent. Her research interest includes different area of Computational Chemistry, Green technology, Nanotechnology, Biotechnology and Environmental science. She is a Life Member of the Indian science congress association. She has reviewed more than 50 papers of various reputed journals and conferences. She is an Editorial board member and Reviewer of various reputed journals like TPI, IJATES, International journal of advance research, Advanced Scientific Research, IJPDS, AJRC and Acta Scientific Computer Sciences. She attended\conducted 28 Seminar\workshop\FDP. She also work as Head of session in TOKYO SUMMIT-II (2020) and TOKYO SUMMIT-III (2021), Turkey (2022) and Turkey 2023.

Christian Kaunert is Professor of International Security at Dublin City University, Ireland. He is also Professor of Policing and Security, as well as Director of the International Centre for Policing and Security at the University of South Wales. In addition, he is Jean Monnet Chair, Director of the Jean Monnet Centre of Excellence and Director of the Jean Monnet Network on EU Counter-Terrorism (www.eucter.net).

Amit Kumar Kundu is Assistant Professor & HOD Department of Chemistry Sripat Singh College, Jiaganj, Murshidabad. Dr. Amit Kumar Kundu is an Assistant professor and HOD at the Department of Chemistry, Sripat Singh College, Jiaganj, Murshidabad. Dr. Kundu did B.Sc. from Krishnagar Government College, University of Kalyani in 2007 and completed MSc. Degree from Department of Chemistry, University of Kalyani in 2009. After that he received his PhD in Analytical Chemistry in 2017 from University of Kalyani. His research mainly focuses on Environmental Analytical Chemistry, carbon nano materials, Graphene, Graphene Oxide-basednano composites Synthesis, Characterization and its application. He has huge experience in Groundwater and surface water onsite sampling, sediment sampling from vadose zone and aquifer. He is also proficient in operating and managing various advanced analytical instruments. To date, he has published near about eighteen peer-reviewed journal papers and he is the author of five book chapters. He received World Environment and Livelihood Award form 'International Foundation for Environment and Ecology'.

Sahil Lal is a Research Scholar at School of Law, Sharda University. He pursued his graduation from National Law University, Lucknow and completed his Master's in Law from Gujarat National Law University, Gandhinagar. Sahil has completed a certificate course on 'AI & Law' from Lund University, Sweden and on 'Cyber Law' from Indian Law Institute, New Delhi. Sahil Lal is a qualified UGC-NET &

JRF candidate. He has a 10+ year Industry Experience and is an active member of Family Dispute Resolution Clinic, Mediation Center, Greater Noida. Sahil has attended 20+ International & National Conferences along with Paper Presentation and has also participated in more than 5+ Faculty Development Programme. Mr. Sahil has more than 50+ publications in various esteemed Journals & Books. Mr. Sahil has a keen interest in area of Artificial Intelligence, Machine Learning, Robotics, Block-Chain Technology and Intellectual Property Rights.

Mukta Vivek Lele has received a Bachelor of Pharmacy from VES College Of Pharmacy, Chembur, Mumbai, Maharashtra. Currently, she is pursuing a Master of Technology in Pharmaceutical Chemistry and Technology from the Institute of Chemical Technology Mumbai, Marathwada Campus, Jalna.

N.P. Bhagya, Associate Professor & HOD, Department of Chemistry, Sai Vidya Institute of Technology, Bengaluru - 560064. She has made research contributions in the fields of photoluminescence, adsorption, and corrosion. She acquired her post-doctoral degree from Kuvempu University and her doctoral degree from Bharathiar University and. She has 15 years of experience in teaching and 8 years of experience in research. She has published research articles in reputed journals. She is a member of the IEEE society.

Vikrant S. Palekar is working as Assistant Professor at Department of Chemistry of Gokhale Education Society's, Arts, Commerce & Science College, Jawhar, Palghar, Maharashtra, India. He has completed his Ph. D. from reputed Institute of Chemistry, Mumbai. He has 11 years of Teaching experience for undergraduate degree programme. He is recognized guide for Ph D (Chemistry) programme. He has good number International Research publication & Book chapter.

Srilatha Rao, Professor, Department of Chemistry, Nitte Meenakshi Institute of Technology, Bengaluru-560 064. She has made her research contributions in the field of plant extracts and energy efficient water purification and desalination. She has published in scholarly journals of international repute. Dr. Rao acquired a doctoral degree and postgraduate degree in Chemistry from the Mangalore University, Karnataka. She has over 23 years of teaching and research experience. She was selected for the prestigious VGST Fellowship, from Vision Group Science & Technology, Government of Karnataka in the year 2018: A research project titled "Synthesis of proteins and its effect on nanomembranes" AICTE, Govt. of India. As an investigator she successfully guided a VTU funded research project. She is a member of the ECSI, IISc Bangalore, India.

Kulbhushan A. Sasane is working as Assistant Professor at Department of Chemistry of Abasaheb Marathe Arts & New Commerce, Science College, Rajapur, Ratnagiri, Maharashtra, India. He has completed his Ph. D. from reputed Institute of Chemistry, Mumbai. He has 08 years of Teaching experience for undergraduate degree programme. He has good number International Research publication & Book chapter.

Trupti Sahebrao Shevkar holds a Bachelor of Pharmacy degree from Sir Dr. M. S. Gosavi College of Pharmaceutical Education and Research, Nashik. She is currently pursuing a Master of Technology from the Department of Pharmaceutical Sciences and Technology at ICT Mumbai, Marathwada Campus. Her area of research focuses on the design of novel molecules using computer-aided drug design and their synthesis for pharmacological activity and molecular docking, as well as the application of artificial intelligence (AI) and machine learning (ML) in drug discovery and development.

Bhupinder Singh working as Professor at Sharda University, India. Also, Honorary Professor in University of South Wales UK and Santo Tomas University Tunja, Colombia. His areas of publications as Smart Healthcare, Medicines, fuzzy logics, artificial intelligence, robotics, machine learning, deep learning, federated learning, IoT, PV Glasses, metaverse and many more. He has 3 books, 139 paper publications, 163 paper presentations in international/national conferences and seminars, participated in more than 40 workshops/FDP's/QIP's, 25 courses from international universities of repute, organized more than 59 events with international and national academicians and industry people's, editor-in-chief and co-editor in journals, developed new courses. He has given talks at international universities, resource person in international conferences such as in Nanyang Technological University Singapore, Tashkent State University of Law Uzbekistan; KIMEP University Kazakhstan, All'ah meh Tabatabi University Iran, the Iranian Association of International Criminal law, Iran and Hague Center for International Law and Investment, The Netherlands, Northumbria University Newcastle UK,

Rutuja Ukhade holds a Bachelor of Pharmacy from the Matoshri College of Pharmacy, Nashik. She is pursuing a Master of Technology in the Department of Pharmaceutical Sciences and Technology at the ICT Mumbai, Marathwada Campus. She is currently working on the design of novel bioactive molecules using computer-aided drug design, molecular docking, and the use of AI and ML in drug discovery, along with the synthesis of novel molecule and their evaluation.

Index

Symbols

A

B

C

D

E

F

H

I

M

N

O

P

S

T

www.ingramcontent.com/pod-product-compliance
Lightning Source LLC
LaVergne TN
LVHW080352011026
849161LV00009B/2273